T0203113

Lecture Notes in Computer Science   14549

Founding Editors

Gerhard Goos
Juris Hartmanis

Editorial Board Members

The series Lecture Notes in Computer Science (LNCS), including its subseries Lecture Notes in Artificial Intelligence (LNAI) and Lecture Notes in Bioinformatics (LNBI), has established itself as a medium for the publication of new developments in computer science and information technology research, teaching, and education.

LNCS enjoys close cooperation with the computer science R & D community, the series counts many renowned academics among its volume editors and paper authors, and collaborates with prestigious societies. Its mission is to serve this international community by providing an invaluable service, mainly focused on the publication of conference and workshop proceedings and postproceedings. LNCS commenced publication in 1973.

Ryuhei Uehara · Katsuhisa Yamanaka ·
Hsu-Chun Yen

Editors

WALCOM: Algorithms and Computation

18th International Conference and Workshops
on Algorithms and Computation, WALCOM
2024, Kanazawa, Japan, March 18–20, 2024
Proceedings

Springer

Editors
Ryuhei Uehara 🆔
Japan Advanced Institute of Science
and Technology
Nomi, Japan

Katsuhisa Yamanaka 🆔
Iwate University
Morioka, Japan

Hsu-Chun Yen 🆔
National Taiwan University
Taipei, Taiwan

ISSN 0302-9743   ISSN 1611-3349 (electronic)
Lecture Notes in Computer Science
ISBN 978-981-97-0565-8   ISBN 978-981-97-0566-5 (eBook)
https://doi.org/10.1007/978-981-97-0566-5

Preface

The 18th International Conference and Workshop on Algorithms and Computation (WALCOM 2024) was held at Kanazawa Bunka Hall, Kanazawa, Japan during March 18–March 20, 2024. The conference covered diverse areas of algorithms and computation, that is, approximation algorithms, algorithmic graph theory and combinatorics, combinatorial algorithms, combinatorial optimization, computational biology, combinatorial reconfiguration, computational complexity, computational geometry, discrete geometry, data structures, experimental algorithm methodologies, graph algorithms, graph drawing, parallel and distributed algorithms, parameterized algorithms, parameterized complexity, network optimization, online algorithms, randomized algorithms, and string algorithms. The conference was sponsored by the Algorithmic Foundations for Social Advancement (AFSA) Grant-in-Aid for Transformative Research Areas, MEXT, Japan and Japan Advanced Institute of Science and Technology. It was also supported by the Special Interest Group for ALgorithms (SIGAL) of the Information Processing Society of Japan (IPSJ), Technical Committees on Theoretical Foundations of Computing (COMP) of the Institute of Electronics, Information and Communication Engineers (IEICE) and the Japan Chapter of the European Association of Theoretical Computer Science (EATCS Japan).

This volume of *Lecture Notes in Computer Science* contains 28 contributed papers that were presented at WALCOM 2024. There were 78 submissions from 25 countries. Each submission was reviewed by at least three Program Committee members with the assistance of external referees. The volume also includes the extended abstracts of two invited talks presented by Shin-ichi Minato and Naomi Nishimura. Two special issues, one of Theoretical Computer Science and one of Journal of Graph Algorithms and Applications, are being prepared for some selected papers among those presented at WALCOM 2024.

We wish to thank all who made this meeting possible: the authors for submitting papers, the Program Committee members and external referees (listed in the following pages) for their excellent work, and our two invited speakers. We acknowledge the Steering Committee members for their continuous encouragement and suggestions. We also wish to express our sincere appreciation to the sponsors, local organizers, Proceedings Committee, and the editors of the *Lecture Notes in Computer Science* series and Springer for their help in publishing this volume. Finally, we thank the EasyChair conference management system, which was very effective in handling the entire reviewing process.

March 2024

Ryuhei Uehara
Katsuhisa Yamanaka
Hsu-Chun Yen

Organization

Steering Committee

Tamal Dey	Ohio State University, USA
Seok-Hee Hong	University of Sydney, Australia
Costas S. Iliopoulos	King's College London, UK
Giuseppe Liotta	University of Perugia, Italy
Petra Mutzel	University of Bonn, Germany
Shin-ichi Nakano	Gunma University, Japan
Subhas C. Nandy	Indian Statistical Institute, India
Md. Saidur Rahman	Bangladesh University of Engineering and Technology, Bangladesh
Ryuhei Uehara	Japan Advanced Institute of Science and Technology, Japan

Program Committee

Sang Won Bae	Kyonggi University, South Korea
Aritra Banik	Indian Institute of Technology Jodhpur, India
Stephane Durocher	University of Manitoba, Canada
Klim Efremenko	Ben Gurion University of the Negev, Israel
Mark Keil	University of Saskatchewan, Canada
Marc van Kreveld	Utrecht University, The Netherlands
Kazuhiro Kurita	Nagoya University, Japan
Chun-Cheng Lin	National Yang Ming Chiao Tung University, Taiwan R. O. C.
Giuseppe Liotta	University of Perugia, Italy
Debajyoti Mondal	University of Saskatchewan, Canada
Shin-Ichi Nakano	Gunma University, Japan
Subhas C. Nandy	Indian Statistical Institute, India
Yoshio Okamoto	University of Electro-Communications, Japan
Yota Otachi	Nagoya University, Japan
Ignaz Rutter	Universität Passau, German
Md. Saidur Rahman	Bangladesh University of Engineering and Technology, Bangladesh
M. Sohel Rahman	Bangladesh University of Engineering and Technology, Bangladesh

Saket Saurabh	Institute of Mathematical Sciences, India
Uéverton Souza	Fluminense Federal University, Brazil
Akira Suzuki	Tohoku University, Japan
Ryuhei Uehara (Co-chair)	Japan Advanced Institute of Science and Technology, Japan
Giovanni Viglietta	University of Aizu, Japan
Sue Whitesides	University of Victoria, Canada
Mingyu Xiao	University of Electronic Science and Technology of China, China
Katsuhisa Yamanaka (Co-chair)	Iwate University, Japan
Hsu-Chun Yen (Co-chair)	National Taiwan University, Taiwan R. O. C.
Guochuan Zhang	Zhejiang University, China

Organizing Committee

Tonan Kamata	Japan Advanced Institute of Science and Technology, Japan
Yota Otachi	Nagoya University, Japan
Ryuhei Uehara	Japan Advanced Institute of Science and Technology, Japan

Additional Reviewers

Ankush Acharyya	Pål Grønås Drange
Abu Reyan Ahmed	Hiroshi Eto
Shareef Ahmed	Simon D. Fink
Hee-Kap Ahn	Fedor Fomin
Taehoon Ahn	Henry Förster
Prashanti Angara	Hiroshi Fujiwara
Toru Araki	Anirban Ghosh
Pradeesha Ashok	Tatsuya Gima
Patricia Bachmann	Erwin Glazenburg
Tian Bai	Petr Golovach
Susobhan Bandopadhyay	Mursalin Habib
Matthias Bentert	Sheikh Azizul Hakim
Sourav Chakraborty	Tesshu Hanaka
Ho-Lin Chen	Tahmid Hasan
Serafino Cicerone	Shinji Imahori
Ikaro Penha Costa	Satyabrata Jana
Gautam K. Das	Lawqueen Kanesh
Minati De	Shuji Kijima

Donggyu Kim
Alex Kitt
Yasuaki Kobayashi
Myroslav Kryven
Anant Kumar
Madhumita Kundu
Chung-Shou Liao
Markus Lohrey
Raul Lopes
Mengfan Ma
Amirhossein Maghghdoust
Andrea Marino
Gopinath Mishra
Sounak Modak
Fabrizio Montecchiani
Joydeep Mukherjee
Miriam Münch
Atsuki Nagao
Yuto Nakashima
Pouria Zamani Nezhad
Rahnuma Islam Nishat
Mateus De Oliveira Oliveira
Giacomo Ortali
Supantha Pandit
Junqiang Peng
Sergey Pupyrev

Johanes Rauch
André van Renssen
Abhishek Sahu
Toshiki Saitoh
Pascal Schweitzer
Sanjay Seetharaman
Sagnik Sen
Zimo Sheng
Bettina Speckmann
Peter Stumpf
Krisztina Szilagyi
Yuma Tamura
Alessandra Tappini
Meng-Tsung Tsai
Anannya Upasana
Mai Ha Vu
Kunihiro Wasa
Alexandra Weinberger
Petra Wolf
Chao Xu
Chenyang Xu
Muhammad Nur Yanhaona
Sang Duk Yoon
Tian-Li Yu
Tom van der Zanden
Jingyang Zhao

Sponsoring Institutions

Algorithmic Foundations for Social Advancement (AFSA) Grant-in-Aid for Transformative Research Areas, MEXT, Japan
Japan Advanced Institute of Science and Technology, Japan
Information Processing Society of Japan (IPSJ), Japan
The Institute of Electronics, Information and Communication Engineers (IEICE), Japan
Japan Chapter of European Association of Theoretical Computer Science (EATCS Japan), Japan

Contents

Recent Research Activities on Algorithmic Foundations for Social Advancement

Shin-ichi Minato(✉)

Kyoto University, Kyoto, Japan
minato@i.kyoto-u.ac.jp

Abstract. Algorithms, the theories, techniques and logical procedures of information processing, perform a key part of the recent sophisticated information society. A five-year nation-wide research project on algorithmic techniques, initiated in 2020, is currently in progress in Japan. This presentation aims to provide an overview of the "AFSA" (Algorithmic Foundations for Social Advancement) project and introduce some selected topics from our recent research activities.

1 Background and Purpose of AFSA Project

With the rapid advancements in the technologies of integrated circuits, optical communication, and storage devices, our personal computing and communication terminals have experienced a remarkable improvement in cost-performance ratios, ranging from thousands to a million times, over recent 30 years. Such rapid technological progress is unprecedented in human history. For instance, although the fundamental algorithmic techniques in deep learning and artificial intelligence were already developed in the 1990s, their full potential has only recently been realized due to a significant increase in computational power and the facilitated acquisition of big data. For the next 10 to 20 years, we expect the further progress in algorithmic technology leveraging such significant improvement in computational power, and it will become a source of competitive advantage to realize social advancement.

For activating the algorithm research communities, we have the following awareness of issues. As shown in Fig. 1, in the early days of computer science, individuals often conducted research on both theory and application. However, in the recent era characterized by advanced information technologies, the gap between theoretical researchers and applied researchers has widened, making it challenging to deeply commit to the both research communities simultaneously. Now it becomes more important to provide the methodologies and software tools contributing to practical applications as a research outcome of algorithmic foundations. Thus, we started to organize a research project to form "Art" layer to bridge Science and Engineering, crossing the gap. Our project logo (Fig. 2) symbolizes this research vision.

Our project aims to develop and organize state-of-the-art techniques on algorithms. The project is named "AFSA" (Algorithmic Foundations for Social Advancement) [13], a five-year national research project started from 2020. The

R. Uehara et al. (Eds.): WALCOM 2024, LNCS 14549, pp. 1–8, 2024.
https://doi.org/10.1007/978-981-97-0566-5_1

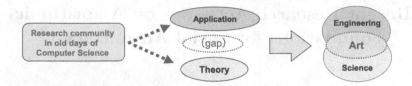

Fig. 1. Research communities in theoretical computer science.

Fig. 2. A visual identity of the project.

results will be provided as open academic resources for many scientists and engineers in various fields, to be utilized for social advancement. Based on the recent drastic progress of computation power, upcoming innovative computation devices, and new concepts from social sense of values, we will reformulate and organize practical computation models to bridge theory and practice. We will also create and organize computational theories and state-of-the-art techniques for algorithms, such as discrete structure manipulation, constraint satisfaction problem solving, enumeration, discrete optimization, quantum computation theory, etc.

2   Organization of the Project

As shown in Fig. 3, our AFSA project consists of six research groups in two categories A and B. The groups in A (A01 and A02) investigate the interface layer to bridge theory and practice, and the groups in B (B01, B02, B03, and B04) investigate specific theories and techniques to support the interface layer. Each research group consists of six or seven PIs (principal investigators). For the application layer, we have a number of external collaborators affiliated with many kinds of research projects in the specific application domains. Those external researchers and engineers communicate with AFSA project members through the interface developed by the research groups A01 and A02. The detailed contents of the six research groups are as shown below.

A01: New Problem Formulation on Next Generation Informatics and Researches on their Algorithms:
Collaborating with researchers in the application layer, this group discusses and formulates a set of new problems to be considered in the future society. We also design efficient algorithms based on a new approach.

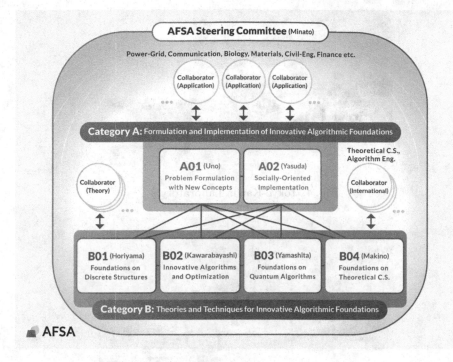

Fig. 3. An organization chart of AFSA project.

A02: Socially-Oriented Algorithm Implementation:
This group implements the algorithms proposed in our project and organizes the algorithmic foundations for social advancement. It provides an interface between theoretical researchers and application engineers.

B01: Algorithmic Foundations Based on Large-Scale Discrete Structures:
By the collaboration of theoretical researchers and application engineers, this group tackles how to deal with exponentially large-scale discrete structures and develops innovative design methodologies of efficient algorithms.

B02: New Computational Models for Algorithms and Discrete Optimization
This group investigates basic research topics in the areas of discrete mathematics, combinatorial optimization, machine learning, etc. to develop efficient algorithms for solving very large-scale problems required in our society.

B03: Creation of Innovative Foundations to Bridge Theory and Practice of Quantum Algorithms
Combining the knowledge of classical computation and new quantum models, this group constructs useful algorithmic foundations to

Fig. 4. Tokyo-Kanda Laboratory.

Fig. 5. Kyoto-Teramachi Laboratory.

implement practically efficient quantum computers connected to conventional systems.

B04: Exploration and Development of the Basic Theory of Algorithms
This group investigates important problems in theoretical computer science, such as performance assurance, preserving fairness and stability, new computation models and design methodologies for social requirements.

We also have publicly selected 17 individual research projects based on call for proposals to work on additional related research topics. They are strongly recommended to collaborate with at least two different research groups in the AFSA project.

To facilitate the collaborative research work, we opened two meeting offices in central Tokyo and Kyoto dedicated for AFSA project activities. The two offices, Tokyo-Kanda Lab. (Fig. 4) and Kyoto-Teramachi Lab. (Fig. 5) are both located nearby the central stations of the two big cities so that many researchers in different universities/institutes can easily access to one of the offices and frequently meet with each other to have research discussions. Unfortunately, due to the overlap of the first year of launching AFSA project and the widespread impact of COVID-19, the project members could not gather closely for discussion. However, we established the two offices as hub centers aiming for the best mix of online and offline activities. Even after the easing of COVID-19 restrictions, it continues to operate effectively as a hybrid activity hub.

3   Recent Research Activities

International Competition on Graph Counting Algorithms

In 2023, the AFSA project conducted International Competition on Graph Counting Algorithms (ICGCA). The graph counting problem is to count the subgraphs satisfying specified constraints on a given graph. The problem belongs to #P-complete, a computationally tough class. Graph counting algorithms are sometimes a key technology to efficiently scan all the subgraphs representing the feasible states of some kinds of societal systems, such as power supply networks [5], communication networks [14], railroad systems [10], electoral districting [9] , and many other systems. In this competition, contestants were asked to count the paths under a length constraint. The benchmark set includes 150 challenging instances, emphasizing on graphs resembling to infrastructure networks.

The competition was held from April to July in 2023. In total eleven solvers were submitted and ranked by the number of benchmarks correctly solved within a time limit. The evaluation results are revealed in the commendation ceremony and symposium at the 22nd annual Forum on Information Technology (FIT 2023), in Osaka, Japan, September 7, 2023. The winning solver, TLDC, was designed based on three fundamental approaches: backtracking search, dynamic programming, and model counting or #SAT (a counting version of Boolean satisfiability). Detailed analyses show that each approach has its own strengths, and one approach is unlikely to dominate the others. The codes and papers of the participating solvers are available at https://afsa.jp/icgca/. A detailed report of this competition will be published as a journal article [6,7].

Our first trial of ICGCA has been finished successfully. We decided to continue this competition next year. The second ICGCA is scheduled to be held from spring to summer in 2024.

Graphillion-2.0: Development of Graph Enumeration Software Tools

Graphillion [4,8] is a software library for very large sets of (vertex-)labeled graphs, based on zero-suppressed binary decision diagrams (ZDDs) [12]. Graphillion was mainly developed by JST ERATO Minato project about ten years ago, but still now well-maintained for solving many kinds of practical problems. Graphillion is implemented as a Python library to encourage easy development of its applications, without introducing significant performance overheads. Graphillion allows us to exhaustively but efficiently search a *graphset* (a set of many subgraphs in a given graph) with complex, even nonconvex, constraints. In addition, we can find top-k optimal graphs from the complex graphset and can also extract common properties among all graphs in the set. Thanks to these features, Graphillion has a variety of applications including graph database, combinatorial optimization, and a graph structure analysis.

In the AFSA project, we are planning to develop an updated version of "Graphillion-2.0" as an interface between algorithmic foundations and applications for social advancement. We are considering various extensions of Graphillion family, to handle extended graph models such as directed graphs (Di-Graphillion), vertex sets of graphs (VertexSet-Graphillion), cost-weighted graphs (Weighted-Graphillion), multi-graphs (Multi-Graphillion), and hyper graphs (Hyper-Graphillion). The development is now ongoing. We also hope to leverage the new insights obtained from the ICGCA competition in our development.

Theoretical Results to be Utilized for Practical Applications

In our project, the research groups in B investigate specific theories and techniques to be utilized for practical applications in future. We already have a number of excellent theoretical results. Some selected topics are briefly summarized as follows.

- In 2021, Kawaharabayashi (group B02) extended the theorem on a graph-cut algorithm of undirected planar graphs to directed ones [11]. This solved a long-standing problem in theoretical computer science, and it expected to contribute to the future development of this research field.
- Hirahara (group B02) successfully achieved to show NP-Hardness of learning programs [3], which was a long-standing problem from 1991. The result is recognized as the *Complexity Result of the Year 2022* [1].
- Hanaka and Ono (group B04) received the Outstanding Paper Award [2] at APDCM 2023, an international workshop in IPDPS 2023, for solving a graph optimization problem inspired by a practical application to the frequency allocation in wireless mobile networks.
- Yamamoto and Shibuya (group B04) developed efficient and highly accurate methods [15] for differentially private statistical genomic analysis. They received the Outstanding Award at IEEE TrustCom2022, a top conference in security and privacy in computing and communications.

4   Concluding Remarks

AFSA project is ongoing until March 2025. This project will lead an active research community where theory and practice meet together. Our expected outputs are not only to produce top conference papers and journal publications, but also to contribute to real-life social problems by collaborating with application research engineers. These algorithmic foundations will be useful for various fields of science and technologies and aim to contribute to social advancement over the long-term.

Acknowledgment. This work was partly supported by JSPS KAKENHI Grant Number JP20H05961, JP20H5962, JP20H05963, JP20H05964, JP20H5965, JP20H05966, JP20H05967.

References

1. Complexity year in review 2022 (2022). https://blog.computationalcomplexity.org/2022/12/
2. Hanaka, T., Ono, H., Sugiyama, K.: Solving distance-constrained labeling problems for small diameter graphs via tsp*. In: IEEE International Parallel and Distributed Processing Symposium, IPDPS 2023 - Workshops, St. Petersburg, FL, USA, 15–19 May 2023, pp. 308–313. IEEE (2023). https://doi.org/10.1109/IPDPSW59300.2023.00059
3. Hirahara, S.: Np-hardness of learning programs and partial MCSP. In: 63rd IEEE Annual Symposium on Foundations of Computer Science, FOCS 2022, Denver, CO, USA, 31 October–3 November 2022, pp. 968–979. IEEE (2022). https://doi.org/10.1109/FOCS54457.2022.00095
4. Inoue, T., Iwashita, H., Kawahara, J., Minato, S.: Graphillion: software library for very large sets of labeled graphs. Int. J. Softw. Tools Technol. Transf. **18**(1), 57–66 (2016). https://doi.org/10.1007/S10009-014-0352-Z
5. Inoue, T., et al.: Distribution loss minimization with guaranteed error bound. IEEE Trans. Smart Grid **5**(1), 102–111 (2014)
6. Inoue, T., Yasuda, N., Nabeshima, H., Nishino, M., Denzumi, S., Minato, S.: International competition on graph counting algorithms 2023. CoRR abs/2309.07381 (2023). https://doi.org/10.48550/ARXIV.2309.07381, https://doi.org/10.48550/arXiv.2309.07381
7. Inoue, T., Yasuda, N., Nabeshima, H., Nishino, M., Denzumi, S., Minato, S.: International competition on graph counting algorithms 2023. IEICE Trans. Fundamentals (2024, to appear)
8. Inoue, T., et al.: Graphillion (2013). http://graphillion.org/
9. Kawahara, J., Horiyama, T., Hotta, K., Minato, S.: Generating all patterns of graph partitions within a disparity bound. In: Proceedings of the 11th International Conference and Workshops on Algorithms and Computation, vol. 10167, pp. 119–131 (2017). https://doi.org/10.1007/978-3-319-53925-6_10
10. Kawahara, J., Saitoh, T., Suzuki, H., Yoshinaka, R.: Solving the longest oneway-ticket problem and enumerating letter graphs by augmenting the two representative approaches with ZDDs. In: Phon-Amnuaisuk, S., Au, T.-W., Omar, S. (eds.) CIIS 2016. AISC, vol. 532, pp. 294–305. Springer, Cham (2017). https://doi.org/10.1007/978-3-319-48517-1_26
11. Kawarabayashi, K., Sidiropoulos, A.: Embeddings of planar quasimetrics into directed $l1$ and polylogarithmic approximation for directed sparsest-cut. In: 62nd IEEE Annual Symposium on Foundations of Computer Science, FOCS 2021, Denver, CO, USA, 7–10 February 2022, pp. 480–491. IEEE (2021). https://doi.org/10.1109/FOCS52979.2021.00055
12. Minato, S.: Zero-suppressed BDDs for set manipulation in combinatorial problems. In: Proceedings of of 30th ACM/IEEE Design Automation Conference (DAC'93), pp. 272–277 (1993)
13. Minato, S., et al.: Creation and organization of innovative algorithmic foundations for social advancement. In: 2020–2024 Grant-in-Aid for Transformative Research Areas, MEXT, Japan (2020). https://afsa.jp/en/

14. Nakamura, K., Inoue, T., Nishino, M., Yasuda, N., Minato, S.: A fast and exact evaluation algorithm for the expected number of connected nodes: an enhanced network reliability measure. In: IEEE INFOCOM 2023 - IEEE Conference on Computer Communications, New York City, NY, USA, 17–20 May 2023, pp. 1–10. IEEE (2023). https://doi.org/10.1109/INFOCOM53939.2023.10228897
15. Yamamoto, A., Shibuya, T.: Efficient and highly accurate differentially private statistical genomic analysis using discrete Fourier transform. In: IEEE International Conference on Trust, Security and Privacy in Computing and Communications, TrustCom 2022, Wuhan, China, 9–11 December 2022, pp. 525–532. IEEE (2022). https://doi.org/10.1109/TRUSTCOM56396.2022.00078

Reasons to Fall (More) in Love with Combinatorial Reconfiguration

Naomi Nishimura[✉]

University of Waterloo, Waterloo, ON, Canada
nishi@uwaterloo.ca

Abstract. The goal of the talk is to give ideas and inspiration to everyone in the audience, whether currently working in combinatorial reconfiguration or new to the area. Organized as a series of reasons to love combinatorial reconfiguration, the presentation will bring non-experts up to speed, setting the stage for a more in-depth investigation. The talk is not intended to be a comprehensive survey of the field, but instead a personal and idiosyncratic tour, full of suggestions for future directions of research.

Keywords: Combinatorial Reconfiguration · Algorithms

1 Introduction

The talk will begin with a demonstration of how reconfiguration can be used to introduce the ideas of research and graphs to a non-technical audience, in the process giving a brief introduction to the reconfiguration framework to those new to the area.

Equipped with all the necessarily terminology for a more technical exploration, we will then cover an eclectic collection of results, ideas, and open questions at varying levels of detail.

In lieu of outlining the entire talk, this abstract provides definitions, resources, and references, as well as a sampling of the topics to be presented.

2 A Primer on Reconfiguration

Stated in its simplest form, the reconfiguration framework, as introduced in 2011 [13], encompasses a large variety of problems that can be characterized as follows:

- a *configuration*, such as a feasible or optimal solution to an instance of an optimization problem, or a snapshot in time of a puzzle, game, or geometric object;

Research supported by the Natural Sciences and Engineering Research Council of Canada.

- a *reconfiguration step* that makes a small change from one configuration into another; and
- a *question*.

Many investigations that fall under this broad umbrella were conducted well before 2011, including work on the 15 puzzle dating at least as far back as 1879 [15]. In recent years, terminology has been standardized, and reconfiguration has been formulated in terms of graphs.

To form a *reconfiguration graph*, we create one node for each possible configuration and create an edge between each pair of nodes representing adjacent configurations, where two configurations are *adjacent* if one can be formed from the other by a single reconfiguration step. The terminology is chosen to allow us to distinguish between a *node* in a reconfiguration graph and a *vertex* in an instance of an optimization problem.

Frequently-studied questions pertain to properties of the graph, such as:

- **Reachability:** Does there exist a path (or *reconfiguration sequence*) from one specified configuration (the *source configuration*) to another specified configuration (the *target configuration*)?
- **Shortest path:** What is the shortest path between a specified pair of configurations?
- **Connectivity:** Is the reconfiguration graph connected?
- **Diameter:** What is the diameter of each connected component of the reconfiguration graph?

Recent work on reconfiguration typically uses a classical problem as a starting point, where configurations are feasible or optimal solutions to an instance of the classical problem. As such, reconfiguration provides an avenue for the study of the solution space of such a problem. For configurations that can be represented as a subset of the vertices of a graph, a solution can be viewed as an assignment of tokens to vertices. Generalizing movements found in sliding block puzzles, such as the 15 puzzle, the reconfiguration step of *token sliding* entails the movement of a token from a vertex to an adjacent free vertex. In a *token jumping* step, a token can be moved to a non-adjacent vertex, and in *token addition and removal* a token can be added to a free vertex or removed from a vertex. References for the steps and more information can be found in the surveys listed in Sect. 3.

Due to the vast range of possible configurations, reconfiguration steps, and questions to consider, unexplored areas of inquiry are easy to find. Much of the work to date has entailed attempts to delineate the boundary between tractable and intractable instances for reachability, with more effort on INDEPENDENT SET, COLOURING, and DOMINATING SET than other classical problems, and more work on token sliding and token jumping (and for COLOURING, the recolouring of a single vertex or the swapping of two colours in a component) than on other types of reconfiguration steps. Results on other combinations of configurations, reconfiguration steps, and questions are sparser.

3   How to Get Started

The list below provides suggestions on how to start (or continue) your exploration of reconfiguration. You can find links to much of the information below, in addition to many more papers, talks, and events, at the site `reconf.wikidot.com`, maintained by Duc A. Hoang.

- Watch Takehiro Ito's excellent talk "Invitation to Reconfiguration", presented at the 16th International Conference and Workshops on Algorithms and Computation, WALCOM 2022.
- Read a survey on reconfiguration:
  - van den Heuvel's "The complexity of change" [12] provides an excellent introduction to the area, with extensive coverage of colouring and generalizations of the 15 puzzle.
  - "Introduction to Reconfiguration" [19] was intended to supplement the previous survey. In the spirit of an annotated bibliography, it attempts to cover all the work on reconfiguration known at the time of writing.
  - Mynhardt and Nasserasr's "Reconfiguration of colourings and dominating sets in graphs" [18] provides in-depth coverage of two of the most well-studied problems.
  - Bousquet, Mouwad, Nishimura, and Siebertz focus on parameterized complexity in "A survey on the parameterized complexity of the independent set and (connected) dominating set reconfiguration problems" [4].
- Look through a list of open problems, such as listed in the report for the Banff International Research Station workshop 22w5090, linked off of the page https://www.birs.ca/events/2022/5-day-workshops/22w5090. The site www.birs.ca also contains a link to the workshop held in 2017 (17w5066); for both the workshops, you can download slides and watch videos of various presentations.
- Participate in a Core challenge, a programming competition aimed at finding practical solutions to combinatorial reconfiguration problems; please see https://core-challenge.github.io/2023/ for information on the 2023 challenge.
- Attend - or organize - a workshop. For information on workshops, sign up for the mailing list at `lists.uwaterloo.ca/mailman/listinfo/reconf`. Past workshops have included talks, mentoring sessions, and open problem sessions.

4   A Sampling of Research Directions

The talk will present specific results and open questions that address ways to expand the range of configurations, reconfiguration steps, and questions to investigate, as well as the relationship among multiple reconfiguration graphs, connections to other areas of research, and applications. Below is a sampling of topics to be covered.

4.1   Extending Configurations

We can enrich the set of problems under consideration by departing even slightly from the standard configuration as a set of tokens on vertices, such as by placing tokens on edges [11] or by adding labels to tokens [6,17].

By being less stringent as to what constitutes a configuration, we expand the realm of tractable solutions to reconfiguration-related problems. Current research includes the investigation of the impact of allowing a bounded number of adjacent vertices in what would otherwise be an independent set [21]. In the *solution discovery framework* [8,9], to be discussed further below, configurations include both feasible and infeasible solutions to problems, allowing one to navigate from an infeasible solution to a reachable feasible solution.

4.2   Extending Questions

Considering only a single reconfiguration graph, there are additional algorithmic and structural questions to consider. Algorithmic questions can be subject to different types of analysis, such as fixed-parameter tractability [4] and approximation [20]. Since a reconfiguration graph is a graph, one can study various ways of characterizing reconfiguration graphs [3], such as considering Hamiltonicity, girth, and whether a reconfiguration graph matches the instance of the problem [2,10].

Instead of providing a source and a target configuration, the optimization variant [14] seeks the best configuration reachable from a given configuration. The area of *solution discovery* extends this idea to consider the reconfiguration graph of both feasible and infeasible solutions to an instance of a source problem (perhaps with constraints imposed on which solutions qualify as configurations), seeking to modify an infeasible solution into a feasible one in a bounded number of steps. Results to date include algorithms, hardness results, differences based on different reconfiguration steps, and the parameterized complexity of such problems [8,9].

In the new area of *resource-focused* reconfiguration, the goal is to minimize the extra resources needed to make reconfiguration possible. As one example, we consider rearranging virtual machines on physical machines in a data centre, for such purposes as maintenance and load-balancing. In addition to asking when such rearrangements are possible, given the original resources, once can also ask how many extra physical machines might be required [16].

4.3   Applications

Reconfiguration not only provides a way of analyzing and characterizing the solution space of a problem, but also captures types of step-by-step changes that occur in real-life problems. Potential uses of reconfiguration include the analysis of gerrymandering [1] and the aforementioned reassignment of virtual machines to physical machines in a data centre [16].

The POWER SUPPLY problem [13] was introduced as a motivation for reconfiguration; here, customers, each with a demand, are assigned to power stations, each with a capacity, with the system functioning when all demands are met and no capacities are exceeded. To ensure that there is no system-wide shutdown when one configuration needs to be changed to another, each move of a customer to a different power station is required to result in a still-functioning system. Practice caught up to theory when reconfiguration was used to solve problems for power stations in Japan [22].

In ongoing work, reconfiguration is being used in preparing atoms for use in quantum simulation. The process of loading atoms into arrays of optical traps results in roughly half of the traps being filled. Reconfiguration algorithms determine how to rearrange atoms into a dense and regular arrangement, such as all locations in a grid, using moving optical tweezers [5–7].

Acknowledgments. I wish to thank all of my co-authors, and all members of this welcoming and supportive research community, for inspiring the title and much of the content of the talk.

References

1. A. Akitaya, H., Korman, M., Korten, O., L. Souvaine, D., D. Tóth, C.: Reconfiguration of connected graph partitions via recombination. In: Calamoneri, T., Corò, F. (eds.) CIAC 2021. LNCS, vol. 12701, pp. 61–74. Springer, Cham (2021). https://doi.org/10.1007/978-3-030-75242-2_4
2. Alikhani, S., Fatehi, D., Klavžar, S.: On the structure of dominating graphs. Graphs Comb. **33**(4), 665–672 (2017)
3. Avis, D., Hoang, D.A.: On reconfiguration graphs of independent sets under token sliding. Graphs Comb. **39**(3) (2023)
4. Bousquet, N., Mouawad, A.E., Nishimura, N., Siebertz, S.: A survey on the parameterized complexity of the independent set and (connected) dominating set reconfiguration problems. CoRR abs/2204.10526 (2022)
5. Cimring, B., et al.: Efficient algorithms to solve atom reconfiguration problems. I. redistribution-reconfiguration algorithm. Phys. Rev. A **108**, 023107 (2023)
6. Cooper, A., Maaz, S., Mouawad, A.E., Nishimura, N.: Parameterized complexity of reconfiguration of atoms. In: Mutzel, P., Rahman, M.S., Slamin (eds.) WALCOM 2022. LNCS, vol. 13174, pp. 263–274. Springer, Cham (2022). https://doi.org/10.1007/978-3-030-96731-4_22
7. El Sabeh, R., et al.: Efficient algorithms to solve atom reconfiguration problems. II. assignment-rerouting-ordering algorithm. Phys. Rev. A **108**, 023108 (2023)
8. Fellows, M.R., et al.: On solution discovery via reconfiguration. In: Gal, K., Nowé, A., Nalepa, G.J., Fairstein, R., Radulescu, R. (eds.) ECAI 2023–26th European Conference on Artificial Intelligence, 30 September–4 October 2023, Kraków, Poland - Including 12th Conference on Prestigious Applications of Intelligent Systems (PAIS 2023). Front. Artif. Intell. Appl. **372**, 700–707. IOS Press (2023)
9. Grobler, M., Maaz, S., Megow, N., Mouawad, A.E., Ramamoorthi, V., Schmand, D., Siebertz, S.: Solution discovery via reconfiguration for problems in P. CoRR abs/2311.13478 (2023)

10. Haas, R., Seyffarth, K.: The k-dominating graph. Graphs Comb. **30**(3), 609–617 (2014)
11. Hanaka, T., et al.: Reconfiguring spanning and induced subgraphs. Theor. Comput. Sci. **806**, 553–566 (2020)
12. van den Heuvel, J.: The complexity of change. Surv. Comb. **2013**(409), 127–160 (2013)
13. Ito, T., et al.: On the complexity of reconfiguration problems. Theor. Comput. Sci. **412**(12–14), 1054–1065 (2011)
14. Ito, T., Mizuta, H., Nishimura, N., Suzuki, A.: Incremental optimization of independent sets under the reconfiguration framework. J. Comb. Optim. **43**(5), 1264–1279 (2022)
15. Johnson, W.W., Story, W.E.: Notes on the "15" puzzle. Am. J. Math. **2**(4), 397–404 (1879)
16. Kam, J., Kamali, S., Miller, A., Nishimura, N.: Reconfiguration of multisets with applications to bin packing. In: Proceedings of the 18th International Conference and Workshops on Algorithms and Computation (2023)
17. Moore, B.R., Nishimura, N., Subramanya, V.: Reconfiguration of graph minors. In: Potapov, I., Spirakis, P.G., Worrell, J. (eds.) 43rd International Symposium on Mathematical Foundations of Computer Science, MFCS 2018, 27–31 August 2018, Liverpool, UK. LIPIcs, vol. 117, pp. 75:1–75:15. Schloss Dagstuhl - Leibniz-Zentrum für Informatik (2018)
18. Mynhardt, C.M., Nasserasr, S.: Reconfiguration of colourings and dominating sets in graphs. In: 50 Years of Combinatorics, Graph Theory, and Computing, chap. 10, pp. 171–191. CRC Press (2019)
19. Nishimura, N.: Introduction to reconfiguration. Algorithms **11**(4), 52 (2018)
20. Ohsaka, N.: Gap preserving reductions between reconfiguration problems. In: Berenbrink, P., Bouyer, P., Dawar, A., Kanté, M.M. (eds.) 40th International Symposium on Theoretical Aspects of Computer Science, STACS 2023, 7–9 March 2023, Hamburg, Germany. LIPIcs, vol. 254, pp. 49:1–49:18. Schloss Dagstuhl - Leibniz-Zentrum für Informatik (2023)
21. Subramanya, V.: Private communication (2023)
22. Sugimura, S., Tanabe, T., Suzuki, A., Zhou, T.I.X.: Method for distribution loss minimization and switching operation procedures with radial network reconfiguration (in Japanese). In: The Papers of Technical Meeting on Power Systems Engineering, IEE Japan, no. 2 (PE-19-083, PSE-19-095), pp. 25–29 (2019)

Plane Multigraphs with One-Bend and Circular-Arc Edges of a Fixed Angle

Csaba D. Tóth[1,2]([✉])[iD]

[1] California State University Northridge, Los Angeles, CA, USA
[2] Tufts University, Medford, MA, USA
csaba.toth@csun.edu

Abstract. For an angle $\alpha \in (0, \pi)$, we consider plane graphs and multigraphs in which the edges are either (i) one-bend polylines with an angle α between the two edge segments, or (ii) circular arcs of central angle $2(\pi - \alpha)$. We derive upper and lower bounds on the maximum density of such graphs in terms of α. As an application, we improve upon bounds for the number of edges in $\alpha AC_1^=$ graphs (i.e., graphs that can be drawn in the plane with one-bend edges such that any two crossing edges meet at angle α). This is the first improvement on the size of $\alpha AC_1^=$ graphs in over a decade.

Keywords: circular arc · one-bend drawing · α-angle crossing drawing

1 Introduction

According to a well-known corollary of Euler's formula, an edge-maximal planar straight-line graph on $n \geq 3$ vertices has at most $3n - 6$ edges, which is attained on any set of $n \geq 3$ points in general position with a triangular convex hull; and at least $2n - 3$ edges, which is attained for n points in convex position. Specifically, on a given set P of n points in the plane in general position, $h \geq 3$ of which are on the convex hull of P, the maximum number of edges of a planar straight-line graph is $M(P) = 3n - h - 3$. This paper explores analogous questions for graphs where the edges are one-bend polylines or circular arcs with a fixed angle. Importantly, there may be multiple edges between a pair of vertices in these drawing styles, and *multigraphs* become relevant.

For an angle $\alpha \in (0, \pi)$, an α-**bend edge** between vertices a and c is a polyline (a, b, c) with one bend at b such that the interior angle of the triangle $\Delta(abc)$ at b is α; and an α-**arc edge** is a circular arc between a and c with central angle $2(\pi - \alpha)$. By the Inscribed Angle Theorem, if e is an α-arc edge between a and c, and b is any interior point of the arc e, then the polyline (a, b, c) is an α-edge; see Fig. 1 for an example.

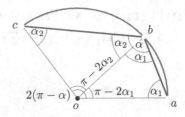

Fig. 1. The relation between an α-arc edge and an α-bend edges.

A simple graph embedded[1] in the plane such that every edge is α-bend (resp., α-arc) is an α-**bend graph** (resp., α-**arc graph**). Similarly, a multigraph embedded in the plane such that every edge is α-bend (resp., α-arc) is an α-**bend multigraph** (resp., α-**arc multigraph**). See Fig. 2 for examples. For a finite set $P \subset \mathbb{R}^2$, denote by $M_a(P, \alpha)$ and $M_b(P, \alpha)$, resp., the maximum number of edges in an α-arc graph and an α-bend graph. Similarly, the maximum number of edges in an α-arc and α-bend multigraph is denoted by $M_a^{\|}(P, \alpha)$ and $M_b^{\|}(P, \alpha)$, respectively. It may be hard to compute $M_a(P, \alpha)$, $M_b(P, \alpha)$, $M_a^{\|}(P, \alpha)$, and $M_b^{\|}(P, \alpha)$, for a given point set P and a given angle $\alpha \in (0, \pi)$; see Problem 1 in Sect. 6.

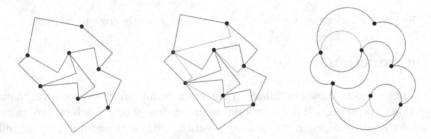

Fig. 2. A $\frac{\pi}{2}$-bend graph, a $\frac{\pi}{2}$-bend multigraph, and a $\frac{\pi}{2}$-arc muligraph

For $n \in \mathbb{N}$, let $M_a(n, \alpha) = \sup_{|P|=n} M_a(P, \alpha)$, and define $M_b(n, \alpha)$, $M_a^{\|}(n, \alpha)$ and $M_b^{\|}(n, \alpha)$ analogously. Our bounds on these quantities are in Table 1.

Table 1. Overview of results for $n \geq 3$ (without assuming general position).

Angle	$\alpha \to 0$	$\alpha \in (0, \frac{\pi}{2}]$	$\alpha \in (\frac{\pi}{2}, \frac{2\pi}{3}]$	$\alpha \in [\frac{5\pi}{5}, \frac{5\pi}{6}]$	$\alpha \in (\frac{2\pi}{3}, \pi)$	$\alpha \to \pi$
$M_a(n, \alpha)$	$2n-3$	$2n-3 \leq . \leq 3n-6$	$3n-6$	$3n-6$	$3n-6$	$3n-6$
$M_b(n, \alpha)$	$3n-6$	$3n-6$	$3n-6$	$3n-6$	$3n-6$	$3n-6$
$M_a^{\|}(n, \alpha)$	$2n-2$	$2n-2 \leq . \leq 4n-6$	$4n-6$	$4n-6$	$6n - O(\sqrt{n})$	$6n-12$
$M_b^{\|}(n, \alpha)$	$4n-6$	$4n-6$	$4n-6$	$6n - O(\sqrt{n})$	$6n - O(\sqrt{n})$	$6n-12$

---

[1] An *embeddng* of a graph into a surface is a continuous injective map of the 1-dimensional simplicial complex formed by the vertices and edges of the graph.

Motivation: RAC and $\alpha AC^=$ Drawings. A **right angle crossing drawing** (or *RAC* drawing, for short) of a graph $G = (V, E)$ is a drawing in which edges are polylines and crossing edges meet at angle $\frac{\pi}{2}$. A **RAC_b** drawing is a RAC drawing where every edge is drawn as a polyline with b bends; and a graph $G = (V, E)$ is a **RAC_b graph** if it admits such a drawing. Didimo et al. [13] proved that a RAC_0-graph on $n \geq 3$ vertices has at most $4n - 10$ edges, and this bound is the best possible; see also [15]. They also showed that every graph is a RAC_3 graph. Angelini et al. [4] proved that an n-vertex RAC_1 graph has at most $5.5n - O(1)$ edges, and this bound is asymptotically tight. Arikushi et al. [7] showed that an n-vertex RAC_2 graph has at most $74.2n$ edges, which was recently improved to $20n$ [20], and is conjectured to be $10n - O(1)$ [5]. Refer to the surveys [12, 14] for an overview on RAC drawings and their relatives.

Dujmović et al. [15] extended the notion of RAC drawings to drawings where the crossing edges meet at an angle greater than some α, for $\alpha \in (0, \frac{\pi}{2})$, and call such drawings α **angle crossing drawings** (αAC **drawings**, for short). They proved that an n-vertex graph with a straight-line αAC drawing has at most $\frac{\pi}{\alpha}(3n - 6)$ edges. Ackerman et al. [2] defined $\alpha AC_b^=$ **graphs**, which are graphs that can be drawn such that the edges are polylines with at most b bends per edge and every crossing occurs exactly at angle α. They proved that n-vertex $\alpha AC_1^=$ and $\alpha AC_2^=$ graphs have $O(n)$ edges for any $\alpha \in (0, \frac{\pi}{2}]$. In an $\alpha AC_1^=$ drawing, each edge is a polyline with two **segments** that are also called **end-segments**. In an $\alpha AC_2^=$ drawing, each edge is a polyline with three segments: Two **end-segments** incident to the vertices, and one **middle segment**.

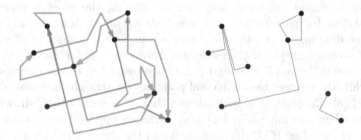

Fig. 3. Left: A RAC_2 drawing of a graph $G = (V, E)$ and the red graph (V, Γ). Right: Perturbation of overlapping red edges yields a $\frac{\pi}{2}$-bend multigraph. (Color figure online)

The main technical tool in the proofs by Ackerman et al. [2] were α-bend graphs, although they did not use this terminology. For example, suppose that we are given an $\alpha AC_2^=$-drawing of a directed graph $G = (V, E)$ in which the edge directions determine a *first* and a *last* end-segment, and suppose that all crossings are between first and last end-segments; see Fig. 3. We can create the multigraph (V, Γ), called the *red graph*, as follows: If the first end-segment s of some edge crosses any other edge, then we create an edge $\gamma(s) \in \Gamma$ as a directed path in the planarization of the $\alpha AC_2^=$-drawing: The path $\gamma(s)$ starts from the (unique) vertex in V incident to s, then follows s until its first crossing with the end-segment s' of some other edge, and then it follows s' to the (unique) vertex in V incident to s'. Each edge of the red graph is an α-bend or a $(\pi - \alpha)$-bend edge. The edges of (V, Γ) do not cross—they may partially overlap, but they can be perturbed to remove overlaps. Thus (V, Γ) is the union of two plane multigraphs: an α-bend and a $(\pi - \alpha)$-bend multigraph.

Ackerman et al. [2] proved that every n-vertex $\alpha AC_1^=$-graph has at most $27n$ edges for $\alpha \in (0, \frac{\pi}{2}]$. We improve this bound to $21n - 36$ for $n \geq 3$ (Theorem 3) using the bound $M_b^{\|}(n, \alpha) \leq 4n - 6$ for $\alpha \in (0, \frac{2\pi}{3}]$ (Theorem 1), applied to a red α-bend or $(\pi - \alpha)$-bend multigraph. This is the first improvement on the size of $\alpha AC_1^=$-graphs in more than a decade.

Further Related Previous Work. Angle constraints in graph drawing have been considered since the 1980s. Vijayan [21] introduced *angle graphs*, which are graphs such that at every vertex, the rotation of incident edges as well as the angles between consecutive edges in the rotation are given, and asked whether a given angle graph can be realized by a straight-line drawing (possibly with crossings). Planarity testing for angle graphs is NP-hard [9,18], but there is a linear-time algorithm for triangulations (with a triangular outer face) [11]. Note that a realization of an angle graph may have crossing edges: Efrat et al. [16] showed that for cycles, one can find a realization with the minimum number of crossings. Importantly, the edges of an angle graph are realized by straight-line segments. In contrast, we consider graphs with one-bend or circular arc edges of a fixed angle, however the angles between adjacent edges are unconstrained.

Circular arcs and one-bend polylines are among the most popular graph drawing styles. In general, both drawing styles allow more flexibility than straight-line drawings. For example, there are universal point sets of size $O(n)$ for planar n-vertex graphs if the edges are drawn as circular arcs [6] or as one-bend polylines [17], even if the bend points are restricted to the universal point set [19], while the current best universal point set for straight-line embeddings is of size $n^2/2$ [8]. Chaplick et al. [10] showed that an n-vertex RAC drawing with circular arc edges can have $4.5n - O(\sqrt{n})$ edges, as opposed to at most $4n - 10$ edges in a straight-line RAC drawing. Refer to the surveys [12,14] for a variety of results on RAC drawings.

One-bend polylines and circular arcs lose most of their competitive advantage against straight-line drawings if the angle α is fixed. A small angle $\alpha > 0$ may even be a disadvantage. Our objective is to obtain quantitative bounds to compare planar straight-line graphs to α-bend and α-arc graphs and multigraphs.

2   Preliminaries

We show that the multiplicity of any edge in a α-bend and an α-arc graph is at most two.

Proposition 1. *For every $\alpha \in (0, \pi)$ and every pair of points $a, c \in \mathbb{R}^2$, a plane α-bend (resp., α-arc) multigraph contains at most two edges between a and c, at most one in each halfplane bounded by the line ac.*

Proof. Let $\alpha \in (0, \pi)$, and let a and c be distinct points in the plane. There are precisely two circular arcs of central angle $\pi - \alpha$ between a and c, which lie in distinct halfplanes bounded by the line ac. Hence there are at most two α-arc edges between a and c. Every α-bend edge between a and c is a polyline (a, b, c) where the bend pointy b is on an α-arc edge between a and c. However, if (a, b_1, c) and (a, b_2, c) are α-bend edges, then b_1 and b_2 lie on distinct α-arcs between a and c, otherwise the two α-bend edges would cross: Indeed, assume w.lo.g. that the points a, b_1, b_2, c are in this order along the same circular arc from a to c. Then the line segments ab_2 and b_1c cross, hence the two α-bend edges cross. □

It is easy to find the maximum number of edges for collinear points.

Proposition 2. *For every set P of $n \geq 3$ collinear points, we have*

1. *$M_b(P, \alpha) = 3n - 6$ and $M_b^{\parallel}(P, \alpha) = 4n - 6$ for all $\alpha \in (0, \pi)$;*
2. *$M_a(P, \alpha) = 3n - 6$ and $M_a^{\parallel}(P, \alpha) = 4n - 6$ for all $\alpha \in [\frac{\pi}{2}, \pi)$;*
3. *$M_a(P, \alpha) = 2n - 3$ and $M_a^{\parallel}(P, \alpha) = 2n - 2$ for all $\alpha \in (0, \frac{\pi}{2})$.*

Proof. Assume w.l.o.g. that $P = \{p_1, \ldots, p_n\}$ is a set of n points on the x-axis sorted by increasing x-coordinates.

Upper Bounds. By definition, α-arc and α-bend graphs are simple planar graphs. By Euler's polyhedron formula, $M_a(P, \alpha) \leq 3n-6$ and $M_b(P, \alpha) \leq 3n-6$. Suppose that multiple edges are allowed. By Proposition 1, the edges in each halfplane form a simple graph. Since all vertices are on the boundary of both halfplanes, the edges in each halfplane form an outerplanar graph each with at most $2n - 3$ edges. This proves $M_a^{\parallel}(P, \alpha) \leq 4n - 6$ and $M_b^{\parallel}(P, \alpha) \leq 4n - 6$ for all $\alpha \in (0, \pi)$.

It remains to consider α-arc graphs and multigraphs for $\alpha \in (0, \frac{\pi}{2})$. Recall that an **$\alpha$-arc edge** is a circular arc between two points with central angle $2(\pi - \alpha)$. Note that $2(\pi - \alpha) \in (\pi, 2\pi)$ for $\alpha \in (0, \frac{\pi}{2})$, thus an α-arc edge is more than a halfcircle. Let $G = (V, E)$ be an α-arc multigraph, and let $G^+ = (V, E^+)$ and $G^- = (V, E^-)$ be the subgraphs formed by the edges in the upper and lower halfplane, respectively. If $i < j < k$ and $\alpha \in (0, \frac{\pi}{2})$, then the α-arc edges $p_i p_j$ and $p_j p_k$ would cross if they are both in the upper halfplane (or both in the lower halfplane). Consequently, in both G^+ and G^-, each vertex is either to the left or to the right of all of its neighbors. It follows that neither G^+ nor G^- can contain cycles, hence they each have at most $n-1$ edges. This implies $M_a^{\parallel}(P, \alpha) \leq 2n-2$ for all $\alpha \in (0, \frac{\pi}{2})$. Note that if G is a simple graph, then G^+ and G^- cannot both contain the edge $p_1 p_n$. Assume w.l.o.g. that G^- does not contain the edge $p_1 p_n$. Then G^- cannot contain any path from p_1 to p_n, consequently it is a forest with at least two components, and so it has at most $n-2$ edges. This yields the upper bound $M_a(P, \alpha) = 2n - 3$ for $\alpha \in (0, \frac{\pi}{2})$.

Lower Bound Constructions. We start with α-arc graphs. For $\alpha \in [\frac{\pi}{2}, \pi)$, an α-arc edge $p_i p_j$ is an x-monotone arc that lies in the vertical strip bounded by the vertical lines through p_i and p_j. Add α-arc edges $p_i p_{i+1}$ for all $i = 1, \ldots, n-1$ and edges $p_1 p_j$ for all $j = 2, \ldots n$ in the upper halfplane; and edge $p_j p_n$ for all $j = 2, \ldots, n-2$ in the lower halfplane. This yields $(n-1)+(n-2)+(n-3) = 3n-6$ edges; see the red edges in Fig. 4 (left). We can augment this construction to an α-arc multigraph by adding α-arc edges $p_i p_{i+1}$ for all $i = 1, \ldots, n-1$ and the edge $p_i p_n$ in the lower halfplane as well. This yields an α-arc multigraph with $(3n-6) + n = 4n-6$ edges; see Fig. 4 (left).

For $\alpha \in (0, \frac{\pi}{2})$, we construct an α-arc graph with $2n-3$ edges as follows. In the upper halfplane, we use the star formed by $p_1 p_i$ for all $i = 2, \ldots, n$; and in the lower halfplane, the star formed by $p_j p_n$ for all $j = 2, \ldots, n-2$. For an α-arc multigraph with $2n-4$ edges, we add the edge $p_1 p_n$ in the lower halfplane as well. Both are subgraphs of the constructions in Fig. 4 (left).

Fig. 4. Left: a $\frac{\pi}{2}$-arc multigraph with $4n-6$ edges; the red edges form a $\frac{\pi}{2}$-arc graph with $3n-6$ edges. Right: a $\frac{\pi}{3}$-bend multigraph with $4n-6$ edges; the red edges form a $\frac{\pi}{3}$-bend graph with $3n-6$ edges. In both examples, $n = 6$.

We can construct an α-bend graph with $3n-6$ edges and an α-bend multigraph with $4n-6$ edges by connecting the same pair of vertices for any $\alpha \in (0, \pi)$. We describe the construction is three steps: (1) Draw all α-arc edges described above (these edges may cross for $\alpha < \pi/2$). (2) For each α-arc $p_i p_j$, create an α-bend polyline (p_i, b, p_j) such that b is the midpoint of the α-arc, which implies that $\Delta(p_i b p_j)$ is an isosceles triangle. Such an α-bend edge $p_i p_j$ is an x-monotone arc for any $\alpha \in (0, \pi)$. Consequently, these α-bend edges do not cross, but adjacent edges may partially overlap. (3) We successively perturb the α-bend edges as follows; see Fig. 4 (right). We perform the perturbation in each halfplane independently. Consider the edges $p_i p_j$ in the upper (resp., lower) halfplane ordered by nonincreasing length. For an edge $p_i p_k$, if both $p_i p_j$ and $p_j p_k$ are edges for some $i < j < k$, then perturb both $p_i p_j$ and $p_j p_k$ by slightly moving their bend points along the corresponding α-arc towards each other (i.e., counterclockwise and clockwise). The perturbation eliminates the overlap between adjacent edges, and yields an α-bend graph with $3n - 6$ edges and an α-bend multigraph with $4n - 6$ edges.   □

Corollary 1. *For every integer $n \geq 3$, we have $M_b(n, \alpha) = 3n - 6$ for all $\alpha \in (0, \pi)$; and $M_a(n, \alpha) = 3n - 6$ for all $\alpha \in [\frac{\pi}{2}, \pi)$.*

3   Asymptotics: Large and Small Angles

In this section, we study the maximum number of edges in α-bend and α-arc graphs as the angle α tends to 0 or π.

3.1   Large Angles

For a sufficiently large α, the α-bend and α-arc edges are similar to straight-line edges.

Proposition 3. *For every $P \subset \mathbb{R}^2$ in general position, there exists a threshold $\alpha_0 \in (0, \pi)$ such that, for every $\alpha \in (\alpha_0, \pi)$, we have $M_a(P, \alpha) = M_b(P, \alpha) = M(P)$ and $M_a^{\parallel}(P, \alpha) = M_b^{\parallel}(P, \alpha) = 2M(P)$.*

Proof. Let G be a planar straight-line graph on P with the maximum number of edges (i.e., with $M(P)$ edges). Then G is a triangulation of the convex hull of P. Let β be the minimum interior angle over all triangular faces of G, and $\alpha_1 = \pi - \beta$. Recall that the bisectors of the interior angles of a triangle T meet the center $c(T)$ of the inscribed circle of T. Subdivide each triangular face T of G into three subtriangles by connecting the corners of T to the center $c(T)$. Then each subtriangle incident to two points in P, and contains an α-arc edge between them for any $\alpha \in (\alpha_1, \pi)$. The outer face of G is the exterior of $\text{conv}(P)$, and it contains an α-arc edge between any two consecutive vertices of $\text{conv}(P)$ for any $\alpha \in (\frac{\pi}{2}, \pi)$. Furthermore, these α-arc edges are pairwise noncrossing. Overall, we can find $2|E(G)| = 2M(P)$ pairwise noncrossing α-arc edges, which form an α-arc multigraph on P. These constructions show that $M_a^{\parallel}(P, \alpha) \geq 2M(P)$ and $M_b^{\parallel}(P, \alpha) \geq 2M(P)$ for $\alpha \in (\alpha_1, \pi)$.

By choosing an arbitrary bend point in each α-arc edge, construct $2|E(G)| = 2M(P)$ pairwise noncrossing α-bend edges. By deleting double edges, we also obtain α-arc and α-bend graphs with $|E(G)| = M(P)$ edges. Consequently, $M_a(P, \alpha) \geq M(P)$ and $M_b(P, \alpha) \geq M(P)$ for $\alpha \in (\alpha_1, \pi)$.

For matching upper bounds, let α_2 be the maximum angle between any two adjacent straight-line edges determined by P. Then, given any α-arc graph HG on P, the convex hull of any edge ab does not contain any other vertices in P. Consequently, we can replace each α-arc edge in H with a straight-line edge, and obtain a planar straight-line graph with $|E(H)|$ edges. This proves $M_a(P,\alpha) \leq M(P)$ and $M_b(P,\alpha) \leq M(P)$ for $\alpha \in (\alpha_2, \pi)$. For multigraphs, up to two parallel edges could be replaced by a straight-line edge by Proposition 1, consequently $M_a^{\parallel}(P,\alpha) \geq 2M(P)$ and $M_b^{\parallel}(P,\alpha) \geq 2M(P)$ for $\alpha \in (\alpha_2, \pi)$.

Overall, we put $\alpha_0 = \max\{\alpha_1, \alpha_2\}$, and then for all $\alpha \in (\alpha_0, \pi)$, we have $M_a(P,\alpha) = M_b(P,\alpha) = M(P)$ and $M_a^{\parallel}(P,\alpha) = M_b^{\parallel}(P,\alpha) = 2M(P)$.   □

3.2 Small Angles

In the other end of the spectrum, for sufficiently small $\alpha > 0$, both $M_b(P,\alpha)$ and $M_b^{\parallel}(P,\alpha)$ have the same behavior as for collinear points.

Proposition 4. *For every set P of n points in the plane, there exists a threshold $\alpha_0 \in (0, \pi)$ such that for all $\alpha \in (0, \alpha_0)$, we have $M_b(P,\alpha) = 3n - 6$ and $M_b^{\parallel}(P,\alpha) = 4n - 6$.*

Proof. By applying a rotation, if necessary, we may assume that the points in P have distinct x-coordinates. Let $P = \{p_0, \ldots, p_{n-1}\}$ be sorted by (increasing) x-coordinate. Let α_0 be the minimum angle between a vertical line and a segment $p_{i-1}p_i$ for $i = 1, \ldots, n-1$. Now for any angle $\alpha \in (0, \alpha_0)$, we can follow the argument in the proof of Proposition 2; see Fig. 5 (left). The x-monotone path $(p_0, \ldots, p_{n-1})$ plays the role of the x-axis: It separates upper and lower edges. We initially draw each α-bend edge $p_i p_j$ so that its two segments have slopes $\pm \cot \frac{\alpha}{2}$, and then perturb overlapping edges as in the proof of Proposition 2.   □

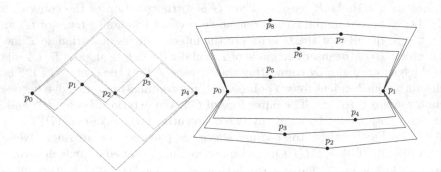

Fig. 5. Left: an α-bend multigraph with $4n - 6$ edges; the red edges form an α-bend graph with $3n - 6$ edges. Right: an α-bend graph with $3n - 7 = M(P)$ edges for a point set P with a quadrilateral convex hull. (Color figure online)

For points in general position and for α-bend graphs, we can take the threshold α_0 in Proposition 4 to be $\frac{\pi}{2}$; see Fig. 5 (right).

Proposition 5. *For every set $P \subset \mathbb{R}^2$ of $n \geq 3$ points in general position, and for every $\alpha \in (0, \frac{\pi}{2}]$, we have $M_b(P,\alpha) \geq 3n - 7$ and $M_b(P,\alpha) \geq M(P)$.*

Proof. Let $\{p_0, p_1\} \subset P$ be a diametric pair of points, which maximizes the pairwise distances in P. By applying a rotation, if necessary, we may assume that the line segment $p_0 p_1$ is horizontal. Since P is in general position, no two points in $P \setminus \{p_0, p_1\}$ have the same y-coordinate. Label the points in P such that p_0 is the leftmost point, p_1 is the rightmost point, and $P \setminus \{p_0, p_1\} = \{p_2, \ldots, p_{n-1}\}$ sorted in increasing y-coordinates; see Fig. 5 (right).

We construct an α-bend graph on P as follows. For all $i \in \{2, \ldots, n-1\}$, add α-bend edges $p_0 p_i$ and $p_i p_1$ such that the edge segments incident to p_i are horizontal, and lie to the left and right of p_i, respectively. Since $\alpha \in (0, \frac{\pi}{2}]$, then the edge segments incident to p_0 lie in the closed halfplane left of p_0; and the edge segments incident to p_1 lies to the right of p_1. Note also that the edge segments incident to p_0 and p_1 may overlap: We perturb these edges to maintain α-bend edges, but eliminate the overlap (as a result, the edge segments incident to p_i are no longer horizontal, but almost horizontal). Add the edge $p_0 p_1$ as well, where the edge segment incident to p_0 is almost horizontal, and the edge segment incident to p_1 is very short. We have added $2(n-2) + 1 = 2n - 3$ edges so far.

For all $i \in \{2, \ldots, n-3\}$, if both p_i and p_{i+1} are on the same side of the horizontal line $p_0 p_1$, then we add an α-bend edge $p_i p_{i+1}$ in the horizontal strip between p_i and p_{i+1}. The edge segment incident to p_i should almost horizontal (but disjoint from the edges $p_0 p_i$ and $p_1 p_i$), and on the same side (left or right) of p_i that contains p_{i+1}. This determines the direction of the edge segment incident to p_{i+1}. There is at most one $i \in \{2, \ldots, n-3\}$ such that p_i and p_{i+1} are on opposite sides of the line $p_0 p_1$, so we add at least $n-4$ edges. We obtain an α-bend graph with $(2n-3) + (n-4) = 3n-7$ edges. This proves that $M(P, \alpha) \geq 3n - 8$.

If the convex hull of P contains 4 or more points, then $M(P) \leq 3n - 7$, consequently $M(P, \alpha) \geq M(P)$. Suppose that the convex hull of is a triangle. Then $p_0 p_1$ is one side of the triangle, and all other points lie on one side of the line $p_0 p_1$. In this case, we add edges $p_i p_{i+1}$ for all $i \in \{2, \ldots, n-3\}$, and the we obtain an α-bend graph with $(2n-3) + (n-3) = 3n - 6 = M(P)$ edges. $\square$

For circular arcs, the number of edges goes down to 2 as α tends to zero.

Proposition 6. *For every finite $P \subset \mathbb{R}^2$ in general position, there exists a threshold $\alpha_0 \in (0, \pi)$ such that, for every $\alpha \in (0, \alpha_0)$, we have $M_a(P, \alpha) = 1$ and $M_a^{\parallel}(P, \alpha) = 2$.*

Proof. Let $P \subset \mathbb{R}^2$ be a set of n points in general position, let W be the set of intersection points of the $\binom{n}{2}$ lines spanned by P. Let $\varepsilon > 0$ be so small that the ε-radius disks centered at the points in W are pairwise disjoint, and let D be a large disk that contains the ε-disks centered at all points in W. Finally, let α_0 be so small that for every point pair $\{p, q\} \subset P$, the intersection of disk D and the circle containing an α_0-arc edge is in the ε-neighborhood of the line spanned by pq. Note that the same property holds for α-arc edges for all $\alpha \in (0, \alpha_0)$.

For the lower bound, it is clear that an α-arc graph can contain any one α-arc edge pq, and an α-arc multigraph can contain any α-arc double-edge pq.

For the upper bound, suppose to the contrary, that $\alpha \in (0, \alpha_0)$ and a plane α-arc graph G contains two α-arc edges p_1q_1 and p_2q_2, where $\{p_1, q_1\} \neq \{p_2, q_2\}$. The circles containing the α-arc edges p_1q_1 and p_2q_2 cross at some point in the ε-neighborhood of the intersection point of lines p_1q_1 and p_2q_2, which is in D. Consequently they also cross at another point, say x, outside of D. The point x lies on both α-arc edges, and so these edges cross, contradicting the assumption that G is a plane graph. $\square$

A combination of Propositions 2 and 6 yields the following.

Corollary 2. *For every finite $P \subset \mathbb{R}^2$, where the maximum number of collinear points is k, there exists a threshold $\alpha_0 \in (0, \pi)$ such that, for every $\alpha \in (0, \alpha_0)$, we have $M_a(P, \alpha) = 2k - 3$ and $M_a^{\parallel}(P, \alpha) = 2k - 2$.*

4   The Size of α-Bend and α-Arc Multigraphs

Let G be an α-bend multigraph. The union of two parallel edges between vertices a and b is a closed curve that contains the line segment ab in its interior by Proposition 1. Let $\overline{G}$ denote the straight-line graph comprising a straight-line edge for each double edge in G. A *lens* of G is the interior of the closed curve formed by two parallel edges. A lens is *empty* if it does not contain any vertex of G. Note that if all double edges in G form empty lenses, then $\overline{G}$ is a planar straight-line graph.

Lemma 1. *For $\alpha \in (0, \frac{2\pi}{3}]$, let G be an α-bend multigraph where every double edge is an empty lens, and let $C = (p_1, \ldots, p_m)$ be a cycle of double edges. Then the edges of G in the (straight-line) polygon $\overline{C}$ do not contain a triangulation of C. (In particular, G does not contain any 3-cycle of double edges, and there is no diagonal in a 4-cycle of double edges.)*

Proof. Suppose, for contradiction, that $C = (p_1, \ldots, p_m)$ is a cycle of double edges and the edges of G in the (straight-line) polygon $\overline{C}$ form a triangulation of C; see Fig. 6. Let pq be an edge in C. Then two closed curves pass through p and q: An empty lens formed by two α-bend edges, and the straight-line cycle $\overline{C}$. These curves can cross only at p and q. The straight-line segment pq is in the interior of the empty lens, and all vertices of $\overline{C}$ are in its exterior. Consequently, one of the α-bend edges between p and q lies in the interior of $\overline{C}$.

Let T be a dual graph of the triangulation of $\overline{C}$, where the nodes correspond to triangles, and two nodes are adjacent if the corresponding triangles share an edge. The dual graph is a tree T. We define an orientation on the edges of T as follows. Consider two adjacent triangles, t_1 and t_2, that share an edge p_ip_j. Direct the edge as (t_1, t_2) if and only if the α-bend edge p_ip_j and triangle t_2 are on the same halfplane of line p_ip_j. Every directed tree contains a sink. Let $t = \Delta(p_ip_jp_k)$ be a sink.

From the discussion above, the straight-line triangle $\Delta(p_ip_jp_k)$ contains three α-bend edges: p_ip_j, p_jp_k, and p_kp_i. The concatenation of these edges is a simple

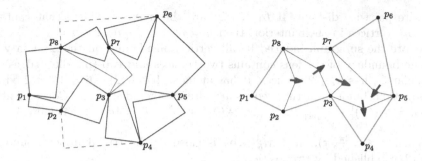

Fig. 6. Left: a cycle of double edges $C = (p_1, \ldots, p_m)$, with two internal and two external diagonals. Right: The corresponding polygon $\overline{C}$, a triangulation of $\overline{C}$ with one-bend edges, and the directions of the dual edges.

hexagon H (formed by three vertices and three bend points). At every bend point the interior angle of H is $2\pi - \alpha > 2\pi - \frac{2\pi}{3} = \frac{4\pi}{3}$. The sum of these three interior angles is greater than 4π. The sum of all interior angles of a hexagon, however, is at most $(6-2)\pi = 4\pi$: a contradiction. $\qquad\square$

Lemma 2. *For $\alpha \in (0, \frac{2\pi}{3}]$, let G be an α-bend multigraph on $n \geq 3$ vertices where every double edge is an empty lens. Then G has at most $4n - 7$ edges.*

Proof. If the double edges do not form any cycle, then there are at most $n - 1$ double edges, hence G has at most $(3n-6)+(n-1) = 4n-7$ edges, as required. Assume now that G contains a cycle of double edges.

Consider the graph $\overline{G}$ (where each straight-line edge represents a double edge in G). By Lemma 1, $\overline{G}$ is triangle-free. By Euler's polyhedron formula, if $\overline{G}$ has f bounded faces, it has $n + f - 1$ edges. That is, G has $n + f - 1$ double edges. By Lemma 1, each bounded face of $\overline{G}$ contains a face of G with 4 or more vertices. By replacing the $n + f - 1$ double edges with single edges, and triangulating each face that has 4 or more vertices, we obtain a triangulation T. Note that T may have parallel edges in the interior and exterior of a face f, but parallel edges cannot form an empty lens, and so T has at most $3n - 6$ edges. Consequently, $|E(G)| - (n + f - 1) + f \leq 3n - 6$, which yields $|E(G)| \leq 4n - 7$, as claimed. $\quad\square$

Theorem 1. *For $\alpha \in (0, \frac{2\pi}{3}]$ and $n \geq 2$, we have $M_b^{\parallel}(n, \alpha) \leq 4n - 6$.*

Proof. We proceed by induction on n. The base case $n = 2$ trivially follows from Proposition 1. Let G be an α-bend graph on $n > 2$ vertices, and assume that the theorem holds for every subgraph of G on fewer than n vertices. If all double edges are empty-lenses, then $|E(G)| \leq 4n - 7$ by Lemma 2. Otherwise, delete all vertices that lie in the interior of a lens (i.e., a cycle formed by double edges), and let G' be the induced subgraph of the remaining $n' \geq 2$ vertices. Then G' has at most $4n' - 7$ edges by Lemma 2 if $n' \geq 3$; and at most $2 = 4n' - 6$ by Proposition 1 if $n' = 2$. In both cases, G' has at most $4n' - 6$ edges.

Note that the lenses of a plane multigraph form a laminar system (i.e., two lenses are either disjoint or one contains the other), and so the maximal lenses of G are pairwise disjoint. If $L_1, \ldots, L_k$ are the maximal lenses that contain $n_1, \ldots n_k$ vertices in their interior, then $n = n' + \sum_{i=1}^{k} n_i$. For $i = 1, \ldots, k$ let G_i denote the subgraph induced by all vertices inside and on the boundary of L_i. The boundary of the lens contains two vertices and two (parallel) edges. In particular G_i has $n_i + 2$ vertices. It has at most $4(n_i + 2) - 6 = 4n_i + 2$ edges by induction, but two of these edges are already included in G'. Consequently, $|E(G)| = |E(G')| + \sum_{i=1}^{k} 4n_i \leq (4n' - 6) + 4 \sum_{i=1}^{k} n_i = 4n - 6$, as claimed. $\quad\square$

When $\alpha \in (\frac{2\pi}{3}, \pi)$, then $M_b^{\parallel}(n, \alpha)$ is close to the trivial upper bound of $6n - 12$, established by Proposition 1.

Proposition 7. *For $\alpha \in (\frac{2\pi}{3}, \pi)$, we have $M_b^{\parallel}(n, \alpha) \geq 6n - O(\sqrt{n})$.*

Proof. Let n points be arranged in a section of a triangular grid. The unit-length edges form a plane graph, with $3n - O(\sqrt{n})$ edges, where all bounded faces are equilateral triangles. We can replace each unit-length edge by two α-bend edges, which form an empty lens, such that the two segments of each edge have equal length. Each equilateral triangle contains three α-bend edges, which are crossing-free. This yields a plane α-bend multigraph with $6n - O(\sqrt{n})$ edges. $\quad\square$

For α-arc multigraphs, Lemma 2, Theorem 1, and Proposition 7 carry over with essentially the same proof, but the angle threshold increases from $\alpha \leq \frac{2\pi}{3}$ to $\alpha \leq \frac{5\pi}{6}$, as a triangle $\Delta(p_1p_2p_3)$ cannot contain three α-arc edges between its vertices for $\alpha \leq \frac{5\pi}{6}$. We summarize the result and omit the details.

Theorem 2. *For $\alpha \in (0, \frac{5\pi}{6}]$ and $n \geq 2$, we have $M_a^{\parallel}(n, \alpha) \leq 4n - 6$. For $\alpha \in (\frac{5\pi}{6}, \pi)$, we have $M_a^{\parallel}(n, \alpha) \geq 6n - O(\sqrt{n})$.*

5   Applications to $\alpha AC_1^{=}$ Graphs

Ackerman et al. [2] proved that every n-vertex $\alpha AC_1^{=}$ graph has at most $27n$ edges, for every $\alpha \in (0, \frac{\pi}{2}]$. More precisely, they proved an upper bound of $24.5n$ if $\alpha \neq \frac{\pi}{3}$ and $27n$ if $\alpha = \frac{\pi}{3}$. We improve these bounds to $18.5n$ and $21n$, resp., using the same general strategy combined with Theorem 1 from Sect. 4. For $\alpha = \frac{\pi}{2}$ (i.e., for RAC$_1$-graphs), however, an asymptotically tight bound of $5.5n - O(1)$ is known [4]. We recall Lemma 2.1 from [2].

Lemma 3 (Ackerman et al. [2]). *Let $\alpha \in (0, \frac{\pi}{2}]$ and let S be a finite set of line segments in the plane such that any two segments may cross only at angle α. Then S can be partitioned into at most three subsets of pairwise noncrossing segments. Moreover, if $\frac{\pi}{\alpha}$ is irrational or if $\frac{\pi}{\alpha} = \frac{p}{q}$, where $\gcd(p, q) = 1$ and p is even, S can be partitioned into at most two subsets of pairwise noncrossing segments.*

The line segments in S may overlap (i.e., intersect in a line segment of positive length). Crossings between line segments is usually defined as follows: Two segments *cross* if they intersect in a single point that lies in the relative interior of both segments. For the application to $\alpha AC_1^=$ graphs, we relax the definition of crossings: Given a set S of line segments and a set V of segment endpoints, two segments in S *cross* if their intersect in a single point that is not in V. The following lemma strengthens Lemma 3, and holds under either notion of crossing.

Lemma 4. *Let $\alpha \in (0, \frac{\pi}{2}]$ and let S be a finite set of line segments in the plane such that any two segments may cross only at angle α. If $\frac{\pi}{\alpha}$ is irrational or if $\frac{\pi}{\alpha} = \frac{p}{q}$, where $\gcd(p,q) = 1$ and $2 \mid p$, then there exists a subset $S' \subset S$ of pairwise noncrossing segments with $|S'| \geq \frac{1}{2}|S|$. Else $\frac{\pi}{\alpha} = \frac{p}{q}$, where $\gcd(p,q) = 1$ and $p = 2k+1$ for some $k \in \mathbb{N}$, and then there exists a subset $S' \subset S$ of pairwise noncrossing segments with $|S'| \geq \frac{k}{2k+1}|S|$.*

Proof. If $\frac{\pi}{\alpha}$ is irrational or if $\frac{\pi}{\alpha} = \frac{p}{q}$, where $\gcd(p,q) = 1$ and $2 \mid p$, then the claim follows directly from Lemma 3.

Assume that $\frac{\pi}{\alpha} = \frac{p}{q}$, where $\gcd(p,q) = 1$ and $p = 2k+1$ is an odd integer. The *direction* of a segment $s \in S$, denoted $\mathrm{dir}(s)$, is the minimum counterclockwise angle from the x-axis to a line parallel to s. Note that $\mathrm{dir}(s) \in [0, \pi)$. Let $D = \{\mathrm{dir}(s) : s \in S\}$, that is, the set of directions of the segments in S. For each direction $d \in D$, let $S(d) = \{s \in S : \mathrm{dir}(s) = d\}$ be the set of segments of direction d.

We define a vertex-weighted graph $G_D = (D, E_D)$, in which two directions $d_1, d_2 \in D$ are joined by an edge if and only if they differ by α; and the weight of a direction $d \in D$ is the cardinality of $S(d)$. Clearly, the maximum degree of a vertex in G_D is at most two, and so G_D is the disjoint union of paths and cycles. Furthermore, if $d_1, d_2 \in D$ are in the same component of G_D, then they differ by a multiple of $\alpha = \frac{q}{p} \cdot \pi$. Since $\gcd(p,q) = 1$, then $m \cdot \alpha \equiv 0 \bmod \pi$ if and only if $m \equiv 0 \bmod p$. Consequently, every cycle in G_D is isomorphic to $C_p = C_{2k+1}$.

Note that if segment $s_1, s_2 \in S$ cross, then $\mathrm{dir}(s_1)$ and $\mathrm{dir}(s_2)$ are adjacent in G_D. We can now construct a subset S' of S. In each connected component H of G_D, we choose a maximum independent set $I(H)$ as follows. (1) If H is a path, then it is 2-colorable, and the weight of one of the color classes is at least half of the weight of H; let $I(H)$ be such a color class. (2) If H is isomorphic to C_{2k+1}, then it has $2k + 1$ maximum independent sets (each containing k vertices), and every vertex lies in precisely k independent sets. By the pigeonhole principle, the weight of an independent set is at least $\frac{k}{2k+1}$ times the weight of H; let $I(H)$ be such an independent set. Let $I \subset D$ be the union of independent sets over all components H of G_D; and note that I is an independent set in G_D, and its weight is at least $\frac{k}{2k+1}$ times the weight of G_D, which is $|S|$. Now let S' be the union of the sets $S(d)$ for all $d \in I$. Clearly, the segments in S' are pairwise disjoint and $|S'| \geq \frac{k}{2k+1}|S|$, as required. $\square$

Corollary 3. *Let $\alpha \in (0, \frac{\pi}{2}]$ and let S be a finite set of line segments in the plane such that any two segments may cross only at angle α. Then there exists a subset $S' \subset S$ of pairwise noncrossing segments with $|S'| \geq |S|/3$.*

Lemma 5. *Let $\alpha \in (0, \frac{\pi}{2}]$, and let $G = (V, E)$ be a graph on $n \geq 3$ vertices that admits an $\alpha AC_1^=$ drawing such that for every edge $e \in E$, both edge segments cross at least one other edge in E. Then*

(i) $|E| \leq 15n - 27$ for all $\alpha \in (0, \frac{\pi}{2}]$;
(ii) $|E| \leq 12n - 18$ for $\alpha \in (0, \frac{\pi}{3})$;
(iii) $|E| \leq 10n - 15$ if $\alpha \in (\frac{\pi}{3}, \frac{\pi}{2})$; and
(iv) $|E| \leq 10n - 18$ if $\frac{\pi}{\alpha}$ is irrational or $\frac{\pi}{\alpha} = \frac{p}{q}$, where $\gcd(p, q) = 1$ and $2 \mid p$.

Proof. (i) Let S be the set of edge segments of all edges in E; hence $|S| = 2|E|$. For each segment $s \in S$, create a directed path $\gamma(s)$ as follows: Start from the (unique) vertex in V incident to s, then follow s until the first crossing with some other segment s', and then follow s' to the (unique) vertex in V incident to s'. Each path $\gamma(s)$ is either an α-bend edge or a $(\pi - \alpha)$-bend edge. We make three observations: (1) For every $s \in S$, the first segment of $\gamma(s)$ is crossing free; (2) if $s_1 \neq s_2$, then $\gamma(s_1) \neq \gamma(s_2)$, since the initial segments of the γ paths are distinct; (3) for $s_1 \neq s_2$, the paths $\gamma(s_1)$ and $\gamma(s_2)$ may correspond to the same undirected polygonal path with opposite directions.

Let $\Gamma = \{\gamma(s) : s \in S\}$, and let (V, Γ) be the *red graph*. Let $\Gamma_1 \subset \Gamma$ be the set of red edges where the second segment crosses some other edge in Γ; and let $\Gamma_2 = \Gamma \setminus \Gamma_1$ be the set of red edges where both segments are crossing-free.

Two edges in Γ_1 cannot follow the same path in opposite directions because the first segment of every red edge is crossing-free. Let S_1 be the set of second segments of the red edges in Γ_1. By Corollary 3, there is a subset $S_1' \subseteq S_1$ of pairwise noncrossing segments of size $|S_1'| \geq \frac{1}{3}|S_1|$. Let Γ_1' be the set of edges in Γ_1 whose second segment lies in S_1', with $|\Gamma_1'| \geq \frac{1}{3}|\Gamma_1|$. If Γ_2 contains two edges that follow the same path in opposite directions, then omit one arbitrarily, and let $\Gamma_2' \subseteq \Gamma_2$ be the remaining edges, with $|\Gamma_2'| \geq \frac{1}{2}|\Gamma_2|$.

Consider the multigraph $(V, \Gamma_1' \cup \Gamma_2')$, and note that $|\Gamma_1' \cup \Gamma_2'| \geq \frac{1}{3}|\Gamma_1| + \frac{1}{2}|\Gamma_2| \geq \min\{\frac{1}{3}, \frac{1}{2}\}(|\Gamma_1| + |\Gamma_2|) = \frac{1}{3}|\Gamma|$. The edges in $\Gamma_1' \cup \Gamma_2'$ are pairwise noncrossing but they may overlap. However, if a segment of $\gamma(s) \in \Gamma_1' \cup \Gamma_2'$ overlaps with another edge in $\Gamma_1' \cup \Gamma_2'$, then the other segment of $\gamma(s)$ is overlap-free. Indeed, the first segments of red edges and the edges in Γ_2' are pairwise nonoverlapping; and the second segments of edges in Γ_1' are pairwise noncrossing (in the strong sense). Consequently, we can partition $\Gamma_1' \cup \Gamma_2'$ into subsets of pairwise overlapping red edges; and we can perturb the edges in each subset to remove overlaps while maintaining the angles between segments. After perturbation, $(V, \Gamma_1' \cup \Gamma_2')$ is a plane multigraph, composed of α-bend and $(\pi - \alpha)$-bend edges.

Let $\Gamma_3 \subset \Gamma_1' \cup \Gamma_2'$ be the set of α-bend edges; and let $\Gamma_4 \subseteq \Gamma_1' \cup \Gamma_2'$ be the set of $(\pi - \alpha)$-bend edges. By Theorem 1, $|\Gamma_3| \leq 4n - 6$. In general, the multiplicity of every edge in (V, Γ_4) is at most two by Proposition 1, and so $|\Gamma_4| \leq 2(3n - 6) = 6n - 12$. Hence we have $|\Gamma| \leq 3|\Gamma_1' \cup \Gamma_2'| = 3|\Gamma_3 \cup \Gamma_4| \leq 3\big((4n - 6) + (6n - 12)\big) = 30n - 54$. Overall, this yields $|E| = \frac{1}{2}|\Gamma| \leq 15n - 27$.

(ii) When $\alpha < \frac{\pi}{3}$, we can apply Theorem 1 for both α-bend and $(\pi - \alpha)$-bend multigraphs. Consequently, $|\Gamma_1' \cup \Gamma_2'| = |\Gamma_3| + |\Gamma_4| \leq 2(4n - 6) = 8n - 12$. This yields $|\Gamma| \leq 3|\Gamma_1' \cup \Gamma_2'| \leq 3(8n - 12) = 24n - 36$, hence $|E| = \frac{1}{2}|\Gamma| \leq 12n - 18$.

(iii) When $\alpha \in [\frac{\pi}{3}, \frac{\pi}{2})$, then $\pi - \alpha < 2\pi/3$, and $|\Gamma_4| \leq 4n - 6$ by Theorem 1. Consequently, $|\Gamma_1' \cup \Gamma_2'| = |\Gamma_3| + |\Gamma_4| \leq 2(4n - 6) = 8n - 12$. If $\frac{\pi}{\alpha} = \frac{p}{q}$ with $\gcd(p,q) = 1$ and $q = 2k + 1$ is an odd integer, then $k \geq 2$ and $|S_1'| \geq \frac{2}{5}|\Gamma_1|$ by Lemma 4. This yields $|\Gamma| \leq 2.5\,|\Gamma_1' \cup \Gamma_2'| \leq 2.5(8n - 12) = 20n - 30$, hence $|E| = \frac{1}{2}|\Gamma| \leq 10n - 15$.

(iv) When $\frac{\pi}{\alpha}$ is irrational or $\frac{\pi}{\alpha} = \frac{p}{q}$ with $\gcd(p,q) = 1$ and $2 \mid p$, then the size of S_1' is bounded by $|S_1'| \geq \frac{1}{2}|\Gamma_1|$ by Lemma 4. This yields $|\Gamma| \leq 2(10n - 18) = 20n - 36$, hence $|E| = \frac{1}{2}|\Gamma| \leq 10n - 18$.   □

Theorem 3. *If $G = (V, E)$ is an $\alpha AC_1^=$ graph with $n \geq 3$ vertices, then*

(i) $|E| \leq 21n - 36$ for $\alpha = \frac{\pi}{3}$;
(ii) $|E| \leq 18.5n - 34$ for $\alpha \in (0, \frac{\pi}{3})$;
(iii) $|E| \leq 16.5n - 31$ if $\alpha \in (\frac{\pi}{3}, \frac{\pi}{2}]$; and
(iv) $|E| \leq 16n - 30$ if $\frac{\pi}{\alpha}$ is irrational or $\frac{\pi}{\alpha} = \frac{p}{q}$, where $\gcd(p,q) = 1$ and $2 \mid p$.

Proof. Let $\alpha \in (0, \frac{\pi}{2}]$; and let $G = (V, E)$ be a graph with $n \geq 4$ vertices with an $\alpha AC_1^=$ drawing. Let $E_1 \subseteq E$ denote the set of edges in E that have at least one crossing-free end segment. Let $G_1 = (V, E_1)$ and $G_2 = (V, E \backslash E_1)$. By Corollary 3, there is a subset $S_1' \subseteq S_1$ of pairwise noncrossing segments of size $|S_1'| \geq \frac{1}{3}|E_1|$. The graph G_1' corresponding to these edges is planar, with at most $3n - 6$ edges. Hence E_1 contains at most $3 \cdot (3n - 6) = 9n - 18$ edges. By Lemma 5(i), G_2 has at most $15n - 18$ edges. Hence, G has at most $(9n - 18) + (15 - 18) = 24n - 36$ edges in general. However, we can improve on this bound for all $\alpha \in (0, \frac{\pi}{2}]$.

(i) When $\alpha = \frac{\pi}{3}$, we can use Lemma 5(ii), which yields at most $(9n - 18) + (12n - 18) = 21n - 36$ edges.

(ii) When $\alpha \in (0, \frac{\pi}{2})$ and $\alpha \neq \frac{\pi}{3}$, then G_1 is quasi-planar (i.e., it does not contain three pairwise crossing edges). Indeed, suppose to the contrary that three edges in G_1 in pairwise cross. As every edge in G_1 has two segments, one of which is crossing-free, then these edges contain three segments that pairwise cross. These segments form a triangle, in which every interior angle is α or $\pi - \alpha$. Since interior angles of a triangle sum to π, it follows that $\alpha = \frac{\pi}{3}$; a contradiction.

It is known [1,3] that a simple quasi-planar n-vertex graph has at most $6.5n - 20$ for $n \geq 4$, hence at most $6.5n - 16$ edges for $n \geq 3$. Combined with Lemma 5(i), G has at most $18.5n - 34$ edges.

(iii) When $\alpha \in (\frac{\pi}{3}, \frac{\pi}{2})$, then again G_1 is quasi-planar, with at most $6.5n - 16$ edges due to [1,3]. Combined with Lemma 5(iii), G has at most $16.5n - 31$ edges.

(iv) When $\frac{\pi}{\alpha}$ is irrational or $\frac{\pi}{\alpha} = \frac{p}{q}$, where $\gcd(p,q) = 1$ and $2 \mid p$. Then we can assume $|S_1'| \geq \frac{1}{2}|E_1|$ by Lemma 4, hence $|E_1| \leq 2 \cdot (3n - 6) = 6n - 12$. Combined with Lemma 5(iv), G has at most $16n - 30$ edges in this case.   □

6   Conclusions and Open Problems

We have introduced α-bend and α-arc graphs and multigraphs for any $\alpha \in (0, \pi)$, and derived upper and lower bounds on the maximum number of edges, $M_a(P, \alpha)$ and $M_b(P, \alpha)$, for a point set P in these drawing styles. However, the computational complexity of the corresponding optimization problems is unknown.

Problem 1. Is it NP-hard to determine $M_a(P, \alpha)$ (resp., $M_b(P, \alpha)$) for a given point set $P \subset \mathbb{R}^2$ and angle $\alpha \in (0, \pi)$? Is it $\exists \mathbb{R}$-hard to determine $M_b(P, \alpha)$?

Intuitively, allowing one-bend edges instead of straight-line edges gives extra flexibility. In several cases, we have shown that an α-bend graph on a point set P has at least as many edges as a straight-line triangulation on P, that is, $M_b(P, \alpha) \geq M(P)$. However, we have also shown that α-arc edges become obstacles as α tends to zero, and $M_a(P, \alpha) < M(P)$ for all sufficiently small $\alpha > 0$. For α-bend edges in general, this remains an open problem.

Problem 2. Does there exist a finite point set $P \subset \mathbb{R}^2$ and an angle $\alpha \in (0, \pi)$ such that $M_b(P, \alpha) < M(P)$?

It is also an open problem to improve the upper or lower bounds for the number of edges in $\alpha AC_1^=$ graphs and $\alpha AC_2^=$ graphs.

Problem 3. Determine the maximum number of edges in an n-vertex $\alpha AC_1^=$ graph and an n-vertex $\alpha AC_2^=$ graph for all angles $\alpha \in (0, \frac{\pi}{2}]$.

Acknowledgments. Research on this paper was partially supported by the NSF award DMS 2154347.

References

1. Ackerman, E.: Quasi-planar graphs. In: Hong, S.H., Tokuyama, T. (eds.) Beyond Planar Graphs: Communications of NII Shonan Meetings, pp. 31–45. Springer, Singapore (2020). https://doi.org/10.1007/978-981-15-6533-5_3
2. Ackerman, E., Fulek, R., Tóth, C.D.: Graphs that admit polyline drawings with few crossing angles. SIAM J. Discret. Math. **26**(1), 305–320 (2012). https://doi.org/10.1137/100819564
3. Ackerman, E., Tardos, G.: On the maximum number of edges in quasi-planar graphs. J. Comb. Theory, Ser. A **114**(3), 563–571 (2007). https://doi.org/10.1016/j.jcta.2006.08.002
4. Angelini, P., Bekos, M.A., Förster, H., Kaufmann, M.: On RAC drawings of graphs with one bend per edge. Theor. Comput. Sci. **828–829**, 42–54 (2020). https://doi.org/10.1016/j.tcs.2020.04.018
5. Angelini, P., Bekos, M.A., Katheder, J., Kaufmann, M., Pfister, M., Ueckerdt, T.: Axis-parallel right angle crossing graphs. In: Proceedings of 31st European Symposium on Algorithms (ESA). LIPIcs, vol. 274, pp. 9:1–9:15. Schloss Dagstuhl (2023). https://doi.org/10.4230/LIPIcs.ESA.2023.9

6. Angelini, P., Eppstein, D., Frati, F., Kaufmann, M., Lazard, S., Mchedlidze, T., Teillaud, M., Wolff, A.: Universal point sets for drawing planar graphs with circular arcs. J. Graph Algorithms Appl. **18**(3), 313–324 (2014). https://doi.org/10.7155/jgaa.00324

7. Arikushi, K., Fulek, R., Keszegh, B., Moric, F., Tóth, C.D.: Graphs that admit right angle crossing drawings. Comput. Geom. **45**(4), 169–177 (2012). https://doi.org/10.1016/j.comgeo.2011.11.008

8. Bannister, M.J., Cheng, Z., Devanny, W.E., Eppstein, D.: Superpatterns and universal point sets. J. Graph Algorithms Appl. **18**(2), 177–209 (2014). https://doi.org/10.7155/jgaa.00318

9. Bekos, M.A., Förster, H., Kaufmann, M.: On smooth orthogonal and octilinear drawings: relations, complexity and Kandinsky drawings. Algorithmica **81**(5), 2046–2071 (2019). https://doi.org/10.1007/s00453-018-0523-5

10. Chaplick, S., Förster, H., Kryven, M., Wolff, A.: Drawing graphs with circular arcs and right-angle crossings. In: Proceedings of 17th Scandinavian Symposium and Workshops on Algorithm Theory (SWAT). LIPIcs, vol. 162, pp. 21:1–21:14. Schloss Dagstuhl (2020). https://doi.org/10.4230/LIPIcs.SWAT.2020.21

11. Di Battista, G., Vismara, L.: Angles of planar triangular graphs. SIAM J. Discret. Math. **9**(3), 349–359 (1996). https://doi.org/10.1137/S0895480194264010

12. Didimo, W.: Right angle crossing drawings of graphs. In: Hong, S., Tokuyama, T. (eds.) Beyond Planar Graphs, Communications of NII Shonan Meetings, pp. 149–169. Springer, Singapore (2020). https://doi.org/10.1007/978-981-15-6533-5_9

13. Didimo, W., Eades, P., Liotta, G.: Drawing graphs with right angle crossings. Theor. Comput. Sci. **412**(39), 5156–5166 (2011). https://doi.org/10.1016/j.tcs.2011.05.025

14. Didimo, W., Liotta, G., Montecchiani, F.: A survey on graph drawing beyond planarity. ACM Comput. Surv. **52**(1), 4:1–4:37 (2019). https://doi.org/10.1145/3301281

15. Dujmović, V., Gudmundsson, J., Morin, P., Wolle, T.: Notes on large angle crossing graphs. Chic. J. Theor. Comput. Sci. **2011** (2011). http://cjtcs.cs.uchicago.edu/articles/CATS2010/4/contents.html

16. Efrat, A., Fulek, R., Kobourov, S.G., Tóth, C.D.: Polygons with prescribed angles in 2D and 3D. J. Graph Algorithms Appl. **26**(3), 363–380 (2022). https://doi.org/10.7155/jgaa.00599

17. Everett, H., Lazard, S., Liotta, G., Wismath, S.K.: Universal sets of n points for one-bend drawings of planar graphs with n vertices. Discret. Comput. Geom. **43**(2), 272–288 (2010). https://doi.org/10.1007/s00454-009-9149-3

18. Garg, A.: New results on drawing angle graphs. Comput. Geom. **9**(1), 43–82 (1998). https://doi.org/10.1016/S0925-7721(97)00016-3

19. Löffler, M., Tóth, C.D.: Linear-size universal point sets for one-bend drawings. In: Di Giacomo, E., Lubiw, A. (eds.) GD 2015. LNCS, vol. 9411, pp. 423–429. Springer, Cham (2015). https://doi.org/10.1007/978-3-319-27261-0_35

20. Tóth, C.D.: On RAC drawings of graphs with two bends per edge. In: Bekos, M.A., Chimani, M. (eds). GD 2023. LNCS, vol. 14465, pp. 69–77. Springer, Cham (2023). https://doi.org/10.48550/arXiv.2308.02663, https://doi.org/10.1007/978-3-031-49272-3_5

21. Vijayan, G.: Geometry of planar graphs with angles. In: Proceedings of 2nd Symposium on Computational Geometry (SoCG), pp. 116–124. ACM Press, New York, NY (1986). https://doi.org/10.1145/10515.10528

Quantum Graph Drawing
[Best Student Paper]

Susanna Caroppo⬛, Giordano Da Lozzo(✉)⬛, and Giuseppe Di Battista⬛

Roma Tre University, Rome, Italy
{susanna.caroppo,giordano.dalozzo,giuseppe.dibattista}@uniroma3.it

Abstract. In this paper, we initiate the study of quantum algorithms in the Graph Drawing research area. We focus on two foundational drawing standards: *2-level drawings* and *book layouts*. Concerning 2-level drawings, we consider the problems of obtaining drawings with the minimum number of crossings, k-planar drawings, quasi-planar drawings, and the problem of removing the minimum number of edges to obtain a 2-level planar graph. Concerning book layouts, we consider the problems of obtaining 1-page book layouts with the minimum number of crossings, book embeddings with the minimum number of pages, and the problem of removing the minimum number of edges to obtain an outerplanar graph. We explore both the quantum circuit and the quantum annealing models of computation. In the *quantum circuit model*, we provide an algorithmic framework based on Grover's quantum search, which allows us to obtain, at least, a quadratic speedup on the best classical exact algorithms for all the considered problems. In the *quantum annealing model*, we perform experiments on the quantum processing unit provided by D-Wave, focusing on the classical 2-level crossing minimization problem, demonstrating that quantum annealing is competitive with respect to classical algorithms.

Keywords: Quantum complexity · Grover's algorithm · QUBO · D-Wave · 2-Level drawings · Book layouts

1 Introduction

We initiate the study of quantum algorithms in the Graph Drawing research area[1].

The Problems. We focus on two foundational graph drawing standards: *2-level drawings* and *book layouts*. In a *2-level drawing* (refer, e.g., to [2, 4, 6, 10, 15, 16,

---

[1] Simultaneously to this research [13], Fukuzawa, Goodrich, and Irani [20] have formulated a model for quantum graph drawing in the circuit model, showing how to use Harrow's quantum algorithm [23] for solving linear systems to compute a so-called Tutte embedding of a 3-connected planar graph.

This research was supported, in part, by MUR of Italy (PRIN Project no. 2022ME9Z78 – NextGRAAL and PRIN Project no. 2022TS4Y3N – EXPAND). The authors are grateful to CINECA and D-Wave Systems Inc. for granting extended access to the D-Wave's Leap™ quantum cloud service to conduct experiments.

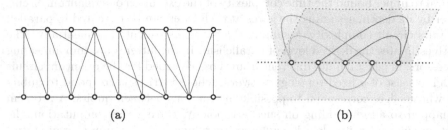

Fig. 1. Examples of graph drawing standards: (a) a 2-level drawing and (b) a 3-page book embedding.

19]), the graph is bipartite, the vertices are placed on two horizontal lines, and the edges are drawn as y-monotone curves; see Fig. 1a. In this drawing standard, we consider the search version of the TWO-LEVEL CROSSING MINIMIZATION (TLCM) problem, where given an integer ρ we seek a 2-level drawing with at most ρ crossings, and of the TWO-LEVEL SKEWNESS (TLS) problem, where given an integer σ we seek to determine a set of σ edges whose removal yields a *2-level planar graph*, i.e., a forest of caterpillars [18]. The minimum value of σ is the *2-level skewness* of the considered graph. We also consider the TWO-LEVEL QUASI PLANARITY (TLQP) problem, where we seek a drawing in which no three edges pairwise cross, i.e., a *quasi-planar* drawing, and the TWO-LEVEL k-PLANARITY (TLKP) problem, where we seek a drawing in which each edge participates to at most k crossings, i.e., a *k-planar* drawing. In a *book layout* (refer, e.g., to [7–9,11,37]), the drawing is constructed using a collection of half-planes, called *pages*, all having the same line, called *spine*, as their boundary; see Fig. 1b. The vertices lie on the spine and each edge is drawn on a page. In this drawing standard, we consider the search version of the ONE-PAGE CROSSING MINIMIZATION (OPCM) problem, where given an integer ρ we seek a 1-page layout with at most ρ crossings; the BOOK THICKNESS (BT) problem, where we search a τ-page layout where the edges in the same page do not cross, i.e., a τ-*page book embedding*; and the BOOK SKEWNESS (BS) problem, where given an integer σ we seek a set of σ edges whose removal yields a graph admitting a 1-*page book embedding*, i.e., it is outerplanar [9]. The minimum value of σ is the *book skewness* of the considered graph.

The Models. We delve into both the *quantum circuit* [29,31] and the *quantum annealing* [28] models of computation. In the former, quantum gates are used to compose a circuit that transforms an input superposition of qubits into an output superposition. The circuit design depends on both the problem and the specific instance being processed. The output superposition is eventually measured, obtaining the solution with a certain probability. The quality of the circuit is measured in terms of its *circuit complexity*, i.e., the number of elementary gates it contains, of its *depth*, i.e., the maximum length of a chain of elementary gates from the input to the output, and of its *width*, i.e., the maximum number of elementary gates "along a cut" separating the input from the output. It is natural to upper bound the time complexity of the execution of a quantum circuit either by its depth, assuming the gates at each layer can be executed in parallel, or by its circuit complexity, assuming the gates are executed sequentially. The width estimates the desired level of parallelism. In the latter, quantum annealing processors, in general quite different from those designed for the quantum circuit model, consist of a fixed-topology network, whose vertices correspond to qubits and whose edges correspond to possible interactions between qubits. A problem is mapped to an embedding on such a topology. During the computation, the solution space of a problem is explored, searching for minimum-energy states, which correspond to, in general approximate, solutions.

Our Contributions. In the quantum circuit model, we first show that the above graph drawing problems can be described by means of quantum circuits.

The first problem that we have to solve is choosing an effective representation for inputs and outputs. Since all problems we tackle require the selection of a permutation of n vertices, a tempting idea is to have as inputs-outputs binary variables explicitly representing the precedence between pairs of vertices. However, this requires to represent the superposition of a number of qubits which is quadratic in n. Hence, we use as inputs-outputs the vertex coordinates, that implicitly represent a permutation and require just $n \log n$ qubits. On the other hand, to solve the above problems, we have to transform the coordinates into orderings. Thus, the first contribution of the paper is a set of efficient quantum methods, that can be of general usage in *Quantum Graph Drawing*, that allow to transform coordinates of vertices into precedence between vertices and vice versa. These methods use a reduced number of "ancilla" qubits. Second, we present an algorithmic framework based on Grover's quantum search [22]. This framework enables us to achieve, at least, a quadratic speedup compared to the best exact classical algorithms for all the problems under consideration. Table 1 overviews our complexity results and compares them with classical algorithms. Within this framework we introduce quantum phase inversion methods composed by building blocks suitable to be combined to solve several types of graph drawing problems. In the quantum annealing model, we focus on the actual processing unit provided by D-Wave, which allows us to perform hybrid computations, which are partly classical and partly quantum. We first show that it is relatively easy to use D-Wave for implementing heuristics for the above problems. Second, we focus on the classical TLCM problem. Through experiments, we demonstrate that quantum annealing exhibits competitiveness when compared to classical algorithms. Figure 7 illustrates our experimental findings, whose raw data are provided in Table 2 of the full version of the paper [13].

State of the Art. Problems TLCM [21], TLQP [3], TLS [34], OPCM [27], BT [35], and BS [36] are NP-complete, whereas the complexity of TLKP is not known. FPT results with respect to the *natural parameters* (total number of crossings ρ, number of crossings per edge k, maximum number of allowed mutually crossing edges, number of pages τ, and number of edges to be removed σ) are known for TLCM [26], TLS [17], and OPCM [5]. Since TLQP and BT are para-NP-hard with respect to their natural parameter, FPT algorithms do not exist for these

Table 1. Results presented in this paper and comparison with exact classic algorithms. FPT algorithms are given, if any, with respect to the *natural parameter*. CC stands for Circuit Complexity. M denotes the number of problem solutions.

Problem	Classic Algorithm Running Time	Upperbound for m	FPT Time	Quantum Oracle Calls	Oracles		
					CC	Depth	Width
TLCM	$2^{n\log n}O(m^2)$	$O(\sqrt[3]{\rho\cdot n^2})$ [4]	$2^{O(\rho)}+n^{O(1)}$ [26]	$\frac{\pi}{4}\sqrt{\frac{2^{n\log n}}{M}}$	$O(m^2)$	$O(n^2)$	$O(m^2)$
TLKP	$2^{n\log n}O(m^2)$	$O(\sqrt{k}\cdot n)$ [4]	-	$\frac{\pi}{4}\sqrt{\frac{2^{n\log n}}{M}}$	$O(m^2)$	$O(m\log^2 m)$	$O(m)$
TLQP	$2^{n\log n}O(m^3)$	$O(n)$ [4]	Para-NP-hard [3]	$\frac{\pi}{4}\sqrt{\frac{2^{n\log n}}{M}}$	$O(m^6)$	$O(m^4)$	$O(m^2)$
TLS	$O(m^\sigma n)$	$O(n+\sigma)$	$2^{O(\sigma^3)}n$ [17]	$\frac{\pi}{4}\sqrt{\frac{2^{n\log n+\sigma\log m}}{M}}$	$O(m^2)$	$O(m)$	$O(m)$
OPCM	$2^{n\log n}O(m^2)$	$O(\sqrt[3]{\rho\cdot n^2})$ [32]	Courcelle's Th. [5]	$\frac{\pi}{4}\sqrt{\frac{2^{n\log n}}{M}}$	$O(n^8)$	$O(n^6)$	$O(m^2)$
BT	$2^{n\log n+m\log\tau}O(\tau n)$	$O(\tau\cdot n)$	Para-NP-hard [27]	$\frac{\pi}{4}\sqrt{\frac{2^{n\log n+m\log\tau}}{M}}$	$O(n^8)$	$O(n^6)$	$O(m)$
BS	$O(m^\sigma n)$	$O(n+\sigma)$	-	$\frac{\pi}{4}\sqrt{\frac{2^{n\log n+\sigma\log m}}{M}}$	$O(n^8)$	$O(n^6)$	$O(m)$

problems (unless $\mathsf{P} = \mathsf{NP}$). Upper bounds for the density of positive instances for all problems are given in Table 1. Let n and m denote the number of vertices and edges of an input graph, respectively. To the best of our knowledge, no exact algorithms for these problems are known that perform asymptotically better than trivial algorithms based on considering each candidate solution and verifying whether it is positive, which can easily be done in polynomial time for each solution. For problems TLCM, TLKP, TLQP, and OPCM, a candidate solution is uniquely identified by one of the possible $n! \in \Theta(2^{n\log n})$ vertex orderings. For problems TLS and BS, together with a vertex ordering, a candidate solution needs also to consider one of the possible $\binom{m}{\sigma} \in O(m^\sigma)$ choices of σ edges to be removed. For problem BT, together with a vertex ordering, a candidate solution needs also to consider one of the τ^m possible page assignments of the edges to τ pages.

Details of omitted and sketched proofs can be found in the full version of the paper [13].

2   Preliminaries

For basic concepts related to graphs and their drawings, we refer the reader to [14,33]. For the standard notation we use for quantum gates and circuits, and for basic concepts about quantum computation, we refer the reader to [29,31].

Let k be a positive integer. To ease the description, we will denote the value $\lceil \log_2 k \rceil$ simply as $\log k$, and the set $\{0, \dots, k-1\}$ as $[k]$. We refer to any of the permutations of the integers in $[k]$ as a k-*permutation*. A k-*set* is a set of size k.

Let X be a ground set. Let X_k be the set of all k-sets of distinct elements of X, i.e., $X_k = \{\{a_1, a_2, \dots, a_k\} | \forall i \neq j : a_i \neq a_j, a_i \in X\}$. A subset S of X_k is *cross-independent* if, for any two k-sets $s_i, s_j \in S$, it holds that $s_i \cap s_j = \emptyset$. To prove the depth bounds of our circuits, we will exploit the following.

Lemma 1. *The set X_k of all k-sets of distinct elements of a set X can be partitioned in $O(\sqrt{k^k} \cdot |X|^{k-1})$ cross-independent sets of size at most $\lfloor \frac{|X|}{k} \rfloor$, if $k < \frac{n+2}{3}$.*

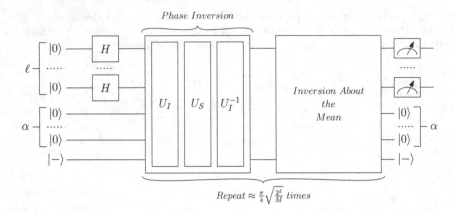

Fig. 2. The quantum graph drawing framework based on Grover's approach.

We introduce the mathematical formulations used in the D-Wave quantum annealing platform. A *constrained binary optimization* (CBO) is the mathematical formulation of an optimization problem, in which the variables are binary. In some cases, we focus on CBO formulations in which the objective function is not defined, and we aim at verifying whether a problem instance only satisfies the given constraints. A *quadratic unconstrained binary optimization* (QUBO) is the mathematical formulation of an optimization problem, in which the variables are binary, the optimization function is quadratic, and there are no constraints.

3   A Quantum Framework for Graph Drawing Problems

In this section, we establish a framework for dealing with several NP-complete graph drawing problems; refer to Fig. 2. The framework is based on Grover's approach for quantum search [22], which builds upon three circuits. The first circuit is a Hadamard gate that builds a uniform superposition of a sequence of qubits representing a potential, possibly not well-formed, solution to the problem. The second circuit exploits an *oracle* to perform the so-called PHASE INVERSION. The third circuit executes the so-called INVERSION ABOUT THE MEAN. The second and the third circuit are executed a number of times which guarantees that a final measure outputs a solution, if any, with high probability.

Theorem 1 (Grover's search [1,22]). *Let P be a search problem whose solutions can be represented using ℓ bits and suppose that there exists a* PHASE INVERSION *circuit for P with $c(\ell)$ circuit complexity and $d(\ell)$ depth. Assume that $c(\ell)$ and $d(\ell)$ are $\Omega(\log \ell)$. Then, there exists a quantum circuit that outputs a solution for P, if any, with $\frac{\pi}{4}\sqrt{\frac{2^\ell}{M}} \cdot O(c(\ell))$ circuit complexity and $\frac{\pi}{4}\sqrt{\frac{2^\ell}{M}} \cdot O(d(\ell))$ depth, where M is the number of solutions of P.*

Let n and m be the number of vertices and edges of an input graph G, respectively. Note that, in all the problems we consider, G admits the sought layout if and only if each of its connected components does. Hence, in the following, we assume that G is connected, and therefore $m \geq n - 1$. During the computation, we will manage a superposition $|\Gamma\rangle = |\Phi\rangle\,|\Psi\rangle\,|\Theta\rangle$, where $|\Phi\rangle$ is a superposition of $n \log n$ qubits, $|\Psi\rangle$ is a superposition of $m \log \tau$ qubits, and $|\Theta\rangle$ is a superposition of $\sigma \log m$ qubits. In particular, for some of the problems, $|\Psi\rangle$ and/or $|\Theta\rangle$ might not be present. We denote by ℓ the value $(n \log n) + (m \log \tau) + (\sigma \log m)$, where the second and/or third terms might be missing.

We denote the set of binary values $\{0, 1\}$ by $\mathbb{B}$. The superposition $|\Phi\rangle = \sum_{\phi \in \mathbb{B}^{n \log n}} c_\phi\,|\phi\rangle$ simultaneously represents all sequences of n natural numbers with values in $[n]$, each corresponding to a binary string ϕ of length $n \log n$. We denote by $\phi[i]$ the i-th natural number contained in ϕ. The purpose of $|\Phi\rangle$ is to represent the position of each vertex of G in a total order. To this aim, observe that, within the superposition $|\Phi\rangle$, all possible states corresponding to assignments of positions from 0 to $n - 1$ for each vertex in G are included. The purpose of $|\Psi\rangle$ is to represent a coloring of the edges of G with colors in $[\tau]$. The purpose of $|\Theta\rangle$ is to represent a subset of the edges of G of size at most σ, each labeled with an integer in $[m]$. For problems TLCM, TLKP, TLQP, and OPCM, we have that $|\Gamma\rangle = |\Phi\rangle$. For problem BT, we have that $|\Gamma\rangle = |\Phi\rangle\,|\Psi\rangle$. Finally, for problems TLS and BS, we have that $|\Gamma\rangle = |\Phi\rangle\,|\Theta\rangle$.

Next, we present an overview of how the superposition $|\Gamma\rangle$ evolves within the three main circuits of the framework; refer to Fig. 2. For space limitations, we will mostly focus only on problem TLCM, and hence we assume that $|\Gamma\rangle = |\Phi\rangle$. In [13], we give details for all the considered problems. We denote by $|0_k\rangle$ the quantum basis state composed of k qubits set to $|0\rangle$.

First, in all problems we study, ℓ qubits set to $|0\rangle$ enter an Hadamard gate that outputs the uniform superposition $|\Gamma\rangle = H^{\otimes \ell}\,|0_\ell\rangle = \frac{1}{\sqrt{2^\ell}} \sum_{\gamma \in \mathbb{B}^\ell} |\gamma\rangle$. Note that, within $|\Gamma\rangle$, all possible solutions of the considered problems are included, if any exist. Second, in Grover's approach, the INVERSION ABOUT THE MEAN circuit is prescribed. Hence, we now focus on the PHASE INVERSION circuit. In the first iteration, it receives as input **(i)** the uniform superposition $|\Gamma\rangle = H^{\otimes \ell}\,|0_\ell\rangle$, **(ii)** α ancilla qubits set to $|0\rangle$, where α depends on the type of problem, and **(iii)** a qubit set to $|-\rangle$. Namely, it receives as input the superposition $|\Gamma\rangle\,|0_\alpha\rangle\,|-\rangle$, where $|\Gamma\rangle = \frac{1}{\sqrt{2^\ell}} \sum_{\gamma \in \mathbb{B}^\ell} |\gamma\rangle$. It outputs the superposition $\frac{1}{\sqrt{2^\ell}} \sum_{\gamma \in \mathbb{B}^\ell} (-1)^{f(\gamma)}\,|\gamma\rangle\,|0_\alpha\rangle\,|-\rangle$, where $f(\gamma) = 1$ if and only if γ represents a valid solution to the considered problem. In general, the PHASE INVERSION circuit receives in input the superposition $|\Gamma\rangle\,|0_\alpha\rangle\,|-\rangle$, where $|\Gamma\rangle = \sum_{\gamma \in \mathbb{B}^\ell} c_\gamma\,|\gamma\rangle$. It outputs the superposition $\sum_{\gamma \in \mathbb{B}^\ell} (-1)^{f(\gamma)} c_\gamma\,|\gamma\rangle\,|0_\alpha\rangle\,|-\rangle$. We remark that the values of the complex coefficients c_γ depend on the iteration.

For each problem we consider, we provide a specific PHASE INVERSION circuit. All such circuits consist of three circuits (see Fig. 2), the first is called INPUT TRANSDUCER and is denoted by U_I, the second is called SOLUTION DETECTOR and is denoted by U_S, and the third is the inverse U_I^{-1} of the INPUT TRANSDUCER. The purpose of the INPUT TRANSDUCER circuits is to "filter out" the states of $|\Gamma\rangle$ that do not correspond to "well-formed" candidate solutions. The purpose of the SOLUTION DETECTOR circuits is to invert the amplitude of the states of $|\Gamma\rangle$ that correspond to positive solutions, if any. The purpose of U_I^{-1} is to restore the state of the ancilla qubits to $|0\rangle$ so that they may be reused in the subsequent iterations of the amplitude-amplification process.

The INPUT TRANSDUCER circuits are described in Sect. 4. The SOLUTION DETECTOR circuits are described in Sect. 5, where the corresponding circuit bounds are also shown (see Table 1). Such bounds and Theorem 1 imply the following.

Theorem 2. *In the quantum circuit model, the TLCM, TLKP, TLQP, TLS, OPCM, BT, and BS problems can be solved with the following sequential (ST) and parallel (PT) time bounds (where M denotes the number of solutions):*

TLCM *ST:* $\sqrt{\frac{2^{n\log n}}{M}}O(m^2)$, *PT:* $\sqrt{\frac{2^{n\log n}}{M}}O(n^2)$. **TLKP** *ST:* $\sqrt{\frac{2^{n\log n}}{M}}O(m^2)$, *PT:* $\sqrt{\frac{2^{n\log n}}{M}}O(m\log^2 m)$. **TLQP** *ST:* $\sqrt{\frac{2^{n\log n}}{M}}O(m^6)$, *PT:* $\sqrt{\frac{2^{n\log n}}{M}}O(m^4)$. **TLS** *ST:* $\sqrt{\frac{2^{n\log n+\sigma\log m}}{M}}O(m^2)$, *PT:* $\sqrt{\frac{2^{n\log n+\sigma\log m}}{M}}O(m)$. **OPCM** *ST:* $\sqrt{\frac{2^{n\log n}}{M}}O(n^8)$, *PT:* $\sqrt{\frac{2^{n\log n}}{M}}O(n^6)$. **BT** *ST:* $\sqrt{\frac{2^{n\log n+m\log\tau}}{M}}O(n^8)$, *PT:* $\sqrt{\frac{2^{n\log n+m\log\tau}}{M}}O(n^6)$. **BS** *ST:* $\sqrt{\frac{2^{n\log n+\sigma\log m}}{M}}O(n^8)$, *PT:* $\sqrt{\frac{2^{n\log n+\sigma\log m}}{M}}O(n^6)$.

4 Input Transducer Circuits

We use two different versions of circuit U_I, depending on the considered problem. Namely, for all problems but for the TLS and the BS problems, circuit U_I consists of just one circuit U_{OT}, called ORDER-TRANSDUCER (see Fig. 3). For problems TLS and BS, circuit U_I executes, in parallel to U_{OT}, another circuit U_{ST}, called SKEWNESS-TRANSDUCER, whose details can be found in the full version.

Let ϕ be a binary string of length $n\log n$, which we interpret as a sequence of n binary integers, each consisting of $\log n$ bits. Recall that, we denote by $\phi[i]$ the i-th binary integer contained in ϕ. Let $|\phi\rangle$ be the basis state corresponding to ϕ.

Lemma 2. *There exists a gate U_{OT} that, when provided with the input superposition $|\phi\rangle\,|0_\alpha\rangle$, where $\alpha = \frac{n}{2}(n-1+\log n)+1$, produces the output superposition $|\phi\rangle\,|f(\phi)\rangle\,|x_{0,1}\rangle\ldots|x_{i,j}\rangle\ldots|x_{n-2,n-1}\rangle\,\big|0_{\frac{n}{2}\log n}\big\rangle$, such that $|x_{i,j}\rangle = 1$ if and only if $\phi[i] < \phi[j]$ and $f(\phi) = 1$ if and only if ϕ represents an n-permutation. Gate U_{OT} has $O(n^2\log n)$ circuit complexity, and $O(n\log n)$ depth and width.*

Proof (Sketch). The gate U_{OT} uses gate U_C, called COLLISION DETECTOR, and gate U_P, called PRECEDENCE-CONSTRUCTOR; see Fig. 3. The purpose of U_C is to compute the superposition $|\phi\rangle\,|f(\phi)\rangle$, where $f(\phi) = 1$ if and only if ϕ represents an n-permutation. The purpose of U_P is to compute the superposition $|\phi\rangle\,|x\rangle$, where $|x\rangle = |x_{0,1}\rangle\ldots|x_{i,j}\rangle\ldots|x_{n-2,n-1}\rangle$ and $x_{i,j} = 1$ if and only if $\phi[i] < \phi[j]$. To achieve $O(n\log n)$ depth, gate U_P exploits Lemma 1 to compare all unordered pairs of integers $\phi[i]$ and $\phi[j]$ in parallel.

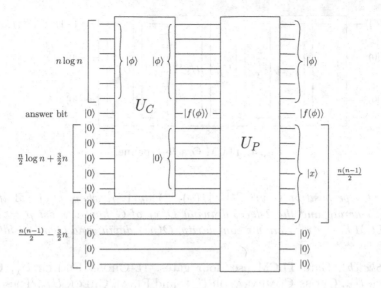

Fig. 3. The ORDER TRANSDUCER gate U_{OT}.

5   Solution Detector Circuits

In this section, we first present the SOLUTION DETECTOR circuit U_S for problem TLCM, and then we sketch the modifications that need to be introduced to address the remaining problems. We refer the reader to the full version for details.

Consider a 2-level drawing of a bipartite graph $G = (U, V, E)$. Let the vertices in U be $u_0, \ldots, u_{|U|-1}$, and the vertices of V be $v_{|U|}, \ldots, v_{|U|+|V|-1}$. Let ℓ_u and ℓ_v be two horizontal lines. We denote by $D(x)$ the 2-level drawing of G defined as follows. Let w_i and w_j be two vertices of G. Suppose that w_i, w_j belong to U (resp. to V). If $x_{i,j} = 1$, then $w_i \prec w_j$ along the horizontal line ℓ_u (resp. ℓ_v), otherwise, $w_j \prec w_i$ along ℓ_u (resp. ℓ_v). Suppose now that $w_i \in U$ and $w_j \in V$, then $x_{i,j} = 0$, which corresponds to the absence of a precedence between such vertices. Note that, if ϕ is not an n-permutation, then $D(x)$ does not correspond to an actual 2-level drawing. In this case, we say that $D(x)$ is *degenerate*.

We will exploit $|x\rangle$ to compute a superposition $|\chi_{0,1}\rangle \cdots |\chi_{i,j}\rangle \cdots |\chi_{m-2,m-1}\rangle$, which we will denote for simplicity by $|\chi\rangle$. We use the values $\chi_{i,j}$ to represent the existence of crossings between pairs of edges in a graph drawing. Namely, for every $0 \le a < b \le m - 1$, consider the value $\chi_{a,b}$, where $e_a = (v_i, v_k)$ and $e_b = (v_j, v_\ell)$. If $D(x)$ corresponds to a 2-level drawing of G, then we have that $\chi_{a,b} = 1$ if e_a and e_b belong to E and cross (i.e., $x_{i,j} \ne x_{k,\ell}$), and $\chi_{a,b} = 0$ if either e_a and e_b belong to E and do not cross (i.e., $x_{i,j} = x_{k,\ell}$) or at least one of e_a and e_b does not belong to E. Note that, the values $\chi_{a,b}$ are completely determined by x.

We call TLCM the SOLUTION DETECTOR circuit for problem TLCM.

Lemma 3. *There exists a gate TLCM that, when provided with the input superposition* $|f(\phi)\rangle |x\rangle |0_h\rangle |-\rangle$, *where* $h = 5\frac{m(m-1)}{2} - \log m - \log(m-1) - 2$, *produces*

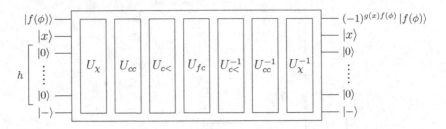

Fig. 4. TLCM Oracle Pipeline.

the output superposition $(-1)^{g(x)f(\phi)} |f(\phi)\rangle |x\rangle |0_h\rangle |-\rangle$, *where* $g(x) = 1$ *if* $D(x)$ *is not degenerate and the 2-level drawing* $D(x)$ *of* G *has at most* ρ *crossings. Gate TLCM has* $O(m^2)$ *circuit complexity,* $O(n^2)$ *depth, and* $O(m^2)$ *width.*

Proof (Sketch). Gate TLCM uses four gates: TL-CROSS FINDER U_χ, CROSS COUNTER U_{cc}, CROSS COMPARATOR $U_{c<}$, and FINAL CHECK U_{fc}. These gates are followed by the inverse gates $U_{c<}^{-1}$, U_{cc}^{-1}, and U_χ^{-1}, whose purpose is to restore the h ancilla qubit to $|0\rangle$ to be reused in the subsequent iterations. See Fig. 4.

TL-CROSS FINDER. The purpose of U_χ is to compute the crossings in $D(x)$ (under the assumption that $D(x)$ is not degenerate); see Fig. 5b. When provided with the input superposition $|x\rangle |0_k\rangle$, where $k = \frac{m(m-1)}{2}$, the gate U_χ outputs the superposition $|x\rangle |\chi\rangle$. Gate U_χ exploits the auxiliary gate U_{cr}, whose purpose is to check if two edges cross; see Fig. 5a. When provided with the input superposition $|x_{i,j}\rangle |x_{k,\ell}\rangle |0\rangle$, gate U_{cr} outputs the superposition $|x_{i,j}\rangle |x_{k,\ell}\rangle |\chi_{a,b}\rangle$, where $e_a = (u_i, v_k)$, $e_b = (u_j, v_\ell)$, and $\chi_{a,b} = x_{i,j} \oplus x_{k,\ell}$ (which is 1 if and only if e_a and e_b cross in $D(x)$). It is implemented using two Toffoli gates with three inputs and outputs. The first one is activated when $x_{i,j} = 1$ and $x_{k,\ell} = 0$. The second one is activated when $x_{i,j} = 0$ and $x_{k,\ell} = 1$. Gate U_{cr} has $O(1)$ circuit complexity, depth, and width. The gate U_χ works as follows. Consider that if two variables $x_{i,j}$ and $x_{k,\ell}$ are compared to determine whether the edges (u_i, v_k) and (u_j, v_ℓ) cross, none of these variables can be compared with another variable at the same time. Therefore, we partition the pairs of such variables using Lemma 1 (with $k = 2$ and $|X| = \frac{n(n-1)}{2}$) into $r \in O(n^2)$ cross-independent sets $S_1, \dots, S_r$ each containing at most $\lceil \frac{n(n-1)}{4} \rceil$ pairs. For $i = 1, \dots, r$, the gate U_χ executes in parallel a gate U_{cr}, for each pair $(x_{i,j}, x_{k,\ell})$ in S_i (see Fig. 5b), in order to output the qubit $|\chi_{a,b}\rangle$. Gate U_χ has $O(n^4)$ circuit complexity, and $O(n^2)$ depth and width.

CROSS COUNTER. The purpose of gate U_{cc} is to count the total number of crossings in the drawing $D(x)$. When provided with the input superposition $|\chi\rangle |0_d\rangle |0_k\rangle$, gate U_{cc} outputs the superposition $|\chi\rangle |\sigma(x)\rangle |0_d\rangle$, where $\sigma(x) = \sum_{e_i, e_j \in E} \chi_{i,j}$ is a binary integer of length d representing the total number of crossings. Gate U_{cc} exploits an auxiliary gate U_{1s}, whose design is given in the full version, that counts the number of 1s contained in a binary string. Note that the number of crossings in $D(x)$ is at most $\frac{m(m-1)}{2}$. Therefore, $\sigma(x)$ can be

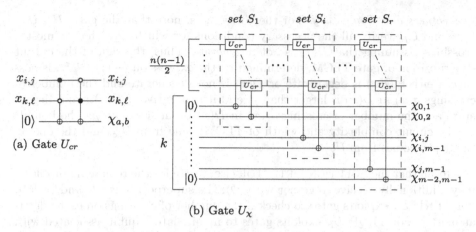

$x_{i,j}$ ———◦——— $x_{i,j}$
$x_{k,\ell}$ ———•——— $x_{k,\ell}$
$|0\rangle$ ———⊕——— $\chi_{a,b}$

(a) Gate U_{cr}

(b) Gate U_χ

Fig. 5. Gate U_{cr} (a) and gate U_χ (b). In (b), it holds $k = \frac{m(m-1)}{2}$.

represented by a binary string of length $d \leq \log m + \log(m-1)$. Gate U_{cc} can be implemented with $O(m^2)$ circuit complexity, $O(\log^2 m)$ depth, and $O(m^2)$ width.

CROSS COMPARATOR. The purpose of gate $U_{c<}$ is to verify if $\sigma(x)$ is less than the allowed number of crossings ρ. If provided with the input superposition $|\sigma(x)\rangle |\rho\rangle |0_h\rangle |0\rangle$, it outputs the superposition $|\sigma(x)\rangle |\rho\rangle |0_h\rangle |g(x)\rangle$, where $g(x) = 1$ if $D(x)$ is not degenerate and $\sigma(x) < \rho$. Gate $U_{c<}$ is an instance of the gate $U_<$, which compares two binary integers (its design is given in the full version). Gate $U_{c<}$ can be implemented with $O(\log m)$ circuit complexity, depth, and width.

FINAL CHECK. The purpose of gate U_{fc} is to check if the current solution is admissible, i.e., if $D(x)$ is not degenerate and the corresponding 2-level drawing of G has at most ρ crossings. See Fig. 6. When provided with the input superposition $|f(\phi)\rangle |g(x)\rangle |-\rangle$, the gate U_{fc} produces the outputs superposition $(-1)^{g(x)f(\phi)} |f(\phi)\rangle |g(x)\rangle |-\rangle$. U_{fc} exploits a Toffoli gate with three inputs and outputs. The control qubits are $|f(\phi)\rangle$ and $|g(x)\rangle$, and the target qubit is $|-\rangle$. If $f(\phi) = g(x) = 1$, the target qubit is transformed into the qubit $-|-\rangle$; otherwise, it leaves unchanged. Gate U_{fc} has $O(1)$ circuit complexity, depth, and width.

$|f(\phi)\rangle$ ———•——— $(-1)^{g(x)f(\phi)} |f(\phi)\rangle$
$|g(x)\rangle$ ———•——— $|g(x)\rangle$
$|-\rangle$ ———⊕——— $|-\rangle$

Fig. 6. Gate U_{fc}.

Correctness and Complexity. For the correctness, note that the gates U_χ, U_{cc}, $U_{c<}$, and U_{fc} verify all the necessary conditions for which $D(x)$ has at most ρ crossings, assuming that $D(x)$ is not degenerate. Thus, the sign of the output superposition of gate TLCM, determined by the expression $(-1)^{g(x)f(\phi)}$, is positive if either $D(x)$ is degenerate or $D(x)$ is not degenerate and the number of crossings $\sigma(x)$ in $D(x)$ is larger than ρ, and it is negative if $D(x)$ is not degenerate and the number of crossings $\sigma(x)$ in $D(x)$ is smaller than ρ. The bounds on the circuit complexity and width of TLCM come from U_{cc}, and the one on the depth comes from U_χ.

Recall that the SOLUTION DETECTOR circuits U_S for the remaining problems may additionally receive, together with $|\Phi\rangle$, the superpositions $|\Psi\rangle$ and/or $|\Theta\rangle$. For TLKP, U_S exploits gates to check if the number of crossings on each edge is at most k. For TLQP, U_S exploits gates to manipulate a qubit associated with a triple of edges. The state of such a qubit is set to $|1\rangle$ if the corresponding three edges mutually cross. For TLS, U_S also receives in input a superposition $|\theta\rangle$ (which identifies a selection of σ edges to be removed) and exploits gates to check if the remaining edges do not cross. For OPCM, U_S exploits gates suited to detect crossing edges in book layouts. For BT, U_S also receives in input a superposition $|\psi\rangle$ (which identifies a page assignment of the edge of G) and exploits gates to check that the edges assigned to the same page do not cross. For BS, U_S only differs from TLS in the gate used to detect crossing edges in book layouts.

6   Exploiting Quantum Annealing for Graph Drawing

In this section, we explore the 2-level problems and the book layout problems, that we have addressed so far from the quantum circuit model perspective, in the context of the quantum annealing model of computation. We pragmatically concentrate on the D-Wave platform, which offers quantum annealing services based on a large-scale quantum annealing solver. To utilize the hybrid facility of D-Wave for solving an optimization problem, there are essentially two ways: Either the problem is provided with its QUBO formulation or it is provided with a CBO formulation with constraints that are at most quadratic. Also, given a CBO formulation, it is quite simple to construct a QUBO formulation. Hence, in the full version, we first provide CBO formulations for the problems introduced in the previous section. Second, we overview a standard method for transforming a CBO formulation into a QUBO formulation.

In what follows, we discuss an experimentation conducted on the quantum annealing platform provided by D-Wave, specifically focusing on TLCM, which has extensive experimental literature compared to other problems considered in this paper. These experiments evaluate the efficiency of D-Wave with respect to well-known classical approaches to the TLCM problem.

We performed our experiments on TLCM using the hybrid solver of D-Wave, which suitably mixes quantum computations with classic tabu-search and simulated annealing heuristics. The obtained results are not guaranteed to be optimal.

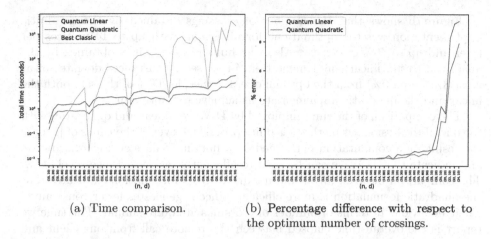

(a) Time comparison.

(b) **Percentage difference with respect to** the optimum number of crossings.

Fig. 7. D-Wave experimentation results: (a) comparison of times and (b) quality of the solutions.

We used the hybrid solved by D-Wave with two CBO formulations: One with linear constrains and one with quadratic constraints. Roughly, D-Wave hybrid solver works as follows. First, it decomposes the problem into parts that fit the quantum processor. The decomposition aims at selecting subsets of variables, and hence sub-problems, maximally contributing to the problem energy. Second, it solves the sub-problems with the quantum processor. Third, it injects the obtained results into the original overall problem that is solved with classical heuristics. These steps can be repeated several times, since an intermediate solution can re-determine the set of variables that contribute the most to the problem energy. An interesting description of the behaviour and of the limitations of D-Wave is presented in [24], although it refers to the quantum processor called Chimera that has been replaced by the new processors called Pegasus and Zephyr.

We compare our results with the figures proposed in [12], where three exact algorithms for TLCM are compared: LIN (the standard linearization approach), JM (the algorithm in [25]), and SDP (the branch-and-bound in [30]). Their experiments were carried out on an Intel Xeon processor with 2.33 GhZ.

Figure 7 illustrate the results of the experimentation on the D-Wave platform. We focused on exactly the same set of graphs used in [12]. Namely, for each value of n, i.e., number of vertices per layer, and for each value of d, i.e., density, we performed 10 experiments on 10 distinct graph instances. Figure 7a reports for each pair n, d, consisting of number of vertices n and density d: (1) the time spent to find the optimum by the fastest classical algorithm between LIN, JM, and SDP, (2) the time spent by our implementation using linear constraints, and (3) the time spent by our implementation using quadratic constraints. All the values are the average computed on the cited 10 instances.

Figure 7b shows that the number of crossings obtained by the quadratic implementation was the optimal one for all graphs with up to 14 vertices per layer and up to 40% of density. Also, the number of crossings obtained by the quadratic (resp., linear) implementation, in all sets of instances deviates of at most 8% (resp., 9%) from the optimal one. Further, for 60% of the sets both the linear and the quadratic implementations achieve the optimum.

The comparison of the time employed by D-Wave (linear and quadratic) with the one of the best exact methods is quite promising, even if the times of [12] are the results of a computation performed on a non-up-to-date classical hardware, and indicate that D-Wave can be used to efficiently tackle instances of TLCM. The comparison between linear and quadratic CBO formulations indicates that the quadratic formulation is more efficient, since it generates fewer constraints. Their behaviour in terms of number of crossings are quite similar. The time we report is the overall time elapsed between the remote call from our client and the reception of the result. The actual time spent on the quantum processor is always between 0.016 and 0.032 sec.

7   Conclusions and Open Problems

We initiate the study of quantum algorithms in the Graph Drawing research area providing a framework that allows us to tackle several classic problems within the 2-level and book layout drawing standards. Our framework, equipped with several quantum circuits of potential interest to the community, builds upon Grover's quantum search approach. It empowers us to achieve, at least, a quadratic speedup compared to the best classical exact algorithms for all the problems under consideration. In addition, we conducted experiments using the D-Wave quantum annealing platform for the TWO-LEVEL CROSSING MIN-IMIZATION problem. Our experiments demonstrated that the platform is highly suitable for addressing graph drawing problems and showcased significant efficiency when compared to the top approaches available for solving such problems. The encounter between Graph Drawing and Quantum Computing is still in its nascent stage, offering a vast array of new and promising problems. Virtually, all graph drawing problems can be explored through the lenses of quantum computation, utilizing both the quantum circuit model and, more pragmatically, quantum annealing platforms.

References

1. Aaronson, S.: Introduction to quantum information science lecture notes, April 2019. https://www.scottaaronson.com/qclec.pdf
2. Ahmed, A.R., et al.: Splitting vertices in 2-layer graph drawings. IEEE Comput. Graph. Appl. **43**(3), 24–35 (2023). https://doi.org/10.1109/MCG.2023.3264244
3. Angelini, P., Da Lozzo, G., Di Battista, G., Frati, F., Patrignani, M.: 2-Level quasi-planarity or how caterpillars climb (SPQR-)trees. In: Marx, D. (ed.) SODA 2021, pp. 2779–2798. SIAM (2021). https://doi.org/10.1137/1.9781611976465.165

4. Angelini, P., Da Lozzo, G., Förster, H., Schneck, T.: 2-Layer k-Planar graphs density, crossing lemma, relationships and pathwidth. Comput. J. (2023). https://doi.org/10.1093/comjnl/bxad038
5. Bannister, M.J., Eppstein, D.: Crossing minimization for 1-page and 2-page drawings of graphs with bounded treewidth. J. Graph Algorithms Appl. **22**(4), 577–606 (2018). https://doi.org/10.7155/jgaa.00479
6. Barth, W., Mutzel, P., Jünger, M.: Simple and efficient bilayer cross counting. J. Graph Algorithms Appl. **8**(2), 179–194 (2004). https://doi.org/10.7155/JGAA.00088
7. Bekos, M.A., Da Lozzo, G., Griesbach, S.M., Gronemann, M., Montecchiani, F., Raftopoulou, C.N.: Book embeddings of k-framed graphs and k-map graphs. Discret. Math. **347**(1), 113690 (2024). https://doi.org/10.1016/J.DISC.2023.113690
8. Bekos, M.A., Gronemann, M., Raftopoulou, C.N.: Two-page book embeddings of 4-planar graphs. Algorithmica **75**(1), 158–185 (2016). https://doi.org/10.1007/S00453-015-0016-8
9. Bernhart, F., Kainen, P.C.: The book thickness of a graph. J. Comb. Theory, Ser. B **27**(3), 320–331 (1979). https://doi.org/10.1016/0095-8956(79)90021-2
10. Binucci, C., et al.: Algorithms and characterizations for 2-layer fan-planarity: from caterpillar to stegosaurus. J. Graph Algorithms Appl. **21**(1), 81–102 (2017). https://doi.org/10.7155/JGAA.00398
11. Binucci, C., Di Giacomo, E., Hossain, M.I., Liotta, G.: 1-page and 2-page drawings with bounded number of crossings per edge. Eur. J. Comb. **68**, 24–37 (2018). https://doi.org/10.1016/J.EJC.2017.07.009
12. Buchheim, C., Wiegele, A., Zheng, L.: Exact algorithms for the quadratic linear ordering problem. INFORMS J. Comput. **22**(1), 168–177 (2010). https://doi.org/10.1287/ijoc.1090.0318
13. Caroppo, S., Da Lozzo, G., Di Battista, G.: Quantum graph drawing. CoRR abs/2307.08371 (2023). https://doi.org/10.48550/arXiv.2307.08371
14. Di Battista, G., Eades, P., Tamassia, R., Tollis, I.G.: Graph Drawing: Algorithms for the Visualization of Graphs. Prentice-Hall, London (1999)
15. Di Giacomo, E., Didimo, W., Eades, P., Liotta, G.: 2-layer right angle crossing drawings. Algorithmica **68**(4), 954–997 (2014). https://doi.org/10.1007/S00453-012-9706-7
16. Diwan, A.A., Roy, B., Ghosh, S.K.: Two-layer drawings of bipartite graphs. Electron. Notes Discret. Math. **61**, 351–357 (2017). https://doi.org/10.1016/J.ENDM.2017.06.059
17. Dujmovic, V., et al.: On the parameterized complexity of layered graph drawing. Algorithmica **52**(2), 267–292 (2008). https://doi.org/10.1007/s00453-007-9151-1
18. Eades, P., McKay, B.D., Wormald, N.C.: On an edge crossing problem. In: 9th Australian Computer Science Conference, ACSC 1986, Proceedings, pp. 327–334 (1986)
19. Eades, P., Whitesides, S.: Drawing graphs in two layers. Theor. Comput. Sci. **131**(2), 361–374 (1994). https://doi.org/10.1016/0304-3975(94)90179-1
20. Fukuzawa, S., Goodrich, M.T., Irani, S.: Quantum Tutte embeddings. CoRR abs/2307.08851 (2023). https://doi.org/10.48550/arXiv.2307.08851
21. Garey, M.R., Johnson, D.S.: Crossing number is NP-complete. SIAM J. Algebr. Discret. Methods **4**(3), 312–316 (1983). https://doi.org/10.1137/0604033
22. Grover, L.K.: A fast quantum mechanical algorithm for database search. In: Miller, G.L. (ed.) STOC 1996, pp. 212–219. ACM (1996). https://doi.org/10.1145/237814.237866

23. Harrow, A.W.: Quantum algorithms for systems of linear equations. In: Encyclopedia of Algorithms, pp. 1680–1683 (2016). https://doi.org/10.1007/978-1-4939-2864-4_771
24. Jünger, M., et al.: Performance of a quantum annealer for ising ground state computations on chimera graphs. CoRR abs/1904.11965 (2019)
25. Jünger, M., Mutzel, P.: 2-layer straightline crossing minimization: performance of exact and heuristic algorithms. J. Graph Algorithms Appl. **1**(1), 1–25 (1997). https://doi.org/10.7155/jgaa.00001
26. Kobayashi, Y., Tamaki, H.: A faster fixed parameter algorithm for two-layer crossing minimization. Inf. Process. Lett. **116**(9), 547–549 (2016). https://doi.org/10.1016/j.ipl.2016.04.012
27. Masuda, S., Kashiwabara, T., Nakajima, K., Fujisawa, T.: On the np-completeness of a computer network layout problem. In: Proceedings of IEEE International Symposium on Circuits and Systems (ISCAS 1987), pp. 292–295 (1987)
28. McGeoch, C.C.: Adiabatic Quantum Computation and Quantum Annealing: Theory and Practice. Synthesis Lectures on Quantum Computing. Morgan & Claypool Publishers (2014). https://doi.org/10.2200/S00585ED1V01Y201407QMC008
29. Nielsen, M.A., Chuang, I.L.: Quantum Computation and Quantum Information, 10th Anniversary edn. Cambridge University Press (2016)
30. Rendl, F., Rinaldi, G., Wiegele, A.: A branch and bound algorithm for max-cut based on combining semidefinite and polyhedral relaxations. In: Fischetti, M., Williamson, D.P. (eds.) IPCO 2007. LNCS, vol. 4513, pp. 295–309. Springer, Heidelberg (2007). https://doi.org/10.1007/978-3-540-72792-7_23
31. Rieffel, E., Polak, W.: Quantum Computing: A Gentle Introduction, 1st edn. The MIT Press, Cambridge (2011)
32. Shahrokhi, F., Sýkora, O., Székely, L.A., Vrt'o, I.: Book embeddings and crossing numbers. In: Mayr, E.W., Schmidt, G., Tinhofer, G. (eds.) WG 1994. LNCS, vol. 903, pp. 256–268. Springer, Heidelberg (1995). https://doi.org/10.1007/3-540-59071-4_53
33. Tamassia, R. (ed.): Handbook on Graph Drawing and Visualization. Chapman and Hall/CRC, Boca Raton (2013)
34. Tan, J., Zhang, L.: The consecutive ones submatrix problem for sparse matrices. Algorithmica **48**(3), 287–299 (2007). https://doi.org/10.1007/s00453-007-0118-z
35. Wigderson, A.: The complexity of the Hamiltonian circuit problem for maximal planar graphs. Technical report TR-298, Princeton University (1982)
36. Yannakakis, M.: Edge-deletion problems. SIAM J. Comput. **10**(2), 297–309 (1981). https://doi.org/10.1137/0210021
37. Yannakakis, M.: Embedding planar graphs in four pages. J. Comput. Syst. Sci. **38**(1), 36–67 (1989). https://doi.org/10.1016/0022-0000(89)90032-9

Simultaneous Drawing of Layered Trees

Julia Katheder[1]([✉])[iD], Stephen G. Kobourov[2][iD], Axel Kuckuk[1][iD],
Maximilian Pfister[1][iD], and Johannes Zink[3][iD]

[1] Wilhelm-Schickard-Institut für Informatik, Universität Tübingen, Tübingen,
Germany
`{julia.katheder,axel.kuckuk,maximilian.pfister}@uni-tuebingen.de`
[2] Department of Computer Science, University of Arizona, Tucson, USA
`kobourov@cs.arizona.edu`
[3] Institut für Informatik, Universität Würzburg, Würzburg, Germany
`zink@informatik.uni-wuerzburg.de`

Abstract. We study the crossing-minimization problem in a layered
graph drawing of planar-embedded rooted trees whose leaves have a given
total order on the first layer, which adheres to the embedding of each
individual tree. The task is then to permute the vertices on the other
layers (respecting the given tree embeddings) in order to minimize the
number of crossings. While this problem is known to be NP-hard for
multiple trees even on just two layers, we describe a dynamic program
running in polynomial time for the restricted case of two trees. If there
are more than two trees, we restrict the number of layers to three, which
allows for a reduction to a shortest-path problem. This way, we achieve
XP-time in the number of trees.

Keywords: layered drawing · tree drawing · crossing-minimization ·
dynamic program · XP-algorithm

1 Introduction

Visualizing hierarchical structures as directed trees is essential for many applications, from software engineering [2] to medical ontologies [1] and phylogenetics
in biology [12]. Phylogenetic trees in particular can serve as an example to illustrate the challenges of working with hierarchical structures, as they are inferred
from large amounts of data using various computational methods [19] and need
to be analyzed and checked for plausibility using domain knowledge [9]. From
a human perspective, visual representations are needed for this purpose. Most
available techniques focus on the visualization of a single tree [7]. However, certain tasks may require working with multiple, possible interrelated trees, such
as the comparison of trees [9,11] or analyzing ambiguous lineages [13]. Graham
and Kennedy [7] provide a survey for drawing multiple trees in this context.

J. Katheder is supported by DFG grant Ka 812-18/2 and
J. Zink is supported by DFG grant Wo 758/11-1.

While there are many different visualization styles for trees (see an overview by Schulz [15]), directed node-link diagrams are the standard. The most common approach to visualize a directed graph as a node-link diagram is the layered drawing approach due to Sugiyama et al. [17]. After assigning vertices to layers, the next step is to permute the vertices on each layer such that the number of crossings is minimized, as crossings negatively affect the readability of a graph drawing [14,18]. However, this problem turns out to be hard even when restricting the number of layers or the type of graphs. For example, if the number of layers is restricted to two, crossing minimization remains NP-hard for general graphs [6], even if the permutation on one layer is fixed [5], known as the *one sided crossing minimization* (OSCM) problem. However, it is known that OSCM is fixed-parameter tractable in the number of crossings, which has first been shown by Dujmovic and Whitesides [4]. For the special case of a single tree on two layers, OSCM can be solved in polynomial time [8] and in the case that both layers are variable, the problem can be reduced to the minimum linear arrangement problem [16], which is polynomial-time solvable [3]. For an arbitrary number of layers, the problem is still NP-hard even for trees [8], however, the obtained trees in the reduction [8] are not drawn upward in the direction from the leaves to a root vertex (and we do not see an obvious way to adjust their construction). With respect to forests, the general case where $k \in \mathcal{O}(n)$ is known to be NP-hard [10] even for $\ell = 2$ layers and trees of maximum degree 4.

Our Contributions. We consider the crossing-minimization problem for an n-vertex forest of k trees whose vertices are assigned to ℓ layers such that all leaves are on the first layer in a fixed total order and the vertices on each of the other layers need to be permuted. In other words, the task is to draw k layered rooted trees whose leaves may interleave simultaneously, while minimizing the number of crossings.

We show that the case of $k = 2$ trees is polynomial-time solvable for arbitrary ℓ using a dynamic program (see Sect. 3). Furthermore, we describe an XP-algorithm[1] in the number k of trees modeling the solution space by a k-dimensional grid graph for $\ell = 3$ layers. Our result generalizes to planar graphs under certain conditions (see Sect. 4). We conclude with the open case of $k \geq 3$ and $\ell \geq 4$ (see Sect. 5).

2   Preliminaries

Let $\mathcal{F}$ be a given forest of k disjoint rooted trees $T_1, \ldots, T_k$ directed towards the roots such that all vertices except for the roots have out-degree 1. For an integer $\ell \geq 2$, let an assignment of the vertices to ℓ layers be given, such that each tree T_i is *drawn upward*, i.e., for any directed edge $(u, v) \in T_i$, we have that the layer of u, denoted by $L(u)$, is strictly less than $L(v)$. This implies that

---

[1]   XP is a parameterized running-time class and an XP-algorithm has a running time in $\mathcal{O}(|I|^{f(k)})$, where $|I|$ is the size of the instance, f a computable function, and k the parameter. Note that every FPT-algorithm is an XP-algorithm but not vice versa.

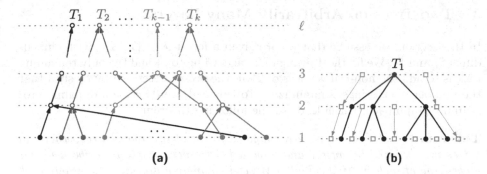

Fig. 1. **(a)** Upward drawing of k disjoint directed rooted trees $T_1, \ldots T_k$ on ℓ layers. As indicated by the filled vertices, the total order $<_1$ of layer 1 is given, while the total orders $<_2, \ldots, <_\ell$ need to be determined. In the following figures, we drop the arrowheads and assume an upward direction. **(b)** Illustration of positions (gray boxes) with respect to T_1 and their respective *ideal positions* indicated by a directed gray arrow from each position p to its ideal position p^*.

if $L(u) = 1$, u is a leaf of T_i. The other way around, we also require that for any leaf v, $L(v) = 1$. Note that the roots of the trees can be placed on different layers, while layer ℓ hosts the root of every tree with height exactly ℓ. We refer to the set of vertices of T_i on layer j as $V_j(T_i)$ and we define the set of all vertices on layer j as $V_j(\mathcal{F}) = V_j(T_1) \cup \cdots \cup V_j(T_k)$.

We further require that the total order $<_1$ of layer 1 (i.e., the order of all leaves) is given as part of the input, with the additional restriction that $<_1$ induces a planar embedding $\mathcal{E}_i$ with respect to each individual tree T_i, that is, there exists an ordering of the (non-leaf) vertices of T_i such that no two edges of T_i cross, see Fig. 1a for an illustration. Since the leaves of each T_i are all on layer 1, the embedding $\mathcal{E}_i$ is unique and implies a total ordering $<_j^i$ of the vertices of T_i on every layer $j \in \{2, \ldots, \ell\}$. Therefore, we henceforth assume that $V_j(T_i)$ appears in the corresponding vertex order $<_j^i$, and if we combine all $<_j^i$ for $i \in \{1, \ldots, k\}$, we obtain a partial order, which we call $\prec_j$.

The task is to find a total order $<_j$ of $V_j(\mathcal{F})$ extending the partial order $\prec_j$ for each $j \in \{2, \ldots, \ell\}$ such that the total number of pairwise edge crossings implied by a corresponding straight-line realization of $\mathcal{F}$ is minimized.

We restrict the notion of an upward drawing even further since we require that for any directed edge $(u, v) \in T_i$, we have that $L(u) + 1 = L(v)$. If our input does not fulfill this requirement, this can be achieved by subdividing edges which span several layers (as commonly done, e.g., in the Sugiyama framework). In the following, we assume that n is the number of vertices after subdivision and let $n_1, \ldots, n_k$ be the number of vertices of $T_1, \ldots, T_k$, respectively. Furthermore, we denote the number of vertices of tree T_i on layer j by $n_{i|j} = |V_j(T_i)|$.

3 Two Trees on Arbitrarily Many Layers

In this section, we assume that we are given a forest $\mathcal{F} = \{T_1, T_2\}$ with embeddings $\mathcal{E}_1$ and $\mathcal{E}_2$. We fix the drawing of T_1 according to $\mathcal{E}_1$ and the only remaining task is to add the non-leaf vertices of T_2 in the order prescribed by $\mathcal{E}_2$ such that the number of crossings is minimized. To this end, we describe a dynamic programming approach, which leads to the following theorem.

Theorem 1. *Let $\mathcal{F}$ be an n-vertex layered forest of two rooted trees, where all leaves are assigned to layer 1 and have a fixed order, which prescribe a planar embedding of each tree individually. We can compute a drawing of $\mathcal{F}$ where each tree is drawn in the prescribed planar embedding with the minimum number of crossings in $\mathcal{O}(n^3)$ time.*

Proof. As stated before, we fix the drawing of T_1 according to $\mathcal{E}_1$. Hence it remains to prove that our dynamic program embeds T_2 according to $\mathcal{E}_2$, which we do in Lemma 1. In Lemma 2, we show that the resulting drawing has the minimum number of crossings. This proves the correct behavior of our algorithm. In Lemma 3, we also show the runtime bound of $\mathcal{O}(n^3)$. $\qquad\square$

Description of the Dynamic Program. Consider some layer $j \in \{2, \ldots, \ell\}$ and index the vertices in $V_j(T_1)$ according to $\mathcal{E}_1$ from left to right by $1, \ldots, n_{1|j}$. In a complete drawing, we define, for a vertex v of T_2, its *position* p on layer j with respect to the index of the closest vertex of T_1 to the left of v. If there is no such vertex, we set $p = 0$. Let $C_v = \{c_1, c_2, \ldots, c_{\text{indeg}(v)}\}$ be the ordered set of children of v in T_2, which lie on layer $j - 1$, where $\text{indeg}(v)$ is the in-degree of v.

For our dynamic program, we define a function o, which has as first parameter a vertex v of T_2 and as second parameter a position p on layer $L(v)$. The value of o shall describe the number of crossings in an optimal partial solution for the drawing of the subtree of T_2 rooted at v and placed at position p. As usual in a dynamic program, we compute a function value once and then save it in a lookup table. Additionally, we save the recursive dependencies that led to a value to reconstruct a drawing in the end. If $j \geq 3$, we define o as follows.

$$o[v, p] = \sum_{i=1}^{|C_v|} \min_{q \in \{0, \ldots, n_{1|j-1}\}} \left(o[c_i, q] + \chi_{j-1}(q, p) \right)$$

where $\chi_x(y, z)$ is a *crossing function* describing the number of crossings an edge between layers x and $x + 1$ admits if its source is arranged at position y (of layer x) and its target is arranged at position z (of layer $x + 1$). If for some c_i, there is more than one position for q resulting in a minimum value of $o[v, p]$, we choose the position q that maximizes $\chi_{j-1}(q, p)$.

For the recursive function o, we add a terminating formulation for the vertices on layer $j = 2$. Recall that for the leaves on layer 1, there is a total order $<_1$ given. Hence, for a vertex $v \in V_2(T_2)$, the position of each child of v is fixed, leading to the following simplified formulation of o, where p_{c_i} is the given position of leaf c_i.

$$o[v,p] = \sum_{i=1}^{|C_v|} \chi_1(p_{c_i}, p)$$

To compute the value $o^\star$ of the dynamic program as a whole, we take the minimum of all values of o for the root r_2 of T_2:

$$o^\star = \min_{p \in \{0,\dots,n_{1|L(r_2)}\}} o[r_2, p].$$

We return a drawing corresponding to $o^\star$, i.e., we specify for each vertex v of T_2 its position with respect to T_1 when computing $o^\star$. Finally, for vertices of T_2 having the same position, we arrange them in the order given by $\mathcal{E}_2$.

Correctness. Next, we prove the correct behavior of our dynamic program by showing that T_2 is embedded according to $\mathcal{E}_2$ (Lemma 1) and by showing that the resulting drawing has the minimum number of crossings (Lemma 2). Mainly because Lemma 1 is rather intricate to prove, we next introduce some more notation and concepts, for which we show four claims that lead to the proofs of these lemmas.

A key observation is that for a position p on layer j, there is precisely one *ideal position* $p^\star$ on layer $j - 1$ such that $\chi_{j-1}(p^\star, p) = 0$ and for two positions p, q with $p < q$ on layer j, the ideal positions $p^\star, q^\star$ on layer $j - 1$ appear in the same order, i.e., $p^\star < q^\star$. (Imagine going down the gap of $\mathcal{E}_1$ where p is located as illustrated in Fig. 1b.) In Claim 1, we formalize another observation regarding ideal positions. Essentially, the claim says the further the endpoints of an edge are away from a pair of position and ideal position, the more crossings occur. For simplicity, we assume henceforth that each of the functions o and χ_j returns ∞ for parameters outside of its domain.

Claim 1. *On a layer $j \in \{2, \dots, L(r_1)\}$, let $p \in \{0, \dots, n_{1|j}\}$ be a position and let $p^\star \in \{0, \dots, n_{1|j-1}\}$ be the ideal position of p on layer $j - 1$. For any $x \in \mathbb{N}_0$, it holds that $\chi_{j-1}(p^\star \pm x, p) = x$ and $\chi_{j-1}(p^\star, p \pm (x+1)) > \chi_{j-1}(p^\star, p \pm x)) \geq x$.*

Proof. Consider an edge (u, v) of T_2 with its endpoints being placed at $p^\star$ and p. We know that $\chi_{j-1}(p^\star, p) = 0$. Now, for every position that we move u (resp. v) to the left or right of $p^\star$ (of p), we change sides with a vertex w of T_1. Because w has exactly (at least) one incident edge going upwards (downwards) to its parent (a child) that we have not crossed before, the number of crossings increases by exactly (at least) 1. □

For a vertex $v \in V_j(T_2)$ on a layer j, we define $P_{\mathrm{opt}}(v)$ as the set of every position p where $o[v,p]$ is minimum. We analyze the properties of P_{opt} in Claim 2.

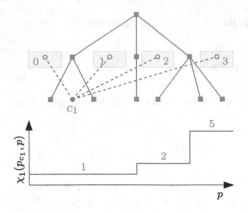

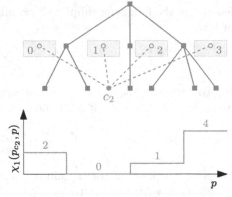

(a) Crossings of the edge (c_1, v) dependent on the position p of v.

(b) Crossings of the edge (c_2, v) dependent on the position p of v.

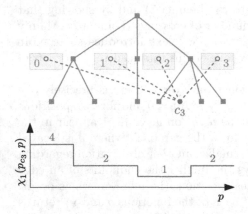

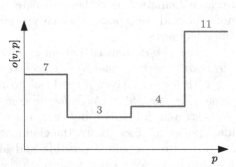

(c) Crossings of the edge (c_3, v) dependent on the position p of v.

(d) Sum of the three crossing functions gives $o[v, p]$.

Fig. 2. Example of a vertex v of $V_2(T_2)$ having three children c_1, c_2, c_3, where the position p_c of a child c and the position p of v determine the value of $o[v, p]$. Here, we perceive χ and o as functions dependent on p.

Claim 2. *For every layer* $j \in \{2, \ldots, L(r_2)\}$, *let* $v_1, v_2, \ldots, v_{n_{2|j}}$ *be the vertices in* $V_j(T_2)$ *in the order of* $\mathcal{E}_2$. *It holds that* $\min \mathrm{P}_{\mathrm{opt}}(v_1) \leq \min \mathrm{P}_{\mathrm{opt}}(v_2) \leq \ldots \leq \min \mathrm{P}_{\mathrm{opt}}(v_{n_{2|j}})$ *and* $\max \mathrm{P}_{\mathrm{opt}}(v_1) \leq \ldots \leq \max \mathrm{P}_{\mathrm{opt}}(v_{n_{2|j}})$.

Further, for every $v \in V_j(T_2)$, $\mathrm{P}_{\mathrm{opt}}(v)$ *is an interval of natural numbers and, for any* $x \in \mathbb{N}_0$, $o[v, \min \mathrm{P}_{\mathrm{opt}}(v) - (x+1)] > o[v, \min \mathrm{P}_{\mathrm{opt}}(v) - x] \geq x$ *and* $o[v, \max \mathrm{P}_{\mathrm{opt}}(v) + (x+1)] > o[v, \max \mathrm{P}_{\mathrm{opt}}(v) + x] \geq x$.

Proof. We show this claim by induction over the layers $j = 2, 3, \ldots$

For $j = 2$ and every $v \in V_j(T_2)$, the children of v have fixed positions and, therefore, $o[v, p]$ only depends on the number of crossings induced by the position $p \in \{0, \ldots, n_{1|j}\}$; see Fig. 2 for an example. We next show that $P_{opt}(v)$ is an interval. Observe that $o[v, p] = \sum_{i=1}^{|C_v|} \chi_1(p_{c_i}, p)$ is a sum of discrete functions (with variable p) where each admits its minimum value for one or two neighboring values of p and apart from at most two values at or around this minimum, all of these functions increase or decrease by the same amount if we add or subtract 1 to p, which follows by the argument presented in the proof of Claim 1. (These functions here are weakly unimodal, i.e., they have a global minimum and they increase monotonously when moving away from that minimum.) Now to find the positions where that sum is minimum, we traverse the values of its domain: we start with $p = 0$. If we increase p by one, then all functions that have not yet reached its minimum decrease, while the functions that had already reached their minimum increase by the same amount. Hence, this sum is minimum in the interval of the domain that has the minima of the single crossing functions equally distributed on the left and on the right side. Furthermore, for each position further to the left or right, the sum increases by at least one. It remains to show that the minima and maxima of $P_{opt}(v_1), P_{opt}(v_2), \ldots, P_{opt}(v_{n_{2|j}})$ increase monotonously. Since the children of the vertices on layer 2 (i.e., the leaves) are ordered, the minima and maxima of all crossing functions are ordered and so are the minima and maxima of the sums.

Now consider $j > 2$. Again $o[v, p]$ is a sum, but now we add, for every child c of v and a position q, $o[c, q]$ and $\chi_{j-1}(q, p)$. The sum of minima is again a sum of unimodal functions with similar properties as before: the $\chi_{j-1}(q, p)$ summands increase and decrease around their minimum as before, while the $o[c, q]$ summands increase and decrease before and after their minimum at least as much as a $\chi_{j-1}(q, p)$ summand due to the induction hypothesis. (They behave like a weighted $\chi_{j-1}(q, p)$ summand.) Hence, if we sum them up, we apply a weighted version of the previous argument to obtain the properties stated in the claim. In particular, the minima and maxima of $P_{opt}(v_1), P_{opt}(v_2), \ldots, P_{opt}(v_{n_{2|j}})$ increase monotonously since the minima and maxima of P_{opt} of the children on layer $j - 1$ are ordered by the induction hypothesis. □

For a vertex u on a layer $j - 1$, we define the *natural position* $p_{nat}(u, p)$ of u with respect to the position p of its parent vertex on layer j as

$$p_{nat}(u, p) = \begin{cases} p^\star, & \text{if } p^\star \in P_{opt}(u) \\ \max P_{opt}(u), & \text{if } p^\star > \max P_{opt}(u) \\ \min P_{opt}(u), & \text{if } p^\star < \min P_{opt}(u). \end{cases}$$

In Claim 3, we describe, for a vertex v, the behavior of the natural positions of v's children and their relationship to $o[v, p]$.

Claim 3. *For a vertex $v \in V_j(T_2)$ on a layer j, let $c_1, \ldots, c_{|C_v|}$ be the children of v. For any position $p \in \{0, \ldots, n_{1|j}\}$, it holds that $p_{nat}(c_1, p) \leq \ldots \leq p_{nat}(c_{|C_v|}, p)$ and $o[v, p] = \sum_{i=1}^{|C_v|} (o[c_i, p_{nat}(c_i, p)] + \chi_{j-1}(p_{nat}(c_i, p), p))$.*

Proof. We partition the children of v into three groups: if for a child c_i (where $i \in \{1, \ldots, |C_v|\}$), $p^\star \in P_{\mathsf{opt}}(c_i)$, we set $q_i = p^\star$. If for a child c_i, $p^\star > \max P_{\mathsf{opt}}(c_i)$, we set $q_i = \max P_{\mathsf{opt}}(c_i)$, and, symmetrically, if $p^\star < \min P_{\mathsf{opt}}(c_i)$, we set $q_i = \min P_{\mathsf{opt}}(c_i)$. By Claim 2, we observe that $q_1 \le \ldots \le q_{|C_v|}$. Since $q_i = p_{\mathsf{nat}}(c_i, p)$, this proves the first part of the claim.

Now for the second part, if $o[v, p] \ne \sum_{i=1}^{|C_v|}(o[c_i, q_i] + \chi_{j-1}(q_i, p))$, then for some i, $o[c_i, q_i] + \chi_{j-1}(q_i, p) > \min_{q' \in \{0, \ldots, n_{1|j-1}\}}(o[c_i, q'] + \chi_{j-1}(q', p))$. Let $\hat{q} \in \{0, \ldots, n_{1|j-1}\}$ be a position such that $o[c_i, q_i] + \chi_{j-1}(q_i, p) > o[c_i, \hat{q}] + \chi_{j-1}(\hat{q}, p)$. Since we know $o[c_i, q_i] \le o[c_i, \hat{q}]$, it follows that $\chi_{j-1}(q_i, p) > \chi_{j-1}(\hat{q}, p)$.

First note that $p^\star \notin P_{\mathsf{opt}}(c_i)$ because otherwise $\chi_{j-1}(q_i, p) < \chi_{j-1}(\hat{q}, p)$. It follows that q_i is the minimum or maximum position of $P_{\mathsf{opt}}(c_i)$ – assume w.l.o.g. that $q_i = \max P_{\mathsf{opt}}(c_i)$. Then, $q_i < \hat{q}$ (and hence $\hat{q} \notin P_{\mathsf{opt}}(c_i)$) because otherwise again $\chi_{j-1}(q_i, p) < \chi_{j-1}(\hat{q}, p)$.

We distinguish two cases. The first case is $q_i < \hat{q} \le p^\star$. By Claim 1, we know that $\chi_{j-1}(q_i, p) - \chi_{j-1}(\hat{q}, p) = \hat{q} - q_i$. By Claim 2, we know that $o[c_i, \hat{q}] - o[c_i, q_i] \ge \hat{q} - q_i$. Hence, we have $o[c_i, \hat{q}] - o[c_i, q_i] \ge \chi_{j-1}(q_i, p) - \chi_{j-1}(\hat{q}, p)$, which we can reformulate as $o[c_i, q_i] + \chi_{j-1}(q_i, p) \le o[c_i, \hat{q}] + \chi_{j-1}(\hat{q}, p)$, which contradicts our initial assumption.

The second case is $q_i < p^\star < \hat{q}$. Now we have $\chi_{j-1}(q_i, p) = p^\star - q_i$ and $\chi_{j-1}(\hat{q}, p) = \hat{q} - p^\star$. If we add up these two equations, we get $\chi_{j-1}(q_i, p) + \chi_{j-1}(\hat{q}, p) = \hat{q} - q_i$. As we still have $o[c_i, \hat{q}] - o[c_i, q_i] \ge \hat{q} - q_i$, we get $o[c_i, q_i] + \chi_{j-1}(q_i, p) \le o[c_i, \hat{q}] - \chi_{j-1}(\hat{q}, p)$, which, of course, also contradicts our initial assumption. $\square$

Now in the last claim, which is Claim 4, we directly investigate the positions that are chosen by our dynamic program as the positions of the children of a vertex – they turn out to be the natural positions.

Claim 4. *For a vertex $v \in V_j(T_2)$ on a layer j and a position $p \in \{0, \ldots, n_{1|j}\}$, the dynamic program selects $p_{\mathsf{nat}}(c_1, p), \ldots, p_{\mathsf{nat}}(c_{|C_v|}, p)$ as the positions of v's children $c_1, \ldots, c_{|C_v|}$ on layer $j - 1$.*

Proof. Recall that, for any $i \in \{1, \ldots, |C_v|\}$, if there is more than one position for c_i resulting in a minimum value of $o[v, p]$, the position of q with the maximum value of $\chi_{j-1}(q, p)$ is used as a tie-breaker rule. If $p_{\mathsf{nat}}(c_i, p) = p^\star$, then $p^\star$ is the only position of c_i that can lead to a minimum value of $o[v, p]$ and our claim is true.

Now, due to symmetry, we assume w.l.o.g. that $p_{\mathsf{nat}}(c_i, p) = \max P_{\mathsf{opt}}(c_i)$. Let $p' \ne p_{\mathsf{nat}}(c_i, p)$ be a position of c_i yielding a minimum value of $o[v, p]$. The position p' cannot lie within $P_{\mathsf{opt}}(c_i)$ by Claim 1 since this would result in a larger number of crossings, while $o[c_i, p'] = o[c_i, p_{\mathsf{nat}}(c_i, p)]$. Hence, $p_{\mathsf{nat}}(c_i, p) < p'$. By Claim 2, $o[c_i, p'] - o[c_i, p_{\mathsf{nat}}(c_i, p)] \ge p' - p_{\mathsf{nat}}(c_i, p)$. This means, that, for each position further to the right of $p_{\mathsf{nat}}(c_i, p)$, the value of the dynamic program for c_i increases by at least one, while the number of crossings according to the function χ_{j-1} increases by exactly one (see Claim 1). Thus, $p_{\mathsf{nat}}(c_i, p)$ is one (of possibly several) position(s) of c_i admitting a minimum value of $o[v, p]$. If p' also admits a minimum value of $o[v, p]$, but $o[c_i, p'] > o[c_i, p_{\mathsf{nat}}(c_i, p)]$, it follows that $\chi_{j-1}(p',p) < \chi_{j-1}(\mathsf{p_{nat}}(c_i,p),p)$. Hence, due to the tie-breaker rule, our algorithm would have selected $\mathsf{p_{nat}}(c_i,p)$ instead of p'. □

Now we have gathered everything to establish the key lemma of this section.

Lemma 1. *The drawing of T_2 is embedded according to $\mathcal{E}_2$.*

Proof. We prove that the vertices of T_2 are ordered according to $<_j^2$ by induction on layer j, starting with the layer of the root r of T_2. On layer $j = L(r_2)$, there is only r, which, of course, cannot contradict $<_j^2$.

Let $j < L(r_2)$ and $v_1, v_2, \ldots v_{n_{2|j+1}}$ be the vertices on layer $j + 1$. Then, by Claims 3 and 4, the children C_{v_i} on layer j have increasing positions respecting $<_j^2$ for every $i \in \{1, \ldots, n_{2|j+1}\}$ and the edges to these children do not cross. It remains to show that no pair of edges between layers j and $j + 1$ without a common endpoint cross. By our induction hypothesis, for two vertices $v_i, v_{i'}$ on layer $j + 1$ with $i < i'$, the position p_i of v_i is not greater than the position $p_{i'}$ of $v_{i'}$. By Claim 1, it follows for the ideal positions of p_i and $p_{i'}$ that $p_i^\star \leq p_{i'}^\star$. By Claim 2, the min and max values of the $\mathsf{P_{opt}}$ of the vertices on layer j are monotonically increasing. Hence, by Claim 4 and the definition of $\mathsf{p_{nat}}$, there are no crossings between edges with target v_i and edges with target $v_{i'}$, as this would contradict $p_i^\star \leq p_{i'}^\star$. Hence, the edges between layer j and $j+1$ are planar, which concludes the induction step. □

After we have now shown that the dynamic program yields a valid solution, i.e., a drawing where both trees are internally crossing-free, it remains to prove that the number of crossings between T_1 and T_2 is minimum.

Lemma 2. *The number of crossings in the computed drawing is minimum.*

Proof. We show by induction over the layers $j = 2, 3, \ldots$ that for a vertex v and a position p, $o[v, p]$ is the minimum number of crossings induced by (T_1 and) the subtree of T_2 rooted at v, which we call T_v, across all drawings of T_v when we place v at position p. For $j = 2$, this is clear as we just sum up the number of crossings induced by the edges to the leaves.

Let $j \geq 3$. By Claim 4, we know that the dynamic program has selected the positions $\mathsf{p_{nat}}(c_1, p) \leq \ldots \leq \mathsf{p_{nat}}(c_{|C_v|}, p)$ for the children $c_1, \ldots, c_{|C_v|}$ of v. By our induction hypothesis, we know that, for each $c \in C_v$, $o[c, \mathsf{p_{nat}}(c, p)]$ corresponds to a drawing of T_c at position $\mathsf{p_{nat}}(c, p)$ with the minimum number of crossings. We add up the number of crossings between layer $j - 1$ and j, and by the formulation of the dynamic program, we know that this is again minimum across all positions of c. □

Running Time. It remains to analyze the running time of our dynamic program.

Lemma 3. *The running time of our algorithm is in $\mathcal{O}(n_1^2 \cdot n_2) \subseteq \mathcal{O}(n^3)$.*

Proof. For a vertex v of $V_j(T_2)$ and a position $p \in \{0, \ldots, n_{1|j}\}$, we can compute $o[v, p]$ by finding, for each child $c \in C_v$ and a position q in a subset of $\{0, \ldots, n_{1|j-1}\}$, the minimum of $o[c, q] + \chi_{j-1}(q, p)$.

The number of children over all steps is in $\mathcal{O}(n_2)$ as T_2 is a tree and the number of positions is in $\mathcal{O}(n_1)$. We can pre-compute and store all values $\chi_j(q,p)$ in $\mathcal{O}(n_1^2)$ time. We have $\mathcal{O}(n_1 n_2)$ entries of $o[v,p]$, which we can compute in overall $\mathcal{O}(n_1^2 n_2) \subseteq \mathcal{O}(n^3)$ time. The optimal root placement can be found in linear time. For the backtracking when constructing the final drawing, we simply store for each entry $o[v,p]$ a pointer to the entries it is based on. □

4 Multiple Trees on Three Layers

In this section, we consider the case that we are given a forest $\mathcal{F} = \{T_1, \ldots, T_k\}$ of k trees spanning (at most) three layers each, and we show the following result.

Theorem 2. *Let $\mathcal{F}$ be an n-vertex layered forest of k rooted trees on three layers, where all leaves are assigned to layer 1 and have a fixed order, which prescribes a planar embedding of each tree individually. We can compute a drawing of $\mathcal{F}$ where each tree is drawn in the prescribed planar embedding with the minimum number of crossings in $\mathcal{O}(n^k)$ time.*

The first property we use to prove Theorem 2 is that the order of roots on layer 3 is fixed, similar to the order of the leaves on layer 1. We can assume this because there are only up to k roots on layer 3, with at most $k!$ ways to arrange them. We simply consider each permutation of the roots on layer 3 individually, and henceforth assume that both total orders $<_1$ and $<_3$ are given, fixing the roots and leaves, and the only remaining task is to compute $<_2$ of the vertices on layer 2 while maintaining their partial order $\prec_2$. Note that if any tree has its root on layer 2, we treat this root like the other vertices of layer 2.

As in Sect. 3, we use the notion of positions and crossing functions, however, we slightly adjust their definitions to better suit the setting of this section. Let σ be a permutation of $V_2(\mathcal{F})$ indexed by $1, 2, \ldots$ and respecting the partial order $\prec_2$. For $i \in \{1, \ldots, k\}$ and some vertex $v \in V_2(\mathcal{F}) \setminus V_2(T_i)$, we denote the *position* (starting at 0) of v within the subsequence of σ consisting of the vertices $V_2(T_i) \cup \{v\}$ by p_i^v.[2] Note that, in a drawing using σ as $<_2$, we can charge every crossing to precisely two vertices of $V_2(\mathcal{F})$ as any crossing occurs between two edges that have two distinct endpoints on layer 2. Now observe that for a vertex $v \in V_2(T_j)$, where $j \in \{1, \ldots, k\}$, the number of crossings charged to v with respect to σ depends only on p_i^v for each $i \in \{1, \ldots, k\} \setminus \{j\}$. Therefore, we introduce the *crossing function* $\chi_i^v(p)$ returning the resulting number of crossings when we insert v at a position $p \in \{0, \ldots, n_{i|2}\}$ into the planar embedding of T_i. The number $\chi_\sigma(v)$ of crossings charged to v when using permutation σ is then $\chi_\sigma(v) = \sum_{i \in \{1, \ldots, k\} \setminus \{j\}} \chi_i^v(p_i^v)$ and the total number $\chi(\sigma)$ of crossings when using permutation σ is then $\chi(\sigma) = \sum_{v \in V_2(\mathcal{F})} \chi_\sigma(v)/2$.

Lemma 4. *For all combinations of $i \in \{1, \ldots, k\}$, $v \in V_2(\mathcal{F}) \setminus V_2(T_i)$, and $p \in \{0, \ldots, n_{i|2}\}$, we can compute every value $\chi_i^v(p)$ in a total of $\mathcal{O}(n^2)$ time.*

---

[2] This is a generalization of the positions introduced in Sect. 3 where all positions were relative to (the given embedding of) T_1.

Proof. First save, for every $v \in V_2(\mathcal{F})$, the star S_v induced by v and v's neighbors in total $\mathcal{O}(n)$ time. Now for a fixed $i \in \{1, \ldots, k\}$, consider the given planar embedding $\mathcal{E}_i$ of T_i. Also fix $v \in V_2(\mathcal{F}) \setminus V_2(T_i)$ and compute $\chi_i^v(0)$ by checking, for every pair of edges of S_v and T_i, if there is a crossing if v is the leftmost vertex on layer 2. Then for $p = 1, \ldots, n_{i|2}$, update $\chi_i^v(p-1)$ to $\chi_i^v(p)$ by checking each pair of edges from S_v and the star around the p-th vertex of $V_2(T_i)$. Over all of these steps, all vertices, and all trees, every pair of edges is considered at most four times, which yields a running time in $\mathcal{O}(n^2)$.   □

Reduction to a Shortest-Path Problem. We now construct a weighted directed acyclic st-graph H (see Fig. 3c) whose st-paths represent precisely all total orders of $V_2(\mathcal{F})$ that respect the vertex orders $<_2^1, \ldots, <_2^k$ given for each tree by its prescribed planar embedding (see Fig. 3a). Moreover, for an st-path π representing a total order σ of $V_2(\mathcal{F})$, the weight of π is twice the number of crossings induced by σ. We let H be the k-dimensional grid graph of side lengths $n_{1|2} \times \cdots \times n_{k|2}$ directed from one corner to an opposite corner. More precisely, H has the node set $\{(x_1, \ldots, x_k) \mid x_1 \in \{0, \ldots, n_{1|2}\}, \ldots, x_k \in \{0, \ldots, n_{k|2}\}\}$ and there is a directed edge from $(x_1, \ldots, x_k)$ to $(y_1, \ldots, y_k)$ if $x_j + 1 = y_j$ for exactly one $j \in \{0, \ldots, k\}$ and $x_i = y_i$ otherwise. Observe that, within H, $(0, \ldots, 0)$ is the unique source and $(n_{1|2}, \ldots, n_{k|2})$ is the unique sink, which we denote by s and t, respectively. We let an edge e from $(x_1, \ldots, x_k)$ to $(y_1, \ldots, y_k)$ where $x_j + 1 = y_j$ represent (i) taking the y_j-th vertex of $V_2(T_j)$, to which we refer as v next, (ii) after having taken x_i vertices of $V_2(T_i)$ for each $i \in \{1, \ldots, k\}$. Thus, we let the weight w_e of e in H be the number of crossings charged to v in this situation, that is, $w_e = \sum_{i \in \{1, \ldots, k\} \setminus \{j\}} \chi_i^v(x_i)$.

Clearly, any st-path π in H has (unweighted) length n_2. If we traverse π, we can think of layer 2 as being empty when we start at s, and then, for each edge of π, we take the corresponding vertex of V_2 and add it to layer 2. Since edge weights equal the number of crossings the corresponding vertices would induce in this situation, finding a lightest st-path in H means finding a crossing-minimal total order of layer 2 (see Fig. 3c). By constructing H (using Lemma 4 to compute the edge weights) and searching for an st-path of minimum weight, we obtain an XP-algorithm in k; see Theorem 2, which we formally prove next.

Proof (of Theorem 2). For $\mathcal{F}$, we fix each order of layer 3 once and compute the corresponding k-dimensional grid graph H. We first argue that the st-paths of H represent precisely the possible total orders of $V_2(\mathcal{F})$ and their weights are twice the number of crossings in the corresponding drawing of $\mathcal{F}$. Thereafter, we argue about the running time.

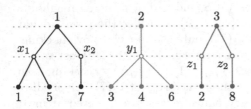

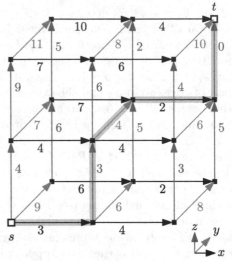

(a) We are given three embedded trees on three layers together with a total order of the leaves and the roots (numbers on the top and the bottom side).

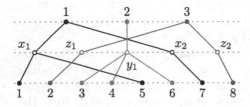

(b) Drawing of the trees from (a) with the minimum number of crossings (six pairwise crossings) where the orders of the leaves and the roots are given. The total order of the vertices on layer 2, the middle layer, corresponds to the shortest st-path highlighted in orange in (c).

(c) Grid graph H with edge weights whose st-paths represent precisely the (allowed) total orders of the vertices on layer 2 of the forest shown in (a). The width in x-dimension is two and represents first choosing vertex x_1 and then vertex x_2 of the first tree. Similarly, the y- and z-dimension represent the vertices of the second and third tree, respectively. The st-path highlighted in orange is the lightest path with weight 12.

Fig. 3. Reducing the problem of finding a layered drawing of k trees on three layers with the minimum number of crossings, where the leaves and the roots are fixed, to a shortest-path problem in a weighted k-dimensional grid graph.

Since we have a k-dimensional grid graph, any st-path in H traverses $n_{1|2}$ edges in the first dimension, $n_{2|2}$ edges in the second dimension, etc. We can interleave the edges of different dimensions arbitrarily to obtain different paths. Hence, each path is equivalent to exactly one total order σ extending the partial order $\prec_2$, which is given by $<_2^1, \ldots, <_2^k$. The number of crossings of a drawing only depends on σ. Every crossing occurs between two edges being incident to precisely two distinct vertices of $V_2(\mathcal{F})$. We *charge* the crossing to these two vertices. Hence, we can add up the numbers of crossings charged to each vertex v (i.e., $\chi_\sigma(v)$) and divide the sum by two. These numbers of crossings charged to the vertices are by definition the edge weights of H. Therefore, each minimum-weight st-path in H is equivalent to a minimum-crossing drawing of $\mathcal{F}$ for the given order of leaves and roots.

It remains to argue about the running time. The number of directed edges of H is upper-bounded by

$$E(H) = \sum_{j=1}^{k} n_{j|2} \prod_{i \in \{1,\dots,k\} \setminus \{j\}} (n_{i|2} + 1)$$

$$\leq k \prod_{i=1}^{k} (n_{i|2} + 1) \leq k \left(\frac{n}{k}\right)^k.$$

To compute the weight of each such edge, we sum up $k - 1$ values of $\chi_i^v(p)$, which we have pre-computed in $\mathcal{O}(n^2)$ time using Lemma 4. Therefore, we can construct H including the assignment of edge weights in $\mathcal{O}(k^2(n/k)^k)$ time and we can find a minimum-weight path in H in $\mathcal{O}(k(n/k)^k)$ time using topological sorting. Recall that we construct a graph H for at most $k!$ permutations of the roots on layer 3. For the final minimum-crossing drawing, we use the permutation of $V_2(\mathcal{F})$ and the permutation of $V_3(\mathcal{F})$ that correspond to the lightest minimum-weight path in any H. Hence, the total running time is in $\mathcal{O}(k!k^2(n/k)^k) \subseteq \mathcal{O}(k^3 \cdot (k-1) \cdot \ldots \cdot 2 \cdot 1/k^k \cdot n^k) \subseteq \mathcal{O}(n^k)$. □

Finally, we remark that our XP-algorithm from Theorem 2 can be generalized in two ways.

Remark 1. By definition, $\prec_2$ has only constraints between vertices of the same tree. We can extend $\prec_2$ by (arbitrarily many) constraints between vertices of different trees and Theorem 2 still holds. This is because we can easily adjust our reduction: say x is the i-th vertex on layer 2 of the first tree, y is the j-th vertex on layer 2 of the second tree, and let the constraint $x \prec_2 y$ be given. Then, in H, we set the weight of every edge representing x and lying in the y-dimension at a position $\geq j$ to ∞. Symmetrically, we set the weight of every edge representing y and lying in the x-dimension at a position $< i$ to ∞. This way, we prevent that a lightest path chooses an edge representing y before it chooses an edge representing x. If and only if there is an st-path with non-infinity weight in H, there is a valid arrangement of the vertices on layer 2.

Remark 2. Requiring trees on the three layers is a stronger restriction than actually needed. For our reduction, we only use the property that the vertex order on layer 3 is fixed, which we achieve by trying all permutations. For this approach, it suffices if layer 3 is sparse. Hence, our result also holds for k planar-embedded graphs provided that on layer 3, there are $\mathcal{O}(k)$ vertices. Moreover, for planar-embedded graphs and an arbitrary number of vertices on layer 3, Theorem 2 holds as well if the total order of vertices on layer 3 is prescribed.

5   Conclusion and Open Problems

In this work, we approach the problem of crossing minimization of layered rooted trees from two directions. First, by describing a cubic-time dynamic program in Theorem 1, keeping the number of trees k small, namely $k = 2$, while allowing an arbitrary number of layers. Inversely, our second result stated in Theorem 2 is an XP-time algorithm for an arbitrary number of trees, restricted to only three layers. Hence, there is a gap between these two results, which has not yet been explored and naturally raises the following open problem. Going one step further, is the case $k = 3$ trees and $\ell = 4$ layers polynomial-time solvable, and if so, for which k and ℓ does it become hard? Moreover, we pose the question of improving the complexity class for the case of $\ell = 3$ and $k > 2$, namely, can we solve the case of three layers in FPT-time in the number k of trees? Alternatively this may be proved to be W[1]-hard. Lastly, note that in our setting, we require that every tree preserves its given planar embedding (imposed by the order of its leaves). It is not clear, whether there exists a solution with less crossings without this restriction, although our current believe is that, in any minimum-crossing solution, all of them are drawn planar.

Acknowledgments. We thank the organizers of the workshop GNV 2022 in Heiligkreuztal for the fruitful atmosphere where some of the ideas of this paper arose. We also thank the anonymous reviewers for their helpful feedback.

References

1. Bodenreider, O.: The unified medical language system (UMLS): Integrating biomedical terminology. Nucleic Acids Res. **32**(suppl_1), 267–270 (2004). https://doi.org/10.1093/nar/gkh061
2. Chawathe, S.S., Rajaraman, A., Garcia-Molina, H., Widom, J.: Change detection in hierarchically structured information. ACM SIGMOD Rec. **25**(2), 493–504 (1996). https://doi.org/10.1145/235968.233366
3. Chung, F.R.K.: On optimal linear arrangements of trees. Comput. Math. Appl. **10**(1), 43–60 (1984). https://doi.org/10.1016/0898-1221(84)90085-3
4. Dujmovic, V., Whitesides, S.: An efficient fixed parameter tractable algorithm for 1-sided crossing minimization. Algorithmica **40**(1), 15–31 (2004). https://doi.org/10.1007/S00453-004-1093-2
5. Eades, P., Wormald, N.C.: Edge crossings in drawings of bipartite graphs. Algorithmica **11**(4), 379–403 (1994). https://doi.org/10.1007/bf01187020
6. Garey, M.R., Johnson, D.S.: Crossing number is NP-complete. SIAM J. Algebraic Discrete Methods **4**(3), 312–316 (1983). https://doi.org/10.1137/0604033
7. Graham, M., Kennedy, J.: A survey of multiple tree visualisation. Inf. Vis. **9**(4), 235–252 (2010). https://doi.org/10.1057/ivs.2009.29
8. Harrigan, M., Healy, P.: k-level crossing minimization Is NP-hard for trees. In: Katoh, N., Kumar, A. (eds.) WALCOM 2011. LNCS, vol. 6552, pp. 70–76. Springer, Heidelberg (2011). https://doi.org/10.1007/978-3-642-19094-0_9
9. Liu, Z., Zhan, S.H., Munzner, T.: Aggregated dendrograms for visual comparison between many phylogenetic trees. IEEE Trans. Visualization Comput. Graph. **26**(9), 2732–2747 (2019). https://doi.org/10.1109/tvcg.2019.2898186
10. Muñoz, X., Unger, W., Vrt'o, I.: One sided crossing minimization is NP-hard for sparse graphs. In: Mutzel, P., Jünger, M., Leipert, S. (eds.) GD 2001. LNCS, vol. 2265, pp. 115–123. Springer, Heidelberg (2002). https://doi.org/10.1007/3-540-45848-4_10

11. Munzner, T., Guimbretiere, F., Tasiran, S., Zhang, L., Zhou, Y.: Treejuxtaposer: scalable tree comparison using Focus+Context with guaranteed visibility. ACM Trans. Graph. **22**(3), 453–462 (2003). https://doi.org/10.1145/1201775.882291
12. Pavlopoulos, G.A., Soldatos, T..G, Barbosa-Silva, A., Schneider, R.: A reference guide for tree analysis and visualization. BioData Mining **3**(1), 1–24 (2010). https://doi.org/10.1186/1756-0381-3-1
13. Puigbò, P., Wolf, Y.I., Koonin, E.V.: Search for a 'tree of life' in the thicket of the phylogenetic forest. J. Biol. **8**(6), 1–17 (2009). https://doi.org/10.1186/jbiol159
14. Purchase, H.C., Carrington, D.A., Allder, J.-A.: Empirical evaluation of aesthetics-based graph layout. Empir. Softw. Eng. **7**(3), 233–255 (2002). https://doi.org/10.1023/A:1016344215610
15. Schulz, H.-J.: Treevis.net: a tree visualization reference. IEEE Comput. Graph. Appl. **31**(6), 11–15 (2011). https://doi.org/10.1109/mcg.2011.103
16. Shahrokhi, F., Sýkora, O., Székely, L.A., Vrto, I.: On bipartite drawings and the linear arrangement problem. SIAM J. Comput. **30**(6), 1773–1789 (2000). https://doi.org/10.1137/S0097539797331671
17. Sugiyama, K., Tagawa, S., Toda, M.: Methods for visual understanding of hierarchical system structures. IEEE Trans. Syst. Man Cybern. **11**(2), 109–125 (1981). https://doi.org/10.1109/TSMC.1981.4308636
18. Ware, C., Purchase, H.C., Colpoys, L., McGill, M.: Cognitive measurements of graph aesthetics. Inf. Vis. **1**(2), 103–110 (2002). https://doi.org/10.1057/palgrave.ivs.9500013
19. Yang, Z., Rannala, B.: Molecular phylogenetics: principles and practice. Nat. Rev. Genet. **13**(5), 303–314 (2012). https://doi.org/10.1038/nrg3186

Fully Dynamic Algorithms for Euclidean Steiner Tree

T-H. Hubert Chan[1] , Gramoz Goranci[2] , Shaofeng H.-C. Jiang[3] ,
Bo Wang[1(✉)] , and Quan Xue[1(✉)]

[1] The University of Hong Kong, Hong Kong, China
{hubert,bwang,qxue}@cs.hku.hk
[2] University of Vienna, Vienna, Austria
gramoz.goranci@univie.ac.at
[3] Peking University, Beijing, China
shaofeng.jiang@pku.edu.cn

Abstract. The Euclidean Steiner tree problem asks to find a min-cost
metric graph that connects a given set of *terminal* points X in $\mathbb{R}^d$, possibly using points not in X which are called Steiner points. Even though
near-linear time $(1+\epsilon)$-approximation was obtained in the offline setting
in seminal works of Arora and Mitchell, efficient dynamic algorithms
for Steiner tree is still open. We give the first algorithm that (implicitly) maintains a $(1+\epsilon)$-approximate solution which is accessed via a set
of tree traversal queries, subject to point insertion and deletions, with
amortized update and query time $O(\text{poly}\log n)$ with high probability.
Our approach is based on an Arora-style geometric dynamic programming, and our main technical contribution is to maintain the DP subproblems in the dynamic setting efficiently. We also need to augment the
DP subproblems to support the tree traversal queries.

Keywords: Steiner Tree · Dynamic Algorithms · Approximation
Schemes

1 Introduction

In the Euclidean Steiner tree problem, a set $X \subset \mathbb{R}^d$ of points called terminals
is given, and the goal is to find the minimum cost metric graph that connects
all points in X. The optimal solution is a tree, and this tree can also use points
other than terminals, called *Steiner points*. Euclidean Steiner tree is a fundamental problem in computational geometry and network optimization, with many
applications in various fields.

In the offline setting, Euclidean Steiner tree is known to be NP-hard [8], even
when the terminals are restricted to lie on two parallel lines [23]. An immediate approximation algorithm is via minimum spanning tree (MST), which gives

---

Full version of this paper is available at [5]

T-H. Hubert Chan was partially supported by the Hong Kong RGC grant 17203122.

$O(1)$-approximation, and MST can be computed in near-linear time in Euclidean $\mathbb{R}^d$. For a better approximation, seminal works by Mitchell [20] and Arora [1] developed PTAS's, i.e., $(1 + \epsilon)$-approximate polynomial time algorithms for the Euclidean Steiner tree, which are based on geometric dynamic programming techniques.

Due to its fundamental importance, Steiner tree has also been studied in the dynamic setting. Dynamic algorithms are algorithms that can handle modifications to the input data efficiently (in terms of running time), such as insertions and deletions. Specifically, the study of dynamic Steiner tree in a graph setting was first introduced by Imase and Waxman [15]. In this setting, the objective is to find a minimum-weight tree that connects all terminals for a given undirected graph with non-negative edge weights. Results for similar settings were also studied in followup papers [9,10,17,19].

Even though an $O(1)$-approximation may be obtained from dynamically maintaining MST efficiently [7,12], we are not aware of any previous work that explicitly studies the dynamic algorithm for the Euclidean Steiner tree, especially whether one can convert the Mitchell/Arora's PTAS into an efficient dynamic algorithm to still achieve $(1 + \epsilon)$-approximation. Technically, the Euclidean setting can differ greatly from the graph setting. In the former, updates typically involve point insertion/deletion, whereas in the latter, updates involve more localized edge insertion/deletion.

1.1 Our Results

In this paper, we investigate the Euclidean Steiner tree problem in the fully dynamic setting, where terminal points in $\mathbb{R}^d$ can be inserted and deleted in each time step. We give the first fully dynamic algorithm that (implicitly) maintains $(1 + \epsilon)$-approximation for the Steiner tree in amortized polylogarithmic time. In particular, the algorithm maintains a) the total weight and b) the root node of this implicit tree. Moreover, it can also answer several tree traversal queries: a) report whether a given point is in the solution tree and b) return neighbors of a given point p in the solution tree, i.e., its parent and the list of its children.

Theorem 1 (Informal; see Theorem 4[1]). *There exists a fully-dynamic algorithm for Euclidean Steiner tree that for any $0 < \epsilon, \delta < 1$ and any operation sequence handles the t-th update in amortized $O(\text{poly}\log(t) \log 1/\delta)$ time for every t, such that with probability at least $1 - \delta$, the maintained Steiner tree is $(1 + \epsilon)$-approximate at all time steps simultaneously. Furthermore, conditioning on the randomness of the maintained solution, the algorithms for accessing the maintained tree are deterministic.*

Note that here we do not consider time cost for pre-computation (see Sect. 4 for the detailed discussion), i.e., constructing an empty solution, and t is the

---

[1] In this informal statement we consider d and ϵ as constants and their dependence are hidden in the big-O.

upper bound of the cardinality of the active set (see Sect. 2 for a formal definition).

Notice that our algorithm can report a $(1+\epsilon)$-approximation to the optimal *value*. This value estimation can be generalized to TSP and related problems, including all solvable problems mentioned in [1] such as k-MST and matching. This is due to the fact that the geometric dynamic programming is defined on a quadtree with height $O(\log n)$, and that an update only requires to change $O(\text{poly}\log n)$ subproblems associated with a leaf-root path of length $O(\log n)$. We give a detailed discussion of this in Sect. 1.2.

However, the same approach cannot readily yield an efficient algorithm for maintaining a *solution* (which is a set of edges). One outstanding issue is that even though only poly $\log n$ subproblems change, the solution that is deduced from the DP subproblems can change drastically. We thus seek for a weaker guarantee, which still offers sufficient types of queries to traverse the implicit tree, instead of explicitly maintaining all edges. Even though one may maintain the solution within $O(1)$-approximation by maintaining MST in poly $\log n$ time per update [7,12], it is an open question to improve the ratio to $1+\epsilon$. Indeed, we remark that maintaining $(1+\epsilon)$-approximate solutions for other related problems, such as TSP, is also open, although it may be possible to achieve a weaker guarantee that can be used to reconstruct the solution via multiple queries, similar to our Theorem 1.

1.2 Technical Overview

Conceptually, our algorithm is based on Arora's geometric DP [1] which gives PTAS's for Euclidean Steiner tree and various related problems in the offline setting, and our main technical contribution is to make it dynamic. For simplicity, we expose the ideas only in 2D, and the input is situated in a discrete grid $[\Delta]^2$, where Δ is a parameter. We start with a brief review of Arora's approach.

Review of Arora's Approach. Let n be the number of points in the input dataset (recalling here we are working with a static dataset), and again for simplicity we assume $\Delta = \text{poly}(n)$. To begin, the algorithm builds a randomly-shifted quadtree on $[2\Delta]^2$. Specifically, the bounding square $[2\Delta]^2$ is evenly divided into four sub-squares of half the edge-length, and this process is done recursively, until reaching a unit square. This recursive process naturally defines a tree whose nodes are the squares. Then crucially, the entire quadtree is randomly shifted independently in each coordinate i by a uniformly random value $v_i \in [-\Delta, 0]$. Indeed, in [1], it was shown that over the randomness of the shift, with constant probability, a "structured" solution that is $(1+\epsilon)$-approximate exists, and finally, such a "structured" solution may be found by a dynamic program. The dynamic program is somewhat involved, but at a high level, it defines poly $\log n$ subproblems for each square in the quadtree, and each subproblem can be evaluated from the information of the poly $\log n$ subproblems associated with the child squares.

Making Arora's Approach Dynamic: Value Estimation. Our main observation is that, for a fixed quadtree, when a point x is inserted or deleted, only subproblems that are associated with squares containing x need to be updated/recomputed (and other subproblems do not change). The number of levels is $O(\log \Delta) = O(\log n)$ and in each level only $O(1)$ squares are involved and need to be recomputed, hence the number of subproblems to be re-evaluated is poly $\log n$. This observation directly leads to an efficient dynamic algorithm that maintains an approximation to the optimal Steiner tree. Moreover, this is very general as it only uses the quadtree structure of DP, hence, our result for value estimation also generalizes to other related problems such as TSP, particularly those solvable in [1].

Other Queries. However, in the abovementioned approach, the solution defined by the DP, which is a set of edges, is *not* guaranteed to have a small change per update. Instead, we maintain an implicit solution that supports the following operations: (i) querying the root, (ii) determining whether a point is Steiner point, and (iii) reporting for a given point p its parent and a list of children. To support these queries, we augment Arora's DP and include additional information in each subproblem. In the original Arora's DP, a subproblem is defined with respect to a square R. Intuitively, if one restricts a global solution to R, then it breaks the solution into several small subtrees. Thus, in Arora's approach, the subproblems associated with R also need to describe the connectivity of these subtrees. Now, in our algorithm, we additionally designate the *root* of each subtree, and encode this in the subproblem. This information about the root is crucial for answering the several types of queries including the root query and children-list query. Since we add this new information and requirement in the subproblems, we need to modify Arora's DP, and eventually we show this can be maintained by enlarging the time complexity by only a poly $\log n$ factor.

1.3   Related Work

Steiner Tree in the Graph Setting. The Steiner tree Problem is a well-known problem in graph theory that involves finding the minimum cost tree spanning a given set of vertices in a graph, i.e., given an undirected graph with edge weights and a subset of vertices, the objective of the problem is to find a tree that connects all of the specified vertices at the lowest possible cost. The tree can also include additional vertices (known as Steiner points) that are not part of the original subset. Steiner trees have been used in various applications, such as VLSI design of microchips [11], multipoint routing [15], transportation networks [6], and phylogenetic tree reconstruction in biology [14]. However, the Steiner tree problem in general graphs is NP-hard [14] and cannot be approximated within a factor of $1 + \epsilon$ for sufficiently small $\epsilon > 0$ unless P = NP [3]. Therefore, efficient approximation algorithms are sought instead of exact algorithms. A series of papers gradually improved the best approximation ratio achievable within polynomial time from 2 to 1.39 for the Steiner tree problem in general graphs [2, 4, 13, 16, 18, 21, 22, 24, 25].

Fully-dynamic Algorithms for Steiner Tree. As noted by [6], the under-standing the dynamic complexity of Steiner Tree problem is an important research question due to its practical applications in transportation and communication networks. In the dynamic setting, a sequence of updates would be made to an underlying point set, and the goal of the dynamic Steiner tree problem is to maintain a solution dynamically and efficiently for each update. Imase and Waxman [15] first introduced this problem, and various papers have explored this direction [9,10,19]. For general graphs, Łacki and Sankowski [17] achieved the current best approximation ratio for the dynamic setting at $(6 + \epsilon)$, utilizing a local search technique. Let D denotes the stretch of the metric induced by G, the time complexity for each addition or removal is $\tilde{O}(\sqrt{n} \log D)$ (note that n is the size of the terminal set). However, there is no polynomial time approximation scheme for the dynamic Steiner tree problem.

2 Preliminaries

Notations. For integer $n \geq 1$, let $[n] := \{1, \ldots, n\}$. In Euclidean space $\mathbb{R}^d$, a metric graph is a graph whose vertex set is a subset of $\mathbb{R}^d$, and the edges are weighted by the Euclidean distance between the end points. For any $x, y \in \mathbb{R}^d$, the distance between them is defined as $\mathrm{dist}(x, y) := \|x - y\|$. For a point set $S \subset \mathbb{R}^d$, let $\mathrm{diam}(S) := \max_{x,y \in S} \mathrm{dist}(x, y)$ be its diameter. When we talk about graphs (e.g., tours, trees) we always refer to a metric graph. For a (metric) graph $G = (V, E)$ where V is the set of vertices and E is the set of edges. The weight of an edge $(x, y) \in E$ is defined as $w(x, y)$. Define the weight of the graph $w(G) = \sum_{(x,y) \in E} w(x, y)$. (Note that in this paper, $w(x, y) = \mathrm{dist}(x, y)$.)

When we talk about a d-dimension hypercube in $\mathbb{R}^d$, we consider both its boundary (its $2d$ facets) and interior. For a set of points X, we will use $\mathrm{OPT}(X)$ to denote the cost of the optimal Steiner tree with respect to X.

Definition 1 (Euclidean Steiner Tree). *Given n terminal points $X \subset \mathbb{R}^d$, find a min-weight graph that connects all points in X. In a solution of Steiner tree, non-terminal points are called Steiner points.*

Implicit Steiner Tree and Queries. As mentioned, our algorithm maintains an (implicit) $(1 + \epsilon)$-approximate Steiner tree along with its cost and the root node. By implicit, we mean we do not maintain an explicit edge set; instead, we provide access to this tree indirectly via queries. The algorithm supports the *membership queries* and the *neighbor queries*. The concrete definitions are given below.

- Membership queries. Given a point u, if u is a node in the implicit Steiner tree, return "Yes"; otherwise, return "No".
- Neighbor queries. Given a point u, if u is not a node in the implicit Steiner tree, return "Null"; if a point u is a node in the underlying Steiner Tree, return its all neighbors in the implicit Steiner tree, i.e., the parent and the list of children of u.

Fully Dynamic Setting. We study the Euclidean Steiner Tree problem in a
dynamic setting. Suppose X is the underlying dataset, and we assume $X \subseteq [\Delta]^d$
for some integer parameter Δ. This is a natural assumption, since in the typical
setting where the set X has at most n points and the aspect ratio of the distances
is $\text{poly}(n)$, the dataset can be discretized and rescaled so that it fits into $[\Delta]^d$ grid
for $\Delta = \text{poly}(n)$. For any time step $t \in \mathbb{N}$, an update operation $\sigma_t \in X \times \{+, -\}$
consists of a point $x_t \in \mathbb{R}^d$ and a flag indicating whether the update is an
insertion $(+)$ or a deletion $(-)$.

Multiple updates can be performed on the same coordinate, involving the
insertion or deletion of multiple points. We track a 'multiplicity' value for each
coordinate, initially set to 0. When a point is inserted, the coordinate's multiplicity increases by 1, and when a point is deleted, the multiplicity decreases by
1. Let X_t be the *active set* at time step t, which contains all the co-ordinates (we
will view them as points) that have non-zero multiplicity. Note that for Steiner
tree, the multiplicity of points do not affect the cost (as they can be connected
with 0-length edges). An update operation is *valid* if a point to be deleted is in
the active set. Queries will be made after each update.

Amortized Time Cost. Suppose A_t is the cost for time step t, the amortized
time cost for t is the average time cost for the first t steps, i.e., $\frac{1}{t} \sum_{i=1}^{t} A_i$.

2.1 Review of Arora's Approach [1]

Randomly-shifted Quadtree. Assume the input is in $[\Delta]^d$. Without loss of generality, we assume Δ is a power of 2. For each $i \in \{1, 2, \ldots, d\}$, pick s_i uniformly at random in $[0, \Delta]$. Take the hypercube $R_{\log \Delta}$ as $[2\Delta]^d$ shifted by $-s_i$
in each coordinate i, and this still contains the dataset. We evenly subdivide
this hypercube into 2^d child clusters, and we continue this process recursively
until the hypercube is of diameter at most 1. This process naturally induces a
2^d-ary tree, i.e., $R_{\log \Delta}$ is the root, and each non-leaf cluster has 2^d child clusters.
For $i \in \{0, \ldots, \log \Delta\}$, a level-$i$ cluster is a hypercube R_i with side-length 2^i.
Denote $\mathcal{R}_i$ as the collection of all such hypercubes with side-length 2^i. Denote
$\mathcal{R} := \cup_{i=0}^{\log \Delta} \mathcal{R}_i$.

(m, r)-Light Graph. For every hypercube R in $\mathcal{R}$, designate its 2^d corners as well
as m evenly placed points on each of R's $2d$ facets as portals. Specifically, portals
in a facet is an orthogonal lattice in the $(d-1)$-dimension hypercube. If the side
length of the hypercube is ℓ and the spacing length between the portals is $\alpha \cdot \ell$,
then we will have $m \leq (\frac{1}{\alpha} + 1)^{d-1}$. For example, in $2D$, a hypercube is a square,
and portals in this case is $4m - 4$ points evenly placed on the 4 boundaries. An
(m, r)-light graph is a geometric graph such that for all R in $\mathcal{R}$, it crosses each
facet in R only via the m portals for at most r times.

Theorem 2 (Structural Property [1] in $\mathbb{R}^d$). *For every $0 < \epsilon < 1$ and
terminal set X in $\mathbb{R}^d$, there is a (random) Steiner tree solution T' defined with*

respect to the randomly-shifted quadtree $\mathcal{R}$ of X, such that T' is ($m := O(d^{\frac{3}{2}} \cdot \frac{\log \Delta}{\epsilon})^{d-1}, r := O(\frac{d^2}{\epsilon})^{d-1}$)-light[2] and

$$\Pr[w(T') \le (1 + \epsilon) \cdot \text{OPT}] \ge \frac{1}{2}.$$

Dynamic Programming (DP). One can find the optimal (m, r)-light solution for Steiner tree in near-linear time using dynamic programming (for m, r guaranteed by Theorem 2). In the original DP for Steiner tree [1], a DP entry is indexed by a tuple (R, A, Π), where

- R is a hypercube in $\mathcal{R}$,
- A is a subset of the *active* portals of R such that A contains at most r portals from each facet in R,
- Π is a partition of A.

If R is a leaf node, then $|A| \le 1$. In the case that R contains a terminal, then $|A| = 1$; if $|A| = 0$, this means that R contains no Steiner point. The value of a subproblem (R, A, Π) is the minimum weight of an (m, r)-light graph G, such that for $A \ne \emptyset$,

- G connects all points $X \cap R$ to some point in A, and
- points in the same part of Π are connected in G.

If $A = \emptyset$, then $X \subseteq R$ and the value is the minimum weight of an (m, r)-light graph G such that G is contained inside R and all terminals in X are connected in G.

3   Static Algorithms for Euclidean Steiner Tree

In this section, we introduce an algorithm that answers the queries in the static setting, as stated in the following Theorem 3. We will show how to make the algorithm fully-dynamic in the next section.

Theorem 3. *There is an algorithm that, for every $0 < \epsilon < 1$, given as input an n-point dataset $X \subset \mathbb{R}^d$, pre-processes X and computes an implicit Steiner tree T' for X in $O(n) \cdot (\log n)^{O(d^{2d} \epsilon^{-d})} \cdot 2^{O(4^d d^{4d} \epsilon^{-2d})}$ time such that T' is $(1 + \epsilon)$-approximate with constant probability. After this pre-processing, the algorithm can deterministically report the weight of T', report the root node of T', answer membership query and answer neighbor query, in $O(1)$, $O(\frac{d^2}{\epsilon})^{d-1}$, $O(2^d \log n)$ and $(\log n)^{O(d^{2d} \epsilon^{-d})} \cdot 2^{O(4^d d^{4d} \epsilon^{-2d})}$ time, respectively.*

We impose the quadtree decomposition with random shifts on the input terminal set X as described in Sect. 2.1. Since there is no inherent parent-child

---

[2] We state the exact dependence of d which was not accurately calculated in [1], see the appendix for the details.

relationship between nodes in a Steiner tree, the global root can be chosen arbitrarily. We assume that the global root is selected based on a predefined priority assigned to all clusters. Specifically, for each level of the quadtree, a deterministic function is used to determine which cluster should contain the global root. The deterministic property ensures that we do not have the issue of failure probability after answering however many queries, and that the failure probability only needs to be analyzed for the error of the implicity Steiner tree

After we construct the quadtree, we can find an $(1+\epsilon)$-approximate solution by applying the original dynamic programming (DP) in Arora's approach, since the original DP can find the optimal (m,r)-light solution and Theorem 2 ensures the existence of a solution that is (m,r)-light and $(1+\epsilon)$-approximate with $\frac{1}{2}$ probability. Note that since $n = \Delta^d$ is an upper bound on the cardinality of X, we analyze the time complexity in terms of n. Also the depth of the quadtree is $O(\log \Delta)$, hence for simplicity, we slightly relax the bounds by using $O(\log n)$ as the depth.

As mentioned, to answer the other types of queries, our plan is to augment the original DP of Arora such that each DP entry is indexed by extra objects and each entry stores a small data structure in addition to a value. The focus is to analyze the running time of computing the augmented DP, as well as the correctness and the time complexity of answering each type of these queries. Then, Theorem 3 would follow from all of these components.

Augmented DP for Steiner Tree. Each entry in the augmented DP is indexed by (R, A, Π, φ), where φ contains additional information for each part S in Π (recalling that Π is the partition of A). Specifically, $\varphi(S)$ specifies (i) whether the connected component containing S inside R includes the global root of the entire Steiner tree, and (ii) the node in S that is the closest to the global root, which we shall call the *root* of S, denoted as $r(S)$. In this context, closeness is defined as the number of edges between a point and the global root. It is worth noting that, since the root of S, namely $r(S)$, is the point that is closest to the global root, it ensures that all points in the connected component containing S (inside R) must connect to the global root via $r(S)$.

Data Structure in Each Entry. For a DP entry indexed by (R, A, Π, φ), in addition to a value as in the original DP, the following information is stored: if R is not a leaf cluster, then for each child cluster R_j (where $j \in [2^d]$) of R, there is a pointer to some DP entry $(R_j, A_j, \Pi_j, \varphi_j)$ in the DP entry.

Consistent Conditions. Moreover, there is a collection E (E would not be stored) of directed edges with vertex set in $\cup_j A_j$ that is *consistent* in the following sense. The collection of entries $\{(R_j, A_j, \Pi_j, \varphi_j) : j \in [2^d]\}$ and the directed edges collection E is consistent with (R, A, Π, φ) if the following conditions hold.

- If we consider each part in $\cup_j \Pi_j$ as a supernode, the edges E will form a forest graph among these supernodes. In this graph, an edge in E must connect components originating from different child clusters, meaning it connects distinct active portals from different child clusters. For the purpose of checking for no cycle, we assume that there is an edge between two parts for every common portal they share. Two parts from two different child clusters can share a common portal that lies on the common face between the two child clusters. The edges in E, along with the edges connecting common portals, will collectively form tree graphs among these supernodes as defined earlier.

- There can be at most one part $S \in \cup_j \Pi_j$ that contains the global root.
- The direction of an edge in E is interpreted as going from a parent to a child. For any part $S \in \cup_j \Pi_j$, the root of S can have both outgoing and incoming edges in E, while all other portals in S can only have outgoing edges in E.
- Several parts in $\cup_j \Pi_j$ are merged to become a single part in Π (the portals lying on the common faces would be excluded since they are no longer portals for parts in Π). This is because either an edge in E connects two different parts, or there is a common portal between two parts.
- If $a \in \cup_j A_j$ and a is not a portal of R, then either a is incident to some edge in E or a is a common portal between two parts from different child clusters. If a is also a portal of R, then a must be in A if a is not incident to any edges in E, otherwise a may or may not appear in A.
- The partition Π of A must be consistent with the aforementioned merging process. (Recall that when several parts are merged, the direction of the edges in E must be consistent with the information in the φ_j's.) Moreover, for the new part $S \in \Pi$, whether S contains the global root or which portal is the root of S must be consistent with φ.

In what follows, we use the same m and r as guaranteed by Theorem 2 for the augmented DP for Steiner tree.

Lemma 1 (Time Complexity of Computing Augmented DP). *The augmented DP for Steiner tree can be solved with time complexity $O(n) \cdot (\log n)^{O(d^{2d}\epsilon^{-d})} \cdot 2^{O(4^d d^{4d}\epsilon^{-2d})}$.*

Lemma 2 (Weight of the Steiner Tree). *After solving the augmented DP for Steiner tree, we can report the weight of the implicit Steiner tree with time complexity $O(1)$, and the value is $(1 + \epsilon)$-approximate with probability $\frac{1}{2}$.*

Lemma 3 (Global Root of the Steiner Tree). *After solving the augmented DP for Steiner Tree, we can report the global root of the Steiner tree with time complexity $O(\frac{d^2}{\epsilon})^{d-1}$.*

Proof. Let $\mathcal{I}$ be the entry whose associated cluster is $R_{\log \Delta}$ in the final solution. $\mathcal{I}$ would point to 2^d subproblems $\mathcal{I}' = \{(R_i, A_i, \Pi_i, \varphi_i)\}_{i=1}^{2^d}$. The global root would belong to exactly one part $S \in \cup_{i=1}^{2^d} \Pi_i$. By checking $\varphi_i(S)$ such that $S \in \Pi_i$ for every $i \in \{1, 2, \ldots, 2^d\}$, exactly one part would report "Yes" and the root of S would be the global root.

There are 2^d subproblems, each subproblem corresponds to a partition and each partition has $O(r) = O(\frac{d^2}{\epsilon})^{d-1}$ parts. We can check φ of each part in $O(1)$, hence the total running time would be $2^d \cdot O(\frac{d^2}{\epsilon})^{d-1} = O(\frac{d^2}{\epsilon})^{d-1}$. $\square$

Lemma 4 (Membership Queries). *After solving the augmented DP for Steiner Tree, we can correctly answer membership queries. The time complexity for answering the query is $O(2^d \log n)$.*

Lemma 5 (Neighbor Queries). *After solving the augmented DP for Steiner Tree, we can correctly answer the neighbor queries. The time complexity for answering the query is $O(\log n)^{O(d^{2d}\epsilon^{-d})} \cdot 2^{O(4^d d^{4d}\epsilon^{-2d})}$.*

4   Dynamic Algorithm

In Sect. 3, we have introduced an algorithm that can answer all types of queries in the static setting. In this section, we will show how to make the algorithm fully-dynamic. Specifically, we propose an algorithm that can efficiently maintain a data structure that supports the insertions/deletions of point as well as the afore-mentioned queries in the dynamic setting, as formally stated in the following Theorem 4. As we mention, we consider update sequences whose data points always lie on a discrete grid $[\Delta]^d$ at any time step. Let $n := [\Delta]^d$. This n is an upper bound for the number of points in the active point set at any time step, and hence we can use n as a parameter for measuring the complexity.

Theorem 4. *There exists a fully-dynamic algorithm that for any $0 < \epsilon, \delta < 1$, and any operation sequence σ whose underlying dataset belongs to $[\Delta]^d$ at any time step, maintains an implicit Steiner tree along with its weight and its root node. This tree is maintained in amortized $O(\log \frac{t}{\delta}) \cdot (\log n)^{O(d^{2d}\epsilon^{-d})} \cdot 2^{O(4^d d^{4d}\epsilon^{-2d})}$ time for the t-th operation ($\forall t \geq 1$), and with probability at least $1 - \delta$, it is $(1 + \epsilon)$-approximate at all time steps simultaneously. The subroutine for answering membership and neighbor queries are deterministic, and run in time $O(2^d \log n)$ and $(\log n)^{O(d^{2d}\epsilon^{-d})} \cdot 2^{O(4^d d^{4d}\epsilon^{-2d})}$, respectively.*

Note that we have already shown in Theorem 3 that all these queries can be answered efficiently after computing the augmented DP. Hence, our proof of Theorem 4 aims to maintain the augmented DP efficiently instead of computing it from scratch after each update.

We mostly focus on the simpler version where a single update can be handled with constant success probability. Then to make the algorithm handle all updates and succeed for all of them simultaneously, a standard amplification is to repeat the algorithm for $\log(1/\delta)$ times, so that the failure probability is reduced to δ. However, the operation sequence can be indefinitely long so it is not possible to set a target failure probability δ in advance, so that the total failure probability is still bounded. Our solution is to rebuild the data structure with a decreased target failure probability whenever sufficiently many updates are preformed.

Next, we would introduce a dynamic data structure wrapping the static algorithm described in Sect. 3 that supports updates.

Data Structure for Dynamic Algorithms. Let Γ_ϵ be a data structure storing a set of tuples that correspond to all subproblems in the augmented dynamic programming described in Sect. 3. We use the same m and r as guaranteed by

Theorem 2 (however, note that the parameters may be different from previous sections since the size of bounding box has changed), i.e., in a tuple (R, A, Π, φ), A is a size of $r := O(\frac{d^2}{\epsilon})^{d-1}$ subset of the $m := O(d^{\frac{3}{2}} \cdot \frac{\log n}{\epsilon})^{d-1}$ portals of R. Γ_ϵ supports the following operations:

- *Initialize()*. Conduct the tree decomposition with random shifts on the bounded area. Then we use the augmented dynamic programming described in Sect. 3 to find the optimal (m, r)-light solution. All subproblems will be identified as a set of tuples (R, A, Π, φ) and will be stored together with their evaluated values. Note each subproblem will have optimal value 0 since there are no points when initializing.
- *Insert(x)*. When we call Insert(x), we find the set of all clusters $\mathcal{R}^x \subset \mathcal{R}$ that contain x. There will be exactly a cluster containing x in each level. Let $R_i^x \in \mathcal{R}^x$ be the level-i cluster that contains x. Note that when a point x is inserted, if it is not the new root of the Steiner Tree, then only the subproblems associated with clusters in $\mathcal{R}^x$ will be affected. If x is the new root, let y be the old root, then the subproblems associated with clusters in $\mathcal{R}^x \cup \mathcal{R}^y$ will be affected. We will update the values of these subproblems in a bottom-up manner. Specifically speaking, the subproblems associated with R_0^k (k can be either x or y) will be updated trivially since k is the only point in the cluster; for the subproblems $\mathcal{I} = (R_i^k, A, \Pi, \varphi)$, where $i \in \{1, 2, \ldots, \log \Delta\}$, we consider all combinations of entry indices I' and directed edges E' that are consistent with $\mathcal{I}$, and update the value of $\mathcal{I}$ to the minimum sum of the DP values of I's and the weight of E' over those combinations.
- *Delete(x)*. Deletion procedure is similar to that of insertion. We need to update the values of all subproblems whose corresponding clusters contain x in a bottom-up approach. If x is the root, let y be the new root. We need to update the values of all subproblems whose corresponding clusters contain y in a bottom-up approach.

Pre-computation. Note that the initialization of Γ_ϵ is independent of σ. Hence, we can construct enough copies of Γ_ϵ by calling *Initialize*() before the start of σ and this part of cost would not count in the time complexity analysis for each update.

Lemma 6 (The Correctness of the Data Structure). *Let $0 < \epsilon < 1$, Γ_ϵ can maintain an implicit Steiner tree that is $1 + \epsilon$-approximate with probability at least $\frac{1}{2}$, and report the weight w and the global root of the implicit Steiner tree. The time complexity for Γ_ϵ to finish a single operation of Initialize(Z) is $O(n) \cdot (\log n)^{O(d^{2d} \epsilon^{-d})} \cdot 2^{O(4^d d^{4d} \epsilon^{-2d})}$; and the time complexity to finish a single operation of Insert(x) or Delete(x) is $(\log n)^{O(d^{2d} \epsilon^{-d})} \cdot 2^{O(4^d d^{4d} \epsilon^{-2d})}$.*

Lemma 7 (Answering the Queries). *By maintaining a copy of Γ_ϵ we can, for any time step t, maintain the weight and the global root of an implicit Steiner tree that is $(1+\epsilon)$-approximation with probability at least $\frac{1}{2}$, and can correctly answer the membership queries and the neighbor queries about the implicit*

Steiner tree, at any time step t. The time complexity for answering the two types of queries are $O(2^d \log n)$ and $(\log n)^{O(d^{2d}\epsilon^{-d})} \cdot 2^{O(4^d d^{4d}\epsilon^{-2d})}$, separately.

Proof. We can construct a copy of Γ_ϵ and call *Initialize* to initialize. Then for each update call *Insert* or *Delete*. According to Lemma 6, the implicit Steiner tree maintained by Γ_ϵ is $(1+\epsilon)$-approximate with probability at least $\frac{1}{2}$, and we can report the weight and the global root of the implicit Steiner tree.

According to Lemma 4 and Lemma 5, we can correctly answer membership queries and parent-child queries by using those methods on Γ_ϵ with time complexity $O(2^d \log n)$ and $(\log n)^{O(d^{2d}\epsilon^{-d})} \cdot 2^{O(4^d d^{4d}\epsilon^{-2d})}$, separately. □

Lemma 8 (High Probability Guarantee by Repetition). *For any $0 < \epsilon, \delta < 1$ and any operation sequence σ with length T, we can maintain an implicit Steiner tree that is $(1+\epsilon)$-approximate with probability at least $1 - \delta$ along with its weight and global root, for all time steps simultaneously by maintaining $\log \frac{T}{\delta}$ independent copies of Γ_ϵ. The time cost for each update is $O(\log \frac{T}{\delta}) \cdot (\log n)^{O(d^{2d}\epsilon^{-d})} \cdot 2^{O(4^d d^{4d}\epsilon^{-2d})}$. We can also correctly answer the membership queries and the neighbor queries with probability 1, at any time step t, with time complexity $O(2^d \log n)$ and $(\log n)^{O(d^{2d}\epsilon^{-d})} \cdot 2^{O(4^d d^{4d}\epsilon^{-2d})}$, separately.*

Proof. We can maintain an $(1+\epsilon)$-approximation to the optimal value with probability at least $\frac{1}{2}$ by one copy of Γ_ϵ according to Lemma 7. Having $\log \frac{T}{\delta}$ independent copies can then guarantee that with probability at least $1-(\frac{1}{2})^{\log \frac{T}{\delta}} = 1-\frac{\delta}{T}$ at least one copy can maintain an $(1 + \epsilon)$-approximation to the optimal value. Then by a union bound, the probability that all time steps succeed simultaneously is $1 - \delta$. Note that each approximate value corresponds to a feasible solution to the Euclidean Steiner tree problem, we hence fix the copy $\tilde{\Gamma}$ with the minimum approximate value to perform further queries.

The time cost comes from two parts: the update (*Insert* and *Delete*), and finding $\tilde{\Gamma}$ after each update. The initialization of $O(\log \frac{T}{\delta})$ copies in the pre-computation. According to Lemma 6, each update takes $(\log n)^{O(d^{2d}\epsilon^{-d})} \cdot 2^{O(4^d d^{4d}\epsilon^{-2d})}$. There are $\log \frac{T}{\delta}$ copies in the algorithm, hence each update would take $\log \frac{T}{\delta} \cdot (\log n)^{O(d^{2d}\epsilon^{-d})} \cdot 2^{O(4^d d^{4d}\epsilon^{-2d})}$ time. After each update, according to Lemma 2, we need $O(\log \frac{T}{\delta})$ time to find $\tilde{\Gamma}$.

Moreover, a global root is maintained in $\tilde{\Gamma}$ according to Lemma 7, and we can also correctly answer the queries by performing those queries on $\tilde{\Gamma}$. Hence the correctness and time complexity for other queries are as desired. □

Proof (Proof of Theorem 4). According to Lemma 7, we can maintain an implicit Steiner tree that is $(1 + \epsilon)$ with probability at least $\frac{1}{2}$, report the weight and the global root of the implicit Steiner tree, and correctly answer the membership queries and the parent-child queries, at any time step t. The time complexity for answering the two types of queries are $O(2^d \log n)$ and $(\log n)^{O(d^{2d}\epsilon^{-d})} \cdot 2^{O(4^d d^{4d}\epsilon^{-2d})}$, separately.

Then we analyze the per time step cost. The time cost comes from the updates (*Insert* and *Delete*) (note that *Initialize* is done in pre-computation). According to Lemma 6, each update would take $(\log n)^{O(d^{2d}\epsilon^{-d})} \cdot 2^{O(4^d d^{4d}\epsilon^{-2d})}$ time.

Partition σ into phases with different lengths indexed from 1. Specifically, the i-th phase would have $2^i n$ time steps. For the i-th phase, we would guarantee that we can maintain an implicit Steiner tree that with probability at least $1 - \frac{\delta}{2^i}$ is $(1 + \epsilon)$-approximate. By a union bound on the phases, we can guarantee that all time steps succeed simultaneously with probability at least $1 - \sum_i \frac{\delta}{2^i} \geq 1 - \delta$. Then we need to maintain $\log \frac{2^i n}{\delta} = O(i + \log n + \log \frac{1}{\delta})$ copies for the i-th phase according to a similar analysis as in the proof of Lemma 8.

Note that we do not need to reconstruct all copies at the beginning of each phase, instead, we just need to add some new copies since Theorem 2 has no dependency on the time step. We will keep inserting all points in the active point set to the new copies for later updates.

We first analyze the reconstruction cost for the copies. The time step t would lie in the $O(\log \frac{t}{n})$-th phase and there are $O(2^{\log \frac{t}{n}} n) = O(t)$ time steps in the phase. The total number of copies of Γ_ϵ needed for the first t steps is $O(\log \frac{t}{\delta})$. For each copy, we may need to insert at most n points for reconstruction. Hence the total cost for this part is $n \cdot O(\log \frac{t}{\delta}) \cdot (\log n)^{O(d^{2d}\epsilon^{-d})} \cdot 2^{O(4^d d^{4d}\epsilon^{-2d})}$.

Then we analyze the update cost for the first t steps. Since there are $\log \frac{2^i n}{\delta} = O(i + \log n + \log \frac{1}{\delta})$ copies in the i-th phase, hence the total update cost would be $\sum_{i=1}^{\log \frac{t}{n}} O((i + \log n + \log \frac{1}{\delta}) \cdot (2^i) \cdot n) \cdot (\log n)^{O(d^{2d}\epsilon^{-d})} \cdot 2^{O(4^d d^{4d}\epsilon^{-2d})} = O(t \log \frac{t}{\delta} \cdot (\log n)^{O(d^{2d}\epsilon^{-d})} \cdot 2^{O(4^d d^{4d}\epsilon^{-2d})}$.

Note that $t > n$. Therefore, the amortized time cost for the time step t is $\log \frac{t}{\delta} \cdot (\log n)^{O(d^{2d}\epsilon^{-d})} \cdot 2^{O(4^d d^{4d}\epsilon^{-2d})}$.

The correctness and time complexity for other queries are also as desired according to Lemma 8.   □

References

1. Arora, S.: Polynomial time approximation schemes for euclidean traveling salesman and other geometric problems. J. ACM **45**(5), 753–782 (1998)
2. Berman, P., Ramaiyer, V.: Improved approximations for the steiner tree problem. J. Algorithms **17**(3), 381–408 (1994)
3. Bern, M., Plassmann, P.: The steiner problem with edge lengths 1 and 2. Inf. Process. Lett. **32**(4), 171–176 (1989)
4. Byrka, J., Grandoni, F., Rothvoß, T., Sanita, L.: An improved lp-based approximation for steiner tree. In: Proceedings of the Forty-Second ACM Symposium on Theory of Computing, pp. 583–592 (2010)
5. Chan, T.H.H., Goranci, G., Jiang, S.H.C., Wang, B., Xue, Q.: Fully dynamic algorithms for euclidean steiner tree. CoRR arxiv (2023)
6. Cheng, X., Du, D.Z.: Steiner trees in industry, vol. 11. Springer Science & Business Media (2013)
7. Eppstein, D.: Dynamic euclidean minimum spanning trees and extrema of binary functions. Discret. Comput. Geom. **13**, 111–122 (1995)

8. Garey, M.R., Graham, R.L., Johnson, D.S.: The complexity of computing steiner minimal trees. SIAM J. Appl. Math. **32**(4), 835–859 (1977)
9. Gu, A., Gupta, A., Kumar, A.: The power of deferral: maintaining a constant-competitive steiner tree online. In: Proceedings of the Forty-Fifth Annual ACM Symposium on Theory of Computing, pp. 525–534 (2013)
10. Gupta, A., Kumar, A.: Online steiner tree with deletions. In: Proceedings of the twenty-fifth annual ACM-SIAM symposium on Discrete algorithms. pp. 455–467. SIAM (2014)
11. Held, S., Korte, B., Rautenbach, D., Vygen, J.: Combinatorial optimization in vlsi design. Combinatorial Optimization, pp. 33–96 (2011)
12. Holm, J., de Lichtenberg, K., Thorup, M.: Poly-logarithmic deterministic fully-dynamic algorithms for connectivity, minimum spanning tree, 2-edge, and biconnectivity. J. ACM **48**(4), 723–760 (2001)
13. Hougardy, S., Prömel, H.J.: A 1.598 approximation algorithm for the steiner problem in graphs. In: Proceedings of the Tenth Annual ACM-SIAM Symposium on Discrete Algorithms, pp. 448–453 (1999)
14. Hwang, F.K., Richards, D.S.: Steiner tree problems. Networks **22**(1), 55–89 (1992)
15. Imase, M., Waxman, B.M.: Dynamic steiner tree problem. SIAM J. Discret. Math. **4**(3), 369–384 (1991)
16. Karpinski, M., Zelikovsky, A.: New approximation algorithms for the steiner tree problems. J. Comb. Optim. **1**, 47–65 (1997)
17. Lacki, J., Ocwieja, J., Pilipczuk, M., Sankowski, P., Zych, A.: The power of dynamic distance oracles: efficient dynamic algorithms for the steiner tree. In: STOC, pp. 11–20. ACM (2015)
18. Matsuyama, A.: An approximate solution for the steiner problem in graphs. Math. Japonica **24**, 573–577 (1980)
19. Megow, N., Skutella, M., Verschae, J., Wiese, A.: The power of recourse for online mst and tsp. SIAM J. Comput. **45**(3), 859–880 (2016)
20. Mitchell, J.S.B.: Guillotine subdivisions approximate polygonal subdivisions: a simple polynomial-time approximation scheme for geometric tsp, k-mst, and related problems. SIAM J. Comput. **28**(4), 1298–1309 (1999)
21. Prömel, H.J., Steger, A.: RNC-approximation algorithms for the steiner problem. In: Reischuk, R., Morvan, M. (eds.) STACS 1997. LNCS, vol. 1200, pp. 559–570. Springer, Heidelberg (1997). https://doi.org/10.1007/BFb0023489
22. Robins, G., Zelikovsky, A.: Tighter bounds for graph steiner tree approximation. SIAM J. Discret. Math. **19**(1), 122–134 (2005)
23. Rubinstein, J.H., Thomas, D.A., Wormald, N.C.: Steiner trees for terminals constrained to curves. SIAM J. Discret. Math. **10**(1), 1–17 (1997)
24. Zelikovsky, A.: Better approximation bounds for the network and euclidean steiner tree problems. Tech. rep., Technical Report CS-96-06, Department of Computer Science, University of (1996)
25. Zelikovsky, A.Z.: An 11/6-approximation algorithm for the network steiner problem. Algorithmica **9**, 463–470 (1993)

The Parameterized Complexity
of Terminal Monitoring Set

N. R. Aravind[(⊠)] and Roopam Saxena[(⊠)]

Department of Computer Science and Engineering IIT Hyderabad, Hyderabad, India
aravind@cse.iith.ac.in, cs18resch11004@iith.ac.in

Abstract. In Terminal Monitoring Set (TMS), the input is an undirected graph $G = (V, E)$, together with a collection T of terminal pairs and the goal is to find a subset S of minimum size that hits a shortest path between every pair of terminals. We show that this problem is W[2]-hard with respect to solution size. On the positive side, we show that TMS is fixed parameter tractable with respect to solution size plus distance to cluster, solution size plus neighborhood diversity, and feedback edge number. For the weighted version of the problem, we obtain a FPT algorithm with respect to vertex cover number, and for a relaxed version of the problem, we show that it is W[1]-hard with respect to solution size plus feedback vertex number.

Keywords: monitoring set · hitting set · hub location · parameterized complexity · fixed parameter tractability

1 Introduction

Consider a communication network and a set of pairs of nodes, say $T = \{\{x_1, y_1\}, \{x_2, y_2\}, \ldots \{x_m, y_m\}\}$ such that every pair in T is communicating through the shortest path between them. Our goal is to monitor this data while deploying monitoring devices at minimum number of nodes in the network. The motivation also comes from incremental deployment of a software defined network over a legacy network by deploying costly smart switches at only a few locations.

Motivated by the above scenarios, we formulate the Terminal Monitoring Set (TMS) problem defined below.

TERMINAL MONITORING SET (TMS):
Input: An instance $I = (G, T, k)$, where $G = (V, E)$ is an undirected graph, $T = \{\{u_1, v_1\}, \ldots, \{u_l, v_l\}\}$ where $u_i, v_i \in V$ and $k \in \mathbb{N}$.
Output: YES, if G contains a set $S \subseteq V$ of size at most k such that for every $i \leq l$, $\exists w \in S : d(u_i, w) + d(w, v_i) = d(u_i, v_i)$; NO otherwise.

R. Uehara et al. (Eds.): WALCOM 2024, LNCS 14549, pp. 76–90, 2024.
https://doi.org/10.1007/978-981-97-0566-5_7

We refer to any set S satisfying the above definition (regardless of its size) as a terminal monitoring set for T. We remark that a vertex can belong to multiple pairs in the list T, but we assume without loss of generality that no terminal pair appears twice. Further, we assume distance between every terminal pair in T is finite (they are reachable by each other, hence belong to a same connected component).

While there are similar problems in the literature, some of which we will discuss in the next section, to the best of our knowledge, the optimization equivalent of TMS has not been studied before.

1.1 Related Work

The closest related problem to TMS in the literature is the (k, r)-center problem, where we are given an undirected graph and it is asked if there exists a vertex set $S \subseteq V(G)$ such that $|S| \leq k$ and for every vertex $v \in V(G) \setminus S$ there exists a vertex $u \in S$ such that distance between u and v is at most r. Optimization of r for a fixed k is studied in [7,13,14,17,18,21], and optimization of k for fixed r is studied in [3,5,8,20]. Recently Katsikarelis, Lampis, Paschos [16] studied parameterized complexity of (k, r)-center with respect to various structural parameters. Benedito, Melo, and Pedrosa [4] studied a problem related to (k, r)-center, under the name of Multiple Allocation k-Hub Center, which is also a closely related problem to TMS.

Another related problem is hub location where packets must travel from each source to its corresponding destination via a small number of hubs. Surveys of hub location can be found in [2,11].

The TMS problem can be formulated as a Hitting Set problem, and thus it is natural that results on Hitting Set are relevant to this work. In particular, we mention the results of [1,12] that use the sunflower lemma to set systems of bounded size. We use similar ideas in some of our algorithmic results.

1.2 Our Results

Theorem 1. *(a) TMS is NP-hard.*
(b) TMS is W[2]-hard with respect to solution size.

On the positive side, we have the following results.

Theorem 2. *TMS admits a FPT algorithm when parameterized by solution size plus distance to cluster.*

Theorem 3. *TMS admits a FPT algorithm when parameterized by solution size plus neighborhood diversity.*

Theorem 4. *Weighted-TMS admits a FPT algorithm when parameterized by vertex cover number.*

Theorem 5. *TMS admits a FPT algorithm when parameterized by the feedback edge number.*

We leave open the parameterized complexity of TMS by feedback vertex number; however we obtain a hardness result for the following relaxation of TMS.

α-RELAXED TERMINAL MONITORING SET (α-RTMS):
Input: An instance $I = (G, \mathcal{T}, k)$, where $G = (V, E)$ is an undirected graph, $\mathcal{T} = \{\{u_1, v_1\}, \dots, \{u_l, v_l\}\}$ where $u_i, v_i \in V$, $k \in \mathbb{N}$ and $\alpha \in \mathbb{Q}_{\geq 0}$.
Output: YES, if G contains a subset S of size at most k such that for every $i \leq l$, $\exists w \in S : d(u_i, w) + d(w, v_i) \leq (1+\alpha) \cdot d(u_i, v_i)$; NO otherwise.

Theorem 6. *For every fixed $0 < \alpha \leq 0.5$, α-RTMS is W[1]-hard with respect to feedback vertex number of the input graph plus solution size.*

2   Preliminaries

We refer [10] for basic graph notations and terminologies. For a graph G, we use $V(G)$ and $E(G)$ are its vertex set and edge set respectively. For $F \subseteq E(G)$, $V(F)$ is the vertex set of F. For $S \subseteq V(G)$, $G[S]$ is the induced sub graph of G on vertex set S, and $G - S$ is the graph $G[V(G) \setminus S]$. A component C of a graph G is a maximally connected subgraph of G. For $F \subseteq E(G)$, $G[F]$ is the sub graph of G with vertex set $V(F)$ and edge set F. We use $v \in G$ instead of $v \in V(G)$ when it is clear that v is a vertex. For a weighted graph G, $w(e)$ is the weight of an edge $e \in E(G)$. For a graph G, $d(u, v)$ is the distance between vertices u and v in G. Given a graph G, and $u, v \in V(G)$, we define $SP_G(u, v)$ to be the set $\{x \in V(G) : d(u, x) + d(x, v) = d(u, v)\}$. When the context is clear, we simply write $SP(u, v)$. Thus, TMS is equivalent to finding a minimum size hitting set for the family $\{SP(u, v) \mid \{u, v\} \in T\}$. For undirected graphs, $SP(u, v)$ and $SP(v, u)$ are the same; and we avoid writing $SP(\{u, v\})$.

We define the **core** of a set family $\mathcal{F}$ to be $\cap_{S \in \mathcal{F}} S$ and denote it by $core(\mathcal{F})$. We say that a collection $\mathcal{F}$ of sets forms a **sunflower** if there is a set C such that $S \cap T = C$ for every distinct pair $S, T \in \mathcal{F}$. Notice that $C = core(\mathcal{F})$ in this case (see [9,12] for details on sunflower, sunflower lemma, and its application to Hitting Set). For details on parameterized complexity and fixed parameter tractability (FPT) we refer to [9,12]. For the details on neighborhood diversity we refer to [19].

HITTING SUBGRAPHS IN A GRAPH (HSG):
Input: An instance $I = (G, \mathcal{V}, k)$, where $G = (V, E)$ is an undirected graph, $\mathcal{V}$ is a collection of subsets of $V(G)$, and $k \in \mathbb{N}$.
Output: YES, if G contains a $S \subseteq V(G)$ of size at most k that hits every set in $\mathcal{V}$; NO otherwise.

If HSG has the constraint that every vertex set in $\mathcal{V}$ induces a simple path in G, then we call the problem HITTING PATHS IN A GRAPH (HPG) [15].

Lemma 1 (folklore, discussed in [15]**).** HITTING PATHS IN A GRAPH *can be solved in polynomial time if the input graph G is a tree.*

From [15] we recall that a graph G is a flower graph if it has a specific vertex z (called its core) such that $G - \{z\}$ is a disjoint union of paths, each such path is called its petal, and no internal vertex of any such path is adjacent to z, assume an arbitrary but distinct ordering $\{R_1, R_2, \ldots R_l\}$ of these petals.

HITTING PATHS IN A FLOWER WITH BUDGETS (HPFB)[15]:
Input: An instance $I = (G, z, P, b)$, where G is a flower graph with core z and petals $R_1, R_2, \ldots R_l$, a set P of simple paths in G, and $b : [l] \to \mathbb{N}_{\geq 1}$.
Question: Is there a set $S \subseteq V(G) \setminus \{z\}$ hitting every path in P such that $|S \cap V(R_i)| = b(i)$ for every $i \in [l]$?.

Lemma 2 ([15]**).** *HPFB is polynomial time solvable.*

3   FPT Algorithms

Our main idea for the first three results is to create an equivalent instance of Hitting Set where the number of sets is a function of k plus the structural parameter. A Hitting Set instance with a set family having m sets over a n-element universe can be solved in $O(2^m \cdot (n+m)^{O(1)})$ time by a standard dynamic programming algorithm (see [9]); this gives the corresponding FPT algorithm for TMS.

Another idea that we use is the following, which we shall call the *standard vertex cover reduction*. This is the reduction applied to obtain a quadratic kernel for vertex cover, in [6].

Observation 1. *Let $\mathcal{F} = \mathcal{F}_1 \cup \mathcal{F}_2$ be a family of sets, such that every set in $\mathcal{F}_2$ is of size 2. Then, given the instance $(\mathcal{F}, k)$ for some integer k, we can replace $\mathcal{F}_2$ by a family $\mathcal{F}_3$, where $|\mathcal{F}_3| = O(k^2)$ such that $(\mathcal{F}, k)$ is a YES instance if and only if $(\mathcal{F}_1 \cup \mathcal{F}_3, k)$ is a YES instance.*

We remark that this can be generalized to subfamilies of bounded size and we indeed do this in a slightly more general way later in this paper (Proposition 1).

On an input instance (G, T, k) of TMS, we will also assume that the following reduction rule has been applied exhaustively, so that it is not applicable to the instance given as input in each algorithm.

Reduction Rule 0: For two distinct pairs $\{u, v\}$ and $\{x, y\}$ in T, if $SP(u, v) \subseteq SP(x, y)$, then remove $\{u, v\}$ from the list of terminal pairs. In the case of equality, remove one of the two pairs arbitrarily.

3.1   Distance to Cluster: Proof of Theorem 2

Let (G, T, k) be the input instance of TMS. We may assume that G is connected (otherwise we find the optimal solution for each connected component separately). We define the auxiliary graph $G_T = (V_T, E_T)$ as $E_T = T$ and $V_T = V(E_T)$. We say that a graph H on a subset $S \subseteq V$ is **core-invariant** if it is a subgraph of G_T and the family $\{SP(u, v) : \{u, v\} \in E(H)\}$ forms a sunflower.

Reduction Rule 1: Given a core-invariant graph H with at least $k + 2$ edges, remove all but $k + 1$ of these pairs from T.

Note that to apply Reduction Rule 1, the subgraph H must be known. The rule is safe because any set of size at most k that hits $(k+1)$ sets in the sunflower must hit the core, and hence hit all the sets $SP(u, v)$ for every edge $\{u, v\}$ in H.

Let $M \subseteq V$ be of cardinality at most q such that every connected component of $G[V \setminus M]$ is a clique. Let these clique components be $C_1, C_2 \ldots, C_r$ and let $C = C_1 \cup C_2 \ldots \cup C_r$. Let $T_0 = \{\{u, v\} \in T \cap E\}$, $T_1 = \{\{u, v\} \in T \setminus T_0 : u \in C, v \in M\}$ and $T_2 = \{\{u, v\} \in T \setminus T_0 : u, v \in C\}$.

We shall first reduce the size of T_2. We fix an arbitrary ordering between both the vertices of every pair $\{u, v\} \in T_2$, and thus $\{u, v\}$ is denoted by (u, v) or (v, u) depending on the ordering. The ordering will not affect $SP(u, v)$ for $\{u, v\}$. For $X \subseteq M \times \{1, 2\} \times M \times \{1, 2\}$, we say that $(u, v) \in T_2$ is of type X if for every shortest path P from u to v, there exists $(x, i, y, j) \in X$ such that x and y are the closest vertices in $V(P) \cap M$ to u and v respectively, and $d(u, x) = i, d(v, y) = j$, and X is the smallest such set (wrt inclusion).

Note that the number of possible types X is at most 2^{4q^2}. Every pair in T_2 is of some type X, and given (u, v) and X, we can verify in polynomial time whether (u, v) is of type X. For each type X, we define an auxiliary graph $H_X = (V_X, E_X)$ with $V_X = \{1, 2, \ldots, r\}$ and $\{i, j\} \in E_X$ if there exists a pair $(u, v) \in T_2$ of type X such that one of its endpoint (either u or v) belong to C_i and the other endpoint belong to C_j. These auxiliary graphs H_X can be computed in polynomial time.

Claim 1. *If $|E_X| > (2(k+2))^3$, then we can find a core-invariant subgraph of G_T with at least $(k+2)$ edges, and hence apply Reduction Rule 1.*

Proof-sketch. If $|E_X| > (2(k+2))^3$, then H_X must contain a vertex of degree at least $(2(k+2))^2$ or (by Vizing's theorem), a matching A of size at least $2(k+2)$.
Case 1: H_X contains a matching A of size at least $2(k+2)$. In this case, for every edge in A, pick exactly one terminal pair from T_2 which corresponds to its construction, let A^* be these picked pairs, every vertex in $V(A^*)$ belongs to a disinct cluster and every pair in A^* is of type X, it follows that $G_T[A^*]$ is core-invariant with $k+2$ edges, and we can apply Reduction Rule 1.
Case 2: H_X contains a vertex i of degree at least $(2(k+2))^2$. For every edge incident on i, pick exactly one terminal pair from T_2 which corresponds to its construction, let B be the set of these picked pairs. Consider the graph $G_T[B]$, there must either be a vertex u in $G_T[B]$ which belong to C_i and has $2(k+2)$ neighbors in $G_T[B]$ each of which belongs to a distinct cluster of $G - M$, or there must be a matching M_B in $G_T[B]$ of size $2(k+2)$ such that for every edge of M_B, its one endpoint is a distinct vertex of C_i and the other endpoint is a vertex of a distinct cluster in $G - M$.
Case 2a: vertex u in $G_T[B]$ belonging to C_i has $2(k+2)$ neighbors in $G_T[B]$ each of which belongs to a distinct cluster of $G - M$. Let U be the set of terminal pairs which correspond to these $2(k+2)$ neighbors of u in $G_T[B]$. Let U_1 be those terminal pairs of U where the first vertex is u (Ex. (u, v)), and let U_2 be those terminal pairs in U where the second vertex is u (Ex. (x, u)). At least one of U_1 or U_2 has at least $k+2$ pairs, let it be U_1. Then every pair in U_1 is of type X and each second vertex in the pair belongs to a distinct cluster. It follows that $G_T[U_1]$ is core-invariant with $(k+2)$ edges. Similar arguments holds when U_2 has at least $k+2$ pairs.
Case 2b: matching M_B in $G_T[B]$ of size $2(k+2)$ such that for every edge of M_B, its one endpoint is a distinct vertex of C_i and the other endpoint is a vertex of a distinct cluster in $G - M$. Let U be the set of terminal pairs which correspond to these $2(k+2)$ edges, let U_1 (resp U_2) be the set of those terminal pairs of U where the first vertex (resp. second vertex) belongs to C_i. Suppose that U_1 has at least $k+2$ pairs. Then every pair in U_1 is of type X and its first vertex is a distinct vertex of C_i and second vertex belongs to a distinct cluster, it follows that $G_T[U_1]$ is core-invariant with $k+2$ edges, similar arguments holds when U_2 has at least $k+2$ pairs. $\qquad\square$

Claim 2. *If there exist $i, j \in [r]$ such that there are more than $(2(k+2))^2$ distinct pairs in T_2 of a same type X with one endpoint in C_i and another endpoint in C_j, then we can find a core-invariant subgraph of G_T with at least $k+2$ edges. Hence, apply Reduction Rule 1.*

The proof of the above claim is similar to that of Claim 1 and hence we skip it. For every type X, we apply Reduction Rule 1 repeatedly while the condition in Claim 1 or the condition in Claim 2 holds.

We now have $|E_X| \le (2(k+2))^3$. Further, for every edge $\{i,j\} \in E_X$, the number of pairs in T_2 with one endpoint in C_i and the other end-point in C_j is at most $(2(k+2))^2$ (because Claim 2 is not applicable). Thus, we obtain a reduced set T_2 such that $|T_2| \le (2(k+2))^5 2^{4q^2}$.

We shall now reduce the size of T_1 in a similar manner. We fix an ordering between both the vertices of every pair $\{u,v\} \in T_1$ such that its first vertex belong to C, and thus $\{u,v\}$ denoted by (u,v) when u belongs to C and v to M. For $X \subseteq M \times \{1,2\}$, we say that $(u,v) \in T_1$ is of type X if for every shortest path P from u to v, there exists $(x,i) \in X$ such that x is the closest vertex in $V(P) \cap M$ to u and $d(u,x) = i$, and X is the smallest such set.

As before, we define an auxiliary bipartite graph $H_X = (V_X, E_X)$, where $V_X = \{1,2,\ldots,r\} \cup M$ and for $i \in [r]$ and $m \in M$, $\{i,m\} \in E_X$ if there exists a pair $(u,m) \in T_1$ of type X such that $u \in C_i$.

Claim 3. *If there is a vertex $m \in M$ with at least $(k+2)$ neighbors in H_X, then we can find a core-invariant subgraph of G_T with at least $(k+2)$ edges and apply Reduction Rule 1.*

Claim 4. *If there exists $i \in [r]$ and $m \in M$ such that there are more than $(k+2)$ distinct pairs in T_1 of a same type X with one endpoint in C_i and other endpoint being vertex m, then we can find a core-invariant subgraph of G_T with at least $k+2$ edges, and apply Reduction Rule 1.*

The above claims are similar to that of Claim 1 and we skip their proof. For every type X, we apply Reduction Rule 1 repeatedly while the condition in Claim 3 or the condition in Claim 4 holds.

We now obtain: $|T_1| \le q(k+2)^2 2^{2q}$.

Finally, we apply Observation 1 to the family $\{SP(u,v) : \{u,v\} \in T_0\}$ so that we obtain a reduced set T_0 such that $|T_0| = O(k^2)$.

Thus, we obtain an instance with the number of terminal pairs are bounded by $O(q(k+2)^5 2^{4q^2})$. We then solve the Hitting Set Instance where the input family is $\{SP(u,v) : \{u,v\} \in T\}$, in time FPT in $q+k$. This completes the proof of Theorem 2.

3.2   Neighborhood Diversity: Proof of Theorem 3

Our main idea is to directly use existing kernels for the hitting set problem with a small modification.

Definition 1. *Given a family $\mathcal{F}$ and a set S, we define the **effective size** of S with respect to $\mathcal{F}$ to be $|S \setminus core(\mathcal{F})|$. We define the **effective size bound** of $\mathcal{F}$ to be the maximum effective size of S wrt $\mathcal{F}$, taken over all $S \in \mathcal{F}$.*

The following result is an adaptation of a well-known application of the sunflower lemma to hitting sets, see e.g.: [1,9,12].

Proposition 1. *There is an algorithm, that given set families $\mathcal{F}_1,\ldots,\mathcal{F}_m$ each with an effective size bound of d, and an integer k, finds set families $\mathcal{C}_1,\ldots,\mathcal{C}_m$ in time polynomial in m,d, and the number of sets such that:*

- $|\mathcal{C}_i| = O(k^d \cdot d!)$;
- $(\cup_i \mathcal{F}_i, k)$ *is a YES instance of hitting set if and only if* $(\cup_i \mathcal{C}_i, k)$ *is a YES instance of hitting set.*

Let (G, T, k) be an input to TMS with G having neighborhood diversity t. We assume that G is connected. Let $V(G) = V_1 \cup \ldots V_t$ where each V_i induces an independent set or clique and such that for every pair i, j, there are either no edges between V_i and V_j or there are all possible edges between them.

For each pair (i, j) such that $1 \le i \le j \le t$, let $T_{i,j} = \{\{u, v\} \in T : u \in V_i, v \in V_j\}$. We define the family $\mathcal{F}_{i,j} = \{SP(u, v) \mid \{u, v\} \in T_{i,j}\}$. Each $\mathcal{F}_{i,j}$ has an effective size bound ≤ 2, where $core(\mathcal{F}_{i,j}) \supseteq (\cup_{S \in \mathcal{F}_{i,j}} S) \setminus (V_i \cup V_j)$. Thus, using Proposition 1, we obtain an equivalent hitting set instance with $O(t^2 k^2)$ sets.

3.3   Weighted TMS and Vertex Cover: Proof of Theorem 4

In this section, we consider TMS where the underlying graph G has positive weights on its edges which satisfy the triangle inequality, i.e. $w(x, y) + w(y, z) \ge w(x, z)$ for all x, y, z. The distance is then the shortest weighted distance and $SP(u, v)$ is defined accordingly.

Let $C = \{v_1, \ldots, v_t\}$ be a given vertex cover and $I = V \setminus C$. Let S be an optimal solution with $k = |S|$. Note that we may assume that $k \le t$; otherwise C itself is a solution of size t.

We first guess a binary matrix M indexed by C and then search for a solution S satisfying the following condition.

$$\forall u, v \in C : S \cap SP(u, v) \ne \emptyset \Leftrightarrow M(u, v) = 1. \tag{1}$$

Note that when $u = v$, the above condition translates to : $v \in S$ if and only if $M(v, v) = 1$. We call such a solution an M-compatible solution.

We shall construct a Hitting Set instance $(\mathcal{F}, l)$ which is a YES instance if and only if (G, T, k) has an M-compatible solution.

To this end, let $C_1 = \{v \in C : M(v, v) = 1\}$ and $C_0 = C \setminus C_1$. Let $S_0 = \cup_{M(u,v)=0} SP(u, v)$. Note that we are looking for a solution S such that $S \cap S_0 = \emptyset$. If $\{u, v\} \in T$ be such that $u, v \in C$ and $M(u, v) = 0$, then no M compatible solution exist. If $S_0 \cap C_1 \ne \emptyset$, then M is inconsistent and we make the next guess. If there exist $u, v \in C$ such that $M(u, v) = 1$ and $SP(u, v) \setminus S_0 = \emptyset$, then no M compatible solution exist and we make the next guess.

Let $\mathcal{F}_1 = \{SP(u, v) \setminus S_0 : u, v \in C, M(u, v) = 1\}$. The family $\mathcal{F}$ will include all the sets of $\mathcal{F}_1$ so that $SP(u, v)$ is hit for every $u, v \in C$ for which $M(u, v) = 1$. Let $T_1 = \{\{u, v\} \in T : \exists x, y \in C : SP(x, y) \subseteq SP(u, v) \text{ and } M(x, y) = 1\}$. Then every hitting set for $\mathcal{F}_1$ also hits $SP(u, v)$ for all terminal pairs $\{u, v\} \in T_1$. Thus, it is sufficient to focus on $T_0 = T \setminus T_1$. We also note that the set T_0 can be computed in polynomial time.

Reduction Rule 2: Let $T_0^* = T_0, l = k$. While there exists a pair $\{u, v\} \in T_0^*$ such that $u \in I, v \in C_0$, do the following:

- Include u into the solution set;
- Let $T_0^* = T_0^* \setminus \{\{x, y\} : u \in SP(x, y)\}$.
- Decrement l.

The above reduction rule is sound because if the solution set does not contain u, then it must contain a vertex from $SP(w, v)$ for some $w \in C$; but this would imply that $M(w, v) = 1$ and hence that $\{u, v\} \in T_1$, a contradiction.

We apply Reduction Rule 2 exhaustively until it can no longer be applied. Let T_0^*, l be the resulting terminal pair set and solution budget respectively.

Let $\{u, v\} \in T_0^*$ be such that $u, v \in I$. We claim that we must include at least one of u, v in the solution. Otherwise S must contain a vertex $z \in C_1$ such that z is neighbor of both u, v and $d(u, v) = w(u, z) + w(z, v)$ or S must intersect $SP(z, y)$ for some $z, y \in C$ such that $d(u, v) = w(u, z) + d(z, y) + w(y, v)$. In either case, it would imply that $\{u, v\} \in T_1$.

Let $\mathcal{F}_2 = \{\{u, v\} \in T_0^* : u, v \in I\}$ and $\mathcal{F} = \mathcal{F}_1 \cup \mathcal{F}_2$. Then, there is a hitting set of $\mathcal{F}$ of size at most l if and only if there is a M-compatible solution for (G, T, k).

Further, by applying Observation 1 to $\mathcal{F}_2$, we can obtain a new collection $\mathcal{F}_3$ replacing $\mathcal{F}_2$ such that $|\mathcal{F}_3| = O(k^2)$ and such that $\mathcal{F}$ has a hitting set of size l if and only if $\mathcal{F}_1 \cup \mathcal{F}_3$ has a hitting set of size l.

We solve the instance $(\mathcal{F}_1 \cup \mathcal{F}_3, l)$; since $|\mathcal{F}_1 \cup \mathcal{F}_3| \leq t^2 + O(k^2) = O(t^2)$, we can solve this instance in time FPT in t. This finishes the proof of Theorem 4.

3.4   Feedback Edge Number: Proof of Theorem 5

Jansen [15] gave an algorithm (which we refer to as Jansen's algorithm) running in time FPT by feedback edge number of the input graph to solve HITTING PATHS IN A GRAPH. In TMS, we need to hit a set of connected subgraphs (union of all the shortest paths between a terminal pair), and we found that Jansen's algorithm solves TMS correctly, and for the proof sketch of Theorem 5, we will recall and discuss Jansen's algorithm [15] in this section.

We may assume that input graph G is connected. We create a clique on 4 vertices $Z = \{z_1, z_2, z_3, z_4\}$ and connect this clique to G by adding an arbitrary edge. This will increase feedback edge number of G by a constant, and it will not change the solution, as no new shortest path between any terminal pair introduced. In the rest of the section we assume that feedback edge number of G is t. Given an instance (G, T, k) of TMS, the following preprocessing is performed.

Preprocessing 1 (adapted from Observation 3 in [15]): While there is a vertex $v \in G$ of degree one.

- If there is a terminal pair $\{v\}$ in T, then put v in solution S, decrease k by one, and remove every terminal pair containing v from T. Otherwise, we replace every terminal pair $\{v, y\}$ with $\{u, y\}$ in T where u is the only neighbor of v in G. Remove v from G.

The above preprocessing is safe, if both vertices of a pair are v then v must be in the solution, else v can be replaced by u in any solution containing v.

Fig. 1. G with minimum degree two, darkened vertices forms $V_{\geq 3}$.

After Preprocessing 1, G has minimum degree two, and we will assume that for every $I = (G, T, k)$, G has minimum degree at least two. For a graph G with degree at least two, let $V_{\geq 3}$ be the set of all the vertices of G with degree at least three (Fig. 1), $G - V_{\geq 3}$ is a disjoint union of paths as $V_{\geq 3} \neq \emptyset$ (we added a clique on Z in G) and G is connected. Let $\mathcal{D}$ be the set of all the components (paths) in $G - V_{\geq 3}$. Observe that every component of $\mathcal{D}$ is connected to rest of the graph by 2 edges, each of which connecting an endpoint of D to a vertex of $V_{\geq 3}$ as every vertex in D has degree two in G (see also [15]). In the remaining part, for the graph in context, we simply use $V_{\geq 3}$ and $\mathcal{D}$ for it as defined above. Given an instance $I = (G, T, k)$ of TMS, the following holds.

Observation 2. *For every* $\{x, y\} \in T$, $SP(x, y)$ *induces a connected graph and every vertex in* $SP(x, y) \setminus \{x, y\}$ *has degree at least two in* $G[SP(x, y)]$.

Proof. It follows from the fact that graph induced by every $SP(x, y)$ is a union of all the shortest path between a terminal pair.   □

Observation 3. *For every* $\{x, y\} \in T$ *and* $D \in \mathcal{D}$ *it holds that: if* $SP(x, y) \subseteq V(D)$, *then* $SP(x, y)$ *induces a sub path of* D *in* G.

Similar to [15], given an instance (G, T, k) of TMS, for every $D \in \mathcal{D}$, let OPT(D) be the minimum size of a terminal monitoring set for $\{\{x, y\} \mid \{x, y\} \in T \wedge SP(x, y) \subseteq V(D)\}$. From Observation 3 and Lemma 1, we have that OPT(D) can be computed in polynomial time for every $D \in \mathcal{D}$.

Lemma 3 (Lemma 5 in [15], stating for TMS). *Given an instance* (G, T, k) *of TMS, there exists a minimum size terminal monitoring set* S' *for* T *such that: for every* $D \in \mathcal{D}$, S' *contains either* OPT(D) *or* OPT(D)+1 *vertices of* D.

Proof of Lemma 3 follows from the proof of [15, Lemma 5] by supplementing it with Observation 2 and Observation 3.

Jansen's algorithm [15] while solving an instance of HPG makes successive guesses (to branch), and if it decides that a guess may lead to a solution within the budget, it constructs an instance of HPFB, and solves it in polynomial time. We found that Jansen's algorithm correctly solves TMS as well, crucially by correctly constructing instances of HPFB for input $I = (G, T, k)$. For completeness, we recall steps of Jansen's algorithm from [15] which are divided into branching and construction of HPFB, we demonstrate them for the input instance $I = (G, T, k)$ of TMS.

Branching (Sect. 3.2 in [15]): Guess $f_v : V_{\geq 3} \to \{0, 1\}$ and $f_d : \mathcal{D} \to \{0, 1\}$. If $k \geq \sum_{v \in V_{\geq 3}} f_v(v) + \sum_{D \in \mathcal{D}} (\text{OPT}(D) + f_d(D))$, then construct an instance of HPFB.

Construction of HPFB (Section 3.2 in [15]): Given f_v, f_d. Create $G_1, \mathcal{D}_1$ as copies of $G, \mathcal{D}$ respectively, and f_d remains same for $\mathcal{D}_1$. Construct $\mathcal{U}_1 = \{SP(x, y) \mid \{x, y\} \in T\}$. Do the following. (1) For every $U \in \mathcal{U}_1$: if U contains a vertex $v \in V_{\geq 3}$ such that $f_v(v) = 1$ or U contains all the vertices of a $D \in \mathcal{D}$ such that $(\text{OPT}(D) + f_d(D)) > 0$, then remove U form $\mathcal{U}_1$. (2) For every $D \in \mathcal{D}_1$ such that $(\text{OPT}(D) + f_d(D)) = 0$: remove D from $\mathcal{D}_1$, remove $V(D)$ from G_1, and remove $V(D)$ from every set $U \in \mathcal{U}_1$. (3) For every $v \in V_{\geq 3}$ such that $f_v(v) = 1$: remove v from G_1. (4) Contract remaining vertices of $V_{\geq 3}$ in G_1 into a single vertex z in G_1, and replace $U \cap V_{\geq 3}$ (if non empty) with z in every $U \in \mathcal{U}_1$. (5) Assign a distinct number in $[\|\mathcal{D}_1\|]$ to every remaining path in $\mathcal{D}_1$, set $b(i) = (\text{OPT}(D) + f_d(D_i))$, where $D_i \in \mathcal{D}_1$. Output $(G_1, z, \mathcal{U}_1, b)$.

For input instance $I = (G, T, k)$ of TMS, for every guess f_v, f_d, and correspondingly constructed $(G_1, z, \mathcal{U}_1, b)$, the following claims hold.

Claim 5 (Claim 5 in [15], stating for (G, T, k)). *G_1 is a flower graph with core z and every vertex set in $\mathcal{U}_1$ induces a simple path in G_1.*

Claim 6 (Claim 6 in [15], stating for (G, T, k)). *The following two statements are equivalent.*

- *There exists a terminal monitoring set S for T such that : $|S \cap V(D)| = \text{OPT}(D) + f_d(D)$ for every $D \in \mathcal{D}$, and $v \in S \Leftrightarrow f_v(v) = 1$ for every $v \in V_{\geq 3}$.*
- *There exists a solution for instance $(G_1, z, \mathcal{U}_1, b)$ of HPFB.*

Proof of Claim 5 and proof of Claim 6 for instance I follow from proof of [15, Claim 5] and proof of [15, Claim 6] respectively by supplementing them with arguments based on Observation 2 and Observation 3. If any constructed instance of HPFB has a solution, then Jansen's algorithm returns YES, otherwise NO [15]. Further, correctness on I follows from Claim 6, Claim 5, and Lemma 3.

Size of $V_{\geq 3}$ and $\mathcal{D}$ can be bounded by $2t$ and $3t$ respectively if $V_{\geq 3} \neq \emptyset$ [15]. We recall from [15] that Jansen's algorithm makes at most 2^{5t} guesses, and its running time is bounded by $2^{5t} \cdot (|V(G)| + |\mathcal{U}|)^{O(1)}$.

4   Hardness Results

4.1   Proof of Theorem 1

We reduce RED-BLUE DOMINATING SET (RBDS) to TMS. In RBDS we are given a bipartite graph $G = (V_B \cup V_R, E)$, and it is asked if there is a vertex set $D \subseteq V_B$

of size at most k such that every vertex in V_R is adjacent to at least one vertex in D. It is known that RBDS is W[2]-hard for parameter solution size k (see [9]).

Let $I = (G = (V_B \cup V_R, E), k)$ be the input instance of RBDS, let $V_R = \{r_1, r_2, \ldots r_n\}$, and let $V_B = \{b_1, b_2, \ldots b_m\}$. We construct an instance I' of TMS as follows. We construct the graph G' as follows. Create a vertex set $V_R' = \{r_1', r_2', \ldots r_n'\}$. For every $i \in [n]$, we connect r_i' to all the neighbors of r_i in V_B, and call the set of all these introduced edges as E'. Essentially we are creating a twin vertex for every $r \in V_R$. We construct terminal set $T = \{\{r_i, r_i'\} \mid i \in [n]\}$. The instance $I' = (G' = (V_B \cup V_R \cup V_R', E \cup E'), T, k)$.

Lemma 4 ($\star^1$). *I is a yes instance of* RED-BLUE DOMINATING SET *if and only if I' is a yes instance of TMS.*

4.2 Proof of Theorem 6

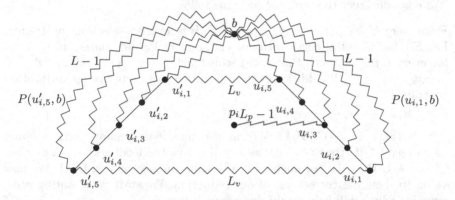

Fig. 2. An example of H_i connected to bridge vertex b, with each path $P(u_{i,j}, b)$ and $P(u_{i,j}', b)$ is of length L and contains $L - 1$ intermediate vertices.

We reduce MULTI COLOR INDEPENDENT SET (W[1]-hard by solution size k [9], and a standard problem in parameterized complexity) to α-RTMS. Let $I = (G, k)$ such that $V(G)$ is partitioned into $\{V_1, V_2, \ldots, V_k\}$, be an instance of MULTI COLOR INDEPENDENT SET, and we need to decide if there exists an independent set S containing exactly one vertex from every V_i, where $i \in [k]$. We may assume that for every $i \in [k]$, $|V_i| = n$, and n is odd. We begin our construction by first defining following values.

For $0 \le \alpha \le 0.5$, we set $L = \lceil \frac{1}{2\alpha} \cdot n \rceil$; $L_v = \lfloor 2 \cdot \alpha \cdot L + 1 - n \rfloor$; $L_p = \lceil \frac{n-1}{\alpha} \rceil$.

The construction of G' is as follows (Fig. 2). We create a vertex b and call it the bridge vertex. For every vertex set V_i in G we construct a gadget H_i as follows.

---

[1] Proofs for claims marked with a $\star$ have been omitted.

- Create a set $U_i' = \{u_{i,j}' \mid v_{i,j} \in V_i\}$ and make a path on these vertices as $(u_{i,1}', u_{i,2}', \ldots u_{i,n}')$. Create another set $U_i = \{u_{i,j} \mid v_{i,j} \in V_i\}$ and create a path $(u_{i,1}, u_{i,2}, \ldots u_{i,n})$ as well. Connect $u_{i,1}'$ to $u_{i,n}$ by creating a path with L_v vertices and connect one endpoint of this path to $u_{i,1}'$ and another to $u_{i,n}$. Similarly, connect $u_{i,n}'$ to $u_{i,1}$ by creating a path on L_v vertices.
- create a vertex p_i, create a path with $L_p - 1$ vertices, and connect one endpoint of this path to p_i and another endpoint with vertex $u_{i,(n+1)/2}$, We denote this path from p_i to $u_{i,(n+1)/2}$ by $P(p_i, u_{i,(n+1)/2})$ which include both p_i and $u_{i,(n+1)/2}$.

For every $i \in [k]$, we connect the vertices of H_i to b as follows.

- For every $i \in [k]$ and $j \in [n]$, connect $u_{i,j}$ to b by creating a path with $L - 1$ intermediate vertices, denote this path $P(u_{i,j}, b)$. Thus, $d(u_{i,j}, b) = L$. Similarly, connect $u_{i,j}'$ with vertex b by creating a path with $L - 1$ intermediate vertices, and denote this path by $P(u_{i,j}', b)$. Thus, $d(u_{i,j}', b) = L$.

We now construct the terminal set T as follows.

- From every H_i we put $\{p_i, u_{i,(n+1)/2}\}$ in T. It is a vertex selection constraints.
- Let $E_{i,j}$ be the set of edges with one endpoint in V_i and another in V_j in G, for every $i, j \in [k]$, such that $i < j$, construct $T_{i,j} = \{\{u_{i,i'}', u_{j,j'}'\} \mid i', j' \in [n] \wedge v_{i,i'} v_{j,j'} \in E_{i,j}\}$. Addition of these terminal pairs acts as edge verification contraints.
- $T = \bigcup_{1 \le i < j \le k} T_{i,j}$.

The instance $I' = (G', T, k)$. We can see that if we remove vertices b and $u_{i,1}$ for every $i \in [k]$ then G' becomes acyclic, thus the feedback vertex number of G' is $k + 1$, thus the parameter FVN + solution size remain $O(k)$. We now move on to show the correctness of our reduction. We start with stating some observation which will help establishing equivalence.

Observation 4. *For every distinct $i, j \in [k]$, for every $x \in U_i' \cup U_i$ and every $y \in U_j' \cup U_j$, the distance $d(x, y) = 2L$.*

Definition 2. *For an $i \in [k]$, we say a vertex $w \in V(G')$ is H_i-representative if the following holds.*

- w belongs to $V(P(p_i, u_{i,(n+1)/2}))$ *(OR)* $d(u_{i,(n+1)/2}, w) \le \frac{n-1}{2}$.

Observation 5. *No vertex in $V(G')$ can be H_i-representative for two or more distinct $i \in [k]$.*

Proof. This is because distance between $u_{i,(n+1)/2}$ and $u_{j,(n+1)/2}$ for every distinct $i, j \in [k]$ is $2L$ and L is at least n. $\qquad \square$

Lemma 5 ($\star$). *If S is a solution to the instance (G', T, k), then for every $i \in [k]$, there is a vertex $w \in S$ such that w is an H_i-representative.*

Definition 3. *A solution set S of I' is nice if and only if the following holds.*

- $|S \cap U_i| = 1$ *for every* $i \in [k]$; *(AND)* $S = \bigcup_{i \in [k]} S \cap U_i$.

Lemma 6 ($\star$). *If there is a solution S for instance $(G', \mathcal{T}, k)$ then there exists a nice solution S^* for instance $(G', \mathcal{T}, k)$.*

Lemma 7 ($\star$). *For every distinct $i, j \in [k]$, and for every $i', j' \in [n]$, the following holds:*

- $d(u'_{i,i'}, u_{i,i'}) + d(u_{i,i'}, u'_{j,j'}) \gtrsim (1 + \alpha) \cdot d(u'_{i,i'}, u'_{j,j'})$.
- $d(u'_{j,j'}, u_{j,j'}) + d(u_{j,j'}, u'_{i,i'}) \gtrsim (1 + \alpha) \cdot d(u'_{i,i'}, u'_{j,j'})$.

Lemma 8 ($\star$). *For every distinct $i, j \in [k]$, and for every distinct $i', i'' \in [n]$ and distinct $j', j'' \in [n]$, the following holds:*

- $d(u'_{i,i'}, u_{i,i''}) + d(u_{i,i''}, u'_{j,j'}) \leq (1 + \alpha) \cdot d(u'_{i,i'}, u'_{j,j'})$.
- $d(u'_{j,j'}, u_{j,j''}) + d(u_{j,j''}, u'_{i,i'}) \leq (1 + \alpha) \cdot d(u'_{i,i'}, u'_{j,j'})$.

Lemma 9 ($\star$). *I is a yes instance of* MULTI COLORED INDEPENDENT SET *if and only if I' is a yes instance of α-RTMS.*

The above lemma finishes the proof of Theorem 6.

References

1. Abu-Khzam, F.N.: A kernelization algorithm for d-hitting set. J. Comput. Syst. Sci. **76**(7), 524–531 (2010). https://doi.org/10.1016/j.jcss.2009.09.002
2. Alumur, S.A., Kara, B.Y.: Network hub location problems: the state of the art. Eur. J. Oper. Res. **190**(1), 1–21 (2008)
3. Barilan, J., Kortsarz, G., Peleg, D.: How to allocate network centers. J. Algorithms **15**(3), 385–415 (1993). https://doi.org/10.1006/jagm.1993.1047
4. Benedito, M.P.L., Melo, L.P., Pedrosa, L.L.C.: A parameterized approximation algorithm for the multiple allocation k-hub center. In: Castañeda, A., Rodríguez-Henríquez, F. (eds.) LATIN 2022: Theoretical Informatics - 15th Latin American Symposium, Guanajuato, Mexico, November 7–11, 2022, Proceedings. Lecture Notes in Computer Science, vol. 13568, pp. 141–156. Springer (2022). https://doi.org/10.1007/978-3-031-20624-5_9
5. Brandstädt, A., Dragan, F.F.: A linear-time algorithm for connected r-domination and steiner tree on distance-hereditary graphs. Networks **31**(3), 177–182 (1998)
6. Buss, J.F., Goldsmith, J.: Nondeterminism within P. SIAM J. Comput. **22**(3), 560–572 (1993). https://doi.org/10.1137/0222038
7. Chechik, S., Peleg, D.: The fault-tolerant capacitated k-center problem. Theor. Comput. Sci. **566**, 12–25 (2015). https://doi.org/10.1016/j.tcs.2014.11.017
8. Coelho, R.S., Moura, P.F.S., Wakabayashi, Y.: The k-hop connected dominating set problem: hardness and polyhedra. Electron. Notes Discret. Math. **50**, 59–64 (2015). https://doi.org/10.1016/j.endm.2015.07.011
9. Cygan, M., et al.: Parameterized Algorithms. Springer (2015)
10. Diestel, R.: Graph Theory, 4th edn. Graduate texts in mathematics, vol. 173. Springer (2012)

11. Farahani, R.Z., Hekmatfar, M., Arabani, A.B., Nikbakhsh, E.: Hub location problems: a review of models, classification, solution techniques, and applications. Comput. Ind. Eng. **64**(4), 1096–1109 (2013)

12. Flum, J., Grohe, M.: Parameterized Complexity Theory. Texts in Theoretical Computer Science. An EATCS Series. Springer (2006). https://doi.org/10.1007/3-540-29953-X

13. Gonzalez, T.F.: Clustering to minimize the maximum intercluster distance. Theor. Comput. Sci. **38**, 293–306 (1985). https://doi.org/10.1016/0304-3975(85)90224-5

14. Hochbaum, D.S., Shmoys, D.B.: A unified approach to approximation algorithms for bottleneck problems. J. ACM **33**(3), 533–550 (1986)

15. Jansen, B.M.P.: On structural parameterizations of hitting set: Hitting paths in graphs using 2-sat. J. Graph Algorithms Appl. **21**(2), 219–243 (2017). https://doi.org/10.7155/jgaa.00413

16. Katsikarelis, I., Lampis, M., Paschos, V.T.: Structural parameters, tight bounds, and approximation for (k, r)-center. Discret. Appl. Math. **264**, 90–117 (2019). https://doi.org/10.1016/j.dam.2018.11.002

17. Khuller, S., Sussmann, Y.J.: The capacitated K-center problem. SIAM J. Discret. Math. **13**(3), 403–418 (2000). https://doi.org/10.1137/S0895480197329776

18. Krumke, S.O.: On a generalization of the p-center problem. Inf. Process. Lett. **56**(2), 67–71 (1995). https://doi.org/10.1016/0020-0190(95)00141-X

19. Lampis, M.: Algorithmic meta-theorems for restrictions of treewidth. Algorithmica **64**(1), 19–37 (2012). https://doi.org/10.1007/S00453-011-9554-X

20. Lokshtanov, D., Misra, N., Philip, G., Ramanujan, M.S., Saurabh, S.: Hardness of r-DOMINATING SET on graphs of diameter $(r + 1)$. In: Gutin, G., Szeider, S. (eds.) IPEC 2013. LNCS, vol. 8246, pp. 255–267. Springer, Cham (2013). https://doi.org/10.1007/978-3-319-03898-8_22

21. Panigrahy, R., Vishwanathan, S.: An o(log* n) approximation algorithm for the asymmetric p-center problem. J. Algorithms **27**(2), 259–268 (1998)

Generating Cyclic 2-Gray Codes for Fibonacci q-Decreasing Words

Dennis Wong[1,2(✉)], Bowie Liu[1], Chan-Tong Lam[1], and Marcus Im[1]

[1] Faculty of Applied Sciences, Macao Polytechnic University, Macao, China
{cwong,p1709065,ctlam,marcusim}@mpu.edu.mo
[2] Department of Computer Science, State University of New York, Incheon, Korea

Abstract. We present a two-stage algorithm for generating cyclic 2-Gray codes for q-decreasing words. In the first step, a simple recursive algorithm is used to generate a cyclic 2-Gray code for q-run constrained words, which are q-decreasing words that start with a 0. Then, by considering the first block of 1 s and concatenating cyclic Gray code listings of q-run constrained words of different length n, we construct the first known cyclic 2-Gray code for q-decreasing words for all positive real numbers q.

Keywords: q-decreasing word · Fibonacci word · Qubonacci word · run-constrained word · Fibonacci sequence · Fibonacci cube · hypercube · Gray code

1 Introduction

A *q-decreasing word* is a binary string in which every maximal factor of the form $0^a 1^b$ satisfies the condition where $a = 0$ or $qa > b$, with q being a positive real number. As an example, the 21 q-decreasing words of length $n = 6$ and $q = 1$ are

$$000000, 000001, 000010, 000011, 000100, 000110, 001000,$$
$$001001, 100000, 100011, 100010, 100100, 100001, 110000, \qquad (1)$$
$$110001, 110010, 111000, 111001, 111100, 111110, 111111.$$

Note that q can be any positive real number, without being restricted to integers. For example, the 31 q-decreasing words of length $n = 5$ and $q = \pi$ are

$$00000, 00001, 00010, 00011, 00100, 00101, 00110, 00111, 01000, 01001, 01010,$$
$$01011, 01100, 01101, 01110, 10000, 10001, 10010, 10011, 10100, 10101, 10110,$$
$$10111, 11000, 11001, 11010, 11011, 11100, 11101, 11110, 11111.$$

The set of q-decreasing words is in bijection with the set of *Fibonacci words* when q is an integer, which are binary strings that avoid the occurrence of 1^{q+1} [2].

The number of q-decreasing words is an interesting topic in combinatorics due to its close relationship with the famous Fibonacci sequence. For example when $q = 1$, the initial six terms of the enumeration of q-decreasing words are 1, 1, 2, 3, 5, 8, 13, which precisely matches the famous Fibonacci sequence. More interestingly, when q is a positive integer, the number of q-decreasing words corresponds to multi-step Fibonacci numbers [10]. For instance when $q = 2$, the initial six terms of the enumeration of q-decreasing words are 1, 1, 2, 5, 8, 15, 28, which aligns with the famous tribonacci number. Similarly when $q = 3$, the number of q-decreasing words follows the tetranacci number, and this property holds for all positive integers q. For more information about multi-step Fibonacci sequences, see [6,7,13]. In general, when q is a positive rational number, Kirgzov [10] derived a linear recurrence formula that enumerates the number of q-decreasing words:

$$Q_n = \sum_{j \in J} Q_{n-j} + Q_{n-(c+d)},$$

where Q_n denotes the number of q-decreasing words of length n, and J denotes the set of powers derived from the model polynomial $P_{q=\frac{c}{d}}(x, x)$. The enumeration sequences for q-decreasing words can be accessed on the Online Encyclopedia of Integer Sequences for various values of q [15].

The study of q-decreasing words and their variations has attracted significant attention from mathematicians [1–4,10,11]. One noteworthy variant that has attracted considerable interest is run-constrained words. The set of run-constrained words is in bijection with the subset of q-decreasing words that consists of q-decreasing words beginning with a 0. These strings have been studied in the context of induced subgraphs of hypercubes [3–5]. Moreover, q-decreasing words have been found to have applications in coding theory, specifically in encoding binary words that avoid consecutive 1 s. This is attributed to the fact that q-decreasing words are in a bijective relationship with Fibonacci words [2]. For more applications of q-decreasing words, see [5,8,9,12,14,16,17,20].

One of the most important aspects of combinatorial generation is to list the instances of a combinatorial object so that consecutive instances differ by a specified *closeness condition* involving a constant amount of change. Lists of this type are called *Gray codes*. This terminology is due to the eponymous *binary reflected Gray code* (BRGC) by Frank Gray, which orders the 2^n binary strings of length n so that consecutive strings differ in one bit. For example, when $n = 5$ the order is

$$
\begin{aligned}
&00000, 00001, 00011, 00010, 00110, 00111, 00101, 00100, \\
&01100, 01101, 01111, 01110, 01010, 01011, 01001, 01000, \\
&11000, 11001, 11011, 11010, 11110, 11111, 11101, 11100, \\
&10100, 10101, 10111, 10110, 10010, 10011, 10001, 10000.
\end{aligned}
\tag{2}
$$

The BRGC listing is a *1-Gray code* in which consecutive strings differ by one bit change. We note that the order is also *cyclic* because the last and first strings also differ by the closeness condition, and this property holds for all n. In this paper, we are focusing on *2-Gray code*, where consecutive strings differ by at most two bit changes.

An interesting problem related to q-decreasing words is thus to discover a Gray code for such strings. There is, however, no 1-Gray code for the set of q-decreasing words for all values of q. To demonstrate this non-existence result, consider the case where $n = 5$

and $q = \frac{2}{3}$. In this scenario, there are a total of twelve q-decreasing words as follows, and among them, seven of them have an even number of 1 s (even parity), while only five of them have an odd number of 1 s (odd parity):

- Even parity: 00001, 00100, 00010, 10000, 11001, 11100, 11111;
- Odd parity: 00000, 10010, 10001, 11000, 11110.

It is well known that in order for a 1-Gray code to exist, the parity difference (the difference between the number of strings with even parity and the number of strings with odd parity) has to be equal to 0, 1, or −1. Consequently, it follows that a 1-Gray code cannot exist for the set of q-decreasing words for all positive real numbers q.

The problem of finding a 1-Gray code for q-decreasing words when q is an integer was initially studied by Eğecioğlu and Iršič [3, 4]. They conjectured that a 1-Gray code exists for q-decreasing words when $q = 1$. Later, Baril, Kirgizov, and Vajnovszki [2] constructed a 1-Gray code for q-decreasing words when $q = 1$, thus proving the conjecture to be correct. They also proved that the binary reflected Gray code induces or filters a 3-Gray code for q-decreasing words for all positive real numbers q [2, 21]. For example, filtering the binary reflected Gray code in listing 2 produces the following 3-Gray code for q-decreasing words of length $n = 5$ and $q = 1$:

$$00000, 00001, 00011, 00010, \overline{00110}, \overline{00111}, \overline{00101}, 00100,$$
$$\overline{01100}, \overline{01101}, \overline{01111}, \overline{01110}, \overline{01010}, \overline{01011}, \overline{01001}, \overline{01000},$$
$$11000, 11001, \overline{11011}, \overline{11010}, 11110, 11111, \overline{11101}, 11100,$$
$$\overline{10100}, \overline{10101}, \overline{10111}, \overline{10110}, 10010, \overline{10011}, 10001, 10000.$$

For more information about Gray codes induced by the binary reflected Gray code, see [18, 19, 21]. Baril et. al further conjectured that a 1-Gray code exists for q-decreasing words when q is a positive integer, however, this conjecture has not yet been proven.

In this paper, we present a two-stage algorithm for generating cyclic 2-Gray codes for q-decreasing words. This is the first known cyclic 2-Gray code for q-decreasing words for all positive real numbers q. It is worth noting that since there is no 1-Gray code available for q-decreasing words for all positive real numbers q (as demonstrated in the previous example when $q = \frac{2}{3}$), our Gray code is optimal.

The rest of the paper is outlined as follows. In Sect. 2, we describe a simple recursive algorithm for generating a cyclic 2-Gray code for q-run constrained words and prove their Gray code property. Then, in Sect. 3, we present an algorithm that leverages our approach for generating q-run constrained words to construct a cyclic 2-Gray code for q-decreasing words for all positive real numbers q.

2 Generating Gray Code for q-Run Constrained Words

In this section, we first describe a simple recursive algorithm for generating a cyclic 2-Gray code for q-run constrained words. A *run-constrained word* is a binary string in which every block (also known as *run*) of 1 s is immediately followed by a strictly longer block of 0 s. For example, the eight run-constrained words of length $n = 6$ are

$$000000, 000100, 001000, 010000, 011000, 100000, 100100, 110000.$$

These strings are used to define Fibonacci-run graphs, which were introduced in [3, 4]. Notably, when we reverse each run-constrained word, they correspond to the q-decreasing words of length $n = 6$ and $q = 1$ starting with a 0 as indicated in listing 1, and this property holds for all positive integers n.

We generalize the concept of run-constrained word to different values of q by introducing the notion of q-run constrained word. A q-*run constrained word* is a binary string in which every maximal factor of the form $0^a 1^b$ satisfies the condition $qa > b$, where q is a positive real number. For example, the eight q-run constrained words of length $n = 6$ and $q = 1$ are

$$000000, 000001, 000010, 000011, 000100, 000110, 001000, 001001.$$

The listing contains exactly the reversal of run-constrained words for length $n = 6$ and $q = 1$. The set is also clearly a subset of q-decreasing words of length $n = 6$ and $q = 1$.

All strings considered in this paper are binary. Our algorithms use a run-length representation for binary strings using a series of blocks which are maximal substrings of the form 0^*1^*. Each block B_i can be represented by two integers (s_i, t_i) corresponding to the number of 0 s and 1 s respectively. For example, the string $\alpha = 000110100011001$ can be represented by $B_1 B_2 B_3 B_4 = (3, 2)(1, 1)(3, 2)(2, 1)$. We first prove the following lemmas.

Lemma 1. *A string* $B_1 B_2 \cdots B_k$ *is a q-run constrained word if and only if B_1 and $B_2 B_3 \cdots B_k$ are both q-run constrained words.*

Proof. The proofs for both directions are straightforward from the definition. □

Lemma 2. *If* $\alpha = b_1 b_2 \cdots b_n$ *is a q-run constrained word of length n, then $\beta = 0 b_1 b_2 \cdots b_n$ is a q-run constrained word of length $n + 1$.*

Proof. Let $\alpha = b_1 b_2 \cdots b_n = B_1 B_2 \cdots B_k$ and $B_1 = (s_1, t_1)$. Since α is a q-run constrained word, clearly each block B_i is of the form $0^a 1^b$ which satisfies the condition $qa > b$. Then we have $\beta = B_1' B_2 \cdots B_k$ with $B_1' = (s_1 + 1, t_1)$, and it is clear that each block in β of the form $0^a 1^b$ also satisfies the condition $qa > b$ ($qs_1 > t_1$ implies $q(s_1 + 1) > t_1$). □

Lemma 3. *If $B_i = (s_i, t_i)$ is a block of length ℓ in a q-run constrained word, then* $\lfloor \frac{\ell}{q+1} \rfloor < s_i \le \ell$ *and* $0 \le t_i < \ell - \lfloor \frac{\ell}{q+1} \rfloor$.

Proof. Assume by contrapositive that $B_i = (s_i, t_i)$ is a block such that $s_i \le \lfloor \frac{\ell}{q+1} \rfloor$. Then, $t_i = \ell - s_i \ge \lfloor \frac{q\ell}{q+1} \rfloor$. Thus, it holds that $qs_i \not> t_i$, and thus B_i cannot be a block of a q-run constrained word. □

Lemma 4. *The shortest possible length ℓ of the first block $B_1 = (s_1, t_1)$ in a q-constrained word is $\ell = \lfloor \frac{1}{q} \rfloor + 2$ when $n \ge \lfloor \frac{1}{q} \rfloor + 2$.*

Proof. For the length ℓ to be minimal, the block must end with a 1; otherwise, its length would be equal to the total length of the word n with $\ell = n \ge \lfloor \frac{1}{q} \rfloor + 2$. Moreover, the number of 1 s of the shortest first block should be minimized to have the shortest possible first block, and thus $t_1 = 1$. Since $qs_1 > t_1$, the smallest possible value of s_1 that satisfies $qs_1 > 1$ is $s_1 = \lfloor \frac{1}{q} \rfloor + 1$. Therefore, $\ell = s_1 + t_1 = \lfloor \frac{1}{q} \rfloor + 2$. □

The main idea of our recursive algorithm is to leverage Lemma 1 by generating a list of all possible first blocks B_1 for q-run constrained words. If the length of B_1 is less than n ($|B_1| < n$), we proceed to fill the remaining part of the string with all possible q-run constrained words of length $n - |B_1|$. We begin by defining $\mathcal{G}_q^\ell$ as a listing of q-run constrained words of length ℓ, each consisting of only a single block. The listing $\mathcal{G}_q^\ell$ starts with the string $0^{\ell-1}1$, and each subsequent string is obtained by progressively changing the last 0 of the string to a 1 until it reaches the string $0^{\lfloor\frac{\ell}{q+1}\rfloor+1}1^{\ell-\lfloor\frac{\ell}{q+1}\rfloor-1}$ (Lemma 3). For example,

- $\mathcal{G}_1^9$: $000000001, 000000011, 000000111, 000001111$;
- $\mathcal{G}_2^7$: $0000001, 0000011, 0000111, 0001111$.

We also use the notation $(\mathcal{G}_q^\ell)^{-1}$ and $\overline{\mathcal{G}}_q^\ell$ to refer to the listing obtained by reversing $\mathcal{G}_q^\ell$, and we use both notations interchangeably. For instance,

- $\overline{\mathcal{G}}_1^9 = (\mathcal{G}_1^9)^{-1}$: $000001111, 000000111, 000000011, 000000001$;
- $\overline{\mathcal{G}}_2^7 = (\mathcal{G}_2^7)^{-1}$: $0001111, 0000111, 0000011, 0000001$.

Similarly, $(\mathcal{G}_q^\ell)^{-k}$ represents the listing obtained by reversing $\mathcal{G}_q^\ell$ k times. Thus, when k is even $(\mathcal{G}_q^\ell)^{-k} = \mathcal{G}_q^\ell$, and when k is odd $(\mathcal{G}_q^\ell)^{-k} = (\mathcal{G}_q^\ell)^{-1}$.

Let $B_1 \cdot \mathcal{L}$ denote the listing $\mathcal{L}$ with the block B_1 prepended to the beginning of each string in $\mathcal{L}$. Furthermore, we use the notation $\prod_{i\in\{1,2,\ldots,k\}} \mathcal{L}_i$ to represent the listing resulting from concatenating the listings $\mathcal{L}_1, \mathcal{L}_2, \ldots, \mathcal{L}_k$ with $\mathcal{L}_1$ appearing before $\mathcal{L}_2$, and $\mathcal{L}_2$ appearing before $\mathcal{L}_3$, and so on. For example, $\prod_{\mathcal{L}\in\{\mathcal{G}_1^9,\mathcal{G}_2^7\}} \mathcal{L}$ is the listing formed by concatenating the listings $\mathcal{G}_1^9$ and $\mathcal{G}_2^7$ as follows:

$000000001, 000000011, 000000111, 000001111, 0000001, 0000011, 0000111, 0001111$.

We proceed to define $\mathcal{R}_q^n$ as our Gray code listing of q-run constrained words of length n. Our recursive definition maintains a variable p which stores the complement of the last bit of the string generated by our algorithm. For instance, if the string just generated by our algorithm is 0000111, then $p = 0$. The listing $\mathcal{R}_q^n$ can be recursively defined as follows:

$$\mathcal{R}_q^n = \prod_{i\in\{0,1,\ldots,n-\lfloor\frac{1}{q}\rfloor-2\}} \left(\prod_{B_1\in(\mathcal{G}_q^{n-i})^{-i}} B_1 \cdot (\mathcal{R}_q^i)^{-p} \right), 0^n.$$

As an example, suppose $q = 1$ and $n = 8$, then we have

- $\prod_{B_1\in\mathcal{G}_1^8} B_1 \cdot (\mathcal{R}_1^0)^{-p} = \mathcal{G}_1^8 = 00000001, 00000011, 00000111$;

- $\prod_{B_1\in\overline{\mathcal{G}}_1^7} B_1 \cdot (\mathcal{R}_1^1)^{-p} = 0000111 \cdot (\mathcal{R}_1^1)^{-0}, 0000011 \cdot (\mathcal{R}_1^1)^{-1}, 0000001 \cdot (\mathcal{R}_1^1)^{-1}$

$= 00001110, 00000110, 00000010;$

Algorithm 1. Recursive algorithm to generate $\mathcal{R}_q^n$ for q-run constrained words.

```
1: function QRUN(n, q)
2:     R ← []
3:     p ← 1
4:     for i from 0 to n − ⌊1/q⌋ − 2 do
5:         if i is even then
6:             for j from 1 to n − i − ⌊(n−i)/(q+1)⌋ − 1 do
7:                 R ← {R, 0^{n−i−j} 1^j ·QRUN(i, q)^{−p}}
8:                 b₁b₂···bₙ ← last string in R
9:                 p ← 1 − bₙ
10:        else
11:            for j from n − i − ⌊(n−i)/(q+1)⌋ − 1 to 1 do
12:                R ← {R, 0^{n−i−j} 1^j ·QRUN(i, q)^{−p}}
13:                b₁b₂···bₙ ← last string in R
14:                p ← 1 − bₙ
15:    R ← {R, 0^n}
16:    return R
```

$$-\prod_{B_1\in\mathcal{G}_1^6} B_1 \cdot (\mathcal{R}_1^2)^{-p} = 000001 \cdot (\mathcal{R}_1^2)^{-1}, 000011 \cdot (\mathcal{R}_1^2)^{-1}$$

$$= 00000100, 00001100;$$

$$-\prod_{B_1\in\overline{\mathcal{G}}_1^5} B_1 \cdot (\mathcal{R}_1^3)^{-p} = 00011 \cdot (\mathcal{R}_1^3)^{-1}, 00001 \cdot (\mathcal{R}_1^3)^{-0}$$

$$= 00011000, 00011001, 00001001, 00001000;$$

$$-\prod_{B_1\in\mathcal{G}_1^4} B_1 \cdot (\mathcal{R}_1^4)^{-p} = 0001 \cdot (\mathcal{R}_1^4)^{-1} = 00010000, 00010010, 00010001;$$

$$-\prod_{B_1\in\overline{\mathcal{G}}_1^3} B_1 \cdot (\mathcal{R}_1^5)^{-p} = 001 \cdot (\mathcal{R}_1^5)^{-0}$$

$$= 00100001, 00100011, 00100010, 00100100, 00100000.$$

As such, the recursive listing for q-run constrained words of $q = 1$ and $n = 8$, that is $\mathcal{R}_1^8$, is as follows:

$$00000001, 00000011, 00000111, 00001110, 00000110, 00000010, 00000100,$$
$$00001100, 00011000, 00011001, 00001001, 00001000, 00010000, 00010010,$$
$$00010001, 00100001, 00100011, 00100010, 00100100, 00100000, 00000000.$$

The listing $\mathcal{R}_q^n$ can be generated recursively with the base cases $\mathcal{R}_q^t = 0^t$ for all integers $0 \le t < \lfloor\frac{1}{q}\rfloor + 2$ (proof of Lemma 4), and using the recursive definition of $\mathcal{R}_q^n$. Pseudocode of the recursive algorithm to generate $\mathcal{R}_q^n$ is given in Algorithm 1.

Corollary 1. *The listing $\mathcal{R}_q^n$ starts with the string $0^{n-1}1$ and ends with the string 0^n when $n \ge \lfloor\frac{1}{q}\rfloor + 2$.*

Proof. If $n \geq \lfloor \frac{1}{q} \rfloor + 2$, then $0^{n-1}1$ is a q-run constrained word by Lemma 4. The first string in $\mathcal{G}_q^n$ is thus $0^{n-1}1$, and the last string in $\mathcal{R}_q^n$ is 0^n by definition. □

Corollary 2. *The listing $\mathcal{R}_q^n$ contains all q-run constrained words of length n.*

Proof. This is a direct application of Lemma 1. □

Lemma 5. *Each consecutive strings in $\mathcal{R}_q^n$ differ from each other by at most two bits.*

Proof. The proof is by induction on n. In the base case when $n < \lfloor \frac{1}{q} \rfloor + 2$, $\mathcal{R}_q^n = \{0^n\}$ where consecutive strings (there is only one string) differ by at most two bits. Inductively, assuming consecutive strings in $\mathcal{R}_q^n$ differ by at most two bits for $n \in \{1, 2, \ldots, k-1\}$, we consider the case when $n = k$ and two consecutive strings α and β in $\mathcal{R}_q^k$, where W.L.O.G. α comes before β. There are four cases:

- α and β have the same first block: α and β are consecutive strings in $B_1 \cdot (\mathcal{R}_q^m)^{-p}$ for some $1 \leq m < k$ with their $n - |B_1|$ suffixes corresponding to two consecutive strings in $\mathcal{R}_q^m$. Thus, α and β differ by at most two bits by induction;
- α and β have a different first block but their first blocks have the same length: α and β are consecutive strings in $\prod_{B_1 \in (\mathcal{G}_q^{n-m})^{-m}} B_1 \cdot (\mathcal{R}_q^m)^{-p}$ with a different B_1 and thus by the definition of $\mathcal{G}_q^{n-m}$, their first blocks differ by one bit. Furthermore by the definition of p, the strings α and β share the same $n - |B_1|$ suffix and thus α and β differ by one bit;
- α and β have their first blocks of different lengths and $\beta \neq 0^n$: Let j be the length of the first block of α, and $\alpha = a_1 a_2 \cdots a_n$ and $\beta = b_1 b_2 \cdots b_n$. Thus $a_{j+1} a_{j+2} \cdots a_n$ is a q-run constrained word of length $n - j$ by Lemma 1. Observe that α and β are consecutive strings that come from different subsequences $\prod_{B_1 \in (\mathcal{G}_q^{n-m})^{-m}} B_1 \cdot (\mathcal{R}_q^m)^{-p}$. By the definition of $\mathcal{G}_q^{n-m}$, the length j prefixes of α and β differ by at most two bits, and $b_j = 0$. Furthermore by Lemma 2, since $a_{j+1} a_{j+2} \cdots a_n$ is a q-run constrained word, $0 a_{j+1} a_{j+2} \cdots a_n$ is also a q-run constrained word. Therefore based on the definition of p, we have $\beta = b_1 b_2 \cdots b_{j-1} 0 a_{j+1} a_{j+2} \cdots a_n$ which implies α and β differ by at most two bits;
- $\beta = 0^n$: The first block B_1 of α is $0^{\lfloor \frac{1}{q} \rfloor + 1} 1$ by Lemma 4. The length $n - \lfloor \frac{1}{q} \rfloor - 2$ suffix of α is either $0^{n - \lfloor \frac{1}{q} \rfloor - 2}$ or $0^{n - \lfloor \frac{1}{q} \rfloor - 3} 1$ by Corollary 1. In either case α differs with $\beta = 0^n$ by at most two bits.

Therefore, by induction, consecutive strings in $\mathcal{R}_q^n$ differ from each other by at most two bits. □

Together, Corollary 1, Corollary 2 and Lemma 5 prove the following theorem.

Theorem 1. *The algorithm* QRUN *generates a list of all q-run constrained words of length n, where q can be any positive real number, in cyclic 2-Gray code order.*

3   Generating Gray Code for q-Decreasing Words

In this section, we leverage our results on q-run constrained words to generate a cyclic 2-Gray code for q-decreasing words. Observe that the set of q-decreasing words can be constructed by taking the union of all strings in the format $1^r\gamma$, where γ is a q-run constrained word of length $n - r$. For instance, consider the set of q-decreasing words for $n = 6$ and $q = 1$ as shown in listing 1. This set can be partitioned into subsets, with each subset containing strings in the format $1^r\gamma$ as follows:

- $r = 0$: $\{000000, 000001, 000010, 000011, 000100, 000110, 001000, 001001\}$;
- $r = 1$: $\{100000, 100011, 100010, 100100, 100001\}$;
- $r = 2$: $\{110000, 110001, 110010\}$;
- $r = 3$: $\{111001, 111000\}$;
- $r = 4$: $\{111100\}$;
- $r = 5$: $\{111110\}$;
- $r = 6$: $\{111111\}$.

Removing the prefix 1^r from the strings of each subset results in a set that contains all possible q-decreasing words of length $n - r$.

The main idea of our algorithm is thus to utilize our recursive algorithm QRUN (Algorithm 1) to generate a cyclic 2-Gray code for q-run constrained words of all possible length $n - r$, with the addition of prepending the corresponding prefix 1^r to each q-run constrained word. Our Gray code listing $\mathcal{Q}_q^n$ for q-decreasing words consists of blocks of subsequences in the form $1^r \cdot (\mathcal{R}_q^{n-r})^{-p}$, where p similarly stores the complement of the last bit of the string just generated by our algorithm, and we initialize $p = 1$.

The Gray code listing $\mathcal{Q}_q^n$ is composed of two parts. The first part begins with blocks of $1^r \cdot (\mathcal{R}_q^{n-r})^{-p}$ with $r = n$, representing a listing consisting solely of the string 1^n. We then decrement r by two until it reaches $r = 0$ when n is even or $r = 1$ when n is odd. For the second part of the listing $\mathcal{Q}_q^n$, we follow a similar process but start with $r = n - 1$, which represents a listing containing the string $1^{n-1}0$. Again, we decrement the value of r by two until it reaches $r = 1$ when n is even or $r = 0$ when n is odd. Finally, we reverse the entire second part of the listing. The formal definition of $\mathcal{Q}_q^n$ is given as follows:

$$\mathcal{Q}_q^n = \begin{cases} \prod_{r\in\{n,n-2,\dots,0\}} 1^r \cdot (\mathcal{R}_q^{n-r})^{-p}, \overline{\prod_{r\in\{n-1,n-3,\dots,1\}} 1^r \cdot (\mathcal{R}_q^{n-r})^{-p}} & \text{if } n \text{ is even;} \\ \prod_{r\in\{n,n-2,\dots,1\}} 1^r \cdot (\mathcal{R}_q^{n-r})^{-p}, \overline{\prod_{r\in\{n-1,n-3,\dots,0\}} 1^r \cdot (\mathcal{R}_q^{n-r})^{-p}} & \text{if } n \text{ is odd.} \end{cases}$$

As an example, consider the case where $n = 6$ and $q = 1$. Then, we have

- $1^6 \cdot (\mathcal{R}_1^0)^{-p} = 111111$;
- $1^4 \cdot (\mathcal{R}_1^2)^{-p} = 1^4 \cdot (\mathcal{R}_1^2)^{-0} = 111100$;
- $1^2 \cdot (\mathcal{R}_1^4)^{-p} = 1^2 \cdot (\mathcal{R}_1^4)^{-1} = 110000, 110010, 110001$;
- $1^0 \cdot (\mathcal{R}_1^6)^{-p} = (\mathcal{R}_1^6)^{-0}$
    $= 000001, 000011, 000110, 000010, 000100, 001000, 001001, 000000$;
- $1^5 \cdot (\mathcal{R}_1^1)^{-p} = 1^5 \cdot (\mathcal{R}_1^1)^{-1} = 111110$.
- $1^3 \cdot (\mathcal{R}_1^3)^{-p} = 1^3 \cdot (\mathcal{R}_1^3)^{-1} = 111000, 111001$;

Algorithm 2. A simple algorithm to generate $\mathcal{Q}_q^n$ for q-decreasing words.

```
1: function QDEC(n, q)
2:     Q ← []
3:     p ← 1
4:     r ← n
5:     while r ≥ 0 do
6:         Q ← {Q, 1^r·QRUN(n − r, q)^{−p}}
7:         b_1 b_2 ··· b_n ← last string in Q
8:         p ← 1 − b_n
9:         r ← r − 2
10:    Q' ← []
11:    p ← 1
12:    r ← n − 1
13:    while r ≥ 0 do
14:        Q' ← {Q', 1^r·QRUN(n − r, q)^{−p}}
15:        b_1 b_2 ··· b_n ← last string in Q'
16:        p ← 1 − b_n
17:        r ← r − 2
18:    Q ← {Q, \overline{Q'}}
19:    return Q
```

- $1^1 \cdot (\mathcal{R}_1^5)^{-p} = 1^1 \cdot (\mathcal{R}_1^5)^{-0} = 100001, 100011, 100010, 100100, 100000;$

As such, the Gray code listing for q-decreasing words for $q = 1$ and $n = 6$, that is $\mathcal{Q}_1^6$, is as follows:

$$111111, 111100, 110000, 110010, 110001, 000001, 000011,$$
$$000110, 000010, 000100, 001000, 001001, 000000, 100000,$$
$$100100, 100010, 100011, 100001, 111001, 111000, 111110.$$

Similarly when $n = 5$ and $q = 1$, we have

- $1^5 \cdot (\mathcal{R}_1^0)^{-p} = 11111;$
- $1^3 \cdot (\mathcal{R}_1^2)^{-p} = 1^3 \cdot (\mathcal{R}_1^2)^{-0} = 11100;$
- $1^1 \cdot (\mathcal{R}_1^4)^{-p} = 1^1 \cdot (\mathcal{R}_1^4)^{-1} = 10000, 10010, 10001;$
- $1^4 \cdot (\mathcal{R}_1^1)^{-p} = 1^4 \cdot (\mathcal{R}_1^1)^{-0} = 11110;$
- $1^2 \cdot (\mathcal{R}_1^3)^{-p} = 1^2 \cdot (\mathcal{R}_1^3)^{-1} = 11000, 11001;$
- $1^0 \cdot (\mathcal{R}_1^5)^{-p} = (\mathcal{R}_1^5)^{-0} = 00001, 00011, 00010, 00100, 00000.$

Similarly, the Gray code listing for q-decreasing words for $q = 1$ and $n = 5$, that is $\mathcal{Q}_1^5$, is as follows:

$$11111, 11100, 10000, 10010, 10001, 00000, 00100,$$
$$00010, 00011, 00001, 11001, 11000, 11110.$$

Observe that consecutive strings in both listings $\mathcal{Q}_1^6$ and $\mathcal{Q}_1^5$ differ by at most two bits. This property holds for all positive integers n and all positive real numbers q. We also note that the Gray code order is cyclic since the last and first strings of the listings also differ by only one bit.

The listing $\mathcal{Q}_q^n$ can be generated by maintaining the first block and utilizing the function QRUN for $\mathcal{R}_q^m$. Pseudocode of the algorithm to generate $\mathcal{Q}_q^n$ is given in Algorithm 2. A complete Python implementation of the algorithm is given in the Appendix.

Theorem 2. *The algorithm* QDEC *generates a list of all q-decreasing words of length n, where q can be any positive real number, in cyclic 2-Gray code order.*

Proof. Clearly, the algorithm QDEC generates all possible q-decreasing words. We now demonstrate that consecutive strings in $\mathcal{Q}_q^n$ differ by at most two bits.

Let $\alpha = 1^x \gamma$ and $\beta = 1^y \rho$ be consecutive strings in $\mathcal{Q}_q^n$ and W.L.O.G. we assume $x \geq y$. Since consecutive strings in $\mathcal{R}_q^n$ differ by at most two bits (Theorem 1), it is evident that consecutive strings in $\mathcal{Q}_q^n$ with $x = y$ also differ by at most two bits.

Now if $x \neq y$, we consider two cases. Recall that $\mathcal{Q}_q^n$ consists of two parts. If $x - y \geq 2$, then by the definition of $\mathcal{Q}_q^n$, we must have $x - y = 2$ and α and β originate from the same part of $\mathcal{Q}_q^n$. Consequently, based on the definition of p, the strings α and β share the same $n - x$ suffix, resulting in a difference of two bits between α and β. On the other hand, if $x - y < 2$, then α and β come from different parts of $\mathcal{Q}_q^n$, implying that $x = 1$ and $y = 0$. Furthermore, the suffixes γ and ρ correspond to either the first or last string generated by the function QRUN. Thus, we have $\gamma \in \{0^{n-x}, 0^{n-x-1}1\}$ and $\rho \in \{0^{n-y}, 0^{n-y-1}1\}$. In either case, α and β differ by at most two bits.

Finally, the first string of $\mathcal{Q}_q^n$ is 1^n, and the last string of $\mathcal{Q}_q^n$ is $1^{n-1}0$, differing by one bit. Hence, $\mathcal{Q}_q^n$ constitutes a cyclic 2-Gray code for q-decreasing words.   □

4   Final Remarks

Efficient algorithms that utilize dynamic programming to generate the same 2-Gray codes for q-decreasing words and q-run constrained words in constant amortized time per string have been developed, and the details will be presented in the full version of the paper.

Acknowledgements. This research is supported by the Macao Polytechnic University research grant (Project code: RP/FCA-02/2022). The research of the first author is also supported by the National Research Foundation of Korea (NRF) grant funded by the Ministry of Science and ICT (MSIT), Korea (No. 2020R1F1A1A01070666).

A part of this work was done while the first author was visiting Kanazawa University in Japan. The first author would like to thank Hiroshi Fujisaki for his hospitality during his stay in Kanazawa.

Appendix: Python code to generate q-decreasing words and q-run constrained words in cyclic 2-Gray code order

```python
import math

def qrun(n, q):
    R = []
    p = 1
    for i in range(0, n-(math.floor(1/q)+2)+1):
        if (i+1)%2: j_range = (1, n-i-(math.floor((n - i)/(q + 1))+1)+1, 1)
        else: j_range = (n-i-(math.floor((n-i)/(q+1))+1), 0, -1)

        for j in range(*j_range):
            if p%2:
                R += ["0"*(n-i-j) + "1"*j + s for s in reversed(qrun(i, q))]
            else:
                R += ["0"*(n-i-j) + "1"*j + s for s in qrun(i, q)]
            p = 1 - int(R[-1][-1])

    R.append("0"*n)
    return R

def qdecreasing(n, q):
    Q = []
    p = 1
    for r in range(n, -1, -2):
        if p%2: Q += ["1"*r + s for s in reversed(qrun(n-r, q))]
        else: Q += ["1"*r + s for s in qrun(n - r, q)]
        p = 1 - int(Q[-1][-1])

    _Q = []
    p = 1
    for r in range(n-1, -1, -2):
        if p%2: _Q += ["1"*r + s for s in reversed(qrun(n-r, q))]
        else: _Q += ["1"*r + s for s in qrun(n - r, q)]
        p = 1 - int(_Q[-1][-1])

    return Q + list(reversed(_Q))

print(' ==========================================')
print(' 1.   q-decreasing words')
print(' 2.   q-run constrained words\n')
print(' ==========================================')
print(' Enter selection #: ')
this_type = int(input())
print('Enter n:')
this_n = int(input())
print('Enter q:')
this_p = float(input())

if this_type == 1: qr = qdecreasing(this_n, this_p)
else: qr = qrun(this_n, this_p)

for curr in qr: print(curr)
print('total: {}'.format(len(qr)))
```

References

1. Baril, J.-L., Kirgizov, S., Vajnovszki, V.: Asymptotic bit frequency in Fibonacci words. Pure Math. Appl. **30**(1), 23–30 (2022)
2. Baril, J.-L., Kirgizov, S., Vajnovszki, V.: Gray codes for Fibonacci q-decreasing words. Theor. Comput. Sci. **927**, 120–132 (2022)
3. Eğecioğlu, O., Iršič, V.: Fibonacci-run graphs I: basic properties. Discret. Appl. Math. **295**, 70–84 (2021)
4. Eğecioğlu, O., Iršič, V.: Fibonacci-run graphs II: degree sequences. Discret. Appl. Math. **300**, 56–71 (2021)
5. Eğecioğlu, O., Klavžar, S., Mollard, M.: Fibonacci Cubes With Applications And Variations. World Scientific Publishing Company (2023)
6. Feinberg, M.: Fibonacci-Tribonacci. Fibonacci Q. **1**(3), 71–74 (1963)
7. Flores, I.: Direct calculation of k-generalized Fibonacci numbers. Fibonacci Q. **5**(3), 259–266 (1967)
8. Goulden, I., Jackson, D.: Combinatorial Enumeration. A Wiley-Interscience Publication. John Wiley & Sons Inc., New York (1983)
9. Hsu, W.-J.: Fibonacci cubes - a new interconnection topology. IEEE Trans. Parallel Distrib. Syst. **4**(1), 3–12 (1993)
10. Kirgizov, S.: q-bonacci words and numbers. Fibonacci Q. **60**(5), 187–195 (2022)
11. Kirgizov, S., Ramírez, J.: Polyominoes and graphs built from Fibonacci words. Fibonacci Q. **60**(5), 196–211 (2022)
12. Knuth, D.: The Art of Computer Programming, vol. 4A. Addison-Wesley Professional, Combinatorial Algorithms (2011)
13. Miles, E.: Generalized Fibonacci numbers and associated matrices. Amer. Math. Monthly **67**(8), 745–752 (1960)
14. Mütze, T.: Combinatorial Gray codes - an updated survey. arXiv Preprint, Feb. 2022. arXiv: arxiv.org/abs/2202.01280
15. OEIS Foundation Inc., The on-line encyclopedia of integer sequences, published electronically at oeis.org (2023)
16. Ruskey, F.: Combinatorial Generation. Book under preparation (2003)
17. Savage, C.: A survey of combinatorial Gray codes. SIAM Rev. **4**, 605–629 (1997)
18. Sawada, J., Williams, A., Wong, D.: Inside the binary reflected Gray code: Flip-swap languages in 2-Gray code order. In: Lecroq, T., Puzynina, S. (eds.) Combinatorics on Words, pp. 172–184, Cham (2021)
19. Sawada, J., Williams, A., Wong, D.: Flip-swap languages in binary reflected Gray code order. Theor. Comput. Sci. **933**, 138–148 (2022)
20. Stanton, D., White, D.: Constructive Combinatorics. Springer Science & Business Media (2012)
21. Vajnovszki, V.: Gray code order for Lyndon words. Discret. Math. Theor. Comput. Sci., 9(2) (2007)

On the Hardness of Gray Code Problems
for Combinatorial Objects

Arturo Merino[1], Namrata[2(✉)], and Aaron Williams[3]

[1] University of Saarland and Max Planck Institute for Informatics, Saarbrücken, Germany
merino@cs.uni-saarland.de
[2] Department of Computer Science, University of Warwick, Coventry, England
namrata@warwick.ac.uk
[3] Department of Computer Science, Williams College, Williamstown, USA
aaron.williams@williams.edu

Abstract. Can a list of binary strings be ordered so that consecutive strings differ in a single bit? Can a list of permutations be ordered so that consecutive permutations differ by a swap? Can a list of non-crossing set partitions be ordered so that consecutive partitions differ by refinement? These are examples of *Gray coding problems*: Can a list of combinatorial objects (of a particular type and size) be ordered so that consecutive objects differ by a flip (of a particular type)? For example, $000, 001, 010, 100$ is a no instance of the first question, while $1234, 1324, 1243$ is a yes instance of the second question due to the order $12\overline{43}, 1\overline{23}4, 13\overline{24}$. We prove that a variety of Gray coding problems are NP-complete using a new tool we call a *Gray code reduction*.

1 Introduction

In a 1947 patent application, Bell Labs engineer Frank Gray devised an order of the 2^n binary strings of length n in which consecutive strings differ by flipping a single bit (i.e., they have Hamming distance one) [10]. He referred to the order as *reflected binary code* due to its recursive structure. Although the order had previously been observed by others, including another Bell Labs engineer George R. Stibitz [35], the order became known as the *binary reflected Gray code (BRGC)*, or simply, the *Gray code*.

While Bell Labs was able to solve their ordering problem several times, similar pursuits are often quite challenging. For example, the well-studied *middle levels conjecture* [2] asked if the same type of ordering exists for the binary strings of length $2k+1$ with either k or $k+1$ copies of 1. Knuth gave this conjecture a difficulty rating of 49/50 [16] before it was settled in the affirmative by Mütze [22], with subsequent work simplifying [24], specializing [18], and generalizing [11,20] the result.

Centuries earlier, bell-ringers developed an order of the $n!$ permutations of $[n] = \{1, 2, \ldots, n\}$ (viewed as strings) where consecutive permutations differ by a *swap* (or *adjacent-transposition*) meaning that two neighboring symbols are exchanged [6]. *Plain changes* was rediscovered independently by Johnson [14], Trotter [37], and Steinhaus [33] in the 1960s for its use in the efficient generation of permutations by computer.

In general, when presented with a combinatorial object and a flip operation, one may ask for an order in which successive objects differ by a flip. Suitable orders are sometimes referred to as *minimal change orders* or *combinatorial Gray codes*. Academic surveys have been written by Savage [29] and more recently Mütze [23], with Ruskey [27] and Knuth [16] devoting extensive textbook coverage to the subject. Despite the long history of the subject, there are still natural Gray code questions that haven't been answered or even posed. For example, in Sect. 4 we'll consider such a question involving non-crossing set partitions.

1.1 Gray Codes and Computational Complexity

When Gray codes are mixed with computational complexity the focus is typically on *generation problems*: How efficiently can a particular order can be generated? For example, Ehrlich's well-known paper [9] provides *loopless algorithms* for the binary reflected Gray code and plain changes using the *shared object model*. In other words, one instance of the object is shared between the generation algorithm and the application, and it is modified in worst-case $O(1)$-time to create the next instance. More recent work has focused on limiting generation algorithms to constant *additional memory* [17,34].

We instead show that there are computationally hard existence problems that underlie the problems solved by Bell Labs engineers, bell-ringers, and many others throughout history. More specifically, we consider *existence problems* like the following:

Q1 Can a list of binary strings be ordered so that consecutive strings differ by a bitflip?
Q2 Can a list of permutations of be ordered so that consecutive strings differ by a swap?

For example, $000, 001, 010, 100$ is a no instance of **Q1**, while $1234, 1324, 1243$ is a yes instance of **Q2** due to the order $12\overline{43}, 1\overline{23}4, 1324$. Note that in these decision problems the type of object and flip operation is fixed, and the input is the list of objects under consideration. To be clear, each object in the list is provided as part of the input, so the size of the input increases along with the number of objects in the list[1].

We refer to these existence problems as *Gray coding problems*, with the connotation that we are trying to *do* something to the list of objects. We consider classic combinatorial objects including binary strings, permutations, combinations, (non-crossing) set partitions, and graphs. In each case, we identify at least one flip operation for which the Gray coding problem is NP-complete (including **Q1** and **Q2**).

1.2 Outline

In Sects. 2–3 we establish that two specific Gray coding problems are NP-complete. Then Sect. 4 introduces our notion of a Gray code reduction. Sections 5–7 use these reductions to obtain additional hardness results for a variety of combinatorial objects. Final remarks are contained in Sect. 8.

---

[1] Conceptually, the input could be described as a subset of the objects. However, subsets of an n-set are often encoded as n-bit incidence vectors, and we want to avoid this misinterpretation.

2   A First NP-Complete Problem

In this section, we discuss a first Gray coding problem that is NP-complete.

2.1   2-Tuple Gray Codes

In the *2-tuple Gray coding problem*, we are given some integer 2-tuples, and we want to decide if we can order the 2-tuples such that consecutive 2-tuples differ only in ± 1 in one of the coordinates. More formally, we use $\mathbb{P}_m$ to denote the permutations of length m and have the following problem.

2TupleGC
Input: A list L of m integer 2-tuples $(a_1, b_1), \ldots, (a_m, b_m) \in \mathbb{N} \times \mathbb{N}$.
Question: Is there a ± 1 Gray code for L? In other words, is there a permutation $\pi \in \mathbb{P}_m$ such that $|a_{\pi(i)} - a_{\pi(i+1)}| + |b_{\pi(i)} - b_{\pi(i+1)}| = 1$ for every $i \in [m-1]$?

For technical reasons, we also consider a version of 2TupleGC where the integers have no gaps between them. We say that a list of integer 2-tuples $(a_1, b_1), \ldots, (a_m, b_m)$ is *continuous* if the set of values that a_i and b_i take for $i \in [m]$ are consecutive integers starting from 1; i.e., $\{a_i \mid i \in [m]\} = [\max_{i \in [m]} a_i]$ and $\{b_i \mid i \in [m]\} = [\max_{i \in I} b_i]$.

2TupleGC'
Input: A continuous list L of m integer 2-tuples $(a_1, b_1), \ldots, (a_m, b_m) \in \mathbb{N} \times \mathbb{N}$.
Question: Is 2TupleGC(L) true?

2.2   Grid Graph Hamiltonicity

Hamilton Path problems have been central to evolution of computational complexity, dating back to Karp's initial list of 21 NP-complete problems [15].

A *grid graph* is a graph, where the vertex set is given by some integer 2-tuples $L = \{(a_1, b_1), \ldots, (a_m, b_m)\}$ and there are edges between all pairs of 2-tuples that differ by 1 on a single coordinate. Since the grid graph is completely defined by the integer 2-tuples, we denote by $grid(L)$ the unique grid graph that has L as vertices.

Of particular relevance to us, is the powerful sharpening of this result by Itai, Papadimitriou, and Szwarcfiter which shows hardness for Hamiltonian paths problems on grid graphs [13]. More formally, the following problem is hard.

GridHamPath
Input: A list L of m integer 2-tuples $(a_1, b_1), \ldots, (a_m, b_m) \in \mathbb{N} \times \mathbb{N}$.
Question: Is there a Hamilton path in $grid(L)$?

Theorem 1 ([13]). GridHamPath *is NP-complete.*

2.3   Hardness Results

The problem GridHamPath can be easily translated into an equivalent 2TupleGC problem. In fact, they are essentially the same problem.

Corollary 1. 2TUPLEGC *is NP-complete.*

Proof. We note that both problems have the same input, and $L = (a_1, b_1), \ldots, (a_m, b_m)$ is a Hamilton path of $grid(L)$ if and only if L is a ± 1 Gray code. □

Furthermore, 2TUPLEGC is hard even when we restrict the input to be continuous. Intuitively, non-continuous inputs give rise to disconnected grid graphs or can be translated to a continuous instance. We have the following theorem.

Theorem 2. 2TUPLEGC' *is NP-complete.*

Proof. It is clear that the problem is in NP, as $\pi \in \mathbb{P}_m$ is a polynomially checkable certificate.

We reduce from 2TUPLEGC. Let $L = (a_1, b_1), \ldots, (a_m, b_m) \in \mathbb{N} \times \mathbb{N}$ be an instance of 2TUPLEGC. We only need to deal with the case of L being non-continuous, as otherwise we simply map L to itself. Let $A = \{a_i \mid i \in I\}$ and $B = \{b_i \mid i \in I\}$. If L is non-continuous, then either (1) there exists a partition of $[m]$ into I and J, and $\alpha \in \mathbb{N}$ such that $a_i < \alpha < a_j$ for every $i \in I$ and $j \in J$, (2) there exists a partition of $[m]$ into I and J, and $\alpha \in \mathbb{N}$ such that $b_i < \alpha < b_j$ for every $i \in I$ and $j \in J$, (3) A and B are the discrete intervals $A = \{\min A, \ldots, \max A\}$ and $B = \{\min B, \ldots, \max B\}$. Furthermore, we can decide if we are in case 1, 2 or 3 in time $O(m \log m)$ by sorting A and B.

If (1) or (2) holds, then for every $i \in I$ and $j \in J$ we have that $|a_i - a_j| + |b_i - b_j| \geq 2$, which implies that L is a no-instance. If (3) holds, we map L to $L' := (a_1', b_1'), \ldots, (a_m', b_m')$ by shifting the instance so that the minimum among the first and second coordinates is one; i.e.,

$$(a_1', b_1'), \ldots, (a_m', b_m') = (1 + a_1 - \min A, 1 + b_1 - \min B), \ldots, (1 + a_m - \min A, 1 + b_m - \min B).$$

Note that the encoding size of L' is at most the encoding size of L and that the mapping can be computed in polynomial time. Furthermore, for every $i, j \in [m]$ we have that

$$|a_i' - a_j'| + |b_i' - b_j'| = |1 + a_i - \min A - (1 + a_j - \min A)| + |1 + b_i - \min B - (1 + b_j - \min B)|$$

$$= |a_i - a_j| + |b_i - b_j|,$$

so L is a yes-instance if and only if L' is a yes-instance. This concludes the proof. □

2.4 Application: Swap Gray Codes for Permutations

Here we show that 2TUPLEGC' can be used as a source problem for establishing the hardness of other Gray coding problems. Consider the following problem.

PERMSWAPGC
Input: A list of m permutations of length n, $\tau_1, \ldots, \tau_m \in \mathbb{P}_n$.
Question: Is there a permutation $\pi \in \mathbb{P}_m$ such that $\tau_{\pi(i)}$ and $\tau_{\pi(i+1)}$ differ in an adjacent transposition for every $i \in [m-1]$?

Theorem 3. PERMSWAPGC *is NP-complete.*

Proof. It is clear that the problem is in NP. To see hardness, we reduce from 2TUPLEGC';
see Fig. 1.

Let $L = (a_1, b_1), \ldots, (a_m, b_m) \in \mathbb{N} \times \mathbb{N}$ be a list of continuous 2-tuples. Let $a = \max_{i \in [m]} a_i$
and $b = \max_{i \in [m]} b_i$. Given a 2-tuple $(x, y) \in [a] \times [b]$, we define a permutation $\tau := \phi(x, y) \in$
$\mathbb{P}_{a+b+2}$ as the unique permutation of length $a + b + 2$ such that $\tau^{-1}(a + b + 1) = x$,
$\tau^{-1}(a + b + 2) = a + y$, and after the removing symbols $(a + b + 1)$ and $(a + b + 2)$ from
τ, we get the identity permutation $1 \cdots (a + b)$.

We map the instance, L to $L' = \phi(a_1, b_1), \ldots, (a_m, b_m)$. Note that every permutation
has encoding size of $(a + b + 2) \log(a + b + 2) \leq (2m + 2) \log(2m + 2)$ and that ϕ can be
implemented in polynomial time.

Note that the permutations produced can only differ on swaps involving either the
symbol $(a+b+1)$ or the symbol $(a+b+2)$. Furthermore, for two permutations produced
by ϕ, say $\tau, \rho \in \phi([a] \times [b])$, they differ if and only if the positions of symbol $(a + b + 1)$
are adjacent or the positions of the symbol $(a + b + 2)$ are adjacent, but not both; i.e.,

$$|\tau^{-1}(a + b + 1) - \rho^{-1}(a + b + 1)| + |\tau^{-1}(a + b + 2) - \rho^{-1}(a + b + 2)| = 1. \quad (1)$$

Finally, if $\tau = \phi(x, y)$ and $\rho = \phi(z, w)$, the left side of (1) is $|x-z| + |y-w|$. Consequently,
L is a yes-instance if and only if L' is a yes-instance, and the theorem follows. □

We will see an alternative proof of Theorem 3 in later subsections.

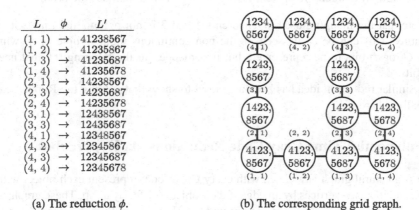

L	ϕ	L'
$(1, 1)$	→	41238567
$(1, 2)$	→	41235867
$(1, 3)$	→	41235687
$(1, 4)$	→	41235678
$(2, 1)$	→	14238567
$(2, 3)$	→	14235687
$(2, 4)$	→	14235678
$(3, 1)$	→	12438567
$(3, 3)$	→	12435687
$(4, 1)$	→	12348567
$(4, 2)$	→	12345867
$(4, 3)$	→	12345687
$(4, 4)$	→	12345678

(a) The reduction ϕ.

(b) The corresponding grid graph.

Fig. 1. The reduction used in Theorem 3, where L is an instance of 2TUPLEGC' and L' has the
corresponding permutations in (a), with the resulting grid graph in (b).

3   A Second Source for NP-Completeness

While 2TUPLEGC' is a useful source problem, we find it convenient to introduce another
NP-complete Gray coding problem for subsequent reductions. Recall that the binary

reflected Gray code lists all 2^n bitstrings so that consecutive strings differ in one bit. Thus, it is natural to ask which bitstrings have bitflip Gray codes (i.e., **Q1** from Sect. 1). We'll show that this BitstringGC problem is hard by a reduction from 2TupleGC'.

BitstringGC
Input: A list of m bitstrings of length n, $x_1, \ldots, x_m \in \{0,1\}^n$.
Question: Is there a permutation $\pi \in \mathbb{P}_m$ such that $x_{\pi(i)}$ and $x_{\pi(i+1)}$ differ in a bitflip for every $i \in [m-1]$?

Theorem 4. BitstringGC *is NP-complete.*

Proof. It is clear that BitstringGC is in NP, as the permutation $\pi \in \mathbb{P}_m$ is a polynomially checkable certificate. Thus, it only remains to prove NP-hardness.

We reduce from 2TupleGC'. Let $L = (a_1, b_1), \ldots, (a_m, b_m) \in \mathbb{N} \times \mathbb{N}$ be a list of continuous 2-tuples. Let $a = \max\limits_{i \in [k]} a_i$ and $b = \max\limits_{i \in [k]} b_i$. For each tuple (a_i, b_i) we define the bitstring $x_i \in \{0,1\}^{a+b}$ as $x_i := 0^{a_i} 1^{a-a_i} 0^{b_i} 1^{b-b_i}$. Since L is continuous, we have $a + b \le 2m$, and consequently, the mapping of the 2TupleGC' instance L to the BitstringGC instance $L' := x_1, \ldots, x_m$ runs in time polynomial in the encoding size of L.

We now show that I is a yes-instance for 2TupleGC' if and only if I' is a yes-instance for BitstringGC. Note that for $i, j \in [k]$ the bitstrings $x_i = 0^{a_i} 1^{a-a_i} 0^{b_i} 1^{b-b_i}$ and $x_j = 0^{a_j} 1^{a-a_j} 0^{b_j} 1^{b-b_j}$ differ in a single bitflip if and only if $|a_i - a_j| + |b_i - b_j| = 1$. Hence, for every permutation $\pi \in \mathbb{P}_m$ and every $i \in [m-1]$ it holds that $x_{\pi(i)}$ and $x_{\pi(i+1)}$ differ in a single bitflip if and only if $|a_i - a_j| + |b_i - b_j| = 1$. This concludes the proof. $\square$

Note that the mapping used in Theorems 4 and 3 is not polynomial time without continuity. This can be easily seen, in the non-continuous input 2-tuple (n, n) which needs $O(\log n)$ bits to be represented, but it is mapped to the bitstring 0^{2n} that needs $O(n)$ bits.

A similar reduction idea has been also used to show that solving the Rubik's cube optimally is hard [5].

4 Polynomial-Time Gray Code Reductions via Hypercubes

There is a natural graph associated with every Gray code: represent each object with a vertex, and join two vertices by an edge if their objects differ by a flip. These graphs are known as *flip graphs*, and a Gray code provides a Hamilton path. For example, the flip graph for bitstrings of length n and bitflips is the n-dimensional hypercube or n-cube.

Let Y be a type of combinatorial object, and Y_m be those objects of size m. In addition, let $F : Y \to Y$ be a type of flip operation acting on objects of type Y without changing their size. In particular, let $F_m : Y_m \to Y_m$ be the flip operation applied to the objects of size m. To prove that the Gray coding problem on Y and F is hard, we will use a new type of reduction defined below.

Definition 1. *A polynomial-time Gray code reduction via hypercubes to Y and F is a poly-time computable function $f : \mathbb{B}_n \to Y_m$ that maps bitstrings of length n to an object of type Y and size m, such that two binary string $b \in \mathbb{B}_n$ and $b' \in \mathbb{B}_n$ differ in a* *single bit if and only if the corresponding objects $f(b) \in Y_m$ and $f(b') \in Y_m$ differ by a flip of type F_m.*

For brevity, we use the term *Gray code reduction* for *polynomial-time Gray code reduction via hypercubes* in the rest of the document. An immediate consequence of Definition 1 is the following remark.

Remark 1. If there is a Gray code reduction from objects Y and flips F, then the flip graph (Y_m, F_m) contains an induced subgraph that is isomorphic to hypercubes of dimension n inside the flip graph of dimension m associated with Y. Moreover, we can efficiently find induced subgraphs of the flip graph that are isomorphic to any induced subgraph of the hypercube.

We now present our main theorem for proving that various Gray coding problems are NP-hard.

Theorem 5. *If there is a Gray code reduction $f : \mathbb{B}_n \to Y_m$ for flips of type F_m, then the following Gray coding problem is NP-hard.*

Gray Coding Problem for Objects Y_m and Flips F_m
Input: *A list L of elemenets in Y_m.*
Question: *Is there a F_m flip Gray code for L?*

Proof. Consider a list of binary strings $B \subseteq \mathbb{B}_n$, and the associated bitflip Gray code problem BITSTRINGGC(L). If there is a Gray code reduction $f : \mathbb{B}_n \to Y_m$ for flips of type F_m, then consider the following list

$$L := \{f(b) \mid b \in B\}. \tag{2}$$

By definition of a Gray code reduction, we know that $b \in \mathbb{B}_n$ and $b' \in \mathbb{B}_n$ differ in a single bit if and only if $f(b) \in Y_m$ and $f(b') \in Y_m$ differ by a flip of type F_m. Therefore, BITSTRINGGC(L) is a yes-instance, if and only if, $GrayCoding(L, F_m)$ is a yes-instance. Also, note that L can be created in polynomial-time with respect to the size of the original input B. Since BITSTRINGGC is NP-hard, we conclude that $GrayCoding(L, F_m)$ is NP-hard.   □

4.1   Example: (Non-crossing) Set Partitions by Refinement

To visualize Theorem 5, let's consider a Gray coding problem that has not been posed in the literature. Let $\mathbb{S}_n^\times$ be the set of *non-crossing set partitions of* $[n]$, which are set partitions in which no pair of subsets cross (i.e., if a and b are in one subset and x and y are in another, then $a < x < b < y$ is not true). Two different set partitions differ by a *refinement* if one can be obtained from the other by splitting a single subset or merging two subsets. The corresponding flip graph is $\mathbb{S}_n^\times$.

NCSETPARTREFGC
Input: A list L of non-crossing partitions from $\mathbb{S}_n^\times$.
Question: Is there a refinement Gray code for L?

A Gray code reduction from BITSTRINGGC to NCSETPARTREFGC problem is shown in Fig. 2. In particular, we map binary strings of length n to non-crossing set partitions of $[n + 1]$ as follows: if $b_i = 0$, then $i + 1$ is a singleton subset, and otherwise $i + 1$ is in the same subset as 1. Thus, toggling b_i is equivalent to moving element $i + 1$ in or out of 1's subset, and this move is a refinement. As a result, the mapping provides an induced subgraph of $S_{n+1}^\times$ that is isomorphic to H_n (as highlighted). Moreover, f can be computed in polynomial-time for each binary string, so we can efficiently find an induced subgraph of $S_{n+1}^\times$ that is isomorphic to any induced subgraph of H_n. Hence, we can conclude that NCSETPARTREFGC is NP-hard. Careful readers may have noticed that hardness also follows for SETPARTREFGC (i.e., NCSETPARTREFGC but on set partitions) since every non-crossing set partition is also a set partition. Both problems are also clearly in NP, so we have the following theorem.

Theorem 6. NCSETPARTREFGC *and* SETPARTREFGC *are NP-complete.*

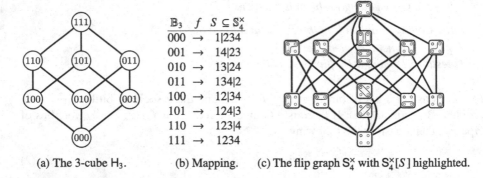

(a) The 3-cube H_3.   (b) Mapping.   (c) The flip graph $S_4^\times$ with $S_4^\times[S]$ highlighted.

Fig. 2. A Gray code reduction from BITSTRINGGC to NCSETPARTREFGC. The one-to-one function $f : B_n \to S_{n+1}^\times$ maps binary strings to non-crossing set partitions in such a way that $b \in B_n$ and $b' \in B_n$ differ by a bit-flip, if and only if, $f(b) \in S_{n+1}^\times$ and $f(b') \in S_{n+1}^\times$ differ by refinement. In (c) we use ∘, ∘, ∘, ∘ for 1, 2, 3, 4 and the non-singleton subsets are surrounded.

In subsequent sections, we visualize our Gray code reductions by illustrating the induced subgraph isomorphic to a hypercube. In other words, we illustrate the list $L = \{f(b) \mid b \in B_n\}$ (i.e., (2) for the full set of binary strings) and argue that the mapping f can be computed in polynomial-time.

For the sake of comparison, the approach taken in Sects. 2.4 and 3 can be defined as a *polynomial-time Gray code reduction via grid graph*, where the source problem was 2TUPLEGC' rather than BITSTRINGGC.

5   Combinations

A *combination* is a subset of $[n]$ of a fixed size k, where $0 \le k \le n$. We denote the set of all combinations of $[n]$ of fixed size k by B_n^k. Commonly, it is represented as a binary string $b_1 b_2 \ldots b_n$ where $b_i := 1$ if i is in the set, otherwise $b_i := 0$. The first transposition Gray code for combinations appeared in [36]. Subsequently, many Gray code results were published in the literature notably [2, 8], and [28].

COMBSWAPGC/COMBTRANSGC/COMBCOMPGC/COMBREVGC

Input: A list L of combinations from $\mathbb{B}_n^k$.

Question: Is there a swap / transposition / substring complement / substring reversal Gray code for L?

Theorem 7. COMBSWAPGC, COMBTRANSGC, COMBCOMPGC, COMBREVGC *are NP-hard.*

Proof. We use a Gray code reduction; see Fig. 3b. For COMBSWAPGC, we define the following list of combinations from $\mathbb{B}_{2n}^n$

$$L := \{c_1 c_2 \ldots c_{2n-1} c_{2n} \mid c_{2i-1} c_{2i} \in \{01, 10\} \text{ for } 1 \le i \le n\}, \tag{3}$$

where each bit b_i is implemented by the pair $c_{2i-1} c_{2i}$. We map each binary string of length n to a combination of length $2n$ with n many 1 s as

$$f(b_1 b_2 \cdots b_n) := c_1 c_2 \ldots c_{2n-1} c_{2n}, \text{ where } c_{2i-1} c_{2i} = 01 \text{ if } b_i = 0 \text{ or } 10 \text{ if } b_i = 1. \tag{4}$$

To understand this construction, note that elements of L can only be modified using the intended swaps within a single pair (i.e., $c_{2i-1} c_{2i}$) rather than between the pair ($c_{2i} c_{2i+1}$). Similar arguments for COMBTRANSGC show that the intended swaps are the only transpositions that can modify members of L.

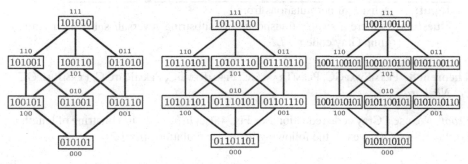

(a) $c_1 c_2 c_3 c_4 c_5 c_6$ from (3) with swaps or transpositions.

(b) $c_1 c_2 1 c_3 c_4 1 c_5 c_6$ from (5) with substring complements.

(c) $c_1 c_2 01 c_3 c_4 01 c_5 c_6$ from (6) with substring reversals.

Fig. 3. Gray code reduction to prove the NP-hardness for combinations. (Color figure online)

The list in (3) does not establish the result for COMBCOMPGC. This is because one substring complement can modify multiple pairs of bits e.g., $\overline{01\,10}$ gives $10\,01$. To avoid this, we insert a padding bit 1 between pairs and use these combinations from $\mathbb{B}_{3n-1}^{2n-1}$

$$L := \{c_1 c_2 1 c_3 c_4 1 \ldots 1 c_{2n-1} c_{2n} \mid c_{2i-1} c_{2i} \in \{01, 10\} \text{ for } 1 \le i \le n\}. \tag{5}$$

The 1 bits of padding prevent any substring complement of length > 2 from modifying elements of L; it is also clear that substring complements of length 1 cannot modify L. As a result, the only valid complements are internal to the pair $(c_{2i-1}c_{2i})$.

Similarly, the list in (5) does not establish the result for COMBREVGC. This is because a single substring reversal can reverse multiple pairs of bits e.g., reversing 01 1 01 gives 10 1 10. To avoid this, we use two bits of padding 01 and the following list of combinations from $\mathbb{B}_{4n-1}^{2n-1}$

$$L := \{c_1 c_2\, 01\, c_3 c_4\, 01\, \ldots\, 01\, c_{2n-1}c_{2n} \mid c_{2i-1}c_{2i} \in \{01, 10\} \text{ for } 1 \le i \le n\}. \tag{6}$$

The 01 padding ensures that the only substring reversals that can modify the elements of L are the intended swaps within the pair $(c_{2i-1}c_{2i})$.

Similar to Eq. 4, we can efficiently map a binary string $b_1 b_2 \ldots b_n$ with combinations of length $3n - 1$ with $2n - 1$ many 1 s and of length $4n - 1$ with $3n - 1$ many 1 s, comprising of padding bits for COMBCOMPGC and COMBREVGC, respectively. $\square$

6   Problems on Permutations Including Pattern-Avoidance

For the set of permutations $\mathbb{P}_n$, we consider an ordering where two consecutive permutations differ by *swaps*. Gray codes then emerged involving *transpositions* [3,31], as well as *prefix-reversals* [26] and *prefix shifts* [4] and [30]. Given a permutation $\pi = p_1 \cdots p_n$ with a substring $p_i \cdots p_j$ where $p_i > p_{i+1} \cdots p_j$, a *right-jump* of the value p_i by $j-i$ steps is a cyclic left rotation of this substring by one position to $p_{i+1} \cdots p_j p_i$. Analogously, we define a *left-jump*. Jump Gray codes were given in [12].

PERMSWAPGC/PERMTRANSGC/PERMREVGC/PERMROTGC/PERMJUMPGC

Input: A list L of permutations from $\mathbb{P}_n$.

Question: Is there a swap / transposition / substring reversal/ substring rotation/ jump Gray code for L?

Theorem 8. PERMSWAPGC, PERMTRANSGC, PERMREVGC, PERMROTGC, PERMJUMPGC *are NP-hard.*

Proof. We use a Gray code reduction; see Fig. 4. Let $b_1 b_2 \cdots b_n$ is a bitstring of length n. For PERMSWAPGC, we use the following list of permutations from $\mathbb{P}_{2n}$

$$L := \{p_1 p_2 \ldots p_{2n-1}p_{2n} \mid p_{2i-1}, p_{2i} \in \{2i - 1, 2i\} \text{ for } 1 \le i \le n\}, \tag{7}$$

where each bit b_i is implemented by the pair (p_{2i-1}, p_{2i}). We map each binary string of length n to a permutation of length $2n$,

$$f(b_1 \cdots b_n) := p_1 \ldots p_{2n-1}p_{2n}, \text{ where } p_{2i-1}p_{2i} = 2i{-}1\, 2i \text{ if } b_i = 0 \text{ or } 2i\, 2i{-}1 \text{ if } b_i = 1. \tag{8}$$

The elements of L can only be modified using the intended swaps within the pair (p_{2i-1}, p_{2i}). Similarly, the intended swaps are the only transpositions, substring reversals, substring rotations, and jumps that can modify the elements of L. Note that unlike Theorem 7, we need not redefine the set L for different operations. $\square$

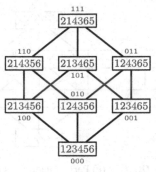 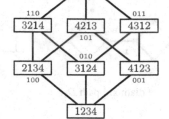

(a) $p_1p_2p_3p_4p_5p_6$ from (7) with swaps. (b) $p_1p_2p_3p_4$ from jumps.

Fig. 4. Gray code reductions to prove the NP-hardness for permutations. (Color figure online)

A *peak* in a permutation $p_1 \cdots p_n$ is a triple $p_{i-1}p_ip_{i+1}$ with $p_{i-1} < p_i > p_{i+1}$. A set of permutations without a peak, also called a *peakless permutation* is denoted by $\mathbb{P}_n^\vee$. We consider the following Gray coding problem on peakless permutations.

PEAKPERMJUMPGC
Input: A list L of permutations from $\mathbb{P}_n^\vee$.
Question: Is there a jump Gray code for L?

Theorem 9. PEAKPERMJUMPGC *is NP-hard.*

Proof. We use a Gray code reduction; see Fig. 4. We define the list of peakless permutations as $\mathbb{P}_n^\vee := \{p_1p_2 \cdots p_n \mid \nexists i \text{ where } p_{i-1} < p_i > p_{i+1}\}$.

We map a binary string $b_2 \cdots b_n$ to a permutation $\pi \in \mathbb{P}_n^\vee$ as follows: we start with 1 and then insert the values $i = 2, \ldots, n$ one by one, either at the leftmost or rightmost position, depending on whether the bit b_i is 1 or 0, respectively. Thus a bitstring of length n maps to a permutation of length $n + 1$ and two permutations in $\mathbb{P}_n^\vee$ differ in a jump if and only if the mapped bitstrings differ in a bitflip. □

For $n \geq k$, let $\pi \in \mathbb{P}_n$ and $\tau \in \mathbb{P}_k$. We say that π *contains* τ, if and only if $\pi = p_1 \cdots p_n$ contains a subpermutation $p_{i_1} \cdots p_{i_k}$ with the same relative order as the elements in τ. Otherwise, p *avoids* τ. We denote $\mathbb{P}_n(\tau)$ as the set of all permutations of length n that avoids τ. Moreover, $\mathbb{P}_n(\tau_1 \wedge \cdots \wedge \tau_\ell)$ is the set of permutations of length n avoiding each of the patterns $\tau_1, \ldots, \tau_\ell$. Gray codes for pattern-avoiding permutations appeared in [1, 7, 12].

Remark 2. Peakless permutations of $[n]$ are $(132 \wedge 231)$-avoiding permutations of $[n]$.

We extend Theorem 9 to multiple permutation patterns consisting of ANDs.

Corollary 2. *For a set of permutation patterns* $\{\tau_1, \tau_2, \ldots, \tau_\ell\}$, *if every* τ_i *contains a peak, then the Gray code problem is NP-hard for on jumps the list of permutations from* $\mathbb{P}_n(\tau_1 \wedge \ldots \wedge \tau_\ell)$.

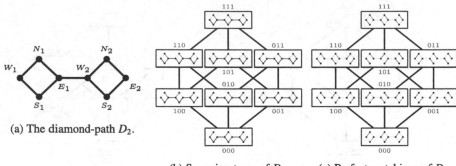

(a) The diamond-path D_2.

(b) Spanning trees of D_3.   (c) Perfect matchings of D_3.

Fig. 5. Gray code reductions for set partitions and graphs. (Color figure online)

7 Problems on Graphs

In this section, we discuss some graph-related problems. Our reductions are based on a particular graph namely, the diamond-path graph. Informally, the n-th diamond-path graph consists of n squares joined through edges. More formally, we define the graph D_n by considering the vertex set $\{N, S, E, W\} \times [n]$, and the edge set $\{N_i E_i \cup N_i W_i \cup S_i E_i \cup S_i W_i \mid i \in [n]\} \cup \{E_i W_{i+1} \mid i \in [n-1]\}$; see Fig. 5a.

7.1 Spanning Trees

A *spanning tree* of a graph G is a connected acyclic subgraph of G. We denote the set of all spanning trees of a fixed graph G by $\mathbb{ST}_G$. We say that two spanning trees T, T' of G differ in an *edge exchange* if they differ in exactly two edges; i.e., there exist edges $e \in T \setminus T'$ and $f \in T' \setminus T$ such that $T = T' + e - f$. Gray codes for spanning trees under edge exchanges have been widely studied from both the combinatorial and computational point of view [19,21,25,32].

SPANNINGTREEGC
Input: A list L of spanning trees from $\mathbb{ST}_{D_n}$.
Question: Is there an edge-exchange Gray code for L?

Theorem 10. SPANNINGTREEGC *is NP-hard.*

Proof (of Theorem 10). We use a Gray code reduction; see Fig. 5b. For bitstrings in $\mathbb{B}_n$, we define the following list of spanning trees of D_n

$$L := \{W_i S_i \mid i \in [n]\} \cup \{E_i S_i \mid i \in [n]\} \cup \{E_i W_{i+1} \mid i \in [n-1] \cup \bigcup_{i:b_i=0} \{W_i N_i\} \cup \bigcup_{i:b_i=1} \{E_i N_i\}.$$

In other words, a spanning tree in L contains all the edges between diamonds, the edges that are incident on S_i vertices, and to join N_i, we use either the edge W_i or E_i, depending on the value of b_i. Therefore, two spanning trees $T_b, T_{b'} \in L$ differ in an edge exchange if and only if $b, b' \in \mathbb{B}_n$ differ in a bitflip. □

7.2 Perfect Matchings

A perfect matching of a graph is a set of edges $M \subseteq E$ such that every vertex is incident to exactly one edge in M. We denote the set of all perfect matchings of a fixed graph G by $\mathbb{PM}_G$. We say that two perfect matchings M, M' of G differ in an *alternating cycle* if their symmetric difference forms a cycle in G. Every graph with a perfect matching has an alternating cycle Gray code for $\mathbb{PM}_G$ that can be efficiently computed [19,25].

PerfectMatchingGC
Input: A list L of perfect matchings from $\mathbb{PM}_{D_n}$.
Question: Is there an alternating cycle Gray code for L?

Theorem 11. PerfectMatchingGC *is NP-hard.*

Proof (of Theorem 11). We use Gray code reductions; see Fig. 5c. For bitstrings in $\mathbb{B}_n$, we define the following list of perfect matchings of D_n

$$L := \bigcup_{i:b_i=0} \{W_i N_i, E_i S_i\} \cup \bigcup_{i:b_i=1} \{E_i N_i, W_i S_i\}.$$

There are two possible choices of perfect matchings for every diamond in D_n. Therefore, two perfect matchings $M_b, M_{b'} \in L$ differ in an alternating cycle if and only if $b, b' \in \mathbb{B}_n$ differ in a bitflip.   □

Theorems 10 and 11 also extend the hardness when we are given the host graph G as input and ask for edge-exchange or alternating cycles Gray codes for lists of spanning trees or perfect matchings, respectively, of G.

8 Final Remarks

We proved that the Gray coding problems are NP-complete for various classical objects. Future work could involve investigating optimization and approximation variants. Furthermore, our techniques apply to many other objects, for example, those involving geometry, that we plan to explore in the full version of this paper.

We note that we were unable to establish NP-hardness for certain subset problems using grid or hypercube reductions. These problems include operations that do not support *independent involutions*. In other words, there is no way to make and unmake multiple local changes, which is a hallmark of hypercube reductions. It is also important to note that some subset problems are poly-time solvable (e.g., those associated with de Bruijn sequences and universal cycles).

References

1. Baril, J.-L.: More restrictive gray codes for some classes of pattern avoiding permutations. Inf. Process. Lett. **109**(14), 799–804 (2009)
2. Buck, M., Wiedemann, D.: Gray codes with restricted density. Discret. Math. **48**(2–3), 163–171 (1984)

3. Compton, R.C., Williamson, S.G.: Doubly adjacent gray codes for the symmetric group. Linear Multilinear Algebra **35**(3–4), 237–293 (1993)
4. Corbett, P.F.: Rotator graphs: an efficient topology for point-to-point multiprocessor networks. IEEE Trans. Parallel Distrib. Syst. **3**(05), 622–626 (1992)
5. Demaine, E.D., Eisenstat, S., Rudoy, M.: Solving the rubik's cube optimally is NP-complete. In: STACS 2018, pp. 24:1–24:13 (2018)
6. Duckworth, R., Stedman, F.: Tintinnalogia: Or, The Art of Ringing. London (1668)
7. Dukes, W.M.B., Flanagan, M.F., Mansour, T., Vajnovszki, V.: Combinatorial gray codes for classes of pattern avoiding permutations. Theor. Comput. Sci. **396**(1–3), 35–49 (2008)
8. Eades, P., McKay, B.D.: An algorithm for generating subsets of fixed size with a strong minimal change property. Inf. Process. Lett. **19**(3), 131–133 (1984)
9. Ehrlich, G.: Loopless algorithms for generating permutations, combinations, and other combinatorial configurations. J. ACM (JACM) **20**(3), 500–513 (1973)
10. Gray, F.: Pulse code communication. United States Patent Number 2632058 (1953)
11. Gregor, P., Mička, O., Mütze, T.: On the central levels problem. J. Comb. Theory. Ser. B **160**, 163–205 (2023)
12. Hartung, E., Hoang, H., Mütze, T., Williams, A.: Combinatorial generation via permutation languages. I. fundamentals. Trans. Am. Math. Soc. **375**(04), 2255–2291 (2022)
13. Itai, A., Papadimitriou, C.H., Szwarcfiter, J.L.: Hamilton paths in grid graphs. SIAM J. Comput. **11**(4), 676–686 (1982)
14. Johnson, S.M.: Generation of permutations by adjacent transposition. Math. Comput. **17**(83), 282–285 (1963)
15. Karp, R.M.: Reducibility among combinatorial problems, complexity of computer computations (re miller and jw thatcher, editors) (1972)
16. Knuth, D.E.: The Art of Computer Programming: Combinatorial Algorithms, Part 1. Addison-Wesley Professional (2011)
17. Lipták, Z., Masillo, F., Navarro, G., Williams, A.: Constant time and space updates for the sigma-tau problem. In: Nardini, F.M., Pisanti, N., Venturini, R. (eds.) SPIRE 2023. LNCS, vol. 14240, pp. 323–330. Springer, Cham (2023). https://doi.org/10.1007/978-3-031-43980-3_26
18. Merino, A., Mička, O., Mütze, T.: On a combinatorial generation problem of knuth. SIAM J. Comput. **51**(3), 379–423 (2022)
19. Merino, A., Mütze, T.: Traversing combinatorial 0/1-polytopes via optimization. arXiv preprint arXiv:2304.08567
20. Merino, A., Mütze, T., Namrata: Kneser graphs are hamiltonian. In: Proceedings of the 55th Annual ACM Symposium on Theory of Computing, pp. 963–970 (2023)
21. Merino, A., Mütze, T., Williams, A.: All your bases are belong to us: listing all bases of a matroid by greedy exchanges. In: 11th International Conference on Fun with Algorithms, FUN 2022 (2022)
22. Mütze, T.: Proof of the middle levels conjecture. Proc. Lond. Math. Soc. **112**(4), 677–713 (2016)
23. Mütze, T.: Combinatorial gray codes-an updated survey. arXiv preprint arXiv:2202.01280 (2022)
24. Mütze, T.: A book proof of the middle levels theorem. arXiv preprint arXiv:2306.13019 (2023)
25. Naddef, D.J., Pulleyblank, W.R.: Hamiltonicity in (0-1)-polyhedra. J. Combin. Theory Ser. B **37**(1), 41–52 (1984)
26. Ord-Smith, R.J.: Algorithm 308: generation of the permutations in pseudo-lexicographic order. Commun. ACM **10**(7), 452 (1967)
27. Ruskey, F.: Combinatorial generation. Preliminary draft, University of Victoria (2003)

28. Ruskey, F., Williams, A.: The coolest way to generate combinations. Discret. Math. **309**(17), 5305–5320 (2009)
29. Savage, C.: A survey of combinatorial gray codes. SIAM Rev. **39**(4), 605–629 (1997)
30. Sawada, J., Williams, A.: Solving the sigma-tau problem. ACM Trans. Algorithms (TALG) **16**(1), 1–17 (2019)
31. Shen, X.S., Williams, A.: A 'hot potato' gray code for permutations. Electron. Notes Discrete Math. **44**, 89–94 (2013)
32. Smith, M.J.: Generating spanning trees (1997)
33. Steinhaus, H.: One hundred problems in elementary mathematics (1979)
34. Stevens, B., Williams, A.: The coolest way to generate binary strings. Theory Comput. Syst. **54**, 551–577 (2014)
35. Stibitz, G.R.: Binary counter. United States Patent Number 2307868 (1943)
36. Tang, D.T., Liu, C.N.: Distance-2 cyclic chaining of constant-weight codes. IEEE Trans. Comput. **100**(2), 176–180 (1973)
37. Trotter, H.F.: Algorithm 115: perm. Commun. ACM **5**(8), 434–435 (1962)

On MAX–SAT with Cardinality Constraint

Fahad Panolan[1] and Hannane Yaghoubizade[2]([✉])

[1] School of Computing, University of Leeds, Leeds, UK
f.panolan@leeds.ac.uk
[2] Department of Mathematical Sciences, Sharif University of Technology,
Tehran, Iran
h.yaghoubizade99@sharif.ir
https://fahadpanolan.github.io/

Abstract. We consider the weighted MAX–SAT problem with an additional constraint that **at most** k variables can be set to true. We call this problem k–WMAX–SAT. This problem admits a $(1 - \frac{1}{e})$-factor approximation algorithm in polynomial time [Sviridenko, Algorithmica 2001] and it is proved that there is no $(1 - \frac{1}{e} + \epsilon)$-factor approximation algorithm in $f(k) \cdot n^{o(k)}$ time for MAXIMUM COVERAGE, the unweighted monotone version of k–WMAX–SAT [Manurangsi, SODA 2020]. Therefore, we study two restricted versions of the problem in the realm of parameterized complexity.

1. When the input is an unweighted 2–CNF formula (the problem is called k–MAX–2SAT), we design an efficient polynomial-size approximate kernelization scheme. That is, we design a polynomial-time algorithm that given a 2–CNF formula ψ and $\epsilon > 0$, compresses the input instance to a 2–CNF formula $\psi^\star$ such that any c-approximate solution of $\psi^\star$ can be converted to a $c(1 - \epsilon)$-approximate solution of ψ in polynomial time.
2. When the input is a planar CNF formula, i.e., the variable-clause incident graph is a planar graph, we show the following results:
   - There is an FPT algorithm for k–WMAX–SAT on planar CNF formulas that runs in $2^{O(k)} \cdot (C + V)$ time.
   - There is a polynomial-time approximation scheme for k–WMAX–SAT on planar CNF formulas that runs in time $2^{O(\frac{1}{\epsilon})} \cdot k \cdot (C + V)$.

The above-mentioned C and V are the number of clauses and variables of the input formula respectively.

Keywords: Parameterized Algorithms · MAX–SAT · MAX–2SAT

1 Introduction

In this paper, we study the well-studied MAX–SAT problem with cardinality constraint. The weighted version of the problem is formally defined as follows.

R. Uehara et al. (Eds.): WALCOM 2024, LNCS 14549, pp. 118–133, 2024.
https://doi.org/10.1007/978-981-97-0566-5_10

WEIGHTED MAX–SAT WITH CARDINALITY CONSTRAINT (k–WMAX–SAT)
Parameter: k
Input: A set of t clauses $\mathcal{C}_{\mathcal{F}} = \{C_1, C_2, \ldots, C_t\}$ of a CNF formula $\mathcal{F}$, a weight function $w : \mathcal{C}_{\mathcal{F}} \rightarrow \mathbb{R}^+$ and a positive integer k.
Objective: Find a subset S of variables such that $|S| \leq k$ and setting variables of S to true and other variables to false, maximizes the **weight** of the satisfied clauses.

---

k–MAX–SAT and its monotone version (a version in which negated literals are not allowed) MAXIMUM COVERAGE are well studied both in the realm of approximation algorithms and parameterized complexity. The input of MAXIMUM COVERAGE is a family $\mathcal{F}$ of subsets of a universe U and a positive integer k. The goal is to find $S_1, S_2, \ldots, S_k \in \mathcal{F}$ that maximizes $|S_1 \cup S_2 \cup \cdots \cup S_k|$.

MAXIMUM COVERAGE, and hence k–MAX–SAT are known to be NP-complete and W[2]-hard because MAXIMUM COVERAGE is a more general case of the DOMINATING SET problem. A simple greedy approximation algorithm for MAXIMUM COVERAGE outputs a $(1 - \frac{1}{e})$-approximate solution, where e is the base of the natural logarithm. This greedy approximation algorithm is essentially optimal for MAXIMUM COVERAGE [7]. Sviridenko [18] obtained a $(1 - \frac{1}{e})$-factor approximation in polynomial time for k–WMAX–SAT. Recently, Manurangsi [14] showed that there is no $f(k) \cdot n^{o(k)}$ time algorithm that can approximate MAXIMUM COVERAGE within a factor of $(1 - \frac{1}{e} + \epsilon)$ for any $\epsilon > 0$ and any function f, assuming Gap Exponential Time Hypothesis (Gap-ETH). Thus, to obtain tractable results for k–WMAX–SAT in the realm of parameterized complexity and approximation algorithms, we need to restrict the input to different classes of formulas. We study cardinality constrained unweighted MAX–SAT when the number of literals in each clause is at most 2. This problem is called k–MAX–2SAT. The problem is formally defined below.

MAX–2SAT WITH CARDINALITY CONSTRAINT (k–MAX–2SAT)
Parameter: k
Input: A set of t clauses $\mathcal{C}_{\mathcal{F}} = \{C_1, C_2, \ldots, C_t\}$ of a 2–CNF formula $\mathcal{F}$ and a positive integer k.
Objective: Find a subset S of variables such that $|S| \leq k$ and setting variables of S to true and other variables to false, maximizes the **number** of the satisfied clauses.

---

k–MAX–2SAT and its monotone version MAX–k–VERTEX COVER (shortly MAX k–VC) are extensively studied [3,9,10,13,16,17]. The best-known polynomial-time approximation ratio for k–MAX–2SAT is 0.75 [9]. Raghavendra and Tan [17] designed an α-approximation algorithm for some $\alpha > 0.92$ that runs in time $n^{\mathsf{poly}(n/k)}$, where n is the number of variables in the input formula [13]. That is, this algorithm runs in polynomial time when k is a constant fraction of V. Assuming Unique Games Conjecture (UGC), it is NP-hard to approximate k–MAX–2SAT with a factor better than 0.929 [1].

The monotone variant of the problem, MAX k–VC gives an interesting connection between approximate kernelization and approximation algorithms. Here, given a graph G, our objective is to find a vertex subset S of size k such that the number of edges in G with at least one endpoint in S is maximized. MAX k–VC is W[1]-hard and Marx [15] designed the first FPT approximation scheme for the problem, where k is the parameter. Lokshtanov et al. [12] showed that, indeed the steps of the algorithm by Marx can be converted to get an efficient polynomial-size approximate kernelization scheme (EPSAKS). We refer to Sect. 2 for the definition of approximate kernelization. Manurangsi [13] improved the kernel size to $O(k/\epsilon)$ and the running time of FPT approximation scheme to $(1/\epsilon)^{O(k)} n^{O(1)}$ for MAX k–VC. Manurangsi applied the algorithm of Raghavendra and Tan [17] for k–MAX–2SAT on the linear size approximate kernel to obtain an approximation factor of 0.92 for MAX k–VC. Approximating MAX k–VC with a factor better than 0.929 is also NP-hard assuming UGC [1]. We prove that k–MAX–2SAT admits an EPSAKS.

Theorem 1. *Given a set of t clauses $C_{\mathcal{F}} = \{C_1, C_2, \ldots, C_t\}$ of a 2–CNF formula $\mathcal{F}$ and a positive integer k, there is an EPSAKS (efficient polynomial-size approximate kernelization scheme) for k–MAX–2SAT such that the size of the output of the reduction algorithm is upper-bounded by $O\left(\frac{k^5}{\epsilon^2}\right)$.*

We also study k–WMAX–SAT when the input is a planar CNF formula, that is, the variable-clause incident graph is a planar graph. Restricting MAX–SAT to planar formulas has been already considered in the realm of approximation algorithms [4,11]. We prove the following result for k–WMAX–SAT on planar CNF formulas.

Theorem 2. *Given a set of t clauses $C_{\mathcal{F}} = \{C_1, C_2, \ldots, C_t\}$ of a **planar** CNF formula $\mathcal{F}$, a weight function $w : C_{\mathcal{F}} \to \mathbb{R}^+$ and a positive integer k, there is an FPT algorithm for k–WMAX–SAT that runs in $O(2^{36k} \cdot k^3 \cdot |C_{\mathcal{F}} \cup V_{\mathcal{F}}|)$ time.*

Khanna and Motwani [11] already designed a PTAS for k–MAX–SAT (the unweighted version) on planar formulas. Using a similar technique, we show that the weighted version k–WMAX–SAT also admits a PTAS.

Theorem 3. *Given a set of t clauses $C_{\mathcal{F}} = \{C_1, C_2, \ldots, C_t\}$ of a **planar** CNF formula $\mathcal{F}$, a weight function $w : C_{\mathcal{F}} \to \mathbb{R}^+$ and a positive integer k, there is a polynomial-time approximation scheme that runs in $O(\frac{1}{\epsilon^2} \cdot 2^{\frac{36}{\epsilon}} \cdot k \cdot |C_{\mathcal{F}} \cup V_{\mathcal{F}}|)$ time and finds $S \subseteq V_{\mathcal{F}}$ such that $|S| \leq k$ and*

$$k\text{–}WMAX\text{–}SAT(C_{\mathcal{F}}, w, k, S) \geq (1 - \epsilon) \cdot OPT(C_{\mathcal{F}}, w, k)$$

Here, $OPT(C_{\mathcal{F}}, w, k)$ is the maximum total weight of clauses in $C_{\mathcal{F}}$ that can be satisfied by an assignment where at most k variables are set to true.

2   Preliminaries

Definition 1 (Conjunctive Normal Form (CNF)). *A formula is said to be in Conjunctive Normal Form (CNF) if it looks like $C_1 \wedge C_2 \wedge \cdots \wedge C_t$ where each $C_i = (l_1 \vee l_2 \vee \cdots \vee l_{t_i})$ is called a clause and each l_i is called a literal. A literal is either a variable, called positive literal, or the negation of a variable, called negative literal.*

A formula is said to be in 2–Conjunctive Normal Form (2–CNF) if it is in CNF and all of its clauses contain 2 literals.

We assume, without loss of generality, that for each variable v, at most one of the v and $\neg v$ is contained in a clause, no literal is repeated in a clause and all clauses are distinct.

For a CNF formula $\mathcal{F}$, the set of clauses and the set of variables appeared in $\mathcal{F}$ are denoted by $\mathcal{C}_{\mathcal{F}} = \{C_1, C_2, \ldots, C_t\}$ and $V_{\mathcal{F}} = \{v_1, v_2, \ldots, v_n\}$, respectively.

2.1   Parameterized Complexity

For a parameterized maximization problem Π and a solution s to the instance (I, k) of Π, we denote the value of s by $\Pi(I, k, s)$, and the task is to find a solution with the maximum possible value. We state the following definitions slightly modified from the Kernelization book [8].

Definition 2 (FPT optimization problem). *A parameterized optimization problem Π is fixed-parameter tractable (FPT) if there is an algorithm (called FPT algorithm) that solves Π, such that the running time of the algorithm on instances of size n with parameter k is upper-bounded by $f(k) \cdot n^{O(1)}$ for a computable function f.*

Definition 3 (α-approximate polynomial-time preprocessing algorithm). *Let $0 < \alpha \leq 1$ be a real number and Π be a parameterized **maximization** problem. An α-approximate polynomial-time preprocessing algorithm $\mathcal{A}$ for Π is a pair of polynomial-time algorithms. The first one is called the **reduction algorithm** $\mathcal{R}_{\mathcal{A}}$, and given an instance (I, k) of Π, it outputs another instance $(I', k') = \mathcal{R}_{\mathcal{A}}(I, k)$. The second algorithm is called the **solution lifting** algorithm. This algorithm takes an instance (I, k) of Π, the output instance (I', k') of the reduction algorithm, and a solution s' to the instance (I', k'). The solution lifting algorithm works in time polynomial in $|I|, k, |I'|, k'$ and $|s'|$, and outputs a solution s to (I, k) such that*

$$\frac{\Pi(I, k, s)}{OPT(I, k)} \geq \alpha \cdot \frac{\Pi(I', k', s')}{OPT(I', k')}$$

Definition 4 (α-approximate kernelization). *An α-approximate kernelization (α-approximate kernel) is an α-approximate polynomial-time preprocessing algorithm $\mathcal{A}$ such that $size_{\mathcal{A}}$ is upper-bounded by a computable function $g : \mathbb{N} \to \mathbb{N}$ where $size_{\mathcal{A}}$ is defined as follows:*

$$size_{\mathcal{A}}(k) = \sup\{|I'| + k' : (I', k')$$
$$= \mathcal{R}_{\mathcal{A}}(I, k) \text{ for any instance } (I, k) \text{ of the problem}\}$$

If the upper-bound $g(\cdot)$ is a polynomial function of k, we say $\mathcal{A}$ is an α-approximate polynomial kernel.

Definition 5 (polynomial-size approximate kernelization scheme (PSAKS)). *A polynomial-size approximate kernelization scheme (PSAKS) for a parameterized maximization problem Π, is a family of $(1-\epsilon)$-approximate polynomial kernels for every $0 < \epsilon < 1$.*

Definition 6 (Efficient PSAKS). *An efficient PSAKS (EPSAKS) is a PSAKS such that for every $(1-\epsilon)$-approximate polynomial kernel $\mathcal{A}$ in that, $size_{\mathcal{A}}(k)$ is upper-bounded by $f(\frac{1}{\epsilon}) \cdot k^c$ for a function f and a constant c independent of I, k and ϵ.*

2.2 Tree Decomposition and Tree-Width

We state the following definitions and lemmas from the Parameterized Algorithms book [5].

Definition 7 (Tree decomposition). *A tree decomposition of a graph G is a pair $\mathcal{T} = (T, \{X_t\}_{t \in V(T)})$, where T is a tree whose every node t is assigned a vertex subset $X_t \subseteq V(G)$, called a bag, such that the following three conditions hold:*

- *Vertex coverage: $\bigcup_{t \in V(T)} X_t = V(G)$, i.e., every vertex of G is in at least one bag.*
- *Edge coverage: For every $uv \in E(G)$, there exists a node t of T such that bag X_t contains both u and v.*
- *Coherence: For every $u \in V(G)$, the set $T_u = \{t \in V(T) : u \in X_t\}$, i.e., the set of nodes whose corresponding bags contain u, induces a connected subtree of T.*

The width of tree decomposition $\mathcal{T} = (T, \{X_t\}_{t \in V(T)})$ equals $\max_{t \in V(T)} |X_t| - 1$.

Definition 8 (Tree-width). *The tree-width of a graph G is the minimum possible width of a tree decomposition of G.*

Definition 9 (Nice tree decomposition). *A tree decomposition $\mathcal{T} = (T, \{X_t\}_{t \in V(T)})$, rooted from $r \in V(T)$, is called nice if the following conditions are satisfied:*

- *$X_r = \emptyset$ and $X_l = \emptyset$ for every leaf l of T.*
- *Every non-leaf node of T is of one of the following three types:*
  - *Introduce node: a node t with exactly one child t' such that $X_t = X_{t'} \cup \{v\}$ for some vertex $v \notin X_{t'}$. We say that v is introduced at t.*

- **Forget node:** *a node t with exactly one child t' such that $X_t = X_{t'} \setminus \{w\}$ for some vertex $w \in X_{t'}$. We say that w is forgotten at t.*
- **Join node:** *a node t with two children t_1, t_2 such that $X_t = X_{t_1} = X_{t_2}$.*

Lemma 1. *If a graph G admits a tree decomposition of width at most d, then it also admits a nice tree decomposition of width at most d. Moreover, given a tree decomposition $\mathcal{T} = (T, \{X_t\}_{t \in V(T)})$ of G of width at most d, one can in time $O(d^2 \cdot \max(|V(T)|, |V(G)|))$ compute a nice tree decomposition of G of width at most d that has $O(d \cdot |V(G)|)$ nodes.*

3 EPSAKS for k–MAX–2SAT with Cardinality Constraint

In this section, we show that k–MAX–2SAT admits an EPSAKS. That is we prove Theorem 1.

There are two main observations used in the algorithm. First, since one can satisfy all clauses containing at least one negative literal by setting all the variables to false, the optimal value is not less than the number of clauses containing negative literals. Second, if a variable v appears positively in many clauses, then one can satisfy all those clauses by setting v true and all the other variables false.

Let $\mathcal{F}$ be a 2–CNF formula with clause set $\mathcal{C}_{\mathcal{F}}$ and variable set $V_{\mathcal{F}}$. For a variable $v \in V_{\mathcal{F}}$, we denote the number of clauses in the form of $(v \vee u)$, $(v \vee \neg u)$, $(\neg v \vee u)$ and $(\neg v \vee \neg u)$ by $d_+^+(v)$, $d_+^-(v)$, $d_-^+(v)$ and $d_-^-(v)$ respectively. For $V \subseteq V_{\mathcal{F}}$ we denote the set of negation of variables in V with $\neg V$, i.e., $\neg V = \{\neg s \mid s \in V\}$. Let $P_{\mathcal{F}} = \{p_1, p_2, \ldots, p_l\}$ be the set of variables that appear only in clauses containing two positive literals, i.e., in the form of $(v \vee u)$, and $N_{\mathcal{F}} = V_{\mathcal{F}} \setminus P_{\mathcal{F}}$. We suppose, without loss of generality, $d_+^+(p_1) \geq d_+^+(p_2) \geq \cdots \geq d_+^+(p_l)$.

We now describe a $(1 - \epsilon)$-approximate polynomial-time preprocessing algorithm $\mathcal{A}_\epsilon$ for an arbitrary ϵ.

Reduction Algorithm $\mathcal{R}_\epsilon$: $\mathcal{R}_\epsilon$ takes the set of clauses $\mathcal{C}_{\mathcal{F}} = \{C_1, C_2, \ldots, C_t\}$ of a 2–CNF formula $\mathcal{F}$ and a parameter k as input. Set λ to be equal to $\frac{4 \cdot \binom{k}{2}}{\epsilon}$. Recall that $P_{\mathcal{F}} = \{p_1, p_2, \ldots, p_l\}$ is the set of variables that appear only in clauses containing two positive literals. Let $\tilde{P}_{\mathcal{F}} = \{p_1, p_2, \ldots, p_{\tilde{l}}\}$ where $\tilde{l} = \min(l, k + k\lambda)$ and $\tilde{\mathcal{C}}_{\mathcal{F}} \subseteq \mathcal{C}_{\mathcal{F}}$ be the set of clauses whose both variables are in $P_{\mathcal{F}} \setminus \tilde{P}_{\mathcal{F}}$. If both of the following requirements are satisfied, $\mathcal{R}_\epsilon$ outputs $(\mathcal{C}_{\mathcal{F}} \setminus \tilde{\mathcal{C}}_{\mathcal{F}}, k)$, otherwise it outputs $(\{C_1\}, k + 1)$.

(R1) There are $< \lambda$ clauses with at least one negative literal.
(R2) $d_+^+(v) < \lambda$ for every variable $v \in V_{\mathcal{F}}$.

Solution Lifting Algorithm $\mathcal{L}_\epsilon$: The algorithm takes $(\mathcal{C}_{\mathcal{F}}, k)$, the output of the reduction algorithm $(\mathcal{C}'_{\mathcal{F}'}, k')$ and a set S' of at most k' variables appeared in $\mathcal{F}'$.

If $k' = k$, $\mathcal{L}_\epsilon$ outputs $S = S'$. Otherwise, let $V_\mathcal{F} = \{v_1, v_2, \ldots, v_n\}$ and without loss of generality suppose

$$d_+^\pm(v_1) - d_-^\pm(v_1) \geq d_+^\pm(v_2) - d_-^\pm(v_2) \geq \cdots \geq d_+^\pm(v_n) - d_-^\pm(v_n)$$

Then the algorithm outputs

$$S = \{v \in \{v_1, v_2, \ldots, v_k\} \mid d_+^\pm(v) - d_-^\pm(v) > 0\}$$

We next show that $\mathcal{A}_\epsilon$ is a $(1-\epsilon)$-approximate polynomial-time preprocessing algorithm. To do so, we need to prove the following lemmas.

Lemma 2. *Suppose $d_+^\pm(v) < \lambda$ for every $v \in V_\mathcal{F}$. Let S^* be an optimal solution for $(\mathcal{C}_\mathcal{F}, k)$ such that $S^* \cap P_\mathcal{F}$ is lexicographically smallest with respect to $p_1, p_2, \ldots, p_l$. Then $(S^* \cap P_\mathcal{F}) \subseteq \tilde{P}_\mathcal{F} = \{p_1, p_2, \ldots, p_{\tilde{l}}\}$.*

Proof. If $\tilde{l} = l$, we have $\tilde{P}_\mathcal{F} = P_\mathcal{F}$. So $(S^* \cap P_\mathcal{F}) \subseteq P_\mathcal{F} = \tilde{P}_\mathcal{F}$ and we are done. So suppose $\tilde{l} = k + k\lambda$ and for the sake of contradiction, suppose there is $p \in (S^* \cap P_\mathcal{F})$ such that $p \notin \tilde{P}_\mathcal{F}$. Define the set A as the following:

$$A = S^* \cup \{v \in V_\mathcal{F} \mid \exists (v \vee u) \in \mathcal{C}_\mathcal{F} : u \in S^*\}$$

Since $|S^*| \leq k$ and $\forall v \in V_\mathcal{F} : d_+^\pm(v) < \lambda$, we have $|A| < k + k\lambda$. Therefore, there is a variable $q \in \{p_1, p_2, \ldots, p_{k+k\lambda}\}$ which is not in A, i.e., $q \in \tilde{P}_\mathcal{F} \setminus A$.

Note that since $p, q \in P_\mathcal{F}$, p and q appear only in clauses with two positive literals, So we have

$$
\begin{aligned}
& k\text{–MAX–2SAT}(\mathcal{C}_\mathcal{F}, k, S^* \setminus \{p\} \cup \{q\}) \\
& \geq k\text{–MAX–2SAT}(\mathcal{C}_\mathcal{F}, k, S^*) - d_+^\pm(p) + d_+^\pm(q) \quad \text{(since } p \in P_\mathcal{F} \text{ and } q \notin A\text{)} \\
& \geq k\text{–MAX–2SAT}(\mathcal{C}_\mathcal{F}, k, S^*) \quad \text{(since } p \notin \tilde{P}_\mathcal{F} \text{ and } q \in \tilde{P}_\mathcal{F}\text{)} \\
& = \text{OPT}(\mathcal{C}_\mathcal{F}, k)
\end{aligned}
$$

Therefore, $S^* \setminus \{p\} \cup \{q\}$ is an optimal solution and since $p \notin \tilde{P}_\mathcal{F}$ but $q \in \tilde{P}_\mathcal{F}$, $(S^* \setminus \{p\} \cup \{q\}) \cap P_\mathcal{F}$ is lexicographically smaller than $S^* \cap P_\mathcal{F}$, which implies a contradiction.

Lemma 3. *If $d_+^\pm(v) < \lambda$ for every $v \in V_\mathcal{F}$, then $OPT(\mathcal{C}_\mathcal{F}, k) = OPT(\mathcal{C}_\mathcal{F} \setminus \tilde{\mathcal{C}}_\mathcal{F}, k)$.*

Proof. Since $(\mathcal{C}_\mathcal{F} \setminus \tilde{\mathcal{C}}_\mathcal{F}) \subseteq \mathcal{C}_\mathcal{F}$, we have $OPT(\mathcal{C}_\mathcal{F}, k) \geq OPT(\mathcal{C}_\mathcal{F} \setminus \tilde{\mathcal{C}}_\mathcal{F}, k)$. For the other direction, let S^* be the optimal solution of $(\mathcal{C}_\mathcal{F}, k)$ described in the Lemma 2. By Lemma 2 we know $S^* \cap (P_\mathcal{F} \setminus \tilde{P}_\mathcal{F}) = \emptyset$ and therefore, by setting only variables of S^* true, none of the clauses with both literals from $P_\mathcal{F} \setminus \tilde{P}_\mathcal{F}$, i.e., clauses in $\tilde{\mathcal{C}}_\mathcal{F}$, gets satisfied. This implies

$$\underbrace{k\text{–MAX–2SAT}(\mathcal{C}_\mathcal{F}, k, S^*)}_{=OPT(\mathcal{C}_\mathcal{F}, k)} = \underbrace{k\text{–MAX–2SAT}(\mathcal{C}_\mathcal{F} \setminus \tilde{\mathcal{C}}_\mathcal{F}, k, S^*)}_{\leq OPT(\mathcal{C}_\mathcal{F} \setminus \tilde{\mathcal{C}}_\mathcal{F}, k)}$$

which proves the lemma.

Lemma 4. $\mathcal{A}_\epsilon$ *is a* $(1-\epsilon)$-*approximate polynomial-time preprocessing algorithm.*

Proof. Clearly, both $\mathcal{R}_\epsilon$ *and* $\mathcal{L}_\epsilon$ *are polynomial algorithms. In the solution lifting algorithm, note that* $\mathcal{C}'_{\mathcal{F}'} \subseteq \mathcal{C}_\mathcal{F}$ *and thus* $S' \subseteq V_\mathcal{F}$. *This implies that the output of* $\mathcal{L}_\epsilon$ *is a subset of* $V_\mathcal{F}$ *with size* $\leq k$ *and therefore a solution to instance* $(\mathcal{C}_\mathcal{F}, k)$ *of* k–MAX–2SAT.

We now show that

$$\frac{k\text{--}MAX\text{--}2SAT(\mathcal{C}_\mathcal{F}, k, S)}{OPT(\mathcal{C}_\mathcal{F}, k)} \geq (1-\epsilon) \cdot \frac{k\text{--}MAX\text{--}2SAT(\mathcal{C}'_{\mathcal{F}'}, k', S')}{OPT(\mathcal{C}'_{\mathcal{F}'}, k')}$$

We consider two cases:

1. *The aforementioned requirements, (R1) and (R2) are satisfied.*
   In this case, $\mathcal{R}_\epsilon$ *outputs* $(\mathcal{C}'_{\mathcal{F}'}, k') = (\mathcal{C}_\mathcal{F} \setminus \tilde{\mathcal{C}}_\mathcal{F}, k)$ *and since* $k = k'$, $\mathcal{L}_\epsilon$ *would output* $S = S'$. *Since* $\mathcal{C}_\mathcal{F} \setminus \tilde{\mathcal{C}}_\mathcal{F} \subseteq \mathcal{C}_\mathcal{F}$, *we have*

$$k\text{--}MAX\text{--}2SAT(\mathcal{C}_\mathcal{F}, k, S') \geq k\text{--}MAX\text{--}2SAT(\mathcal{C}_\mathcal{F} \setminus \tilde{\mathcal{C}}_\mathcal{F}, k, S')$$

   And by Lemma 3 we get

$$\frac{k\text{--}MAX\text{--}2SAT(\mathcal{C}_\mathcal{F}, k, S')}{OPT(\mathcal{C}_\mathcal{F}, k)} \geq \frac{k\text{--}MAX\text{--}2SAT(\mathcal{C}_\mathcal{F} \setminus \tilde{\mathcal{C}}_\mathcal{F}, k, S')}{OPT(\mathcal{C}_\mathcal{F} \setminus \tilde{\mathcal{C}}_\mathcal{F}, k)}$$

$$\geq (1-\epsilon) \cdot \frac{k\text{--}MAX\text{--}2SAT(\mathcal{C}_\mathcal{F} \setminus \tilde{\mathcal{C}}_\mathcal{F}, k, S')}{OPT(\mathcal{C}_\mathcal{F} \setminus \tilde{\mathcal{C}}_\mathcal{F}, k)}$$

   Which completes the proof for the first case.
2. *At least one of the requirements, (R1) and (R2) is not satisfied.*
   If (R1) is not satisfied we have k–MAX–2SAT$(\mathcal{C}_\mathcal{F}, k, \emptyset) \geq \lambda$. *If (R2) is not satisfied, there is a variable* $v \in V_\mathcal{F}$ *such that* $d_+^+(v) \geq \lambda$, *thus* k–MAX–2SAT$(\mathcal{C}_\mathcal{F}, k, \{v\}) \geq \lambda$. *Therefore, in this case* $OPT(\mathcal{C}_\mathcal{F}, k) \geq \lambda$. *Note that for any solution* S:

$$k\text{--}MAX\text{--}2SAT(\mathcal{C}_\mathcal{F}, k, S) = \sum_{v \in V_\mathcal{F}} d_-^+(v) - |\{(\neg v \vee u) \mid v \in S, u \in V_\mathcal{F} \setminus S\}|$$

$$+ \frac{\sum_{v \in V_\mathcal{F}} d_-^-(v)}{2} - |\{(\neg v \vee \neg u) \mid v, u \in S\}|$$

$$+ \sum_{v \in S} d_+^+(v) - |\{(v \vee u) \mid v, u \in S\}|$$

And also:

$$|\{(\neg v \vee u) \mid v \in S, u \in V_\mathcal{F} \setminus S\}| = \left(\sum_{v \in S} d_-^+(v) - |\{(\neg v \vee u) \mid v, u \in S\}| \right)$$

Which implies:

$$k\text{-}MAX\text{-}2SAT(\mathcal{C}_{\mathcal{F}}, k, S) = \sum_{v \in V_{\mathcal{F}}} d_-^+(v) - \left(\sum_{v \in S} d_-^+(v) - |\{(\neg v \lor u) \mid v, u \in S\}| \right)$$

$$+ \frac{\sum_{v \in V_{\mathcal{F}}} d_-^-(v)}{2} - |\{(\neg v \lor \neg u) \mid v, u \in S\}|$$

$$+ \sum_{v \in S} d_+^+(v) - |\{(v \lor u) \mid v, u \in S\}| \tag{1}$$

And since $|S| \le k$ and all clauses are distinct, we have:

$$|\{(\neg v \lor \neg u)|v, u \in S\}| \quad , \quad |\{(v \lor u)|v, u \in S\}| \quad \le \quad \binom{k}{2}$$

Therefore, considering Eq. (1) we have:

$$k\text{-}MAX\text{-}2SAT(\mathcal{C}_{\mathcal{F}}, k, S) \ge \sum_{v \in V_{\mathcal{F}}} d_-^+(v) - \left(\sum_{v \in S} d_-^+(v) - |\{(\neg v \lor u) \mid v, u \in S\}| \right)$$

$$+ \frac{\sum_{v \in V_{\mathcal{F}}} d_-^-(v)}{2} - \binom{k}{2} + \sum_{v \in S} d_+^+(v) - \binom{k}{2}$$

$$= \sum_{v \in V_{\mathcal{F}}} d_-^+(v) + |\{(\neg v \lor u) \mid v, u \in S\}|$$

$$+ \frac{\sum_{v \in V_{\mathcal{F}}} d_-^-(v)}{2} + \left(\sum_{v \in S} d_+^+(v) - \sum_{v \in S} d_-^+(v) \right) - 2 \cdot \binom{k}{2} \tag{2}$$

Note that in this case $\mathcal{R}_\epsilon$ outputs $(\mathcal{C}_{\mathcal{F}'}', k') = (\{C_1\}, k+1)$ and since $k \ne k'$, $\mathcal{L}_\epsilon$ outputs $S = \{v \in \{v_1, v_2, \ldots, v_k\} | d_+^+(v) - d_-^+(v) > 0\}$. Let $S^* \subseteq V_{\mathcal{F}}$ be an optimal solution to $(\mathcal{C}_{\mathcal{F}}, k)$. Then we have:

$$\sum_{v \in S} d_+^+(v) - d_-^+(v) \ge \sum_{v \in S^*} d_+^+(v) - d_-^+(v)$$

And considering inequality (2):

$$k\text{-}MAX\text{-}2SAT(\mathcal{C}_{\mathcal{F}}, k, S)$$

$$\ge \sum_{v \in V_{\mathcal{F}}} d_-^+(v) + |\{(\neg v \lor u)|v, u \in S\}|$$

$$+ \frac{\sum_{v \in V_{\mathcal{F}}} d_-^-(v)}{2} + \left(\sum_{v \in S^*} d_+^+(v) - \sum_{v \in S^*} d_-^+(v) \right) - 2 \cdot \binom{k}{2}$$

$$= \sum_{v \in V_{\mathcal{F}}} d_-^+(v) - \left(\sum_{v \in S^*} d_-^+(v) - |\{(\neg v \lor u)|v, u \in S^*\}| \right)$$

$$+ (|\{(\neg v \lor u)|v, u \in S\}| - |\{(\neg v \lor u)|v, u \in S^*\}|)$$

$$+ \frac{\sum_{v \in V_{\mathcal{F}}} d_-^-(v)}{2} + \sum_{v \in S^*} d_+^+(v) - 2 \cdot \binom{k}{2}$$

$$\geq k\text{-}MAX\text{-}2SAT(\mathcal{C}_{\mathcal{F}}, k, S^*) + |\{(\neg v \vee u)|v, u \in S\}|$$

$$- |\{(\neg v \vee u)|v, u \in S^*\}| - 2 \cdot \binom{k}{2} \qquad \text{(By Eq. 1)}$$

Plugging $|\{(\neg v \vee u)|v, u \in S^*\}| \leq 2 \cdot \binom{k}{2}$ into the above inequality, we get:

$$k\text{-}MAX\text{-}2SAT(\mathcal{C}_{\mathcal{F}}, k, S) \geq k\text{-}MAX\text{-}2SAT(\mathcal{C}_{\mathcal{F}}, k, S^*) - 4 \cdot \binom{k}{2}$$

$$= OPT(\mathcal{C}_{\mathcal{F}}, k) - \epsilon\lambda \qquad \text{(since } S^* \text{ is an optimal solution and } \lambda = \frac{4 \cdot \binom{k}{2}}{\epsilon})$$

Finally, as $OPT(\mathcal{C}_{\mathcal{F}}, k) \geq \lambda$ we have:

$$k\text{-}MAX\text{-}2SAT(\mathcal{C}_{\mathcal{F}}, k, S) \geq (1 - \epsilon) \cdot OPT(\mathcal{C}_{\mathcal{F}}, k)$$

Which implies $\frac{k\text{-}MAX\text{-}2SAT(\mathcal{C}_{\mathcal{F}},k,S)}{OPT(\mathcal{C}_{\mathcal{F}},k)} \geq (1 - \epsilon) \geq (1 - \epsilon) \cdot \frac{MAX\text{-}2SAT(\mathcal{C}'_{\mathcal{F}'},k',S')}{OPT(\mathcal{C}'_{\mathcal{F}'},k')}$ and proves the second case.

The next lemma states an upper-bound for $size_{\mathcal{A}_\epsilon}(k)$.

Lemma 5. $size_{\mathcal{A}_\epsilon}(k)$ is of $O\left(\frac{k^5}{\epsilon^2}\right)$ where $size_{\mathcal{A}_\epsilon}(k)$ is defined in Definition 4.

Proof. Note that $\mathcal{R}_\epsilon$ returns either $(\{C_1\}, k + 1)$ or $(\mathcal{C}_{\mathcal{F}} \setminus \tilde{\mathcal{C}}_{\mathcal{F}}, k)$. In the first case $size_{\mathcal{A}_\epsilon}(k)$ is of $O(1)$ and so we need to only consider the case of returning $(\mathcal{C}_{\mathcal{F}} \setminus \tilde{\mathcal{C}}_{\mathcal{F}}, k)$. In this case, (R1) and (R2) are satisfied. Since (R1) is satisfied, there are less than 2λ variables that appear in at least one clause with at least one negative literal, i.e., $|N_{\mathcal{F}}| < 2\lambda$. Therefore, $|N_{\mathcal{F}} \cup \tilde{P}_{\mathcal{F}}| \leq 2\lambda + \tilde{l} \leq 2\lambda + k\lambda + k = O(k\lambda)$. (R1) and (R2) together imply that $d_+^+(v) + d_-^-(v) + d_+^-(v) + d_-^-(v) < d_+^+(v) + \lambda < 2\lambda$ which means every variable $v \in V_{\mathcal{F}}$ appears in less than 2λ clauses of $\mathcal{F}$. Therefore, $|\mathcal{C}_{\mathcal{F}} \setminus \tilde{\mathcal{C}}_{\mathcal{F}}|$ is less than $2\lambda \cdot |N_{\mathcal{F}} \cup \tilde{P}_{\mathcal{F}}| = O(k\lambda^2) = O\left(\frac{k^5}{\epsilon^2}\right)$.

We finally prove Theorem 1. For convenience, we restate the theorem here.

Theorem 1. Given a set of t clauses $\mathcal{C}_{\mathcal{F}} = \{C_1, C_2, \ldots, C_t\}$ of a 2-CNF formula $\mathcal{F}$ and a positive integer k, there is an EPSAKS (efficient polynomial-size approximate kernelization scheme) for $k\text{-}MAX\text{-}2SAT$ such that the size of the output of the reduction algorithm is upper-bounded by $O\left(\frac{k^5}{\epsilon^2}\right)$.

Proof. According to Definition 6, the proof is directly derived from Lemma 4 and Lemma 5.

4 k–WMAX–SAT with Cardinality Constraint on Planar Formulas

In this section, we present an FPT algorithm as well as a PTAS (Polynomial-time approximation scheme) for k–WMAX–SAT on a special family of sparse CNF formulas that we will refer to as *planar formulas*. We now describe this family of formulas.

For a CNF formula $\mathcal{F}$, let $G_{\mathcal{F}} = (\mathcal{C}_{\mathcal{F}} \cup V_{\mathcal{F}}, E_- \cup E_+)$ be a bipartite graph such that $(C_i, v_j) \in E_+$ if C_i contains v_j and $(C_i, v_j) \in E_-$ if C_i contains $\neg v_j$. We call $\mathcal{F}$ a *planar CNF formula* if $G_{\mathcal{F}}$ is a planar graph.

Both algorithms presented in this section are designed using Baker's technique [2] and dynamic programming on tree decomposition. First, we need the following lemmas.

Lemma 6 (Eppstein [6]). *Let planar graph G have diameter d. Then G has tree-width at most $3d - 2$, and a tree-decomposition of G with such a width can be found in time $O(d \cdot |V(G)|)$.*

Lemma 7. *Let $\mathcal{F}$ be a planar CNF formula. Then there is an algorithm with running time $O(2^{3d} \cdot kd \cdot |\mathcal{C}_{\mathcal{F}} \cup V_{\mathcal{F}}|)$ that takes $\mathcal{C}_{\mathcal{F}} = \{C_1, C_2, \ldots, C_t\}$, a weight function $w : \mathcal{C}_{\mathcal{F}} \to \mathbb{R}^+$, a positive integer k, and a tree decomposition of $G_{\mathcal{F}}$ of width at most d with $O(d \cdot |V(G_{\mathcal{F}})|)$ nodes as input and solves k–WMAX–SAT, i.e., finds $S \subseteq V_{\mathcal{F}}$ such that $|S| \le k$ and setting variables of S to true and other variables to false maximizes the weight of the satisfied clauses.*

Proof. First, we construct a nice tree decomposition $\mathcal{T} = (T, \{X_t\}_{t \in V(T)})$ of width at most d with $O(d \cdot |V(G_{\mathcal{F}})|)$ nodes in time $O(d^3 \cdot |V(G_{\mathcal{F}})|)$ using Lemma 1. Then, we use a dynamic programming routine.

For each $t \in V(T)$ let $V_t \subseteq V(G_{\mathcal{F}}) = \mathcal{C}_{\mathcal{F}} \cup V_{\mathcal{F}}$ be the union of all the bags present in the subtree of T rooted at t, including X_t. For each $t \in V(T)$, $S \subseteq (X_t \cap V_{\mathcal{F}})$, $C \subseteq (X_t \cap \mathcal{C}_{\mathcal{F}})$ and $0 \le i \le k$ define the following:

$$dp[t, S, C, i] := \begin{array}{l} \textit{Maximum possible weight of satisfied clauses in } V_t \textit{ if we set at} \\ \textit{most } i \textit{ variable from } V_t \textit{ to true, set other variables of } V_t \textit{ to} \\ \textit{false and ignore variables of } V_{\mathcal{F}} \setminus V_t \textit{ such that } \hat{S} \cap X_t = S \textit{ and} \\ \hat{C} \cap X_t = C \textit{ where } \hat{S} \textit{ is the set of true variables and } \hat{C} \textit{ is the} \\ \textit{set of satisfied clauses in } V_t. \end{array}$$

If we manage to compute values of dp, then since $X_r = \emptyset$, where r is the root of T, the answer would be $dp[r, \emptyset, \emptyset, k]$ and we can fill the dp array in a bottom-up manner and in the following way:

- **Leaf node:** *If t is a leaf, $X_t = \emptyset$ and we have $dp[t, \emptyset, \emptyset, i] = 0$ for all $0 \le i \le k$. So in this case, filling each cell of dp takes $O(1)$ time.*
- **Introduce node:** *If t is an introduce node with child t' that $X_t = X_{t'} \cup \{v\}$, we consider two cases and fill the entries $dp[t, S, C, i]$ in the following way.*
  1. *$v \in V_{\mathcal{F}}$, i.e., v is a variable. Then $C' \subseteq C$ might be the set of satisfied clauses of $X_{t'}$, if it satisfies one of the two below conditions:*

(C1) All clauses in $C \setminus C'$ contain a positive literal of v, i.e., Setting v to true satisfies all clauses in $C \setminus C'$.

(C2) All clauses in $C \setminus C'$ contain a negative literal of v, i.e., setting v to false satisfies all clauses in $C \setminus C'$.

So we can compute $dp[t, S, C, i]$ as follows

$$\begin{cases} \max_{C' \text{ satisfies (C1)}} dp[t', S \setminus \{v\}, C', i-1] + w(C \setminus C') \text{ if } v \in S \\ \max_{C' \text{ satisfies (C2)}} dp[t', S, C', i] + w(C \setminus C') \hspace{1.2cm} \text{ if } v \notin S \end{cases}$$

So in this case, filling one cell of dp takes $O(2^d)$ time.

2. *$v \in C_{\mathcal{F}}$, i.e., v is a clause. Note that because of edge coverage and coherence properties, $Var(v) \cap V_t = Var(v) \cap X_t$ where $Var(v)$ is the set of variables present in the clause v, either as a positive or negative literal. So, there are two possibilities:*

(P1) $v \in C$ and v gets satisfied by setting all variables of S to true, $X_t \setminus S$ to false and ignoring variables of $V_{\mathcal{F}} \setminus X_t$.

(P2) $v \notin C$ and v is not satisfied by setting all variables of S to true, $X_t \setminus S$ to false and ignoring variables of $V_{\mathcal{F}} \setminus X_t$.

Therefore, we have:

$$dp[t, S, C, i] = \begin{cases} dp[t', S, C \setminus \{v\}, i] + w(v) \text{ if (P1) is true} \\ dp[t', S, C, i] \hspace{1.8cm} \text{if (P2) is true} \\ INVALID \hspace{2.2cm} \text{otherwise} \end{cases}$$

So, in this case filling one cell of dp takes $O(1)$ time.

Overall we can fill $dp[t, S, C, i]$ for an introduce node t in $O(2^d)$ time.

- **Forget node:** *If t is a forget node with child t' that $X_t = X_{t'} \setminus \{w\}$, we again consider two cases:*

  1. *$w \in V_{\mathcal{F}}$, i.e., w is a variable. Note that w is either set to true or false and therefore:*

  $$dp[t, S, C, i] = \max \begin{cases} dp[t', S, C, i] \hspace{1.2cm} \text{setting } w \text{ to false} \\ dp[t', S \cup \{w\}, C, i] \text{ setting } w \text{ to true} \end{cases}$$

  2. *$w \in C_{\mathcal{F}}$, i.e., w is a clause.*

  $$dp[t, S, C, i] = \max \begin{cases} dp[t', S, C, i] \hspace{1.2cm} w \text{ does not get satisfied} \\ dp[t', S, C \cup \{w\}, i] \hspace{0.1cm} w \text{ gets satisfied} \end{cases}$$

Note that in this case filling one cell of dp takes $O(1)$ time.

- **Join node:** *If t is a join node with children t_1 and t_2 that $X_t = X_{t_1} = X_{t_2}$, we consider all possibilities of S_1, S_2 and C_1, C_2, and compute the value of $dp[t, S, C, i]$ by:*

$$\max_{S_1 \cup S_2 = S, \ C_1 \cup C_2 = C, \ |S_1| \leq j \leq i} \begin{pmatrix} dp[t_1, S_1, C_1, j] \\ + dp[t_2, S_2, C_2, i - j + |S_1 \cap S_2|] \\ - w(C_1 \cap C_2) \end{pmatrix}$$

So in the case of join nodes, we can compute the value of each cell of dp in $O(2^{2d})$, because of 2^d possibilities for $S_1 \cup C_1$ and at most 2^d possibilities for $S_2 \cup C_2$.

The total number of array's cells is $O(|V(T)| \cdot 2^d \cdot k)$ and we can fill each cell in time $O(2^{2d})$, since by Lemma 1 $|V(T)| = O(d \cdot |V(G_{\mathcal{F}})|) = O(d \cdot |C_{\mathcal{F}} \cup V_{\mathcal{F}}|)$ we can fill all the cells in time $O(2^{3d} \cdot kd \cdot |C_{\mathcal{F}} \cup V_{\mathcal{F}}|)$. Again by Lemma 1, constructing T is done in time $O(d^3 \cdot |C_{\mathcal{F}} \cup V_{\mathcal{F}}|$ which gives us the overall runtime of $O(2^{3d} \cdot kd \cdot |C_{\mathcal{F}} \cup V_{\mathcal{F}}|)$.

Finally, using the standard technique of backlinks, i.e., memorizing for every cell of dp how its value was obtained, we can find an optimal solution, i.e., a subset $S \subseteq V_{\mathcal{F}}$ such that $|S| \leq k$ and setting its variables to true maximizes the weight of the satisfied clauses, within the same running time.

4.1 FPT Algorithm

Here, we use Lemma 6 and Lemma 7 to show that k–WMAX–SAT on planar formulas is FPT. That is we prove Theorem 2. For convenience, we restate the theorem here.

Theorem 2. *Given a set of t clauses $C_{\mathcal{F}} = \{C_1, C_2, \ldots, C_t\}$ of a **planar** CNF formula $\mathcal{F}$, a weight function $w : C_{\mathcal{F}} \to \mathbb{R}^+$ and a positive integer k, there is an FPT algorithm for k–WMAX–SAT that runs in $O(2^{36k} \cdot k^3 \cdot |C_{\mathcal{F}} \cup V_{\mathcal{F}}|)$ time.*

Proof. Construct $G_{\mathcal{F}}$ and without loss of generality suppose the graph is connected. Then, do a breadth-first search (BFS) on the graph starting from an arbitrary variable. Since $G_{\mathcal{F}}$ is bipartite the first level would contain variables, the second level would contain clauses, the third level would contain variables, etc.

If the number of levels is more than $2k$, for each $0 \leq i$ label the level $2i + 1$, which contains variables, with $[i \mod (k+1)]$. Note that since the number of levels is at least $2k + 1$, we would use all the $k + 1$ different labels and therefore there should be a label that all of its variables are set to false in the optimal answer. We consider all the $k + 1$ possibilities for this label and each time, set variables of one of the $k + 1$ labels, say label l, to false. This makes some clauses satisfied, then we remove variables with label l and also satisfied clauses to get a new graph $G_{\mathcal{F},l}$. Each connected component of $G_{\mathcal{F},l}$ would contain at most $2k + 1$ levels and therefore its diameter is at most $4k$. Using Lemma 6 a tree decomposition of $G_{\mathcal{F},l}$ with width at most $12k$ can be found in time $O(k \cdot |V_{\mathcal{F}} \cup C_{\mathcal{F}}|)$, and thus with $O(k \cdot |V_{\mathcal{F}} \cup C_{\mathcal{F}}|)$ nodes. Then using Lemma 7 we can solve k–WMAX–SAT on the CNF formula induced by $G_{\mathcal{F},l}$ in time $O(2^{36k} \cdot k^2 \cdot |C_{\mathcal{F}} \cup V_{\mathcal{F}}|)$. By doing so for every label $0 \leq l < k + 1$, we can find the optimal solution in time $O(2^{36k} \cdot k^3 \cdot |C_{\mathcal{F}} \cup V_{\mathcal{F}}|)$.

If the number of levels is at most $2k$, we can use Lemma 6 and Lemma 7 on $G_{\mathcal{F}}$ directly.

4.2 Polynomial-Time Approximation Scheme

Now, we prove Theorem 3. For convenience, we restate the theorem here.

Theorem 3. *Given a set of t clauses $\mathcal{C}_{\mathcal{F}} = \{C_1, C_2, \dots, C_t\}$ of a **planar** CNF formula $\mathcal{F}$, a weight function $w : \mathcal{C}_{\mathcal{F}} \to \mathbb{R}^+$ and a positive integer k, there is a polynomial-time approximation scheme that runs in $O(\frac{1}{\epsilon^2} \cdot 2^{\frac{36}{\epsilon}} \cdot k \cdot |\mathcal{C}_{\mathcal{F}} \cup V_{\mathcal{F}}|)$ time and finds $S \subseteq V_{\mathcal{F}}$ such that $|S| \leq k$ and*

$$k\text{-}WMAX\text{-}SAT(\mathcal{C}_{\mathcal{F}}, w, k, S) \geq (1 - \epsilon) \cdot OPT(\mathcal{C}_{\mathcal{F}}, w, k)$$

Proof. Fix an arbitrary $0 < \epsilon \leq 1$, let $d = \lceil \frac{1}{\epsilon} \rceil$ and suppose $S^* \subseteq V_{\mathcal{F}}$ is an optimal solution to k–WMAX–SAT on $(\mathcal{F}, w, k)$, i.e., $|S^*| \leq k$ and setting variables of S^* to true maximizes the weight of the satisfied clauses. Also, let $\mathcal{C}^*$ be the set of clauses that get satisfied by setting variables of S^* to true. Construct $G_{\mathcal{F}}$ and without loss of generality suppose the graph is connected. Then, do a breadth-first search (BFS) on the graph starting from an arbitrary clause.

If the number of levels is at least $2d$, for each $0 \leq i$ label the level $2i + 1$, which contains clauses, with $[i \mod d]$. Let $\mathcal{C}_{\mathcal{F},l}$ be the set of all clauses with label l. Note that since the number of levels is at least $2d$, we would use all the d different labels and therefore there should be a label l^* such that $w(\mathcal{C}^* \cap \mathcal{C}_{\mathcal{F},l}) \leq \frac{w(\mathcal{C}^*)}{d} = \frac{OPT(\mathcal{C}_{\mathcal{F}}, w, k)}{d}$.

We consider all the d possibilities for l^* and each time remove clauses with one of the labels, say label l, to get a new graph $G_{\mathcal{F},l}$. Each connected component of $G_{\mathcal{F},l}$ contains at most $2d$ levels, and therefore its diameter is at most $4d$.

Using Lemma 6 a tree decomposition of $G_{\mathcal{F},l}$ with width at most $12d$ can be found in time $O(d \cdot |V_{\mathcal{F}} \cup \mathcal{C}_{\mathcal{F}}|)$ and thus with $O(d \cdot |V_{\mathcal{F}} \cup \mathcal{C}_{\mathcal{F}}|)$ nodes. Then using Lemma 7 we can solve k–WMAX–SAT on the CNF formula induced by $G_{\mathcal{F},l}$ in time $O(2^{36d} \cdot kd \cdot |\mathcal{C}_{\mathcal{F}} \cup V_{\mathcal{F}}|)$. Let S_l be the optimal solution of k–WMAX–SAT on the CNF formula induced by $G_{\mathcal{F},l}$ and let k–WMAX–SAT$(\mathcal{C}, w, k, S)$ be the weight of satisfied clauses in $\mathcal{C} \subseteq \mathcal{C}_{\mathcal{F}}$ if we set variables of S to true. Then we have the following for every label $0 \leq l < d$:

$$
\begin{aligned}
k\text{-WMAX-SAT}(\mathcal{C}_{\mathcal{F}}, w, k, S_l) &\geq k\text{-WMAX-SAT}(\mathcal{C}_{\mathcal{F}} \setminus \mathcal{C}_{\mathcal{F},l}, w, k, S_l) \\
&\geq k\text{-WMAX-SATT}(\mathcal{C}_{\mathcal{F}} \setminus \mathcal{C}_{\mathcal{F},l}, w, k, S^*) \\
&= k\text{-WMAX-SAT}(\mathcal{C}_{\mathcal{F}}, w, k, S^*) - w(\mathcal{C}^* \cap \mathcal{C}_{\mathcal{F},l}) \\
&= OPT(\mathcal{C}_{\mathcal{F}}, w, k) - w(\mathcal{C}^* \cap \mathcal{C}_{\mathcal{F},l})
\end{aligned}
$$

And for l^* we also have:

$$
\begin{aligned}
k\text{-WMAX-SAT}(\mathcal{C}_{\mathcal{F}}, w, k, S_{l^*}) &\geq OPT(\mathcal{C}_{\mathcal{F}}, w, k) - w(\mathcal{C}^* \cap \mathcal{C}_{\mathcal{F},l^*}) \\
&\geq OPT(\mathcal{C}_{\mathcal{F}}, w, k) - \frac{OPT(\mathcal{C}_{\mathcal{F}}, w, k)}{d} \\
&\geq (1 - \epsilon) \cdot OPT(\mathcal{C}_{\mathcal{F}}, w, k)
\end{aligned}
$$

Therefore, by finding S_l for every label $0 \leq l < d$, we can find the optimal solution in time $O(\frac{1}{\epsilon^2} \cdot 2^{\frac{36}{\epsilon}} \cdot k \cdot |\mathcal{C}_{\mathcal{F}} \cup V_{\mathcal{F}}|)$.

5   Conclusion

In this work, we showed that k–MAX–2SAT admits an EPSAKS of size $O(\frac{k^5}{\epsilon^2})$. As the monotone variant of the problem, MAXIMUM k–VERTEX COVER, admits an EPSAKS of size $O(\frac{k}{\epsilon})$ [13], which also works for weighted graphs, is it possible to improve the kernel size for k–MAX–2SAT or design an EPSAKS for its weighted version?

We also showed that k–WMAX–SAT on planar graphs admits an FPT algorithm as well as a PTAS. Does this problem also admit a kernelization?

References

1. Austrin, P., Stankovic, A.: Global cardinality constraints make approximating some max-2-csps harder. In: Achlioptas, D., Végh, L.A. (eds.) Approximation, Randomization, and Combinatorial Optimization. Algorithms and Techniques, APPROX/RANDOM 2019, Massachusetts Institute of Technology, Cambridge, MA, USA, 20–22 September 2019. LIPIcs, vol. 145, pp. 24:1–24:17. Schloss Dagstuhl - Leibniz-Zentrum für Informatik (2019). https://doi.org/10.4230/LIPIcs.APPROX-RANDOM.2019.24
2. Baker, B.S.: Approximation algorithms for np-complete problems on planar graphs. J. ACM **41**(1), 153–180 (1994). https://doi.org/10.1145/174644.174650
3. Bläser, M., Manthey, B.: Improved approximation algorithms for max-2SAT with cardinality constraint. In: Bose, P., Morin, P. (eds.) ISAAC 2002. LNCS, vol. 2518, pp. 187–198. Springer, Heidelberg (2002). https://doi.org/10.1007/3-540-36136-7_17
4. Crescenzi, P., Trevisan, L.: Max np-completeness made easy. Theor. Comput. Sci. **225**(1–2), 65–79 (1999). https://doi.org/10.1016/S0304-3975(98)00200-X
5. Cygan, M., et al.: Parameterized Algorithms. Springer, Cham (2015). https://doi.org/10.1007/978-3-319-21275-3
6. Eppstein, D.: Subgraph isomorphism in planar graphs and related problems. In: Proceedings of the Sixth Annual ACM-SIAM Symposium on Discrete Algorithms, SODA 1995, p. 632–640. Society for Industrial and Applied Mathematics, USA (1995)
7. Feige, U.: A threshold of ln n for approximating set cover. J. ACM **45**(4), 634–652 (1998). https://doi.org/10.1145/285055.285059
8. Fomin, F.V., Lokshtanov, D., Saurabh, S., Zehavi, M.: Kernelization: Theory of Parameterized Preprocessing. Cambridge University Press, Cambridge (2019). https://doi.org/10.1017/9781107415157
9. Hofmeister, T.: An approximation algorithm for MAX-2-SAT with cardinality constraint. In: Di Battista, G., Zwick, U. (eds.) ESA 2003. LNCS, vol. 2832, pp. 301–312. Springer, Heidelberg (2003). https://doi.org/10.1007/978-3-540-39658-1_29
10. Jain, P., et al.: Parameterized approximation scheme for biclique-free max k-weight SAT and max coverage, pp. 3713–3733. https://doi.org/10.1137/1.9781611977554.ch143. https://epubs.siam.org/doi/abs/10.1137/1.9781611977554.ch143
11. Khanna, S., Motwani, R.: Towards a syntactic characterization of ptas. In: Proceedings of the Twenty-Eighth Annual ACM Symposium on Theory of Computing, STOC 1996, pp. 329–337. Association for Computing Machinery, New York (1996). https://doi.org/10.1145/237814.237979

12. Lokshtanov, D., Panolan, F., Ramanujan, M.S., Saurabh, S.: Lossy kernelization. In: Hatami, H., McKenzie, P., King, V. (eds.) Proceedings of the 49th Annual ACM SIGACT Symposium on Theory of Computing, STOC 2017, Montreal, QC, Canada, 19–23 June 2017, pp. 224–237. ACM (2017). https://doi.org/10.1145/3055399.3055456

13. Manurangsi, P.: A note on max k-vertex cover: faster fpt-as, smaller approximate kernel and improved approximation. In: Fineman, J.T., Mitzenmacher, M. (eds.) 2nd Symposium on Simplicity in Algorithms, SOSA 2019, San Diego, CA, USA, 8–9 January 2019. OASIcs, vol. 69, pp. 15:1–15:21. Schloss Dagstuhl - Leibniz-Zentrum für Informatik (2019). https://doi.org/10.4230/OASIcs.SOSA.2019.15

14. Manurangsi, P.: Tight running time lower bounds for strong inapproximability of maximum k-coverage, unique set cover and related problems (via t-wise agreement testing theorem). In: Chawla, S. (ed.) Proceedings of the 2020 ACM-SIAM Symposium on Discrete Algorithms, SODA 2020, Salt Lake City, UT, USA, 5–8 January 2020, pp. 62–81. SIAM (2020). https://doi.org/10.1137/1.9781611975994.5

15. Marx, D.: Parameterized complexity and approximation algorithms. Comput. J. **51**(1), 60–78 (2008). https://doi.org/10.1093/comjnl/bxm048

16. Panolan, F., Yaghoubizade, H.: Partial vertex cover on graphs of bounded degeneracy. In: Kulikov, A.S., Raskhodnikova, S. (eds.) Computer Science - Theory and Applications - 17th International Computer Science Symposium in Russia, CSR 2022, Virtual Event, 29 June–1 July 2022, Proceedings. Lecture Notes in Computer Science, vol. 13296, pp. 289–301. Springer, Heidelberg (2022). https://doi.org/10.1007/978-3-031-09574-0_18

17. Raghavendra, P., Tan, N.: Approximating csps with global cardinality constraints using SDP hierarchies. In: Rabani, Y. (ed.) Proceedings of the Twenty-Third Annual ACM-SIAM Symposium on Discrete Algorithms, SODA 2012, Kyoto, Japan, 17–19 January 2012, pp. 373–387. SIAM (2012). https://doi.org/10.1137/1.9781611973099.33

18. Sviridenko, M.: Best possible approximation algorithm for MAX SAT with cardinality constraint. Algorithmica **30**(3), 398–405 (2001). https://doi.org/10.1007/s00453-001-0019-5

Minimizing Corners in Colored Rectilinear Grids

Thomas Depian[1], Alexander Dobler[1], Christoph Kern[1], and Jules Wulms[2(✉)]

[1] TU Wien, Vienna, Austria
{e11807882,e11904675}@student.tuwien.ac.at, adobler@ac.tuwien.ac.at
[2] TU Eindhoven, Eindhoven, Netherlands
j.j.h.m.wulms@tue.nl

Abstract. Given a rectilinear grid G, in which cells are either assigned a single color, out of k possible colors, or remain white, can we color white grid cells of G to minimize the total number of corners of the resulting colored rectilinear polygons in G? We show how this problem relates to hypergraph visualization, prove that it is NP-hard even for $k = 2$, and present an exact dynamic programming algorithm. Together with a set of simple kernelization rules, this leads to an FPT-algorithm in the number of colored cells of the input. We additionally provide an XP-algorithm in the solution size, and a polynomial $\mathcal{O}(OPT)$-approximation algorithm.

Keywords: Shape complexity · Rectilinear polygons · Set visualization

1 Introduction

Hypergraphs are a prominent way of modeling set systems. In a hypergraph, vertices correspond to set elements and hyperedges represent sets. To gain insight into the structure of hypergraphs, many hypergraph visualizations have been developed. These visualizations can be roughly subdivided into area-based visualizations [12,15,17], resembling the traditional Euler and Venn diagrams [3], edge-based techniques [1,10,12], where set elements are connected by links, and matrix-based approaches [14,16], in which columns and rows represent vertices and hyperedges. The surveys by Alsallakh et al. [2] and Fischer et al. [8] consider state-of-the-art set visualizations and their classification in more detail.

Most area- and edge-based hypergraph visualizations represent the vertices as points in the plane, and visualize hyperedges as either regions or connections, respectively. These hyperedges usually intersect at common vertices to convey membership, while other intersections are considered to violate *planarity* [11], a well-established quality criterion for graph drawings [13].

Other visualization techniques completely prevent visual intersections between hyperedge representations, such as the grid-based visualization introduced by van Goethem et al. [9]. In this visual encoding, hyperedges are represented by disjoint (connected) polygons and each vertex corresponds to a cell in

This work has been supported by the Vienna Science and Technology Fund (WWTF) under grant [10.47379/ICT19035].

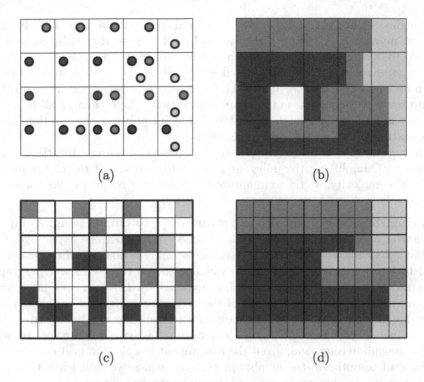

Fig. 1. (a) A grid with colors assigned to grid cells as in [9]. (b) A disjoint polygon visualization for Fig. 1a. (c) A subdivision of the grid in Fig. 1a; colors assigned in Fig. 1a are now assigned to a single cell after subdivision. (d) A coloring of Fig. 1c that minimizes the number of corners of the colored polygons.

a rectilinear grid. Membership is conveyed by overlap of such a polygon with a grid cell representing a vertex. In their setting, van Goethem et al. allow each hyperedge to overlap only those grid cells corresponding to incident vertices and all such cells must be overlapped, see Figs. 1a and 1b. Their input consists of a grid, in which each grid cell is assigned a subset of colors, and van Goethem et al. show how to test whether a disjoint polygon representation of two hyperedges can be realized, given such a grid and an assignment of colors to grid cells. They also prove that, if such a representation exists, the complexity within each grid cell can be bounded. Here, complexity refers to how many times a grid cell is intersected by hyperedge polygons.

In this paper, we further study these disjoint polygon hypergraph visualizations. A question left open in [9] is whether it can be beneficial to make use of empty grid cells, not assigned to any hyperedge: Can coloring such cells allow for more grids to result in valid visualizations, or can coloring these white cells reduce the visual complexity? We work towards answering the latter question, and to do so, we weaken multiple requirements with respect to the original setting. First, we allow hyperedge polygons to also overlap white/empty grid cells not assigned to any set. Second, we consider more than two sets, namely any constant number $k \geq 2$. And third, we no longer require that each set is represented by a single (connected) polygon. The former two adaptations create a more general problem, in which unused grid cells can be used as well. To deal with more than two sets, we allow each hyperedge representation to be broken up into multiple polygons via the latter adaptation. These changes allow us to represent more hypergraphs in a grid-based disjoint polygon visualization, even in the restricted case of only two hyperedges.

Thus, we study the *shape complexity* of polygons in the visualization: With the intent of simplifying the polygons representing sets, and thereby reducing the visual complexity, we try to minimize *the number of corners of the polygons*, and call this problem MINCORNER COMPLEXITY.

Adapting the Hypergraph Visualization of [9] to Our Setting. As input, we consider a grid in which each cell is assigned to at most one hyperedge. We say that cells assigned to a hyperedge are *colored*. This does not directly correspond to the original setting of [9], in which each cell represents a hypergraph vertex incident with potentially multiple hyperedges. However, our input can be obtained by subdividing the grid of the original setting, and assigning each hyperedge incident with a vertex to a distinct subcell (see Fig. 1c and 1d). This leads to two subproblems: finding a good/optimal assignment of colors to grid cells in the subdivision, and, given the assignment of colors to grid cells, coloring the grid to optimize the number of corners (which we called MINCORNER COMPLEXITY). In this paper, we consider only the latter subproblem, and we leave the former question as an open problem. Notice that, especially in sparsely colored grids, solving MINCORNER COMPLEXITY will likely lead to few polygons per hyperedge, similar to the goal of the original setting: A single polygon often has fewer corners than the sum of corners of the colored grid cells it encompasses. Thus, we tangentially still work towards colored grids with few polygons, sometimes even achieving the goal of the original setting.

Contributions. We formally define MINCORNER COMPLEXITY in Sect. 2 and introduce the necessary terminology to work towards our results. In Sect. 3 we show that the decision version of MINCORNER COMPLEXITY is NP-complete. Section 4 presents an exact dynamic programming algorithm with exponential running time and a polynomial-time $\mathcal{O}(OPT)$-approximation algorithm. This approximation is closer to optimal when the optimum is small. By introducing a set of simple kernelization rules, we show that MINCORNER COMPLEXITY is fixed-parameter tractable with respect to the number of colored cells, and give an XP-algorithm with respect to the number of corners in the solution in Sect. 5. We conclude the paper with future research directions in Sect. 6.

The proofs for statements marked with ($\star$) can be found in the full version [7].

2   Preliminaries

We denote the set $\{1, 2, \ldots, n\}$ by $[n]$ and $\{n, n+1, \ldots, m\}$ by $[n, m]$. Let $\mathcal{C} = [k]$ be a set of k (non-white) *colors*, encoded as integers. We define the color 0 to represent white and denote with $\mathcal{C}_0 := \mathcal{C} \cup \{0\}$ the set of colors including white. A matrix $G \in \mathcal{C}_0^{m \times n}$ represents a colored $m \times n$ *grid*, which we call a *coloring*. With $G_{i,\cdot}$, $i \in [m]$, we will address the i-th row of the coloring, with $G_{\cdot,j}$ the j-th column, and with $G_{i,j}$, $i \in [m]$, $j \in [n]$ one of its cells. Since we often address rows, we will use G_i if there is no risk of confusion. To address all rows from i to j, $i \le j$, in G, we use $G_{[i,j],\cdot}$, and analogously for columns $G_{\cdot,[i,j]}$, and to access a sub-grid we write $G_{[i_1,i_2],[j_1,j_2]}$. Throughout this paper, we implicitly assume that i and j are in the correct domain. We call a cell $G_{i,j}$ *colored* if $G_{i,j} \ne 0$, and otherwise we say that it is *white*. Let $\mathcal{K}$ be a non-empty subset of $\mathcal{C}_0$. A coloring H is a valid $\mathcal{K}$-*extension* of G, if it respects the color of the colored cells in G, i.e., for all colored cells $G_{i,j}$ we have $H_{i,j} = G_{i,j}$ and for the white cells we have $H_{i,j} \in \mathcal{K}$. If $\mathcal{K}$ is clear from context, it will be omitted. We denote with $\Gamma_{\mathcal{K}}(G)$ the set of all valid $\mathcal{K}$-extensions and use $\Gamma(G)$ as a shorthand for $\Gamma_{\mathcal{C}_0}(G)$.

Problem Description. As explained in the introduction, we aim to find extensions with few corners. Roughly speaking, a corner is a 90°- or 270°-angled bend of a color in the coloring, which always occurs at a center point of a 2×2-region of the grid. Let $\delta_c : \mathcal{C}_0^{2 \times 2} \to \mathbb{Z}_0^+$ be a function that counts the number of corners of a color $c \in \mathcal{C}$, i.e., the c-*corners*, at the center of such a 2×2-region. When counting c-corners, we can treat all other cells as white. We observe the following distinct scenarios (disregarding symmetries) in a 2×2-region.

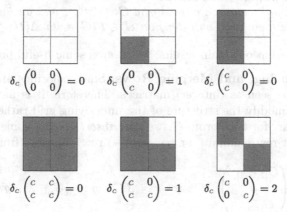

$$\delta_c \begin{pmatrix} 0 & 0 \\ 0 & 0 \end{pmatrix} = 0 \qquad \delta_c \begin{pmatrix} 0 & 0 \\ c & 0 \end{pmatrix} = 1 \qquad \delta_c \begin{pmatrix} c & 0 \\ c & 0 \end{pmatrix} = 0$$

$$\delta_c \begin{pmatrix} c & c \\ c & c \end{pmatrix} = 0 \qquad \delta_c \begin{pmatrix} c & 0 \\ c & c \end{pmatrix} = 1 \qquad \delta_c \begin{pmatrix} c & 0 \\ 0 & c \end{pmatrix} = 2$$

Above cases lead to the following definition of δ_c.

$$\delta_c(C) = |C_{1,1}^{=c} + C_{2,2}^{=c} - C_{1,2}^{=c} - C_{2,1}^{=c}|, \text{ where } C_{i,j}^{=c} := \begin{cases} 1 & \text{if } C_{i,j} = c \\ 0 & \text{otherwise} \end{cases}$$

In order to count all corners of a color c on the entire grid, we iterate over all 2×2-regions of the grid and sum up the number of corners of c. To ensure we do not miss corners on the boundary of the grid, we enlarge the $m \times n$ grid G in each direction by a row/column of white cells, resulting in the $(m + 2) \times (n + 2)$ grid G^P. This is equivalent to initializing G^P as a white $(m + 2) \times (n + 2)$ grid and setting $G^P_{[1,m],[1,n]} = G$. Observe that the added white rows/columns are in the row/column with index 0 and $m + 1/n + 1$. We use this slight abuse of notation to ease our arguments, as this preserves the row and column indices of G in G^P. We denote by g^P the analogous operation for a row g: We add one white cell left and right of g, such that $g^P = (G^P)_i$ for appropriate G and i.

Let $\Delta_c : C_0^{m \times n} \to \mathbb{Z}_0^+$ be a function that counts all corners of color $c \in C$ of a coloring G of the $m \times n$ grid. Formally, we can define Δ_c as follows:

$$\Delta_c(G) = \sum_{i=0}^{m} \sum_{j=0}^{n} \delta_c \left(G^P_{[i,i+1],[j,j+1]} \right).$$

The total number of corners, $\Delta : C_0^{m \times n} \to \mathbb{Z}_0^+$, is the sum of all non-white corners, i.e., $\Delta(G) = \sum_{c \in C} \Delta_c(G)$. To count the number of corners, of a particular color or in total, between two (consecutive) rows g and h, we use $\overline{\Delta_c}(g, h)$ and $\overline{\Delta}(g, h)$, respectively. With this we can formally define MINCORNER COMPLEXITY, with optimal extensions $\Gamma^*(G)$, and its decision variant CORNER COMPLEXITY.

Definition 1. MINCORNER COMPLEXITY
Given: *A set C of k colors and a colored grid $G \in C_0^{m \times n}$.*
Compute: *Extension $H^* \in \Gamma^*(G)$ s.t. $\Delta(H^*) \leq \Delta(H')$ for all $H' \in \Gamma(G)$.*

Definition 2. CORNER COMPLEXITY
Given: *A set C of k colors, a colored grid $G \in C_0^{m \times n}$, and an integer ℓ.*
Question: *Does there exist an extension $H \in \Gamma(G)$ with $\Delta(H) \leq \ell$?*

Before we present our main results, we discuss some useful properties of Δ.

Adding, Removing, and Merging Rows. Since colorings are defined on a grid, they can be seen as (integer) matrices. Therefore, it is natural to define operations that modify the structure of the underlying grid rather than its colors. In particular, for a coloring G, we can *insert* ($\succ/\curlyvee$) some other colored row/column g at row/column i, or *remove* ($-$) row/column i from G:

$$G \succ_i g := \begin{pmatrix} G_{[1,i-1],\cdot} \\ g \\ G_{[i,m],\cdot} \end{pmatrix}, \qquad G \curlyvee_i g := \left(G_{\cdot,[1,i-1]} \; g \; G_{\cdot,[i,m]} \right),$$

$$G_{(i,\cdot)-} := \begin{pmatrix} G_{[1,i-1],\cdot} \\ G_{[i+1,m],\cdot} \end{pmatrix}, \qquad G_{(\cdot,i)-} := \left(G_{\cdot,[1,i-1]} \; G_{\cdot,[i+1,m]} \right).$$

Lemma 1 ($\star$). *Let $G \in C_0^{m \times n}$, $g \in C_0^n$, and $i \in [m]$, then $\Delta(G) \leq \Delta(G \succ_i g)$.*

Lemma 2 ($\star$). *Let $G \in C_0^{m \times n}$ and $i \in [m]$, then $\Delta(G_{(i,\cdot)-}) \leq \Delta(G)$.*

Lemmata analogous to Lemma 1 and Lemma 2 can be proved for columns.

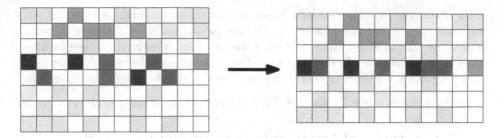

Fig. 2. The result of merging the two highlighted rows.

Finally, one can also *merge* ($\oplus$) the colorings of two adjacent rows (see Fig. 2). The merge operator $\oplus$ for row colorings $g, h \in \mathcal{C}_0^n$ is defined as

$$(g \oplus h)_i := \begin{cases} g_i & \text{if } g_i = h_i \vee h_i = 0 \\ h_i & \text{if } g_i = 0 \end{cases} , \text{ for } i \in [n].$$

Observe that $g \oplus h$ is undefined if we have $g_i \neq h_i$ and neither of g_i or h_i is 0, for some $i \in [n]$. In that case, we set $g \oplus h = \bot$. Additionally, we define that $g \oplus \bot = \bot \oplus g = \bot$, and $\Delta(\bot) = \infty$, and observe the following property.

Property 1. $\oplus$ is commutative and associative; the white row 0 is the identity.

Let $\left(G_{[i,j]}\right)^\oplus$, abbreviated as $G_{[i,j]}^\oplus$, denote the row coloring after consecutively merging rows i to j of G. We use $G^\oplus$ as a shorthand for $G_{[1,m]}^\oplus$ and define

$$G_{[i,j]}^\oplus := \begin{cases} G_i & \text{if } i = j \\ G_{[i,j-1]}^\oplus \oplus G_j & \text{otherwise.} \end{cases}$$

Bounds on the Number of Corners. While finding a minimum-corner extension of a grid G is NP-hard (see Sect. 3), we can prove bounds on the number of corners between two consecutive rows and in (extensions of) G in general.

First, it is easy to see that there are no corners between two identical row colorings. However, distinct rows have at least 2 corners between them.

Property 2. For a row coloring $g \in \mathcal{C}_0^n$ it holds that $\overline{\Delta}(g, g) = 0$.

Lemma 3 ($\star$). *For a coloring $G \in \mathcal{C}_0^{m \times n}$, if $G_i \neq G_{i+1}$ then $\overline{\Delta}(G_i, G_{i+1}) \geq 2$.*

Second, as a consequence of Lemma 1, we can bound the number of corners in G from below by considering the number of corners within an arbitrary row.

Lemma 4 ($\star$). *For any coloring $G \in \mathcal{C}_0^{m \times n}$ and row G_i, $\Delta(G_i) \leq \Delta(G)$ holds.*

Finally, we want to argue about the number of corners of a single row. Let $\eta : \mathcal{C}_0^n \rightarrow \mathcal{C}_0^n$ be a function that, for a given row coloring g, extends the coloring by doing one sweep over the cells of g from left to right and coloring each white cell the same color as the previous cell in the row. Additionally, we define $\eta(\bot) = \bot$. Intuitively, $\Delta(\eta(g)) \leq \Delta(g)$ holds since the number of colored rectangles in g never increases, but might decrease when two rectangles of the same color merge. This property generalizes further: For a single row g, $\eta(g)$ is an optimal extension.

Property 3 For any $g \in \mathcal{C}_0^n$ and $h \in \Gamma(g)$, it holds that $\Delta(\eta(g)) \leq \Delta(h)$.

3   Computational Complexity of CORNER COMPLEXITY

In this section, we show that CORNER COMPLEXITY, the decision variant of MINCORNER COMPLEXITY, is NP-complete. Whilst NP-membership follows from the corner-counting function, we show NP-hardness using a series of reductions. The base problem for our reduction is RESTRICTED PLANAR MONOTONE 3-BOUNDED 3-SAT (see Sect. 3.1), a variant of 3-SAT. The centerpiece of this section is the reduction of the aforementioned problem to RESTRICTED c-CORNER COMPLEXITY, a restricted variant of CORNER COMPLEXITY (Sect. 3.2). The final step is to reduce to CORNER COMPLEXITY. The reduction effectively uses only two colors, c and c', which we sometimes call (b)lue and (r)ed, respectively.

3.1   RESTRICTED PLANAR MONOTONE (RPM) 3-BOUNDED 3-SAT

An instance of RPM 3-BOUNDED 3-SAT is a monotone Boolean formula φ in 3-CNF over variables $\mathcal{X} = \{x_1, \ldots, x_n\}$: Each clause of φ has only *positive* or only *negative* literals, forming the sets $\mathcal{P}$ and $\mathcal{N}$ of positive and negative clauses, respectively. Furthermore, φ is *3-bounded*: each variable appears in at most three positive and in at most three negative clauses. Let the graph $\mathcal{G}(\varphi)$ be the incidence graph of φ. We require that $\mathcal{G}(\varphi)$ has a *restricted planar rectilinear embedding*. This means that we can embed $\mathcal{G}(\varphi)$ on a rectilinear grid of polynomial size in the plane [5, Section 3], separating the positive from the negative clauses on different sides of the variables. See Fig. 6 for a typical restricted planar rectilinear embedding of $\mathcal{G}(\varphi)$ [4].

Definition 3 RPM 3-BOUNDED 3-SAT
Given: *A monotone Boolean 3-bounded formula φ and a restricted planar rectilinear embedding of the associated incidence graph $\mathcal{G}(\varphi)$.*
Question: *Is φ satisfiable?*

Darmann and Döcker [6] showed that this problem is NP-complete (even when a variable may appear in at most p positive and at most q negative clauses).

3.2   Via RESTRICTED c-CORNER COMPLEXITY To CORNER COMPLEXITY

RESTRICTED c-CORNER COMPLEXITY is a restricted variant of CORNER COMPLEXITY that uses only two distinct colors c and c'. Only color c can be used to find an extension with at most ℓ c-corners, and all grid corners must be c'-colored.

Definition 4 RESTRICTED c-CORNER COMPLEXITY
Given: *Coloring $G \in \{c, c'\}_0^{m \times n}$, with c'-colored grid corners, and an integer ℓ.*
Question: *Does there exist a valid extension $H \in \Gamma_{\{c,0\}}(G)$ s.t. $\Delta_c(H) \leq \ell$?*

Since we can only color white cells in c or leave them white, the white cells can be used to connect c-colored cells into larger shapes to reduce the number of c-corners. The c'-colored cells can be seen as obstacles for those connections.

To show NP-hardness of RESTRICTED c-CORNER COMPLEXITY we reduce from RPM 3-BOUNDED 3-SAT, and first create variable and clause gadgets.

Variable Gadget. Figure 3a shows the layout of the variable gadget, consisting of two 3×3-checkerboard-like patterns on the top-left and bottom-right quadrants with pathways between them over the other two quadrants. In the top and bottom row are three white cells each, which we refer to as *outlets*. These act as the connection points to the clause gadgets. As each variable occurs at most three times in both positive and negative clauses, three outlets suffice. When considering various $\{b, 0\}$-extensions over the variable gadget, we observe that inside a 3×3-checkerboard-like pattern, it is beneficial to connect the blue cells in rows or columns to reduce the number of blue corners. We can further reduce corners by connecting a row from one 3×3-checkerboard-like pattern with a column from the other, using the pathways (see Fig. 3). Then, at least one side will not have colored outlets in a minimum $\{b, 0\}$-extension. Due to the constant size of the gadget, we can prove this by enumerating all $\{b, 0\}$-extensions.

Lemma 5 $(\star)$. *Any minimum c-corner extension $H^* \in \Gamma_{\{c,0\}}^*(G_{x_i})$ of variable gadget G_{x_i} has (1) $\Delta_c(H^*) = 18$, and (2) colored outlets on at most one side.*

We want to emphasize two minimum extensions that represent the true and false states of a variable (see Fig. 3). While other minimum extensions exist, they can always be replaced by the true or false extensions.

Clause Gadget. Figure 4 shows the layout of a clause gadget: One blue cell with a line of white cells to its right. These cells are surrounded by red cells except for outlets at the bottom (positive clause) or top (negative clause), one for each clause literal. Each outlet is connected by a vertical pathway to an outlet of the corresponding variable gadget. Any minimum blue corner extension of a clause

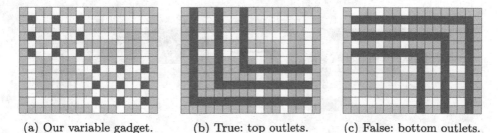

(a) Our variable gadget.   (b) True: top outlets.   (c) False: bottom outlets.

Fig. 3. A variable gadget and its true and false state, coloring different outlets.

Fig. 4. A positive clause gadget.

gadget contributes at most four corners, as it can leave all white cells white. If the outlet of a variable gadget is colored, we can extend the clause coloring as in Fig. 5 to reduce the number of corners by two. We cannot eliminate more than two blue corners, as the two blue corners left of the initial blue cell always remain. Lastly, we cannot remove any corner if no outlet is colored.

Lemma 6 (⋆). *Any minimum c-corner extension $H^* \in \Gamma^*_{\{c,0\}}(G_C)$ of a clause gadget G_C contributes either (1) two c-corners if it is connected to at least one colored outlet of a variable gadget, or (2) it contributes four c-corners.*

Complete Construction. For a given instance $(\varphi, \mathcal{G}(\varphi))$ of RPM 3-BOUNDED 3-SAT we construct a coloring G for a bounded grid as shown in Fig. 6.

To do so, we first create a variable gadget for each variable $x_i \in \mathcal{X}$ and place it at the rectangular vertex representing x_i in $\mathcal{G}(\varphi)$. Next, we create a clause gadget for each clause $C \in \varphi$ and place it at the position of C in $\mathcal{G}(\varphi)$. The gadgets determine the size of our grid and we color the remaining area red while ensuring that the vertical pathways between clause gadgets and variable gadgets remain white. This process takes polynomial time and results in a polynomial-sized grid with grid corners colored red. The outcome is a valid instance $(G, 18n + 2m)$ of RESTRICTED c-CORNER COMPLEXITY, for $c = (b)$lue and $c' = (r)$ed.

To show that this yields a correct reduction, we observe in one direction that, starting with a truth assignment T over $\mathcal{X}$ that satisfies every clause in φ, we can color the variable gadgets according to the states in Fig. 3. As φ is satisfied, we have for each clause C a variable x_i that satisfies C. We then color the pathway from the single blue cell of G_C to the outlet of G_{x_i}. Due to Lemma 5, each variable gadget admits 18 blue corners. As at least one outlet is colored for each clause gadget, Case (1) of Lemma 6 applies and thus each clause gadget contributes two additional blue corners, resulting in $18n + 2m$ corners.

For the other direction, Lemma 5 tells us that in any witness extension the variable gadget has at least 18 blue corners and blue outlets at most on one side.

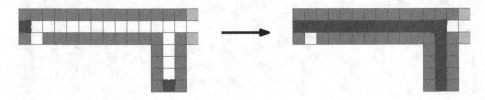

Fig. 5. Connecting the blue clause gadget cell with a blue variable gadget outlet. (Color figure online)

Thus, each clause gadget may contribute only two additional blue corners, which due to Lemma 6, occurs only if each clause gadget is connected to a colored outlet of a variable gadget. Therefore, we can create a satisfying truth assignment T for φ that mimics the variable gadget state for x_i by setting $T(x_i) = $ true, if G_{x_i} has a colored outlet at the top (or no outlet at all), and $T(x_i) = $ false otherwise.

Theorem 1 ($\star$). RESTRICTED c-CORNER COMPLEXITY *is NP-hard.*

To complete the overall reduction, we introduce the notion of *internal* corners, which are all corners *except* those at the corner of the grid in G. We denote with $\Delta_c^- : C_0^{m \times n} \to \mathbb{Z}_0^+$ the number of internal corners for a color $c \in C$, which is

$$\Delta_c^-(G) := \Delta_c(G) - \sum_{\substack{i \in \{0,m\} \\ j \in \{0,n\}}} \delta_c \left(G^P{}_{[i,i+1],[j,j+1]} \right).$$

Here, Δ^- is defined analogously to Δ. For two colors c and c', we can see that in any $\{c, c'\}$-extension of G every internal c-corner is also an internal c'-corner and vice versa, since there are no white cells. This brings us to Property 4, which we use to prove Lemma 7: The number of corners of a coloring G does not increase if we color all white cells with the color that has more corners in G.

Property 4. Any $\{c, c'\}$-extension H of $G \in \{c, c'\}_0^{m \times n}$ has $\Delta_c^-(H) = \Delta_{c'}^-(H)$.

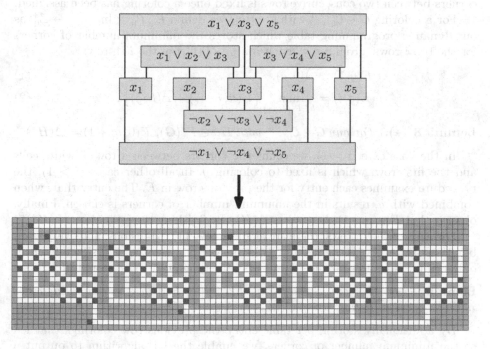

Fig. 6. Construction of coloring G from a monotone 3-bounded formula φ.

Lemma 7 ($\star$). *Let $G \in \{c, c'\}_0^{m \times n}$ be a coloring in which all grid corners are colored in c'. There exists an extension $H \in \Gamma_{\{c,c'\}}(G)$ s.t. $\Delta(H) \leq \Delta(G)$.*

To show NP-hardness of CORNER COMPLEXITY, we make use of Lemma 7. Take an instance (G, ℓ) of RESTRICTED c-CORNER COMPLEXITY and define the instance $(G, 2\ell + 4)$ of CORNER COMPLEXITY. Correctness follows by pairing off the internal c- and c'-corners in a completely colored grid (Property 4) in addition to the four c'-corners we have in RESTRICTED c-CORNER COMPLEXITY at the corners of G. For NP-membership, notice that Δ is evaluated in polynomial time.

Theorem 2 ($\star$). CORNER COMPLEXITY *is NP-complete, even for $k = 2$.*

4   Computing Low-Complexity Extensions

Despite the hardness result in Sect. 3, our goal is still to compute extensions with few corners. While the arithmetically simple corner counting function naturally lends itself to integer linear programming to find an optimal solution (see the full version [7]), in this section, we focus on dynamic programming (DP).

4.1   Exact Dynamic Programming Algorithm

A core observation utilized by our exact DP-algorithm is that once a 2×2-region is assigned a fixed coloring, the number of corners at the center of the 2×2-region will not change again. This property can be scaled up for rows: The number of corners between two consecutive rows is fixed once a coloring has been assigned.

For a coloring $G \in \mathcal{C}_0^{m \times n}$ with $\mathcal{C} = [k]$, we define $E : \mathcal{C}_0^n \times [m+1] \to \mathbb{Z}_0^+$ as our dynamic programming table which stores the minimum number of corners for the top i rows, given a row extension $g \in \Gamma(G_i)$ of the i-th row:

$$E(g, 1) = \overline{\Delta}(0, g), \tag{1}$$

$$E(g, i) = \min_{h \in \Gamma(G_{i-1})} \left\{ E(h, i-1) + \overline{\Delta}(h, g) \right\}. \tag{2}$$

Lemma 8 ($\star$). *For any $G \in \mathcal{C}_0^{m \times n}$ with $H^* \in \Gamma^*(G)$, $E(0, m+1) = \Delta(H^*)$.*

In the base case ($i = 1$), we count the corners between a row of white cells and the first row, which is fixed to coloring g. In all other cases ($i > 1$), the procedure examines each entry for the previous row in E. The entry that, when combined with g, results in the minimum number of corners is chosen. Finally, observe that $E(0, m+1)$ then stores $\Delta(H^*)$ by finishing the evaluation of $(H^*)^P$.

For each of the m rows, there are at most $(k+1)^n$ extensions, and every combination of two rows is checked in $\mathcal{O}(n)$ time per pair. As the recursion references only the previous row, at most two rows at a time need to be stored in E.

Lemma 9 ($\star$). *For any $G \in \mathcal{C}_0^{m \times n}$ with $\mathcal{C} = [k]$, $E(0, m+1)$ can be computed in $\mathcal{O}((k+1)^{2n} mn)$ time using $\mathcal{O}((k+1)^n)$ space.*

By additionally storing per table entry the previous row colorings that led to the minimum number of corners, we enable the DP-algorithm to output a minimum corner extension. This increases space usage to $\mathcal{O}((k+1)^n mn)$.

4.2   Approximation Algorithm

Using an alternative dynamic programming algorithm we can approximate the optimal solution in polynomial time. We leverage the observation that in an optimal extension $H^* \in \Gamma^*(G)$, there are often identical consecutive rows.

Let H' represent the coloring obtained by retaining only one representative row from each set of identical consecutive rows in H^*. By Lemma 2, we know that removing rows does not increase the number of corners. In H', any two consecutive rows cannot be identical, and thus, there must be at least two corners between any pair of consecutive rows in accordance to Lemma 3. Consequently, H' may only have at most $\frac{1}{2}\Delta(H^*)$ rows in total. Furthermore, each row by itself can have only $\Delta(H^*)$ corners, as per Lemma 4.

The primary objective of our approximation algorithm is to identify a partition of all rows into sets of consecutive rows that can be merged. By the arguments above, we will find a partition into at most $\frac{1}{2}\Delta(H^*)$ sets, each of which consists of mergeable rows. Rows in such a set can hence be colored identically, with at most $\Delta(H^*)$ corners, leading to at most $\frac{1}{2}\Delta(H^*)^2$ corners in total.

Let $G \in \mathcal{C}_0^{m \times n}$ be a coloring of $\mathcal{C} = [k]$. We define $A : [0, m] \to \mathbb{Z}_0^+$ as our dynamic programming table, which for each row $i \in [m]$ stores an approximation of the number of corners for a minimum corner extension of rows 1 to i:

$$A(0) = 0, \tag{3}$$

$$A(i) = \min_{j \in [0, i-1]} \left\{ A(j) + \Delta \left(\eta \left(G^{\oplus}_{[j+1, i]} \right) \right) \right\}. \tag{4}$$

Lemma 10 ($\star$).   For $G \in \mathcal{C}_0^{m \times n}$ with $H^* \in \Gamma^*(G)$, $\Delta(H^*) \leq A(m) \leq \frac{1}{2}\left(\Delta(H^*)\right)^2$.

Each entry in the dynamic programming table can be computed in $\mathcal{O}(mn)$ time by iteratively merging the rows inside the min-function of Eq. 4.

Lemma 11 ($\star$).   For any $G \in \mathcal{C}_0^{m \times n}$, $A(m)$ can be computed in $\mathcal{O}(m^2 n)$ time using $\mathcal{O}(n + m)$ additional space.

5   Parameterized Complexity

We now investigate the complexity of MINCORNER COMPLEXITY with respect to the number of colored cells, and to the number of corners in the solution.

5.1   FPT in the Number of Colored Cells due to Kernelization

We propose a kernelization procedure for our problem that involves the exhaustive application of the following two kernelization rules on a coloring $G \in \mathcal{C}_0^{m \times n}$.

Rule 1: If there is an empty row or column, remove it from the grid.
Rule 2: If there are two consecutive rows or columns that only contain cells of a singular color $c \in [k]$ and white (0), merge them.

We denote the resulting coloring by G'. To show that both rules are *safe*, we prove that the number of corners in optimal solutions of G and G' are equal.

To show the safety of Rule 1, we can utilize Lemma 2, which states that removing rows does not increase corners, and Property 2, which observes that there are no corners between two identical rows. Safety of Rule 2 can be shown by similar, slightly more sophisticated arguments.

Lemma 12 ($\star$). *Kernelization rules 1 and 2 are safe.*

Trivially, each rule application takes polynomial time and can be applied at most a polynomial amount of times. Thus, the entire kernelization procedure runs in polynomial time. Furthermore, the size of the kernel can be expressed by a polynomial on the number of colored cells in G. Specifically, Rule 1 ensures that every row/column has a colored cell. Furthermore, let $c^* \in C$ be the color that has the most rows and columns where it occurs as a singular color (besides white), and let $C' = C \backslash \{c^*\}$. If there are r C'-colored cells, then the rows/columns containing these cells form boundaries between the rows/columns with c^*-colored cells merged by Rule 2: For two colors, the kernel size depends on only one color.

Lemma 13 ($\star$). *Exhaustively applying Rules 1 and 2 on a coloring $G \in C_0^{m \times n}$, with r C'-colored cells, results in a kernel of size at most $\mathcal{O}(r) \times \mathcal{O}(r)$.*

By applying the exact DP-algorithm from Sect. 4.1 to the obtained kernel, we show that MINCORNER COMPLEXITY is FPT in the number of colored cells of G.

5.2 XP in the Solution Size

We construct an XP-algorithm, which decides, for a given coloring $G \in C_0^{m \times n}$ with $C = [k]$, and an integer ℓ as parameter, whether there exists an extension $H \in \Gamma(G)$ such that $\Delta(H) \leq \ell$. The algorithm is a modification of the algorithm presented in Sect. 4.1. Utilizing Lemma 4, we generate only row extensions for each row of G, which, by themselves, will not admit more than ℓ corners.

Lemma 14 ($\star$). *Let $\Gamma'(g)$ be the set of all possible extensions of $g \in C_0^n$ of $C = [k]$ such that $\Delta(h) \leq \ell$ for each $h \in \Gamma'(g)$. Then $|\Gamma'(g)| \leq (n(k+1))^{\mathcal{O}(\ell)}$.*

This leads to the following running time and space requirement.

Lemma 15 ($\star$). *Deciding whether $G \in C_0^{m \times n}$ of $C = [k]$ admits an extension H with $\Delta(H) \leq \ell$ can be done in $(n(k+1))^{\mathcal{O}(\ell)} m$ time using $(n(k+1))^{\mathcal{O}(\ell)}$ space.*

This solves the problem in XP-time: $(nm)^{\lambda(\ell)}$ for some computable function λ.

6   Conclusion

We studied the combinatorial properties of grid-based hypergraph visualizations with disjoint polygons, by trying to minimizing the visual complexity. We assumed as input an assignment of at most one hyperedge per grid cell, which differs from the standard mapping between set elements and grid cells in [9]. We leave finding such an assignment, that minimizes the number of polygon corners, as an open problem. Furthermore, when representing a hyperedge by multiple polygons, a natural optimization goal is to minimize the number of polygons per hyperedge. While minimizing shape complexity may incidentally result in few polygons, the complexity of minimizing the number of polygons remains open.

Acknowledgements. The authors would like to thank anonymous referees for their careful reviews and pointing us to [6].

References

1. Alper, B., Riche, N., Ramos, G., Czerwinski, M.: Design study of linesets, a novel set visualization technique. IEEE Trans. Vis. Comput. Graph. **17**(12), 2259–2267 (2011)
2. Alsallakh, B., Micallef, L., Aigner, W., Hauser, H., Miksch, S., Rodgers, P.: The state-of-the-art of set visualization. Comput. Graph. Forum **35**(1), 234–260 (2016)
3. Baron, M.: A note on the historical development of logic diagrams: Leibniz, Euler and Venn. Math. Gaz. **53**(384), 113–125 (1969)
4. de Berg, M., Khosravi, A.: Optimal binary space partitions in the plane. In: Proceedings 16th COCOON, pp. 216–225 (2010)
5. Cabello, S., Demaine, E., Rote, G.: Planar embeddings of graphs with specified edge lengths. In: Proceedings of the 11th GD, pp. 283–294 (2003)
6. Darmann, A., Döcker, J.: On simplified NP-complete variants of monotone 3-SAT. Discret. Appl. Math. **292**, 45–58 (2021)
7. Depian, T., Dobler, A., Kern, C., Wulms, J.: Minimizing corners in colored rectilinear grids. arXiv (2023)
8. Fischer, M., Frings, A., Keim, D., Seebacher, D.: Towards a survey on static and dynamic hypergraph visualizations. In: 2021 IEEE VIS - Short Papers, pp. 81–85 (2021)
9. van Goethem, A., Kostitsyna, I., van Kreveld, M., Meulemans, W., Sondag, M., Wulms, J.: The painter's problem: covering a grid with colored connected polygons. In: Proceedings of the 25th GD, pp. 492–505 (2017)
10. Jacobsen, B., Wallinger, M., Kobourov, S., Nöllenburg, M.: Metrosets: visualizing sets as metro maps. IEEE Trans. Vis. Comput. Graph. **27**(2), 1257–1267 (2021)
11. Johnson, D., Pollak, H.: Hypergraph planarity and the complexity of drawing Venn diagrams. J. Graph Theory **11**(3), 309–325 (1987)
12. Meulemans, W., Riche, N., Speckmann, B., Alper, B., Dwyer, T.: KelpFusion: a hybrid set visualization technique. IEEE Trans. Vis. Comput. Graph. **19**(11), 1846–1858 (2013)
13. Purchase, H.: Metrics for graph drawing aesthetics. J. Vis. Lang. Comput. **13**(5), 501–516 (2002)

14. Rodgers, P., Stapleton, G., Chapman, P.: Visualizing sets with linear diagrams. ACM Trans. Comput. Hum. Interact. **22**(6), 27:1–27:39 (2015)
15. Rottmann, P., Wallinger, M., Bonerath, A., Gedicke, S., Nöllenburg, M., Haunert, J.: MosaicSets: embedding set systems into grid graphs. IEEE Trans. Vis. Comput. Graph. **29**(1), 875–885 (2023)
16. Wallinger, M., Dobler, A., Nöllenburg, M.: Linsets.zip: compressing linear set diagrams. IEEE Trans. Vis. Comput. Graph. **29**(6), 2875–2887 (2023)
17. Wang, Y., et al.: F2-bubbles: faithful bubble set construction and flexible editing. IEEE Trans. Vis. Comput. Graph. **28**(1), 422–432 (2022)

On Layered Area-Proportional Rectangle Contact Representations

Carolina Haase(✉)🆔 and Philipp Kindermann🆔

Universität Trier, Trier, Germany
{haasec,kindermann}@uni-trier.de

Abstract. Semantic word clouds visualize the semantic relatedness between the words of a text by placing pairs of related words close to each other. Formally, the problem of drawing semantic word clouds corresponds to drawing a rectangle contact representation of a graph whose vertices correlate to the words to be displayed and whose edges indicate that two words are semantically related. The goal is to maximize the number of realized contacts while avoiding any false adjacencies. We consider a variant of this problem that restricts input graphs to be layered and all rectangles to be of equal height, called MAXIMUM LAYERED CONTACT REPRESENTATION OF WORD NETWORKS or MAX-LAYERED-CROWN, as well as the variant MAX-INTLAYEREDCROWN, which restricts the problem to only rectangles of integer width and the placement of those rectangles to integer coordinates.

We classify the corresponding decision problem k-INTLAYERED-CROWN as NP-complete even for triangulated graphs and k-LAYERED-CROWN as NP-complete for planar graphs. We introduce three algorithms: a 1/2-approximation for MAX-LAYEREDCROWN of triangulated graphs, and a PTAS and an XP algorithm for MAX-INTLAYEREDCROWN with rectangle width polynomial in n.

1 Introduction

Word clouds can be used to visualize the importance of (key-)words in a given text. Usually, words will be scaled according to their frequency and, in case of semantic word clouds, arranged in such a way that closely related words are placed closer together than words that are unrelated. There are multiple tools like Wordle[1] [12], which was launched in 2008 by Jonathan Feinberg, that allow for automized drawing of classical word clouds, i.e., word clouds that disregard semantic relatedness; see Fig. 1 for an example.

However, classical word clouds have certain disadvantages, as they are frequently misinterpreted. This has been analyzed in a survey conducted by Viegas et al. [12]: different colors and positioning of words give the impression to bear

---

[1] At the time of writing, the tool (usually found at http://www.wordle.net/) is not available, but the creator states on their website (https://mrfeinberg.com/) that they have "hopes to bring it back to life".

R. Uehara et al. (Eds.): WALCOM 2024, LNCS 14549, pp. 149–162, 2024.
https://doi.org/10.1007/978-981-97-0566-5_12

Fig. 1. Randomly arranged word cloud (left) and semantic word cloud (right), generated using the first chapter of "Alice's Adventures in Wonderland" by Lewis Carroll.

meaning, even if they don't. For this reason, it makes sense to pay special attention to *semantic* word clouds, which resolve these shortcomings by placing related words closely together and sometimes using color to indicate, for example, clusters of semantically related words. Semantic relatedness, in this case, can be measured by how often two words occur together in the same sentence [3].

Tools to generate semantic word clouds are, however, not as widely available. One such tool can be found online at http://wordcloud.cs.arizona.edu that implements different algorithms for semantic word clouds [2,3,5]. A semantic word cloud generated by the tool is shown in Fig. 1. In the given example, the placement of words was calculated using cosine similarity. Compared to the classical word cloud generated using the same tool, with the same coloring for clusters, but a greedy, randomized approach to place words, the advantages of arranging words semantically become quite clear.

Problem Statement. To formalize the problem of drawing semantic word clouds, Barth et al. [2] introduced the problem CONTACT REPRESENTATION OF WORD NETWORKS (CROWN). Given a graph $G = (V, E)$, where every vertex v_i of G corresponds to a word of width w_i and height h_i, and every (weighted) edge between two vertices indicates the level of semantic relatedness between the corresponding words, the goal is to draw a contact representation where each vertex v_i is drawn as an axis-aligned rectangle of width w_i and height h_i such that bounding boxes of semantically related words touch.

In this paper, we consider a more restricted variant of the problem, which we will call (MAX-)LAYEREDCROWN, that has been introduced by Nöllenburg et al. [10]. Here, the input graph is planar and the vertices are assigned to layers. Furthermore, all bounding boxes have the same height. The goal is to maximize the number of contacts between semantically related words, while words that are not semantically related are not allowed to touch.

More formally, the problem is defined as follows. Let $G = (V, E)$ be a planar vertex-weighted *layered graph* with L layers, i.e., each vertex is assigned to one of L layers. The order of vertices within a layer is fixed, i.e., each vertex $v_{i,j}$ can be identified by its layer $1 \leq i \leq L$ and its position j within the layer. Edges can only exist between neighboring vertices $v_{i,j}, v_{i,j+1}$, on the same layer and

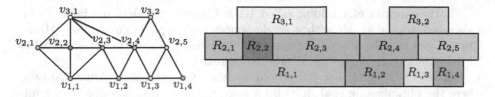

Fig. 2. Internally triangulated graph with 3 layers (left) and a contact representation (right).

between vertices on adjacent layers. Like Nöllenburg et al., we consider the case that the edges are unweighted. To each vertex v we assign an axis-aligned unit-height rectangle $R(v)$ with width $w(v)$, given by the weight of the vertex. We will also use the notation $R_{i,j} = R(v_{i,j})$ and $w_{i,j} = w(v_{i,j})$; see Fig. 2. The goal is to calculate the position $x_{i,j}$ for each vertex $v_{i,j}$, where $x_{i,j}$ denotes the x-coordinate of the bottom left corner of $R_{i,j}$, in such a way that rectangles do not overlap except on their boundaries. We call such an assignment a *representation*. Two rectangles $R(v)$ and $R(u)$ touch if their intersection is a line segment of length $\varepsilon > 0$. In this case, we say that $R(v)$ and $R(u)$ are in *contact*. An edge $\{v, u\}$ is *realized* if R_v and R_u are in contact. We call a contact *horizontal* if R_v and R_u are neighbors on the same layer and *vertical* if R_v and R_u are on adjacent layers. Contacts between rectangles whose vertices are not adjacent are not allowed and are called *false adjacencies*. Representations with false adjacencies are *invalid*; otherwise, they are *valid*. Gaps between vertices $v_{i,j}, v_{i,j+1}$ on the same layer are allowed.

The maximization problem MAXIMUM LAYERED CONTACT REPRESENTATION OF WORD NETWORKS (MAX-LAYEREDCROWN) is to find a valid representation for a given graph G such that the number of realized contacts is maximized. The respective decision problem LAYERED CONTACT REPRESENTATION OF WORD NETWORKS (k-LAYEREDCROWN) is to decide whether there exists a valid contact representation that realizes at least k contacts. Many fonts are monospaced, i.e., all letters and characters occupy the same amount of horizontal space. Thus, we also consider the further restriction that rectangles may only be of integer width and may only be placed with their lower left corner on integer coordinates. This implies that two rectangles are in contact if and only if the intersection of their boundaries is a line segment of positive integer length. We call those problems MAX-INTLAYEREDCROWN and k-INTLAYEREDCROWN.

For information about graph drawing and parameterized complexity in general, we refer to books [4, 7, 8, 11].

Related Work. Barth et al. [2] have shown that CROWN is strongly NP-hard even when restricted to trees and weakly NP-hard even when restricted to stars, but can be solved in linear time on irreducible triangulations. They also provided constant-factor approximation algorithms for several graph classes like stars, trees, and planar graphs. These were improved by Bekos et al. [5] and partially implemented and compared to other algorithms by Barth et al [3].

Another variant of CROWN, called HIER-CROWN, restricts the input to be a directed acyclic graph with a single source and a plane embedding. HIER-CROWN can be solved in polynomial time, but can be shown to become weakly NP-complete if rectangles are allowed to be rotated [2].

Barth et al. [2] further introduced another variant called AREA-CROWN, where the optimization goal shifts from maximizing rectangle contacts to minimizing the area of a bounding box containing the contact representation. They show that this problem is NP-hard, even if restricted to paths.

Nöllenburg et al. [10] introduced MAX-LAYEREDCROWN, but they only considered triangulated graphs. They gave a linear-time algorithm for triangulated graphs with only 2 layers and proposed an ILP-formulation for triangulated graphs with more than 2 layers. They further showed how to solve AREA-LAYEREDCROWN in polynomial time with a flow formulation.

Espenant and Mondal [9] study STREAMTABLES, where one seeks to visualize a matrix such that each cell is drawn as a rectangle of a specified area, cells in the same row have uniform height and align horizontally, while maximizing contacts and/or minimizing excess area. Their model is similar to LAYEREDCROWN on grids, but false adjacencies are not forbidden, point contacts count as realized edges, and rows can generally be permuted.

Our Contribution. In this work, we study the computational complexity of INTLAYEREDCROWN and algorithms for MAX-LAYEREDCROWN and MAX-INT-LAYEREDCROWN. In Sect. 2, we classify k-INTLAYEREDCROWN as an NP-complete problem even for triangulated graphs, using a reduction from PLANAR MONOTONE 3-SAT. We will then adjust the proof to show NP-completeness for k-LAYEREDCROWN for planar graphs. In Sect. 3, we present a $1/2$-approximation for MAX-LAYEREDCROWN on triangulated graphs (Sect. 3.1) and formulate a dynamic program for MAX-INTLAYEREDCROWN that is an XP algorithm if the maximum rectangle width is polynomial in n (Sect. 3.2). Finally, we combine the ideas of the two algorithms to formulate a polynomial-time approximation scheme for MAX-INTLAYEREDCROWN if the maximum rectangle width is polynomial in n (Sect. 3.3). We conclude with a list of research questions in Sect. 4.

2   NP-Completeness of k-INTLAYEREDCROWN

In this section, we prove that k-INTLAYEREDCROWN is NP-complete. We first show that k-INTLAYEREDCROWN lies in NP.

Lemma 1. k-INTLAYEREDCROWN *lies in NP.*

Proof. For a given contact representation of a layered graph G, one can verify in polynomial time if the representation is valid and whether at least k contacts are realized. Thereby, k-INTLAYEREDCROWN is a member of the class NP.   □

We prove NP-hardness by reducing from PLANAR MONOTONE 3-SAT, which is NP-complete [6]. Let B be a boolean formula in conjunctive normal form

(CNF) and $X = \{x_1, \ldots, x_n\}$ its variable set. That is, $B = C_1 \wedge C_2 \wedge \cdots \wedge C_m$ is a conjunction of clauses C_i, where a clause is a disjunction of literals and a literal is defined as either x or $\bar{x}$ for a variable $x \in X$. In PLANAR MONOTONE 3-SAT, all clauses consist of at most three literals and are either *positive* (they only contain positive literals) or *negative* (they only contain negative literals), and the variable-clause incidence graph can be drawn such that (i) it is crossing-free; (ii) all variable vertices lie on the x-axis; (iii) all positive clause vertices lie above the x-axis; and (iv) all negative clause vertices lie below the x-axis.

We construct a vertex-weighted layered graph G whose contact representation closely resembles the rectilinear representation of B. To this end, we use gadgets to represent variables and clauses, as well as an additional gadget to split/duplicate variable values. Just as in the rectilinear representation, vertices representing variable gadgets are aligned horizontally, and positive clauses are drawn above, while negative clauses are drawn below the variable gadgets. The goal is for G to have a valid contact representation if and only if B is satisfiable. We choose k as the maximum number of possible contacts in our construction.

Variable Gadget. A variable gadget consists of five vertices $v_{l_1}, v_{l_2}, v_m, v_{r_2}, v_{r_1}$ that each have a rectangle width of 1 on layer 1, as well as three vertices u_l, u_m, u_r on layer 0. The rectangles $R(u_l)$ and $R(u_r)$ both have width 2, $R(u_m)$ has width 1; see Fig. 3. As edges between the layers we add $u_l v_{l_1}, u_l v_{l_2}, u_l v_m, u_l v_{r_2}$, $u_m v_{r_2}, u_r v_{r_1}$, and $u_r v_{r_2}$. Note that there is no edge between v_m and u_m, and the corresponding rectangles are therefore not allowed to touch. We want to use this to create a gap in each layer, which will allow us to assign opposite variable values above and below the gadget, thus realizing the notion of positive clauses above and negative clauses below the variable gadgets.

For the gadget to work as intended we need additional walls on either side. Walls are constructed from three rectangles of width 1 per layer. Edges are added in such a way that moving any wall rectangle to either side reduces realized contacts by at least one and/or introduces false adjacencies; see Fig. 4a.

To determine variable values, we add vertices v_x and u_x of rectangle width 3 to layers 2 and -1, respectively, with edges to all vertices of the variable gadget and the innermost wall vertices on the adjacent layers. Since u_m and v_m are not allowed to touch, they split the rectangles on layers 0 and 1 into two blocks of rectangles of width 3 and 2, respectively. To maximize contacts, both v_x and u_x have to realize vertical contacts to the larger block of width 3 and a horizontal contact to a wall vertex. Since the blocks of width 3 on layers 0 and 1 are in contact with opposite walls, so are v_x and u_x. We interpret a variable assignment as follows: if v_x realizes contacts to v_{l_1} and v_{l_2}, and u_x realizes a contact to u_r, the assigned value of the variable is `true`, otherwise `false`; see Figs. 3 to 3c.

Note that u_l could also realize contacts to v_{l_2}, v_m, u_m instead of v_{l_1}, v_{l_2} and a wall vertex; see Fig. 3d. However, this does not change the position of u_x and can therefore be disregarded. The same holds for vertices v_{r_2}, v_{r_1}, which could be moved to the left by one without changing the number of realized contacts; see Fig. 3e. Every other valid placement of vertices results in the variable gadget to be wider and thus realize less contacts; see for example Fig. 3f.

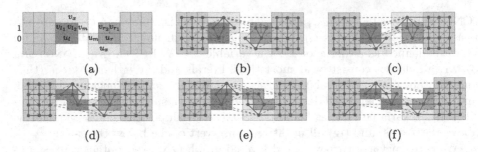

Fig. 3. (a) Contact representation and (b) underlying graph for a variable gadget with variable assignment `true`; (c) variable gadget with variable assignment `false`; (d, e) alternative representations with the same number of realized contacts; (f) valid representation realizing fewer contacts.

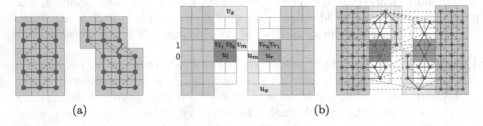

Fig. 4. (a) Moving wall vertices leads to false adjacencies (red curve) and (b) propagating variable values. (Color figure online)

In order to use variable values within multiple clauses, we will have to propagate them; see Fig. 4b. We do so by adding alternating rows of five vertices with rectangle width 1 and rows of three vertices of width 2, 1, and 2, essentially repeating the pattern we used for the variable gadget. The difference is that this time the middle rectangles have edges to their counterparts in adjacent rows and are therefore allowed to touch. Thus, the gap stays as assigned by the variable gadget. We can proceed to add vertices v_x and u_x as before.

Clause Gadget. Let C be a clause that contains variables x_a, x_b, x_c in B. Recall that all clauses above the variable layer are positive while all clauses below the variable layer are negative, and the variable gadgets propagate the positive variable assignment to the top and the negated variable assignment to the bottom.

Assume that x_a, x_b, x_c occur in this order. To determine whether a clause is satisfied, we use a *slider* vertex v_s. The slider shall realize 4 contacts if the variable assignments satisfy C, and 3 contacts otherwise. The slider has rectangle width 2 and can therefore only be in contact with one variable gadget at a time.

We describe the clause gadget for the case that C is a negative clause; see Fig. 5. The other case is symmetric. Suppose that the propagation of the variable assignments for x_a, x_b, x_c ends with vertices u_a, u_b, u_c on layer i. On layer $i - 1$, we place v_s and continue the outermost walls with two vertices of rectangle

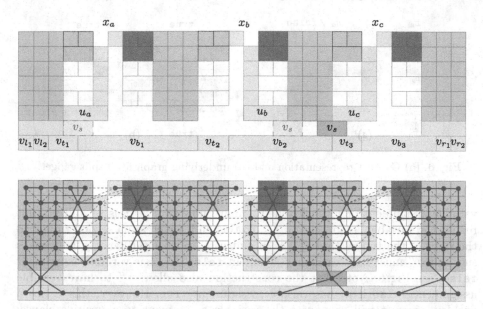

Fig. 5. Contact representation (top) and underlying graph (bottom) for a clause gadget, including multiple examples of placements for v_s. Unrealized edges between v_s and vertices of adjacent layers are omitted for readability.

width 3. On layer $i - 2$, we add vertices $v_{l_1}, v_{l_2}, v_{t_1}, v_{b_1}, v_{t_2}, v_{b_2}, v_{t_3}, v_{b_3}, v_{r_1}, v_{r_2}$ in this order to close the bottom of the gadget. The rectangles $R_{l_1}, R_{l_2}, R_{r_1}, R_{r_2}$ have width 1; $R_{t_1}, R_{t_2}, R_{t_3}$ have width 2. The width of $R_{b_1}, R_{b_2}, R_{b_3}$ is set such that the remaining space is filled and $v_{t_1}, v_{t_2}, v_{t_3}$ are each placed on the leftmost position underneath a variable gadget, i.e., on the side of the positive-valued variable propagation. Edges exist from v_s to most vertices of the gadget on adjacent layers such that the triangulation is preserved and v_s can be placed freely along the whole width of the gadget. For the exact edges, refer to Fig. 5.

The only ways for v_s to realize four contacts are the following. (i) it touches v_{t_1} and v_{b_1} at the bottom, the wall at the left, and u_a at the top, if x_a has a negative variable assignment; (ii) it touches v_{b_1} and v_{t_2} at the bottom, u_b and the wall left of u_b at the top, if x_b has a negative variable assignment; or (iii) it touches v_{b_2} and v_{t_3} at the bottom, u_c and the wall left of u_c at the top, if x_c has a negative variable assignment. Thus, v_s only realizes four contacts if the variable assignment satisfies C.

Split Gadget. To duplicate variable values that occur in multiple clauses, we use a split gadget; see Fig. 6. Let x_a be a variable such that its variable assignment ends at a vertex v_a. Recall that all clauses above the variable layer are positive while all clauses below the variable layer are negative. Assume that v_a lies above the variable layer, so the variable assignment has to be propagated to a positive clause; the other case is symmetric. In the split gadget, we want to split the

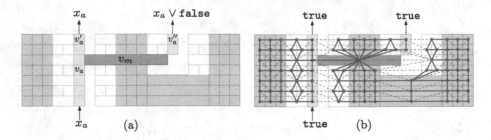

Fig. 6. (a) Contact representation and (b) underlying graph for a split gadget.

variable assignment of x_a such that there are now two vertices v_a' and v_a'' that realize the variable assignment of x_a. To this end, we create a second tunnel to the right of the tunnel that v_a lies in and use a horizontal bar v_m that makes sure that v_a'' must have variable assignment `false` if v_a has variable assignment `false`; see Fig. 6. Note that the construction also allows v_a'' to have variable assignment `false` if v_a has variable assignment `true`. However, this is not a problem since it will propagate the variable assignment to a positive clause; hence, this cannot satisfy a clause that should be unsatisfied due to the variable assignment.

Combining the Gadgets. Combining these gadgets, we construct a vertex-weighted layered graph G for the planar monotone boolean formula B. Let h be the number of layers of G, and let w be the minimum width of G (i.e., the sum of rectangle widths among all layers). Obviously, any layered contact representation of G has at most $w \cdot h$ contacts. To make sure that the representation of G has to be drawn inside a designated bounding box of width w and height h, we add a *frame* around G consisting of walls of width $w \cdot h$ on the left and right and $w \cdot h$ stacked rectangles that span the whole width of G at the top and bottom, creating a graph G^+. Moving any rectangle of G outside of the designated bounding box also moves parts of the frame and thus removes at least $w \cdot h$ contacts. We choose the number of desired contacts k as the number of contacts that would be realized if every single gadget maximizes its number of contacts. A full example can be seen in Fig. 7.

Assume that we have a solution for B. For each variable, we draw the corresponding variable gadget of G^+ such that it represents the variable assignment of the solution, and we propagate the variable assignments along the tunnels and split gadgets. Since the variable assignment satisfies all clauses, we can place v_s at each clause such that it has 4 contacts, thus maximizing the number of contacts at every gadget and obtaining k contacts in total.

For the other direction, assume that we have a drawing of G^+ that realizes k contacts. From each variable gadget, we can read the corresponding variable assignment. Since each clause gadget must have v_s in a position such that it has four contacts (otherwise, there cannot be k contacts in total), every clause has a satisfied literal. Together with Lemma 1, this proves the following theorem.

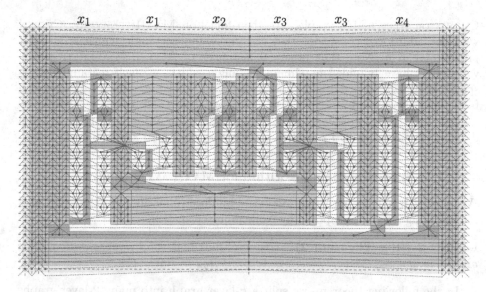

Fig. 7. Contact representation for the boolean formula B with variable set $\{x_1, x_2, x_3, x_4\}$, clauses $\{\overline{x_1}, \overline{x_2}, \overline{x_3}\}$, $\{\overline{x_1}, \overline{x_3}, \overline{x_4}\}$ and $\{x_1, x_3, x_4\}$ and variable assignment $x_1 = \texttt{false}, x_2 = \texttt{true}, x_3 = \texttt{true}, x_4 = \texttt{false}$ (Edges between v_s and above/below layer left out for readability purposes)..

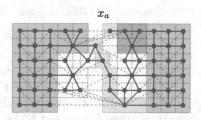

Fig. 8. If non-integral positions are allowed, then variable assignments may flip.

Theorem 1. k-INTLAYEREDCROWN *is NP-complete for internally triangulated graphs.*

Note that the proof cannot be immediately extended to k-LAYEREDCROWN, as placing rectangles on non-integer positions might lead to situations where a variable assignment flips; see Fig. 8. However, if we drop the requirement that the graph is triangulated, then we can adjust the construction by removing unwanted contacts from the graph, which leads to the following theorem.

Theorem 2. k-LAYEREDCROWN *is NP-complete.*

3   Parameterized and Approximation Algorithms

In this section, we provide parameterized and approximation algorithms. As a warmup (Sect. 3.1), we first describe a 1/2-approximation for MAX-LAYERED-

CROWN on triangulated graphs. We then focus on MAX-INTLAYEREDCROWN with the additional constraint that the maximum rectangle width is at most polynomial in n. Note that practical instances of MAX-INTLAYEREDCROWN will always have bounded maximum rectangle width, as each rectangle corresponds to a word, and words have an upper limit of letters in most languages (in fact, the longest word in an English dictionary, has 45 letters: *pneumonoultra-microscopicsilicovolcanoconiosis*). We first describe an XP-algorithm based on a dynamic program (Sect. 3.2), which we then use to obtain a PTAS (Sect. 3.3).

3.1 1/2-Approximation Algorithm for MAX-LAYEREDCROWN

We show that a 1/2-approximation exists by describing an algorithm that uses the following Lemma, proposed by Nöllenburg et al. [10].

Lemma 2 ([10], **Theorem 2**). *A contact-maximal valid representation for a given triangulated 2-layer graph can be computed in linear time.*

In the following theorem, we split a k-layer graph into many 2-layer graphs and solve these optimally with Lemma 2. Half of these 2-layer graphs are vertex-disjoint, so their optimal solutions can be combined to a valid solution of the input graph.

Theorem 3. MAX-LAYEREDCROWN *on triangulated graphs admits a 1/2-approximation in linear time.*

Proof. Let G be an L-layered graph. For $i = 1, \ldots, L-1$, let A_i be the subgraph of G induced by the vertices on layers i and $i + 1$. We construct two groups of subgraphs $G_{\text{even}} = \bigcup_{i \text{ even}} A_i$ and $G_{\text{odd}} = \bigcup_{i \text{ odd}} A_i$; see Fig. 9.

We solve every subgraph A_i, $i = 1, \ldots, L - 1$ optimally using Lemma 2. Let ALG_i be the number of contacts realized for A_i. Let Γ^* be an optimal drawing of G that realizes OPT contacts, and let OPT_i be the number of contacts realized for A_i in Γ^*. Since the 2-layer algorithm yields an optimal solution, it holds that $\text{ALG}_i \geq \text{OPT}_i$ for $i = 1, \ldots, L-1$, so $\sum_{i=1}^{L-1} \text{ALG}_i \geq \sum_{i=1}^{L-1} \text{OPT}_i \geq \text{OPT}$. Note that any two subgraphs $A_i, A_j \in G_{\text{even}}$ are vertex-disjoint. Hence, we can obtain a valid solution for G_{even} with $\text{ALG}_{\text{even}} = \sum_{i: A_i \in G_{\text{even}}} \text{ALG}_i$ contacts by combining the computed solutions for the corresponding subgraphs. Analogously, we can obtain a valid solution for G_{odd} with $\text{ALG}_{\text{odd}} = \sum_{i: A_i \in G_{\text{odd}}} \text{ALG}_i$ contacts. We get a 1/2-approximation by choosing the contacts realized by the instances corresponding to the larger of both sums: $\max\{\text{ALG}^{G_{\text{even}}}, \text{ALG}^{G_{\text{odd}}}\} \geq \text{OPT}/2$.

For the running time, note that every vertex lies in at most two subgraphs, and Lemma 2 solves each subgraph optimally in time linear in its size. □

3.2 XP-Algorithm for MAX-INTLAYEREDCROWN

We now use a dynamic programming approach to solve MAX-INTLAYERED-CROWN with bounded maximum rectangle width optimally (Fig. 10).

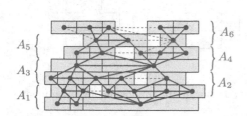

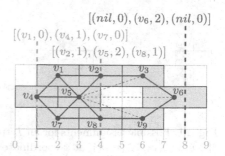

Fig. 9. 2-layered sub-graphs split into two groups G_{odd} (blue) and G_{even} (green). (Color figure online)

Fig. 10. Three cuts and respective k-tuples of an optimal solution of a 3-layered graph.

Theorem 4. MAX-INTLAYEREDCROWN *is solvable in time* $\mathcal{O}(nW)^L$, *where W is the maximum rectangle width. If $W \in \text{poly}(n)$,* MAX-INTLAYEREDCROWN *lies in XP when parameterized by the number of layers L of the input graph.*

Proof. Let G be an L-layered vertex-weighted graph with maximum weight W. We want to define subproblems based on vertical cuts through integer x-coordinates; see Sect. 10. At each such cut through any valid representation, we can obtain the following information: (i) Which vertex has been cut at each layer (if any)? (ii) At what length was the corresponding rectangle cut (i.e., how much of the rectangle has already been drawn on the left of the cut)? (iii) If no vertex has been cut, which vertex will be drawn next on the specific layer?

To formalize this, we use a tuple (v, l) for each layer, where v denotes the vertex that is being cut and l denotes the length of the rectangle to the left of the cut. The tuple $(v, 0)$ indicates that v is next in line but has not yet been placed, while $(nil, 0)$ means that there is no more vertex to be drawn on the corresponding layer. For every possible cut, we therefore obtain an L-tuple $[(v_1, l_1), (v_2, l_2), \ldots, (v_L, l_L)]$. As each rectangle has at most width W, there are no more than $((n + 1) \cdot (W + 1))^L$ such L-tuples. We store in an L-dimensional table D for each L-tuple the maximum number of contacts that can be achieved to the *right* of the corresponding cut.

We set $D[(nil, 0), \ldots, (nil, 0)] = 0$, which corresponds to the right boundary of the drawing. Consider any L-tuple T and its corresponding cut. To calculate $D[T]$, we have to look at each cut T' through a solution one coordinate to the right. Consider any layer i and the corresponding tuple $T_i = (v_i, l_i)$; see Fig. 11. If T cuts through the middle of v_i, i.e., $0 < l_i < w(v_i)$, then this rectangle has to continue, i.e., $T_i' = (v_i, l_i + 1)$. If T cuts through no vertex, i.e., $l_i = 0$, then we can either place v_i, i.e., $T_i' = (v_i, 1)$, or not place it yet, i.e., $T_i' = (v_i, 0)$. Finally, if T touches the right side of v_i, i.e., $l_i = w(v_i)$, then we can either immediately place the next vertex v_i' (if it exists), i.e., $T_i' = (v_i', 1)$, or not place it yet, i.e., $T_i' = (v_i', 0)$. Doing this for every layer, we can find each possible next cut. For each such cut, we calculate whether it is feasible, i.e., whether the newly

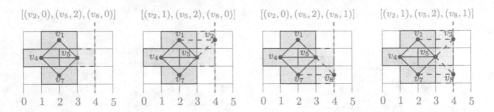

Fig. 11. Possible assignments for an L-tuple at x-coordinate 4 following an assignment of $[(v_1, 2), (v_5, 1), (v_7, 2)]$ at x-coordinate 3.

placed vertices have any false adjacencies. If it is not feasible, then we discard it; otherwise, we count how many edges are realized by the newly placed vertices, and thus calculate $D[T]$ from $D[T']$. We can obtain the optimum solution for G from $D[(v_1, 0), \ldots, (v_L, 0)]$, where v_i is the leftmost vertex of layer $i = 1, \ldots, L$.

All in all, this leaves us with at most $(n \cdot (W + 1) + 1)^L$ different table entries that each take $\mathcal{O}(2^L)$ time to be calculated. The algorithm thus runs in $\mathcal{O}(nW)^L$ time. To obtain the solution instead of the number of contacts, we can use an additional lookup table in the same time. □

3.3   PTAS for MAX-INTLAYEREDCROWN

In the following, we use Baker's technique [1] to combine the ideas of the previously described 1/2-approximation (Sect. 3.1) and dynamic program (Sect. 3.2).

Lemma 3. *For every integer $\ell > 0$, MAX-INTLAYEREDCROWN admits a $(1 - \frac{1}{\ell})$-approximation in $\mathcal{O}(nW)^{\ell+1}$ time, where W is the maximum rectangle width.*

Proof. If $\ell \geq L$, then we can solve the problem optimally in $\mathcal{O}(nW)^L$ time using Theorem 4. Otherwise, similar to Theorem 3, we split the graph into multiple subgraphs of ℓ layers each, which we will then solve using the dynamic program described in Theorem 4. We assume that L is evenly divisible by ℓ; otherwise, we add empty dummy layers to the top, increasing L by at most factor 2. For technical reasons, we treat layer 0 to be the same as layer L.

For $i = 1, \ldots, L$, let A_i be the subgraph of G induced by the vertices on the ℓ layers $i, \ldots, i + \ell - 1 \mod L$. We can solve each of these subgraphs optimally using Theorem 4 in $\mathcal{O}(nW)^\ell$ time. Since $L \in \mathcal{O}(n)$, this takes $\mathcal{O}(nW)^{\ell+1}$ time in total. Let ALG_i be the number of contacts for A_i obtained this way.

Let Γ^* be an optimal representation of G that realizes OPT contacts, let OPT_i be the number of horizontal contacts realized for each layer i in Γ^*, and let $\mathrm{OPT}_{i,i+1}$ denote the number of vertical contacts between layers i and $i + 1$ in Γ^*. Since we solved A_i optimally, we have

$$\mathrm{ALG}_i \geq \sum_{j=i}^{(i+\ell-2) \bmod L} (\mathrm{OPT}_j + \mathrm{OPT}_{j,j+1}) + \mathrm{OPT}_{i+\ell-1}.$$

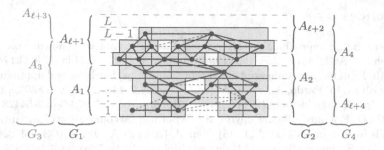

Fig. 12. An L-layered graph with one dummy layer split into L subgraphs of ℓ layers, partitioned into ℓ groups, for $L = 8$ and $\ell = 4$.

Horizontal contacts of each layer are covered by ℓ subgraphs and vertical contacts between pairs $\{i, i+1\}$ of layers are covered by $\ell-1$ subgraphs. Therefore,

$$\sum_{i=1}^{L} \mathrm{ALG}_i \geq \ell \sum_{j=1}^{L} \mathrm{OPT}_j + (\ell-1) \sum_{j=1}^{L-1} \mathrm{OPT}_{j,j+1} \geq (\ell-1)\,\mathrm{OPT}$$

We then partition these subgraphs into ℓ groups $G_1 \ldots, G_\ell$ such that $G_i = A_i \cup A_{i+\ell} \cup A_{i+2\ell} \cup \ldots \cup A_{i+L-\ell}$; see Fig. 12. Note that the subgraphs in a group are vertex-disjoint, so combining the optimum solutions for $A_i, A_{i+\ell}, \ldots, A_{i+L-\ell}$ gives an optimum solution for G_i with $\mathrm{ALG}_{G_i} = \sum_{j=0}^{L/\ell-1} \mathrm{ALG}_{i+j\ell}$ contacts. Further, every subgraph lies in exactly one group, so $\sum_{i=1}^{\ell} \mathrm{ALG}_{G_i} = \sum_{i=1}^{L} \mathrm{ALG}_i$.
We now choose $1 \leq j \leq \ell$ such that $\mathrm{ALG}_{G_j} = \max_{i=1}^{\ell} \mathrm{ALG}_{G_i}$. Then,

$$\mathrm{ALG}_{G_j} = \max_{i=1}^{\ell} \mathrm{ALG}_{G_i} \geq \frac{1}{\ell} \sum_{i=1}^{L} \mathrm{ALG}_{G_i} \geq \left(1 - \frac{1}{\ell}\right) \mathrm{OPT}.$$

□

For any $\varepsilon > 0$, by choosing $\ell = \lceil 1/\varepsilon \rceil$, Lemma 3 provides a PTAS if $W \in \mathrm{poly}(n)$.

Theorem 5. *For every $\varepsilon > 0$, MAX-INTLAYEREDCROWN admits a $(1-\varepsilon)$-approximation in $\mathcal{O}(nW)^{1+\lceil \frac{1}{\varepsilon} \rceil}$ time, where W is the maximum rectangle width.*

4 Conclusion

We have proved that k-INTLAYEREDCROWN and k-LAYEREDCROWN are NP-complete, and provided an XP-algorithm parameterized by the number of layers and a PTAS for MAX-INTLAYEREDCROWN when rectangle widths are polynomial in n. Several interesting problems remain open, for example: (i) Is there an FPT-algorithm parameterized by the number of layers for MAX-INTLAYERED-CROWN? (ii) Is there a PTAS for MAX-INTLAYEREDCROWN for which the running time does not depend on the maximum rectangle width? (iii) What can we do if rectangles can have different (integer) heights, thus spanning more than one layer?

References

1. Baker, B.S.: Approximation algorithms for NP-complete problems on planar graphs. J. ACM **41**(1), 153–180 (1994). https://doi.org/10.1145/174644.174650
2. Barth, L., et al.: Semantic word cloud representations: hardness and approximation algorithms. In: Pardo, A., Viola, A. (eds.) LATIN 2014. LNCS, vol. 8392, pp. 514–525. Springer, Heidelberg (2014). https://doi.org/10.1007/978-3-642-54423-1_45
3. Barth, L., Kobourov, S.G., Pupyrev, S.: Experimental comparison of semantic word clouds. In: Gudmundsson, J., Katajainen, J. (eds.) SEA 2014. LNCS, vol. 8504, pp. 247–258. Springer, Cham (2014). https://doi.org/10.1007/978-3-319-07959-2_21
4. Battista, G.D., Eades, P., Tamassia, R., Tollis, I.G.: Graph Drawing: Algorithms for the Visualization of Graphs. Prentice-Hall (1999)
5. Bekos, M.A., et al.: Improved approximation algorithms for box contact representations. Algorithmica **77**(3), 902–920 (2017). https://doi.org/10.1007/s00453-016-0121-3
6. de Berg, M., Khosravi, A.: Optimal binary space partitions for segments in the plane. Int. J. Comput. Geom. Appl. **22**(3), 187–206 (2012). https://doi.org/10.1142/S0218195912500045
7. Cygan, M., et al.: Parameterized Algorithms. Springer, Cham (2015). https://doi.org/10.1007/978-3-319-21275-3
8. Downey, R.G., Fellows, M.R.: Fundamentals of Parameterized Complexity. TCS, Springer, London (2013). https://doi.org/10.1007/978-1-4471-5559-1
9. Espenant, J., Mondal, D.: StreamTable: an area proportional visualization for tables with flowing streams. In: Mutzel, P., Rahman, M.S., Slamin (eds.) WALCOM: Algorithms and Computation. WALCOM 2022. LNCS, vol. 13174. Springer, Cham (2022). https://doi.org/10.1007/978-3-030-96731-4_9
10. Nöllenburg, M., Villedieu, A., Wulms, J.: Layered area-proportional rectangle contact representations. In: Purchase, H.C., Rutter, I. (eds.) GD 2021. LNCS, vol. 12868, pp. 318–326. Springer, Cham (2021). https://doi.org/10.1007/978-3-030-92931-2_23
11. Tamassia, R. (ed.): Handbook on Graph Drawing and Visualization. Chapman and Hall/CRC (2013). https://cs.brown.edu/people/rtamassi/gdhandbook
12. Viegas, F., Wattenberg, M., Feinberg, J.: Participatory visualization with wordle. IEEE Trans. Visual Comput. Graphics **15**(6), 1137–1144 (2009). https://doi.org/10.1109/tvcg.2009.171

Short Flip Sequences to Untangle
Segments in the Plane

Guilherme D. da Fonseca[1]([✉])[iD], Yan Gerard[2][iD], and Bastien Rivier[2][iD]

[1] Aix-Marseille Université and LIS, Marseille, France
guilherme.fonseca@lis-lab.fr
[2] Université Clermont Auvergne and LIMOS, Clermont-Ferrand, France
{yan.gerard,bastien.rivier}@uca.fr

Abstract. A (multi)set of segments in the plane may form a TSP tour, a matching, a tree, or any multigraph. If two segments cross, then we can reduce the total length with the following *flip* operation. We *remove* a pair of crossing segments, and *insert* a pair of non-crossing segments, while keeping the same vertex degrees. The goal of this paper is to devise efficient strategies to flip the segments in order to obtain crossing-free segments after a small number of flips. Linear and near-linear bounds on the number of flips were only known for segments with endpoints in convex position. We generalize these results, proving linear and near-linear bounds for cases with endpoints that are not in convex position. Our results are proved in a general setting that applies to multiple problems, using multigraphs and the distinction between removal and insertion choices when performing a flip.

Keywords: Planar geometry · Reconfiguration · Matching · Euclidean TSP

1 Introduction

The *Euclidean Travelling Salesman Problem (TSP)* is one of the most studied geometric optimization problems. We are given a set P of points in the plane and the goal is to find a tour S of minimum length. While the optimal solution has no crossing segments, essentially all approximation algorithms, heuristics, and PTASs may produce solutions S with crossings. Given S, the only procedure known to obtain a solution S' without crossings and of shorter length is to perform a flip operation. In our case, a *flip* consists of *removing* a pair of crossing segments, and then *inserting* a pair of non-crossing segments preserving a tour (and consequently reducing its length). Flips are performed in *sequence* until a crossing-free tour is obtained, in a procedure called *untangle*.

The same flip operation may be applied in other settings. More precisely, a *flip* consists of removing a pair of crossing segments s_1, s_2 and inserting a

This work is supported by the French ANR PRC grant ADDS (ANR-19-CE48-0005). Full version: https://arxiv.org/abs/2307.00853.

pair of segments s_1', s_2' in a way that s_1, s_1', s_2, s_2' forms a cycle and a certain graph *property* is preserved. In the case of TSP tours, the property is being a Hamiltonian cycle. Other properties have also been studied, such as spanning trees, perfect matchings, and multigraphs. Notice that flips preserve the degrees of all vertices and multiple copies of the same edge may appear when we perform a flip on certain graphs.

When the goal is to obtain a crossing-free TSP tour, we are allowed to *choose* which pair of crossing segments to remove in order to perform fewer flips, which we call *removal choice* (Fig. 1(a)). Notice that, in a tour, choosing which pair of crossing edges we remove defines which pair of crossing edges we insert. However, this is not the case for matchings and multigraphs. There, we are also allowed to choose which pair of segments to insert among two possibilities, which we call *insertion choice* (Fig. 1(b)).

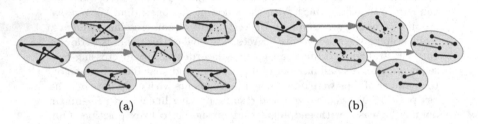

Fig. 1. (a) Three untangle sequences for a tour with different *removal choices*. (b) Three untangle sequences for a matching with different *insertion choices*. We highlight the segments removed and inserted at each flip.

Using removal or insertion choices to obtain shorter flip sequences has not been explicitly studied before and opens several new questions, while unifying the solution to multiple reconfiguration problems. Next, we describe previous work according to which choices are used. Throughout, P denotes the set of points and n the number of segments.

Using No Choice: The *length* (i.e. the number of flips) of any untangle sequence for a TSP tour is $\mathcal{O}(n^3)$ [22] and it is easy to construct $\Omega(n^2)$ examples. The same proof has been rediscovered in the context of matchings [9] after 35 years. If P is in convex position, then the number of crossings decreases at each flip, which gives a tight bound of $\Theta(n^2)$. If all points except the endpoints of t segments are in convex position, then the authors [15] recently showed a bound of $\mathcal{O}(tn^2)$.

Using Only Insertion Choice: It is possible to untangle a matching using only insertion choice and $\mathcal{O}(n^2)$ flips [9]. Let σ be the *spread* of P, that is, the ratio between the maximum and minimum distances among points in P. Using insertion choice, it is also possible to untangle a matching using $\mathcal{O}(n\sigma)$ flips [7].

Using Only Removal Choice: If P is in convex position, then by using $\mathcal{O}(n)$ flips we can untangle a TSP tour [25,28], as well as a red-blue matching [7], while the best known bound for trees is $\mathcal{O}(n \log n)$ [7]. If instead of convex position, we have colinear red points in a red-blue matching, then $\mathcal{O}(n^2)$ flips suffice [7,12].

Using Both Removal and Insertion Choices: If P is in convex position, then by using $\mathcal{O}(n)$ flips we can untangle a matching [7].

1.1   New Results

Previous results are usually stated for a single graph property. Using choices, we are able to state the results in a more general setting. Proofs that use insertion choice are unlikely to generalize to red-blue matchings, TSP tours, or trees, where insertion choice is not available (still, they may hold for both non-bipartite matchings and multigraphs). In contrast, bounds for multigraphs using only removal choice apply to all these cases. Previously, we only knew linear or near-linear bounds when the points P are in convex position and removal choice is available. The goal of the paper is to obtain linear and near-linear bounds to as many cases as possible, considering near-convex configurations as well as removal and insertion choices.

Let $P = C \cup T$ where C is in convex position and the points of T are outside the convex hull of C, unless otherwise specified. Let S be a multiset of n segments with endpoints P and t be the number of segments with at least one endpoint in T. We prove the following results to untangle S, and some are summarized in Table 1.

Using only insertion choice (Section 2): If $T = \emptyset$, then $\mathcal{O}(n \log n)$ flips suffice. If T is separated from C by two parallel lines, then $\mathcal{O}(tn \log n)$ flips suffice.

Using only removal choice (Section 3): If $|T| \leq 2$ and $t = \mathcal{O}(1)$, then $\mathcal{O}(n \log n)$ flips suffice. In this case, our results hold with the points T being anywhere with respect to the convex hull of C, however, if both points are outside, then S must be a matching. As the bound for $|T| \leq 1$ holds for trees, it is useful to compare it against the $\mathcal{O}(n \log n)$ bound for trees in the convex case from [7] that strongly uses the fact that S forms a tree. The $\mathcal{O}(\log n)$ factor is not present for the special cases of TSP tours and red-blue matchings.

Using both removal and insertion choices (Section 4): If T is separated from C by two parallel lines, then $\mathcal{O}(tn)$ flips suffice. If T is anywhere outside the convex hull of C and S is a matching, then $\mathcal{O}(t^3 n)$ flips suffice.

In a matching or TSP tour, we have $t = \mathcal{O}(|T|)$ and $n = \mathcal{O}(|P|)$, however in a tree, t can be as high as $\mathcal{O}(|T|^2)$. In a multigraph t and n can be much larger than $|T|$ and $|P|$. The theorems describe more precise bounds as functions of all these parameters. For simplicity, the introduction only shows bounds in terms of n and t.

1.2 Related Reconfiguration Problems

Combinatorial reconfiguration studies the step-by-step transition from one solution to another, for a given combinatorial problem. Many reconfiguration problems are presented in [18]. We give a brief overview of reconfiguration among line segments using alternative flip operations.

Table 1. Upper bounds to different versions of the problem with points having $\mathcal{O}(1)$ degree. The letter R corresponds to removal choice, I to insertion choice, and $\emptyset$ to no choice. New results are highlighted in yellow with the theorem number in parenthesis and tight bounds are bold.

Property:	Matching				TSP, Red-Blue	
Choices:	RI	I	R	$\emptyset$	R	$\emptyset$
Convex	n [7]	$n \log n$ (Thm 1)	$n \log n$ (Thm 3)	n^2	n [7,25,28]	n^2
$\|T\| = 1$	n (Thm 8)	$n \log n$ (Thm 2)	$n \log n$ (Thm 4)	n^2 [15]	n (Thm 4)	n^2 [15]
$\|T\| = 2$	n (Thm 9)	n^2 [9]	$n \log n$ (Thm 5)	n^2 [15]	n (Thm 5)	n^2 [15]
separated	tn (Thm 8)	$tn \log n$ (Thm 2)	tn^2 [15]			
$C \cup T$	$t^3 n$ (Thm 9)	n^2 [9]	tn^2 [15]			

The *2OPT flip* is not restricted to crossing segments. It removes and inserts pairs of segments (the four segments forming a cycle) as the total length decreases. In contrast to flips among crossing segments, the number of 2OPT flips performed may be exponential [13].

It is possible to relax the flip definition even further to all operations that replace two segments by two others forming a cycle [5,6,8,10,14,27]. This definition has also been considered for multigraphs [16,17,20].

Another type of flip consists of removing a single segment and inserting another one. Such flips are widely studied for triangulations [3,19,21,23,24,26]. They have also been considered for non-crossing trees [1] and paths. It is possible to reconfigure any two non-crossing paths if the points are in convex position [4, 11] or if there is *one* point inside the convex hull [2].

1.3 Preliminaries

Throughout, we consider *multigraphs* (P, S) whose vertices P (called *endpoints*) are points in the plane and edges S are a multiset of line *segments*. We assume that the endpoints are in *general position* and that the two endpoints of a segment are distinct. Given two (possibly equal) sets P_1, P_2 of endpoints, we say that a segment is a $P_1 P_2$-*segment* if one endpoint is in P_1 and the other is in P_2. Similarly, we say that a segment is a P_1-*segment* if at least one endpoint is in P_1.

We say that two segments *cross* if they intersect at a single point that is not an endpoint of either segment. We say that a line *crosses* a segment if they intersect at a single point that is not an endpoint of the segment. We say that a segment or a line h *separates* a set of points P if P can be partitioned into two non-empty sets P_1, P_2 such that every segment $p_1 p_2$ with $p_1 \in P_1, p_2 \in P_2$ crosses h. Several proofs in this paper use the following two lemmas from previous papers.

Lemma 1 ([22]).   *Given a multiset S of segments and a line ℓ, let the* line potential *of ℓ, denoted $\lambda(\ell)$, be the number of segments in S crossing ℓ. Then, $\lambda(\ell)$ never increases at a flip.*

Lemma 2 ([9]).   *Consider a partition $S = \bigcup_i S_i$ of the multiset S of segments and let P_i be the set of endpoints of S_i. If no segment of $\binom{P_i}{2}$ crosses a segment of $\binom{P_j}{2}$ for $i \neq j$, then the sequences of flips in each S_i are independent.*

We say that a segment s is *uncrossable* if for any two endpoints p_1, p_2, we have that $p_1 p_2$ do not cross s. Lemma 2 implies that an uncrossable segment cannot be flipped.

Our bounds often have terms like $\mathcal{O}(tn)$ and $\mathcal{O}(n \log |C|)$ that would incorrectly become 0 if t or $\log |C|$ is 0. In order to avoid this problem, factors in the $\mathcal{O}$ notation should be made at least 1. For example, the aforementioned bounds should be respectively interpreted as $\mathcal{O}((1 + t)n)$ and $\mathcal{O}(n \log(2 + |C|))$.

2   Insertion Choice

In this section, we show how to untangle a multigraph using only insertion choice, that is, our strategies do not choose which pair of crossing segments is removed, but only which pair of segments with the same endpoints is subsequently inserted. We start with the convex case, followed by points outside the convex separated by two parallel lines.

2.1   Convex

Let $P = C = \{p_1, \ldots, p_{|C|}\}$ be a set of points in convex position sorted in counterclockwise order along the convex hull boundary (Fig. 2(a)). Given a segment $p_a p_b$, we define the *depth* $\delta(p_a p_b) = |b - a|$. This definition resembles but is not the same as the depth used in [7]. We use the depth to prove the following theorem.

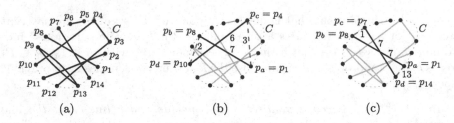

Fig. 2. (a) A multigraph (C, S) with $|C| = 14$ points in convex position and $n = 9$ segments. (b) Insertion choice for Cases 1 and 2 of the proof of Theorem 1. (c) Insertion choice for Case 3.

Theorem 1. *Every multigraph (C, S) with C in convex position has an untangle sequence of length $\mathcal{O}(n \log |C|) = \mathcal{O}(n \log n)$ using only insertion choice, where $n = |S|$.*

Proof. Let the potential function $\phi(S) = \prod_{s \in S} \delta(s)$. As $\delta(s) \in \{1, \ldots, |C| - 1\}$, we have that $\phi(S)$ is integer, positive, and at most $|C|^n$. Next, we show that for any flipped pair of segments $p_a p_b, p_c p_d$ there exists an insertion choice that multiplies $\phi(S)$ by a factor of at most $3/4$, and the theorem follows.

Consider a flip of a segment $p_a p_b$ with a segment $p_c p_d$ and assume without loss of generality that $a < c < b < d$. The contribution of the pair of segments $p_a p_b, p_c p_d$ to the potential $\phi(S)$ is the factor $f = \delta(p_a p_b)\delta(p_c p_d)$. Let f' be the factor corresponding to the pair of inserted segments.

Case 1: If $\delta(p_a p_c) \leq \delta(p_c p_b)$, then we insert the segments $p_a p_c$ and $p_b p_d$ and we get $f' = \delta(p_a p_c)\delta(p_b p_d)$ (Fig. 2(b)). We notice $\delta(p_a p_b) = \delta(p_a p_c) + \delta(p_c p_b)$. It follows $\delta(p_a p_c) \leq \delta(p_a p_b)/2$ and we have $\delta(p_b p_d) \leq \delta(p_c p_d)$ and then $f' \leq f/2$.

Case 2: If $\delta(p_b p_d) \leq \delta(p_c p_b)$, then we insert the same segments $p_a p_c$ and $p_b p_d$ as previously. We have $\delta(p_a p_c) \leq \delta(p_a p_b)$ and $\delta(p_b p_d) \leq \delta(p_c p_d)/2$, which gives $f' \leq f/2$.

Case 3: If (i) $\delta(p_a p_c) > \delta(p_c p_b)$ and (ii) $\delta(p_b p_d) > \delta(p_c p_b)$, then we insert the segments $p_a p_d$ and $p_c p_b$ (Fig. 2(c)). The contribution of the new pair of segments is $f' = \delta(p_a p_d)\delta(p_c p_b)$. We introduce the coefficients $x = \frac{\delta(p_a p_c)}{\delta(p_c p_b)}$ and $y = \frac{\delta(p_b p_d)}{\delta(p_c p_b)}$ so that $\delta(p_a p_c) = x\delta(p_c p_b)$ and $\delta(p_b p_d) = y\delta(p_c p_b)$. It follows that $\delta(p_a p_b) = (1+x)\delta(p_c p_b)$, $\delta(p_c p_d) = (1+y)\delta(p_c p_b)$ and $\delta(p_a p_d) = (1+x+y)\delta(p_c p_b)$. The ratio f'/f is equal to a function $g(x, y) = \frac{1+x+y}{(1+x)(1+y)}$. Due to (i) and (ii), we have that $x \geq 1$ and $y \geq 1$. In other words, we can upper bound the ratio f'/f by the maximum of the function $g(x, y)$ with $x, y \geq 1$. It is easy to show that the function $g(x, y)$ is decreasing with both x and y. Then its maximum is obtained for $x = y = 1$ and it is equal to $3/4$, showing that $f' \leq 3f/4$.   □

2.2   Separated by Two Parallel Lines

In this section, we prove the following theorem, which is a generalization of Theorem 1.

Theorem 2. *Consider a multigraph (P, S) with $P = C \cup T_1 \cup T_2$ where C is in convex position and there exist two horizontal lines ℓ_1, ℓ_2, with T_1 above ℓ_1 above C above ℓ_2 above T_2. Let $T = T_1 \cup T_2$, $n = |S|$, and t be the number of T-segments. There exists an untangle sequence of length $\mathcal{O}(t|P|\log|C|+n\log|C|) = \mathcal{O}(tn\log n)$ using only insertion choice.*

Proof. We start by describing the insertion choice for flips involving at least one point in T. Let $p_1, \ldots, p_{|P|}$ be the points P sorted vertically from top to bottom. Consider a flip involving the points p_a, p_b, p_c, p_d with $a < b < c < d$. The insertion choice is to create the segments $p_a p_b$ and $p_c p_d$. As in [9], we define the potential η of a segment $p_i p_j$ as $\eta(p_i p_j) = |i - j|$. Notice that η is an integer between 1 and $|P| - 1$. We define $\eta_T(S)$ as the sum of $\eta(p_i p_j)$ for $p_i p_j \in S$ with p_i or p_j in T. Notice that $0 < \eta_T(S) < t|P|$. It is easy to verify that any flip involving a point in T decreases $\eta_T(S)$ and other flips do not change $\eta_T(S)$. Hence, the number of flips involving at least one point in T is $\mathcal{O}(t|P|)$.

For the flips involving only points of C, we use the same choice as in the proof of Theorem 1. The potential function

$$\phi(S) = \prod_{p_i p_j \in S \, : \, p_i \in C \text{ and } p_j \in C} \delta(p_i p_j)$$

is at most $|C|^n$ and decreases by a factor of at most $3/4$ at every flip that involves only points of C.

However, $\phi(S)$ may increase by a factor of $\mathcal{O}(|C|^2)$ when performing a flip that involves a point in T. As such flips only happen $\mathcal{O}(t|P|)$ times, the total increase is at most a factor of $|C|^{\mathcal{O}(t|P|)}$.

Concluding, the number of flips involving only points in C is at most

$$\log_{4/3}\left(|C|^{\mathcal{O}(n)}|C|^{\mathcal{O}(t|P|)}\right) = \mathcal{O}(n\log|C| + t|P|\log|C|).$$

$\square$

3 Removal Choice

In this section, we show how to untangle a multigraph using only removal choice. We start with the convex case, followed by 1 point inside or outside the convex hull of C, then 2 points outside the convex hull of C, 2 points inside the convex hull of C, and 1 point inside and 1 outside the convex hull of C. As only removal choice is used, all results also apply to red-blue matchings, TSP tours, and trees.

3.1 Convex

Let $P = C = \{p_1, \ldots, p_{|C|}\}$ be a set of points in convex position sorted in counterclockwise order along the convex hull boundary and consider a set of segments S with endpoints P. Given a segment $p_a p_b$ and assuming without loss of generality that $a < b$, we define the *crossing depth* $\delta_\times(p_a p_b)$ as the number of points in $p_{a+1}, \ldots, p_{b-1}$ that are an endpoint of a segment in S that crosses any other segment in S (not necessarily $p_a p_b$). We use the crossing depth to prove the following theorem, which implies a simpler and more general proof of the $\mathcal{O}(n \log n)$ bound for trees [7].

Theorem 3. *Every multigraph (C, S) with C in convex position has an untangle sequence of length $\mathcal{O}(n \log |C|) = \mathcal{O}(n \log n)$ using only removal choice, where $n = |S|$.*

Proof. We repeat the following procedure until there are no more crossings. Let $p_a p_b \in S$ be a segment *with crossings* (hence, crossing depth at least one) and $a < b$ minimizing $\delta_\times(p_a p_b)$ (Fig. 3(a)). Let $q_1, \ldots, q_{\delta_\times(p_a p_b)}$ be the points defining $\delta_\times(p_a p_b)$ in order and let $i = \lceil \delta_\times(p_a p_b)/2 \rceil$. Since $p_a p_b$ has minimum crossing depth, the point q_i is the endpoint of segment $q_i p_c$ that crosses $p_a p_b$. When flipping $q_i p_c$ and $p_a p_b$, we obtain a segment s (either $s = q_i p_a$ or $s = q_i p_b$) with $\delta_\times(s)$ at most half of the original value of $\delta_\times(p_a p_b)$ (Fig. 3(b,c)). Hence, this operation always divides the value of the smallest positive crossing depth by at least two. As the crossing depth is an integer smaller than $|C|$, after performing this operation $\mathcal{O}(\log |C|)$ times, it produces a segment of crossing depth 0. As the segments of crossing depth 0 can no longer participate in a flip, the claimed bound follows. □

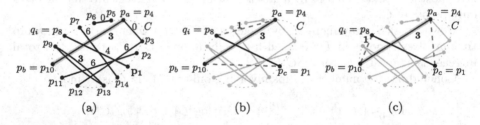

Fig. 3. Proof of Theorem 3. (a) The segments of a convex multigraph are labeled with the crossing depth. (b,c) Two possible pairs of inserted segments, with one segment of the pair having crossing depth $\lfloor \frac{3}{2} \rfloor = 1$.

3.2   One Point Inside or Outside a Convex Region

In this section, we prove Theorem 4. In the case of TSP tours [25,28] and red-blue matchings [8], the preprocessing to untangle CC-segments takes $\mathcal{O}(n)$ flips. However, in the case of trees [8] and in general (Theorem 3), the best bound known is $\mathcal{O}(n \log n)$. We first state a lemma used to prove Theorem 4.

Lemma 3. *Consider a set C of points in convex position, and a multiset S of n crossing-free segments with endpoints in C. Consider the multiset $S \cup \{s\}$ where s is an extra segment with one endpoint in C and one endpoint q anywhere in the plane. There exists an untangle sequence for $S \cup \{s\}$ of length $\mathcal{O}(n)$ using only removal choice.*

Proof. Iteratively flip the segment qp_1 with the segment $p_2p_3 \in S$ crossing qp_1 the farthest from q. This flip inserts a CC-segment p_1p_2, which is impossible to flip again, because the line p_1p_2 is crossing free. The flip does not create any crossing between CC-segments.   □

We are now ready to state and prove the theorem.

Theorem 4. *Consider a multigraph (P, S) with $P = C \cup T$ where C is in convex position, where $T = \{q\}$, and such that there is no crossing pair of CC-segments (possibly after a preprocessing for the convex case). Let $n = |S|$ and t be the number of T-segments. There exists an untangle sequence of length $\mathcal{O}(tn)$ using only removal choice.*

Proof. For each segment s with endpoint q with crossing, we apply Lemma 3 to s and the CC-segments crossing s. Once a segment s incident to q is crossing free, it is impossible to flip it again as we fall in one of the following cases. Let ℓ be the line containing s.

Case 1: If ℓ is crossing free, then it splits the multigraph in three partitions: the segments on one side of ℓ, the segments on the other side of ℓ, and the segment s itself.

Case 2: If ℓ is not crossing free and q is outside the convex hull of C, then s is *uncrossable*.

Case 3: If q is inside the convex hull of C, then introducing a crossing on s would require that q lies in the interior of the convex quadrilateral whose diagonals are the two segments removed by a flip. The procedure excludes this possibility by ensuring that there are no crossing pair of CC-segments, and, therefore, that one of the removed segment already has q as an endpoint.

Therefore, we need at most n flips for each of the t segments incident to q. □

3.3 Two Points Not in Convex Position

In this section, we consider the case of two points that are not in convex position. We present the proof for a matching with two points outside a convex region. The remaining cases hold for general multigraphs but the proofs are long and technical. These proofs are presented in the full version.

Theorem 5. *Consider a matching (P, S) with $P = C \cup T$ where C is in convex position, where $T = \{q, q'\}$, and such that there is no crossing pair of CC-segments (possibly after a preprocessing for the convex case). Let $n = |S|$. There exists an untangle sequence of length $\mathcal{O}(n)$ using only removal choice.*

Proof. Throughout this proof, we partition the TT-segments respectively the CT-segments into two types: *TTI-segments* and *CTI-segments* if they intersect the interior of the convex hull of C and *TTO-segments* and *CTO-segments* otherwise. Next, we describe the removal choices of an untangle sequence of a matching such as the one shown in Fig. 4(a). We decompose this sequence into two phases.

Phase 1: handle a TT-segment. If the TT-segment $s = qq'$ is in S and crosses no segment in S, then S is crossing free. If the TT-segment $s = qq'$ is in S and crosses at least one segment in S (necessarily a CC-segment), then we flip $s = qq'$ with any segment, say s'. The line containing s' now splits S into one matching with q and another matching with q'. Figures 4(b) and (c) show the splitting line in the two possible insertion cases. We untangle each of these two matchings using $\mathcal{O}(n)$ flips by Theorem 4. Figure 4(d) shows an example of the matching S at the end of Phase 1.

Phase 2: handle CT-segments. We remove an arbitrary CT-segment s, say the segment incident to q, from S. We then untangle S using $\mathcal{O}(n)$ flips by Theorem 4, and insert the segment s back in S afterwards. Notice that all crossings are now on s.

While s', the segment of S that crosses s the farthest away from q, is a CC-segment, we flip s and s' and we set s to be the newly inserted CT-segment incident to q. By Lemma 3, at most $\mathcal{O}(n)$ flips are performed in this loop.

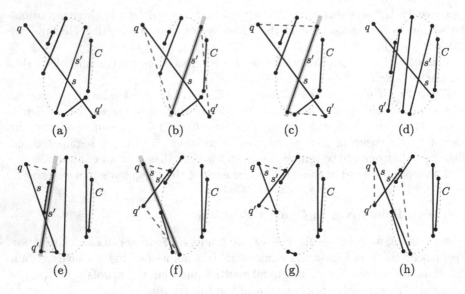

Fig. 4. Illustrations for the proof of Theorem 5. (a) Example of the matching S before Phase 1. (b) & (c) Illustration of the two insertion cases of Phase 1. In both insertion cases, the highlighted line slpits the matching. (d) Example of the matching S before Phase 2. (e) & (f) & (g) The three sub-cases of Insertion case 1 of Phase 2. (h) Example of the Insertion case 2 of Phase 2.

At the end of the loop, either s is crossing free, or s' is adjacent to q'. Then, we also flip s and s'.

Insertion case 1: If two CT-segments are inserted, then we examine the two following three cases.

If s' is a CTI-segment, then the line containing s' now splits S into one matching with q and another matching with q' (Fig. 4(e)). Again, we untangle each of these two matchings using $\mathcal{O}(n)$ flips by Theorem 4.

If s' is a CTO-segment and s is a CTI-segment, then s' is the only segment crossing s before the flip. Thus, after the flip, the line containing s splits S into one matching with q and another matching with q' (Fig. 4(f)). Again, we untangle each of these two matchings using $\mathcal{O}(n)$ flips by Theorem 4.

If both s and s' are CTO-segments, then S is crossing free after the flip (Fig. 4(g)).

Insertion case 2: If the TT-segment qq' is inserted (Fig. 4(h)), then we apply Phase 1 to untangle S using $\mathcal{O}(n)$ flips.

In total, we have used $\mathcal{O}(n)$ flips to untangle S.   □

In the full version, we prove the following two theorems that handle the remaining cases of two points that are not in convex position.

Theorem 6. *Consider a multigraph (P, S) with $P = C \cup T$ where C is in convex position, the points of T are inside the convex hull of C, and $T = \{q, q'\}$. Let $n = |S|$ and t be the number of T-segments. There exists an untangle sequence of length $\mathcal{O}(d_{conv}(n) + tn)$ using only removal choice, where $d_{conv}(n)$ is the number of flips to untangle any multiset of at most n segments with endpoints in convex position.*

Proof. (Sketch) The untangle sequence is decomposed in five phases. At the end of each phase, one more type of crossings is removed, and types of crossings removed in the previous phases are not present, even if they may temporarily appear during the phase. *Phase 1.* We untangle the CT-segments using $\mathcal{O}(d_{\text{conv}}(n))$ flips. *Phase 2.* We untangle the CC-segments using $\mathcal{O}(d_{\text{conv}}(n))$ flips. *Phase 3.* We remove the crossings between CT-segments and CC-segments that cross qq' using $\mathcal{O}(tn)$ flips. *Phase 4.* We remove the remaining crossings between CT-segments and CC-segments using $\mathcal{O}(tn)$ flips. *Phase 5.* We remove the remaining crossings, which are between qq' and CC-segments, using $\mathcal{O}(d_{\text{conv}}(n))$ flips.   □

Theorem 7. *Consider a multigraph (P, S) with $P = C \cup T$ where C is in convex position, and $T = \{q, q'\}$ such that q is outside the convex hull of C and q' is inside the convex hull of C. Let $n = |S|$ and t be the number of T-segments. There exists an untangle sequence of length $\mathcal{O}(d_{conv}(n) + \delta(q)\delta(q')n) = \mathcal{O}(d_{conv}(n) + t^2n)$ using only removal choice, where $d_{conv}(n)$ is the number of flips to untangle any multiset of at most n segments with endpoints in convex position.*

Proof. (Sketch) The untangle sequence is decomposed in four phases. *Phase 1.* We untangle the CC-segments using $d_{\text{conv}}(n)$ flips. *Phase 2.* We remove the crossings between Cq'-segments and CC-segments using $\mathcal{O}(tn)$ flips (Theorem 4). *Phase 3.* We remove the crossings between Cq-segments using $\mathcal{O}(t^2n)$ flips. *Phase 4.* We remove the crossings between qq' and CC-segments, using $\mathcal{O}(tn)$ flips.   □

4   Removal and Insertion Choices

In this section, we show how to untangle a matching or a multigraph using both removal and insertion choices. We start with the case of points outside the convex separated by two parallel lines. Afterwards, we prove an important lemma and apply it to untangle a matching with points outside the convex.

4.1   Separated by Two Parallel Lines

We start with the simpler case in which T is separated from C by two parallel lines. In this case, our bound of $\mathcal{O}(n + t|P|)$ interpolates the tight convex bound of $\mathcal{O}(n)$ from [7] and the $\mathcal{O}(t|P|)$ bound from [9] for t arbitrary segments.

Theorem 8. *Consider a multigraph (P, S) with $P = C \cup T_1 \cup T_2$ where C is in convex position and there exist two horizontal lines ℓ_1, ℓ_2, with T_1 above ℓ_1 above C above ℓ_2 above T_2. Let $n = |S|$, $T = T_1 \cup T_2$, and t be the number of T-segments. There exists an untangle sequence of length $\mathcal{O}(n + t|P|) = \mathcal{O}(tn)$ using both removal and insertion choices.*

Proof. The algorithm runs in two phases.

Phase 1. We use removal choice to perform the flips involving a point in T. At the end of the first phase, there can only be crossings among segments with all endpoints in C. The insertion choice for the first phase is the following. Let $p_1, \ldots, p_{|P|}$ be the points P sorted vertically from top to bottom. Consider a flip involving the points p_a, p_b, p_c, p_d with $a < b < c < d$. The insertion choice is to create the segments $p_a p_b$ and $p_c p_d$. As in [9], we define the potential η of a segment $p_i p_j$ as $\eta(p_i p_j) = |i - j|$. Notice that η is an integer from 1 to $|P| - 1$. We define $\eta(S)$ as the sum of $\eta(p_i p_j)$ for $p_i p_j \in S$ with p_i or p_j in T. Notice that $0 < \eta(S) < t|P|$. It is easy to verify that any flip involving a point in T decreases $\eta(S)$. Hence, the number of flips in Phase 1 is $\mathcal{O}(t|P|)$.

Phase 2. Since T is outside the convex hull of C, flips between segments with all endpoints in C cannot create crossings with the other segments, which are guaranteed to be crossing free at this point. Hence, it suffices to run an algorithm to untangle a convex set with removal and insertion choice from [7], which performs $\mathcal{O}(n)$ flips.   □

4.2   Liberating a Line

In this section, we prove the following key lemma, which we use next. The lemma only applies to matchings and it is easy to find a counter-example for multisets (S consisting of n copies of a single segment that crosses pq).

Lemma 4. *Consider a matching S of n segments with endpoints C in convex position, and a segment pq separating C. Using $\mathcal{O}(n)$ flips with removal and insertion choices on the initial set $S \cup \{pq\}$, we obtain a set of segments that do not cross the line pq.*

Proof. For each flip performed in the subroutine described hereafter, at least one of the inserted segments does not cross the line pq and is removed from S (see Fig. 5).

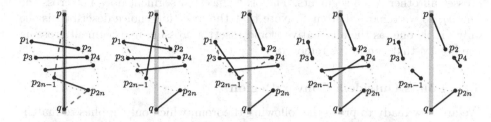

Fig. 5. An untangle sequence of the subroutine to liberate the line pq (with $n = 4$).

Preprocessing. First, we remove from S the segments that do not intersect the line pq, as they are irrelevant. Second, anytime two segments in S cross, we flip them choosing to insert the pair of segments not crossing the line pq. One such flip removes two segments from S. Let p_1p_2 (respectively $p_{2n-1}p_{2n}$) be the segment in S whose intersection point with pq is the closest from p (respectively q). Without loss of generality, assume that the points p_1 and p_{2n-1} are on the same side of the line pq.

First Flip. Elementary geometry yields that at least one of the segments among pp_{2n-1}, qp_1, qp_2 intersects all the segments of S (see full version).

Without loss of generality, assume that pp_{2n-1} is such a segment, i.e., that pp_{2n-1} crosses all segments of $S \setminus \{p_{2n-1}p_{2n}\}$. We choose to remove the segments pq and $p_{2n-1}p_{2n}$, and we choose to insert the segments pp_{2n-1} and qp_{2n}. As the segment qp_{2n} does not cross the line pq, we remove it from S.

Second Flip. We choose to flip the segments pp_{2n-1} and p_1p_2. If n is odd, we choose to insert the pair of segments pp_1, p_2p_{2n-1}. If n is even, we insert the segments pp_2, p_1p_{2n-1}.

By convexity, one of the inserted segment (the one with endpoints in C) crosses all other $n - 2$ segments. The other inserted segment (the one with p as one of its endpoints) does not cross the line pq, so we remove it from S. Note that the condition on the parity of n is there only to ensure that the last segment $p_{2n-3}p_{2n-2}$ is dealt with at the last flip.

Remaining Flips. We describe the third flip. The remaining flips are performed similarly. Let s be the previously inserted segment. Let p_3p_4 be the segment in S whose intersection point with pq is the closest from p. Without loss of generality, assume that p_3 is on the same side of the line pq as p_1 and p_{2n-1}.

We choose to flip s with p_3p_4. If $s = p_2p_{2n-1}$, we choose to insert the pair of segments p_2p_4, p_3p_{2n-1}. If $s = p_1p_{2n-1}$, we choose to insert the pair of segments p_1p_3, p_4p_{2n-1}.

By convexity, one inserted segment (the one with p_{2n-1} as an endpoint) crosses all other $n - 3$ segments. The other inserted segment does not cross the line pq, so we remove it from S. Note that the insertion choice described is the only viable one, as the alternative would insert a crossing-free segment crossing the line pq that cannot be removed.   □

4.3   Points Outside a Convex Region

We are now ready to prove the following theorem, which only applies to matchings because it uses Lemma 4.

Theorem 9. *Consider a matching S consisting of n segments with endpoints $P = C \cup T$ where C is in convex position and T is outside the convex hull of C. Let $t = |T|$. There exists an untangle sequence of length $\mathcal{O}(t^3n)$ using both removal and insertion choices.*

Proof. Throughout this proof, we partition the TT-segments into two types: TTI-*segment* if it intersects the interior of the convex hull of C and TTO-*segment* otherwise.

TT-segments. At any time during the untangle procedure, if there is a TTI-segment s that crosses more than t segments, we apply Lemma 4 to liberate s from every CC-segment using $\mathcal{O}(n)$ flips. Let ℓ be the line containing s. Since $\lambda(\ell)$ cannot increase (Lemma 1), $\lambda(\ell) < t$ after Lemma 4, and there are $\mathcal{O}(t^2)$ different TTI-segments, it follows that Lemma 4 is applied $\mathcal{O}(t^2)$ times, performing a total $\mathcal{O}(t^2n)$ flips. As the number of times s is inserted and removed differ by at most 1 and $\lambda(\ell)$ decreases at each flip that removes s, it follows that s participates in $\mathcal{O}(t)$ flips. As there are $\mathcal{O}(t^2)$ different TTI-segments, the total number of flips involving TTI-segments is $\mathcal{O}(t^3)$.

We define a set L of $\mathcal{O}(t)$ lines as follows. For each point $q \in T$, we have two lines $\ell_1, \ell_2 \in L$ that are the two tangents of the convex hull of C that pass through q. As the lines $\ell \in L$ do not separate C, the potential $\lambda(\ell) = \mathcal{O}(t)$. When flipping a TTO-segment q_1q_2 with another segment q_3p with $q_3 \in T$ (p may be in T or in C), we make the insertion choice of creating a TTO-segment q_1q_3 such that there exists a line $\ell \in L$ whose potential $\lambda(\ell)$ decreases. It is easy to verify that ℓ always exist (see full version). Hence, the number of flips involving TTO-segments is $\mathcal{O}(t^2)$ and the number of flips involving TT-segments in general is $\mathcal{O}(t^3)$.

All except pairs of CC-segments. We keep flipping segments that are not both CC-segments with the following insertion choices. Whenever we flip two CT-segments, we make the insertion choice of creating a TT-segment. Hence, as the number of flips involving TT-segments is $\mathcal{O}(t^3)$, so is the number of flips of two CT-segments.

Whenever we flip a CT-segment p_1q with $q \in T$ and a CC-segment p_3p4, we make the following insertion choice. Let $v(q)$ be a vector such that the dot product $v(q) \cdot q < v(q) \cdot p$ for all $p \in C$, that is, v is orthogonal to a line ℓ separating q from C and pointing towards C. We define the potential $\eta(p_x q)$ of a segment with $p_x \in C$ and $q \in T$ as the number of points $p \in C$ such that $v(q) \cdot p < v(q) \cdot p_x$, that is the number of points in C before p_x in direction v. We choose to insert the segment $p_x q$ that minimizes $\eta(p_x q)$ for $x = \{1,2\}$. Let $\eta(S)$ be the sum of $\eta(p_x q)$ for all CT-segments $p_x q$ in S. It is easy to see that $\eta(S)$ is $\mathcal{O}(t|C|)$ and decreases at each flip involving a CT-segment (not counting the flips inside Lemma 4).

There are two situation in which $\eta(S)$ may increase. One is when Lemma 4 is applied, which happens $\mathcal{O}(t^2)$ times. Another one is when a TT-segment and a CC-segment flip, creating two CT-segments, which happens $\mathcal{O}(t^3)$ times. At each of these two situations, $\eta(S)$ increases by $\mathcal{O}(|C|)$. Consequently, the number of flips between a CT-segment and a CC-segment is $\mathcal{O}(t^3|C|) = \mathcal{O}(t^3 n)$.

CC-segments. By removal choice, we choose to flip the pairs of CC-segments last (except for the ones flipped in Lemma 4). As T is outside the convex hull of C, flipping two CC-segments does not create crossings with other segments (Lemma 2). Hence, we apply the algorithm from [7] to untangle the remaining segments using $\mathcal{O}(n)$ flips. □

References

1. Aichholzer, O., et al.: Reconfiguration of non-crossing spanning trees. arXiv preprint (2022). https://arxiv.org/abs/2206.03879
2. Aichholzer, O., et al.: Flipping plane spanning paths. In: International Conference and Workshops on Algorithms and Computation (WALCOM) (2023). https://doi.org/10.1007/978-3-031-27051-2_5, https://arxiv.org/abs/2202.10831
3. Aichholzer, O., Mulzer, W., Pilz, A.: Flip distance between triangulations of a simple polygon is NP-complete. Discrete Comput. Geom. **54**(2), 368–389 (2015). https://doi.org/10.1007/s00454-015-9709-7
4. Akl, S.G., Islam, M.K., Meijer, H.: On planar path transformation. Inf. Process. Lett. **104**(2), 59–64 (2007). https://doi.org/10.1016/j.ipl.2007.05.009
5. Bereg, S., Ito, H.: Transforming graphs with the same degree sequence. In: Computational Geometry and Graph Theory, pp. 25–32 (2008). https://doi.org/10.1007/978-3-540-89550-3_3
6. Bereg, S., Ito, H.: Transforming graphs with the same graphic sequence. J. Inf. Process. **25**, 627–633 (2017). https://doi.org/10.2197/ipsjjip.25.627
7. Biniaz, A., Maheshwari, A., Smid, M.: Flip distance to some plane configurations. Comput. Geom. **81**, 12–21 (2019). https://doi.org/10.1016/j.comgeo.2019.01.008, https://arxiv.org/abs/1905.00791
8. Bonamy, M., et al.: The perfect matching reconfiguration problem. In: 44th International Symposium on Mathematical Foundations of Computer Science (MFCS). LIPIcs, vol. 138, pp. 80:1–80:14 (2019). https://doi.org/10.4230/LIPIcs.MFCS.2019.80
9. Bonnet, É., Miltzow, T.: Flip distance to a non-crossing perfect matching. arXiv 1601.05989 (2016). http://arxiv.org/abs/1601.05989
10. Bousquet, N., Joffard, A.: Approximating shortest connected graph transformation for trees. In: Theory and Practice of Computer Science, pp. 76–87 (2020). https://doi.org/10.1007/978-3-030-38919-2_7

11. Chang, J.M., Wu, R.Y.: On the diameter of geometric path graphs of points in convex position. Inf. Process. Lett. **109**(8), 409–413 (2009). https://doi.org/10.1016/j.ipl.2008.12.017

12. Das, A.K., Das, S., da Fonseca, G.D., Gerard, Y., Rivier, B.: Complexity results on untangling red-blue matchings. Comput. Geom. **111**, 101974 (2023). https://doi.org/10.1016/j.comgeo.2022.101974, https://arxiv.org/abs/2202.11857

13. Englert, M., Röglin, H., Vöcking, B.: Worst case and probabilistic analysis of the 2-Opt algorithm for the TSP. Algorithmica **68**(1), 190–264 (2014). https://doi.org/10.1007/s00453-013-9801-4

14. Erdős, P.L., Király, Z., Miklós, I.: On the swap-distances of different realizations of a graphical degree sequence. Comb. Probab. Comput. **22**(3), 366–383 (2013). https://doi.org/10.1017/S0963548313000096

15. da Fonseca, G.D., Gerard, Y., Rivier, B.: On the longest flip sequence to untangle segments in the plane. In: Lin, CC., Lin, B.M.T., Liotta, G. (eds.) WALCOM: Algorithms and Computation. WALCOM 2023. LNCS, vol. 13973. Springer, Cham (2023). https://doi.org/10.1007/978-3-031-27051-2_10

16. Hakimi, S.L.: On realizability of a set of integers as degrees of the vertices of a linear graph. I. J. Soc. Ind. Appl. Math. **10**(3), 496–506 (1962)

17. Hakimi, S.L.: On realizability of a set of integers as degrees of the vertices of a linear graph II. uniqueness. J. Soc. Ind. Appl. Math. **11**(1), 135–147 (1963)

18. van den Heuvel, J.: The complexity of change. Surv. Comb. **409**, 127–160 (2013)

19. Hurtado, F., Noy, M., Urrutia, J.: Flipping edges in triangulations. Discrete Comput. Geom. **22**(3), 333–346 (1999)

20. Joffard, A.: Graph domination and reconfiguration problems. Ph.D. thesis, Université Claude Bernard Lyon 1 (2020)

21. Lawson, C.L.: Transforming triangulations. Discret. Math. **3**(4), 365–372 (1972)

22. van Leeuwen, J., Schoone, A.A.: Untangling a traveling salesman tour in the plane. In: 7th Workshop on Graph-Theoretic Concepts in Computer Science (1981)

23. Lubiw, A., Pathak, V.: Flip distance between two triangulations of a point set is NP-complete. Comput. Geom. **49**, 17–23 (2015). https://doi.org/10.1016/j.comgeo.2014.11.001

24. Nishimura, N.: Introduction to reconfiguration. Algorithms **11**(4), 52 (2018). https://doi.org/10.3390/a11040052

25. Oda, Y., Watanabe, M.: The number of flips required to obtain non-crossing convex cycles. In: Kyoto International Conference on Computational Geometry and Graph Theory, pp. 155–165 (2007)

26. Pilz, A.: Flip distance between triangulations of a planar point set is APX-hard. Comput. Geom. **47**(5), 589–604 (2014). https://doi.org/10.1016/j.comgeo.2014.01.001

27. Will, T.G.: Switching distance between graphs with the same degrees. SIAM J. Discret. Math. **12**(3), 298–306 (1999). https://doi.org/10.1137/S0895480197331156

28. Wu, R., Chang, J., Lin, J.: On the maximum switching number to obtain non-crossing convex cycles. In: 26th Workshop on Combinatorial Mathematics and Computation Theory, pp. 266–273 (2009)

Improved Approximation Algorithms for Cycle and Path Packings

Jingyang Zhao[ID] and Mingyu Xiao[(✉)][ID]

University of Electronic Science and Technology of China, Chengdu, China
myxiao@gmail.com

Abstract. Given an edge-weighted (metric/general) complete graph with n vertices, the maximum weight (metric/general) k-cycle/path packing problem is to find a set of $\frac{n}{k}$ vertex-disjoint k-cycles/paths such that the total weight is maximized. In this paper, we consider approximation algorithms. For metric k-cycle packing, we improve the previous approximation ratio from 3/5 to 7/10 for $k = 5$, and from $7/8 \cdot (1 - 1/k)^2$ for $k > 5$ to $(7/8 - 0.125/k)(1 - 1/k)$ for constant odd $k > 5$ and to $7/8 \cdot (1 - 1/k + \frac{1}{k(k-1)})$ for even $k > 5$. For metric k-path packing, we improve the approximation ratio from $7/8 \cdot (1 - 1/k)$ to $\frac{27k^2 - 48k + 16}{32k^2 - 36k - 24}$ for even $10 \geq k \geq 6$. For the case of $k = 4$, we improve the approximation ratio from 3/4 to 5/6 for metric 4-cycle packing, from 2/3 to 3/4 for general 4-cycle packing, and from 3/4 to 14/17 for metric 4-path packing.

Keywords: Approximation algorithms · Cycle packing · Path packing

1 Introduction

In a graph with n vertices, a *k-cycle/path packing* is a set of $\frac{n}{k}$ vertex-disjoint k-cycles/paths (i.e., a simple cycle/path on k different vertices) covering all vertices. For an edge-weighted complete graph, every edge has a non-negative weight. Moreover, it is called a *metric* graph if the weight satisfies the triangle inequality; Otherwise, it is called a *general* graph. Given a (metric/general) graph, the maximum weight (metric/general) k-cycle/path packing problem (kCP/kPP) is to find a k-cycle/path packing such that the total weight of the k-cycles/paths in the packing is maximized.

When $k = n$, kCP becomes the well-known maximum weight traveling salesman problem (MAX TSP). One may obtain approximation algorithms of kCP and kPP by using approximation algorithms of MAX TSP. In the following, we let α (resp., β) denote the current-best approximation ratio of MAX TSP on metric (resp., general) graphs. We have $\alpha = 7/8$ [19] and $\beta = 4/5$ [9].

1.1 Related Work

For $k = 2$, kCP and kPP are equivalent with the maximum weight perfect matching problem, which can be solved in $O(n^3)$ time [10,20]. For $k \geq 3$, metric kCP

© The Author(s), under exclusive license to Springer Nature Singapore Pte Ltd. 2024
R. Uehara et al. (Eds.): WALCOM 2024, LNCS 14549, pp. 179–193, 2024.
https://doi.org/10.1007/978-981-97-0566-5_14

and kPP become NP-hard [17], and general kCP and kPP become APX-hard even on $\{0,1\}$-weighted graphs (i.e., a complete graph with edge weights 0 and 1) [22]. There is a large number of contributions on approximation algorithms.

General kCP. For $k = 3$, Hassin and Rubinstein [13,14] proposed a randomized $(0.518 - \varepsilon)$-approximation algorithm, Chen et al. [7,8] proposed an improved randomized $(0.523 - \varepsilon)$-approximation algorithm, and Van Zuylen [32] proposed a deterministic algorithm with the same approximation ratio. For lager k, Li and Yu [21] proposed a 2/3-approximation algorithm for $k = 4$ and a $\beta \cdot (1 - 1/k)^2$-approximation algorithm for $k \geq 5$. On $\{0,1\}$-weighted graphs, Bar-Noy et al. [2] gave a 3/5-approximation algorithm for $k = 3$. Note that Berman and Karpinski [4] gave a 6/7-approximation algorithm for the *Maximum Path Cover Problem*, which seeks a set of node disjoint paths such that the number of edges in all the paths is maximal. Their algorithm could be used to obtain a $(6/7 - \varepsilon)$-approximation algorithm for general kCP and kPP with $k = n$ on $\{0,1\}$-weighted graphs.

Metric kCP. For $k = 3$, Hassin et al. [15] firstly gave a deterministic 2/3-approximation algorithm and Chen et al. [5] proposed a randomized $(0.66768 - \varepsilon)$-approximation algorithm. For lager k, Li and Yu [21] proposed a 3/4-approximation algorithm for $k = 4$, a 3/5-approximation algorithm for $k = 5$, and an $\alpha \cdot (1 - 1/k)^2$-approximation algorithm for $k \geq 6$.

General kPP. For $k = 3$, Hassin and Rubinstein [13] proposed a randomized $(0.5223 - \varepsilon)$-approximation algorithm, Chen et al. [27] proposed a deterministic $(0.5265 - \varepsilon)$-approximation algorithm, and Bar-Noy et al. [2] proposed an improved 7/12-approximation algorithm. For lager k, Hassin and Rubinstein [11] proposed a 3/4-approximation algorithm for $k = 4$, and a $\beta \cdot (1 - 1/k)$-approximation algorithm for $k \geq 5$. On $\{0,1\}$-weighted graphs, Hassin and Schneider [16] gave a 0.55-approximation algorithm for $k = 3$ and the ratio was improved to 3/4 [2].

Metric kPP. Li and Yu [21] proposed a 3/4-approximation algorithm for $k = 3$, a 3/4-approximation algorithm for $k = 5$, and an $\alpha \cdot (1 - 1/k)$-approximation algorithm for $k \geq 6$. The best-known result for $k = 4$ is still 3/4 due to the general 4PP, by Hassin and Rubinstein [11]. On $\{1,2\}$-weighted graphs, there is a 9/10-approximation algorithm for $k = 4$ [23].

General/metric kCP and kPP can be seen as a special case of the weighted k-set packing problem, which admits an approximation ratio of $\frac{1}{k-1} - \varepsilon$ [1], $\frac{2}{k+1} - \varepsilon$ [3], and $\frac{2}{k+1-1/31850496} - \varepsilon$ [24]. Recently, these results have been further improved (see [25,26,28]). They can be used to obtain a $1/1.786 \approx 0.559$-approximation ratio for general 3CP [28].

1.2   Our Results

We study approximation algorithms for metric/general kCP and kPP.

Firstly, we consider metric kCP. We propose a $(7/8 - 0.125/k)(1 - 1/k)$-approximation algorithm for constant odd k and a $7/8 \cdot (1 - 1/k + \frac{1}{k(k-1)})$-approximation algorithm for even k, which improve the best-known approximation ratio of $3/5$ for $k = 5$ [21] and $7/8 \cdot (1 - 1/k)^2$ for $k \geq 6$ [21]. Moreover, we propose an algorithm based on the maximum weight matching, which can further improve the approximation ratio from $17/25$ to $7/10$ for $k = 5$. An illustration of the improved results for metric kCP with $k \geq 5$ can be seen in Table 1.

Table 1. Improved approximation ratios for metric kCP with $k \geq 5$

Metric kCP	5	6	7	8
Previous Ratio [21]	0.600	0.607	0.642	0.669
Our Ratio	**0.700**	**0.758**	**0.734**	**0.781**

Secondly, we consider metric kPP. We propose a $\frac{27k^2-48k+16}{32k^2-36k-24}$-approximation algorithm for even $10 \geq k \geq 6$, which improves the best-known approximation ratio of $7/8 \cdot (1 - 1/k)$ [11]. An illustration of the improved results for metric kPP with even $10 \geq k \geq 6$ can be seen in Table 2.

Table 2. Improved approximation ratios for metric kPP with even $10 \geq k \geq 6$

Metric kPP	6	8	10
Previous Ratio [11]	0.729	0.765	0.787
Our Ratio	**0.767**	**0.783**	**0.794**

At last, we focus on the case of $k = 4$ for metric/general kCP and kPP. For metric 4CP, we propose a $5/6$-approximation algorithm, improving the best-known ratio $3/4$ [21], and as a corollary, we also give a $7/8$-approximation algorithm on $(1, 2)$-weighted graphs. For general 4CP, we propose a $3/4$-approximation algorithm, improving the best-known ratio $2/3$ [21]. For metric 4PP, we propose a $14/17$-approximation algorithm, improving the best-known ratio $3/4$ [11]. An illustration of the improved results for the case of $k = 4$ can be seen in Table 3.

Table 3. Improved results for the case of $k = 4$

	Metric Graphs	General Graphs
4CP	$3/4$ [21] → **5/6**	$2/3$ [21] → **3/4**
4PP	$3/4$ [11] → **14/17**	$3/4$ [11]

Due to limited space, the proofs of lemmas and theorems marked with "*" were omitted and they can be found in the full version of this paper [31].

1.3 Paper Organization

The remaining parts of the paper are organized as follows. In Sect. 2, we introduce basic notations. In Sect. 3, we consider metric kCP. In Sect. 3.1, we present
a better reduction from metric kCP to metric TSP, which has already led to an
improved ratio for $k \geq 5$. In Sect. 3.2, by using some properties of the current-
best approximation algorithm for metric TSP, we obtain a further improved
ratio. In Sect. 3.2, we consider a simple algorithm based on matching with a better ratio for $k = 5$. In Sect. 4, we consider metric kPP and propose an improved
algorithm for even $10 \geq k \geq 6$. Note that metric kPP is harder to improve, unlike
metric kCP. In Sect. 5, we propose non-trivial algorithms for metric/general kCP
and kPP with $k = 4$. In Sect. 5.1, we obtain a better algorithm for general 4CP.
In Sect. 5.2, we obtain a better algorithm for metric 4CP. In Sect. 5.3, we obtain a
better approximation algorithm for metric 4PP. Finally, we make the concluding
remarks in Sect. 6.

2 Preliminaries

We use $G = (V, E)$ to denote an undirected complete graph with n vertices such
that $n \bmod k = 0$. There is a non-negative weight function $w : E \to \mathbb{R}_{\geq 0}$ on the
edges in E. For an edge $uv \in E$, we use $w(u, v)$ to denote its weight. A graph
is called a *metric* graph if the weight function satisfies the triangle inequality;
Otherwise, it is called a *general* graph. For any weight function $w : X \to \mathbb{R}_{\geq 0}$,
we define $w(Y) = \sum_{x \in Y} w(x)$ for any $Y \subseteq X$.

Two subgraphs or subsets of edges of a graph are *vertex-disjoint* if they
do not appear a common vertex. We only consider simple paths and simple
cycles with more than two vertices. The *length* of a path/cycle is the number
of vertices it contains. A *cycle packing* is a set of vertex-disjoint cycles such
that the length of each cycle is at least three and all vertices in the graph are
covered. Given a cycle packing $\mathcal{C}$, we use $l(\mathcal{C})$ to denote the minimum length
of cycles in $\mathcal{C}$. We also use $\mathcal{C}^*$ to denote the maximum weight cycle packing.
A path (resp., cycle) on k different vertices $\{v_1, v_2, \ldots, v_k\}$ is called a k-*path*
(resp., k-*cycle*), denoted by $v_1 v_2 \cdots v_k$ (resp., $v_1 v_2 \cdots v_k v_1$). A k-*path packing*
(resp., k-*cycle packing*) in graph G is a set of vertex-disjoint n/k k-paths (resp.,
k-cycles) such that all vertices in the graph are covered. Note that we can obtain
a k-cycle packing by completing every k-path of a k-path packing. Let $\mathcal{P}_k^*$ (resp.,
$\mathcal{C}_k^*$) denote the maximum weight k-path packing (resp., k-cycle packing). We can
get $w(\mathcal{C}^*) \geq w(\mathcal{C}_k^*)$ for $k \geq 3$.

A 2-path packing is called a *matching* of size $n/2$. The maximum weight
matching of size $n/2$ is denoted by $\mathcal{M}^*$. An n-cycle is called a *Hamiltonian*
cycle. MAX TSP is to find a maximum weight Hamiltonian cycle. We simply use
general/metric TSP to denote MAX TSP in general/metric graphs. We use H^*
to denote the maximum weight Hamiltonian cycle. For a k-path $P = v_1 v_2 \cdots v_k$
where k is even, we define $\widetilde{w}(P) = \sum_{i=1}^{k/2} w(v_{2j-1}, v_{2j})$.

3 Approximation Algorithms for Metric kCP

In this section, we improve the approximation ratio for metric kCP with $k \geq 5$. We will first present a better black-box reduction from metric kCP to metric TSP, which is sufficient to improve the previous ratio for $k \geq 5$. Then, based on the approximation algorithm for metric TSP, we prove an improved approximation ratio. Finally, we consider a matching-based algorithm that can further improve the ratio of metric 5CP.

3.1 A Better Black-Box

Given an α-approximation algorithm for metric TSP, Li and Yu [21] proposed an $\alpha \cdot (1 - 1/k)^2$-approximation algorithm for metric kCP. We will show that the ratio can be improved to $\alpha \cdot (1 - 0.5/k)(1 - 1/k)$. Moreover, for even k, the ratio can be further improved to $\alpha \cdot (1 - 0.5/k)(1 - 1/k + \frac{1}{k(k-1)})$. We first consider a simple algorithm, denoted by Algorithm 1, which mainly contains three following steps.

Step 1. Obtain a Hamiltonian cycle H using an α-approximation algorithm for metric TSP;

Step 2. Get a k-path packing $\mathcal{P}_k$ with $w(\mathcal{P}_k) \geq (1 - 1/k)w(H)$ from H: we can obtain a k-path packing by deleting one edge per k edges from H; since there are $(1 - 1/k)n$ edges in $\mathcal{P}_k$ and n edges in H, if we carefully choose the initial edge, we can make sure that the weight of $\mathcal{P}_k$ is at least $(1 - 1/k)n \cdot (1/n) \cdot w(H)$, i.e., on average each edge has a weight of at least $(1/n) \cdot w(H)$.

Step 3. Obtain a k-cycle packing $\mathcal{C}_k$ by completing the k-path packing $\mathcal{P}_k$.

To analyze the approximation quality, we use the path patching technique, which has been used in some papers [12,18,19].

Lemma 1 ([12,18]). *Let G be a metric graph. Given a cycle packing $\mathcal{C}$, there is a polynomial-time algorithm to generate a Hamiltonian cycle H such that $w(H) \geq (1 - 0.5/l(\mathcal{C}))w(\mathcal{C})$.*

Since the length of every k-cycle in the maximum weight k-cycle packing $\mathcal{C}_k^*$ equals to k, we have $l(\mathcal{C}_k^*) = k$. By Lemma 1, we have the following lemma.

Lemma 2. $w(H^*) \geq (1 - 0.5/k)w(\mathcal{C}_k^*)$.

Theorem 1. *Given an α-approximation algorithm for metric TSP, Algorithm 1 is a polynomial-time $\alpha \cdot (1 - 0.5/k)(1 - 1/k)$-approximation algorithm for metric kCP.*

Proof. By the algorithm, we can easily get that $w(\mathcal{C}_k) \geq w(\mathcal{P}_k) \geq (1 - 1/k)w(H) \geq \alpha \cdot (1 - 1/k)w(H^*)$. By Lemma 2, we have $w(\mathcal{C}_k) \geq \alpha \cdot (1 - 0.5/k)(1 - 1/k)w(\mathcal{C}_k^*)$. Therefore, the algorithm achieves an approximation ratio of $\alpha \cdot (1 - 0.5/k)(1 - 1/k)$ for metric kCP. $\qquad\square$

Next, we propose an improved $\alpha \cdot (1 - 0.5/k)(1 - 1/k + \frac{1}{k(k-1)})$-approximation algorithm for even k, denoted by Algorithm 2. The previous two steps of Algorithm 2 are the same as Algorithm 1. However, Algorithm 2 will obtain a better k-cycle packing in Step 3:

New Step 3. For each k-path $P_i = v_{i1}v_{i2}\cdots v_{ik} \in \mathcal{P}_k$, we obtain $k - 1$ k-cycles $\{C_{i1}, \ldots, C_{i(k-1)}\}$ where $C_{ij} = v_{i1}v_{i2}\cdots v_{ij}v_{ik}v_{i(k-1)}\cdots v_{i(j+1)}v_{i1}$ (See Fig. 1 for an illustration); let C_{ij_i} denote the maximum weight cycle from these cycles; return a k-cycle packing $\mathcal{C}_k = \{C_{ij_i}\}_{i=1}^{n/k}$.

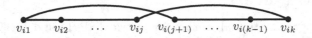

$$v_{i1} \quad v_{i2} \quad \cdots \quad v_{ij} \quad v_{i(j+1)} \quad \cdots \quad v_{i(k-1)} \quad v_{ik}$$

Fig. 1. An illustration of the k-cycle C_{ij} obtained from P_i, where $j \in \{1, 2, \ldots, k-1\}$

Lemma 3. *It holds that* $w(\mathcal{C}_k) \geq \frac{k-2}{k-1}w(\mathcal{P}_k) + \frac{2}{k-1}\widetilde{w}(\mathcal{P}_k)$.

Proof. Since C_{ij_i} is the maximum weight cycle from these cycles, we have

$$w(C_{ij_i}) \geq \frac{1}{k-1}\sum_{j=1}^{k-1} w(C_{ij})$$

$$= \frac{1}{k-1}\sum_{j=1}^{k-1}(w(P_i) + w(v_{i1}, v_{i(j+1)}) + w(v_{ij}, v_{ik}) - w(v_{ij}, v_{i(j+1)}))$$

$$= \frac{1}{k-1}\left((k-1)w(P_i) + \sum_{j=1}^{k-1}(w(v_{i1}, v_{i(j+1)}) + w(v_{ij}, v_{ik})) - w(P_i)\right)$$

$$= \frac{1}{k-1}\left((k-2)w(P_i) + \sum_{j=1}^{k-1}(w(v_{i1}, v_{i(j+1)}) + w(v_{ij}, v_{ik}))\right).$$

By the triangle inequality, we can get that

$$\sum_{j=1}^{k-1} w(v_{i1}, v_{i(j+1)}) = w(v_{i1}, v_{i2}) + \sum_{j=2}^{k/2}(w(v_{i1}, v_{i(2j-1)}) + w(v_{i1}, v_{i(2j)}))$$

$$\geq w(v_{i1}, v_{i2}) + \sum_{j=2}^{k/2} w(v_{i(2j-1)}, v_{i(2j)})$$

$$= \sum_{j=1}^{k/2} w(v_{i(2j-1)}, v_{i(2j)})$$

$$= \widetilde{w}(P_i).$$

Similarly, we can get $\sum_{j=1}^{k-1} w(v_{ij}, v_{ik}) \geq \widetilde{w}(P_i)$. Hence,

$$w(C_{ij_i}) \geq \frac{1}{k-1}\left((k-2)w(P_i) + \sum_{j=1}^{k-1}(w(v_{i1}, v_{i(j+1)}) + w(v_{ij}, v_{ik}))\right)$$

$$\geq \frac{(k-2)w(P_i) + 2\widetilde{w}(P_i)}{k-1}.$$

By doing this for all k-paths in $\mathcal{P}_k$, we can get a k-cycle packing $\mathcal{C}_k$ such that $w(\mathcal{C}_k) \geq \frac{(k-2)w(\mathcal{P}_k)+2\widetilde{w}(\mathcal{P}_k)}{k-1}$. □

Theorem 2. *Given an α-approximation algorithm for metric TSP, for metric kCP with even k, Algorithm 2 is a polynomial-time $\alpha \cdot (1-0.5/k)(1-1/k+\frac{1}{k(k-1)})$-approximation algorithm.*

Proof. Recall that all k-paths in $\mathcal{P}_k$ are obtained from the α-approximate Hamiltonian cycle H. By deleting one edge per k edges from a Hamiltonian cycle H and choosing the initial edge carefully, we can get a k-path packing $\mathcal{P}_k$ such that

$$(k-2)w(\mathcal{P}_k) + 2\widetilde{w}(\mathcal{P}_k) \geq \frac{(k-2)(k-1)+k}{k}w(H) = \frac{(k-1)^2+1}{k}w(H)$$

since $(k-2)w(\mathcal{P}_k)+2\widetilde{w}(\mathcal{P}_k)$ contains the weight of $\frac{n(k-2)(k-1)+nk}{k}$ (multi-)edges in H. By Lemma 3, we can obtain a k-cycle packing $\mathcal{C}_k$ such that

$$w(\mathcal{C}_k) \geq \frac{(k-2)w(\mathcal{P}_k)+2\widetilde{w}(\mathcal{P}_k)}{k-1}$$

$$\geq \frac{(k-1)^2+1}{k(k-1)}w(H)$$

$$= \left(1 - 1/k + \frac{1}{k(k-1)}\right)w(H).$$

Since $w(H) \geq \alpha \cdot w(H^*) \geq \alpha \cdot (1 - 0.5/k)w(\mathcal{C}_k^*)$ by Lemma 2, we have $w(\mathcal{C}_k) \geq \alpha \cdot (1 - 0.5/k)(1 - 1/k + \frac{1}{k(k-1)})w(\mathcal{C}_k^*)$. □

Note that for metric TSP there is a randomized $(7/8 - O(1/\sqrt{n}))$-approximation algorithm [12], a deterministic $(7/8 - O(1/\sqrt[3]{n}))$-approximation algorithm [6], and a deterministic 7/8-approximation algorithm [19]. By Theorem 2, we obtain an approximation ratio of $7/8 \cdot (1 - 0.5/k)(1 - 1/k)$ for metric kCP with odd k, and $7/8 \cdot (1 - 0.5/k)(1 - 1/k + \frac{1}{k(k-1)})$ for metric kCP with even k.

3.2 A Further Improvement

In this subsection, we show that the approximation ratio of Algorithm 2 can be further improved based on the properties of the 7/8-approximation algorithm for metric TSP [19]. We recall the following result.

Lemma 4 ([19]). *Let G be a metric graph with even n. There is a polynomial-time algorithm to get a Hamiltonian cycle H with $w(H) \geq \frac{5}{8}w(\mathcal{C}^*) + \frac{1}{2}w(\mathcal{M}^*)$.*

For any k-cycle packing with k being even or Hamiltonian cycle with an even number of vertices, the edges can be decomposed into two edge-disjoint matchings of size $n/2$. We can get the following bounds.

Lemma 5. *It holds that $w(\mathcal{M}^*) \geq \frac{1}{2}w(\mathcal{C}_k^*)$ for even k and $w(\mathcal{M}^*) \geq \frac{1}{2}w(H^*)$ for even n.*

Note that for metric kCP with even k, the number of vertices is always even since it satisfies $n \bmod k = 0$. But for odd k, the number may be odd, and then there may not exist a matching of size $n/2$. Since we mainly consider the improvements for constant k, for the case of odd k and n, we can first use $n^{O(k)} = n^{O(1)}$ time to enumerate a k-cycle in $\mathcal{C}_k^*$, and then consider an approximate k-cycle packing in the rest graph. The approximation ratio preserves. Hence, we may assume that n is even for the case of constant k.

Theorem 3 (*).   *For metric kCP, there is a $(7/8 - 0.125/k)(1 - 1/k)$-approximation algorithm for constant odd k and a $7/8 \cdot (1 - 1/k + \frac{1}{k(k-1)})$-approximation algorithm for even k.*

3.3   An Improved Algorithm Based on Matching

Consider metric kCP with odd k. By deleting the least weighted edge from every k-cycle in $\mathcal{C}_k^*$, we can get a k-path packing $\mathcal{P}_k$ with $w(\mathcal{P}_k) \geq (1 - 1/k)w(\mathcal{C}_k^*)$. Note that $\mathcal{P}_k$ can be decomposed into two edge-disjoint matchings of size $p := (n/k) \cdot (k-1)/2$. Let $\mathcal{M}_p^*$ be the maximum weight matching of size p, which can be computed in polynomial time [10,20]. Then, we can get $2w(\mathcal{M}_p^*) \geq w(\mathcal{P}_k) \geq (1 - 1/k)w(\mathcal{C}_k^*)$. Note that there are also n/k *isolated vertices* not covered by $\mathcal{M}_p^*$. Next, we construct a k-cycle packing using $\mathcal{M}_p^*$ with the isolated vertices. The algorithm, denoted by Algorithm 3, is shown as follows.

> **Step 1.** Arbitrarily partition the p edges of $\mathcal{M}_p^*$ into n/k sets with the same size, denoted by $\mathcal{S}_1, \mathcal{S}_2, \ldots, \mathcal{S}_{n/k}$. Note that each edge set contains $m := (k-1)/2$ edges. For each of the n/k edge sets, arbitrarily assign an isolated vertex.
> **Step 2.** Consider an arbitrary edge set $\mathcal{S}_i = \{e_1, e_2, \ldots, e_m\}$ with the isolated vertex v. Assume w.o.l.g. that $w(e_1) \geq w(e_m) \geq w(e_i)$ for $2 \leq i < m$, i.e., $w(e_1) + w(e_m) \geq (2/m)w(\mathcal{S}_i)$. Orient each edge e_i uniformly at random from the two choices. Let t_i (resp., h_i) denote the tail (resp., head) vertex of e_i. Construct a k-cycle C_i such that $C_i = vt_1h_1t_2h_2 \cdots t_mh_mv$.
> **Step 3.** Get a k-cycle packing $\mathcal{C}_k$ by packing the k-cycles from the edge sets and the isolated vertices.

Algorithm 3 can be derandomized efficiently by conditional expectations [29].
Next, we analyze the expected weight of $C_i = vt_1h_1t_2h_2 \cdots t_mh_mv$, obtained from the edge set $\mathcal{S}_i$ and the isolated vertex v.

Lemma 6. *It holds that* $\mathbb{E}[w(v,t_1)] \geq \frac{1}{2}w(e_1)$, $\mathbb{E}[w(v,h_m)] \geq \frac{1}{2}w(e_m)$, *and* $\mathbb{E}[w(h_i,t_{i+1})] \geq \frac{1}{4}(w(e_i)+w(e_{i+1}))$ *for* $1 \leq i < m$.

Proof. Consider $\mathbb{E}[w(v,t_1)]$. Since we orient the edge e_1 uniformly at random, each vertex of e_1 has a probability of $1/2$ being t_1. Hence, we can get $\mathbb{E}[w(v,t_1)] = \frac{1}{2}\sum_{u \in e_1} w(v,u) \geq \frac{1}{2}w(e_1)$ by the triangle inequality. Similarly, we can get $\mathbb{E}[w(v,h_m)] \geq \frac{1}{2}w(e_m)$.

Consider $\mathbb{E}[w(h_i,t_{i+1})]$. We can get $\mathbb{E}[w(h_i,t_{i+1})] = \frac{1}{4}\sum_{u \in e_i}\sum_{w \in e_{i+1}} w(u,w)$. Let $e_i = u'u''$ and $e_{i+1} = o'o''$. By the triangle inequality, we can get that

$$\sum_{u \in e_i}\sum_{w \in e_{i+1}} w(u,w) = w(u',o') + w(u',o'') + w(u'',o') + w(u'',o'')$$

$$\geq w(o',o'') + w(u',u'')$$

$$= w(e_i) + w(e_{i+1}).$$

Therefore, $\mathbb{E}[w(h_i,t_{i+1})] \geq \frac{1}{4}(w(e_i)+w(e_{i+1}))$ for $1 \leq i < m$. $\qquad\square$

Lemma 7. *It holds that* $\mathbb{E}[w(C_i)] \geq \frac{3m+1}{2m}w(\mathcal{S}_i)$.

Proof. Note that

$$w(C_i) = w(v,t_1) + w(v,h_m) + \sum_{i=1}^{m-1}(w(t_i,h_i) + w(h_i,t_{i+1}))$$

$$= w(\mathcal{S}_i) + w(v,t_1) + w(v,h_m) + \sum_{i=1}^{m-1} w(h_i,t_{i+1}).$$

We can get that

$$\mathbb{E}[w(C_i)] \geq w(\mathcal{S}_i) + \frac{1}{2}(w(e_1)+w(e_m)) + \frac{1}{4}\sum_{i=1}^{m-1}(w(e_i)+w(e_{i+1}))$$

$$= w(\mathcal{S}_i) + \frac{1}{2}(w(e_1)+w(e_m)) + \frac{1}{2}w(\mathcal{S}_i) - \frac{1}{4}(w(e_1)+w(e_m))$$

$$= \frac{3}{2}w(\mathcal{S}_i) + \frac{1}{4}(w(e_1)+w(e_m))$$

$$\geq \left(\frac{3}{2}+\frac{1}{2m}\right)w(\mathcal{S}_i)$$

$$= \frac{3m+1}{2m}w(\mathcal{S}_i),$$

where the first inequality follows from Lemma 6, and the second from $w(e_1) + w(e_m) \geq (2/m)w(\mathcal{S}_i)$ by the algorithm. $\qquad\square$

Theorem 4. *For metric kCP with odd k, Algorithm 3 is a polynomial-time $(3/4 - 0.25/k)$-approximation algorithm.*

Proof. Recall that $2w(\mathcal{M}_p^*) \geq (1 - 1/k)w(\mathcal{C}_k^*)$ and $\mathcal{M}_p^* = \bigcup_{i=1}^{n/k} \mathcal{S}_i$. Using a derandomization based on conditional expectations [29], by Lemma 7, we can get that

$$w(\mathcal{C}_k) \geq \sum_{i=1}^{n/k} \frac{3m+1}{2m} w(\mathcal{S}_i) = \frac{3m+1}{2m} w(\mathcal{M}_p^*) \geq \frac{3m+1}{4m} \left(1 - \frac{1}{k}\right) w(\mathcal{C}_k^*).$$

Since $m = (k-1)/2$, we can get an approximation ratio of $\frac{3m+1}{4m}(1 - \frac{1}{k}) = 3/4 - 0.25/k$. □

By Theorem 4, we obtain a 7/10-approximation algorithm for metric 5CP, which improves the previous ratio 17/25 in Theorem 3, and the ratio 3/5 in [21].

Corollary 1. *For metric* 5*CP, Algorithm 3 is a 7/10-approximation algorithm.*

4 Approximation Algorithms for Metric kPP

In this section, we consider metric kPP. Using a reduction from metric kPP to metric TSP, metric kPP admits a $7/8 \cdot (1 - 1/k)$-approximation algorithm [11]. Note that, unlike metric kCP, it is not easy to construct a better black box to improve the ratio. However, we will combine the properties of the 7/8-approximation algorithm for metric TSP with an algorithm based on matching to obtain a better approximation ratio for even $6 \leq k \leq 10$. Next, we assume that k is even.

The first algorithm, denoted by Algorithm 4, is to use the reduction from metric kPP to metric TSP [11].

Step 1. Obtain a Hamiltonian cycle H using the 7/8-approximation algorithm for metric TSP [19];
Step 2. Get a k-path packing $\mathcal{P}_k$ with $w(\mathcal{P}_k) \geq (1-1/k)w(H)$ from H using the same method in Step 2 of Algorithm 1.

For every $P_i = v_{i1}v_{i2}\cdots v_{ik} \in \mathcal{P}_k^*$, let $\mathcal{E}_i' = \{v_{i(2j-1)}v_{i(2j)}\}_{j=1}^{k/2}$ and $\mathcal{E}_i'' = \{v_{i(2j)}v_{i(2j+1)}\}_{j=1}^{(k-2)/2}$. Then, we can obtain a matching $\mathcal{M}_{n/2} = \bigcup_i \mathcal{E}_i'$ of size $n/2$ and a matching $\mathcal{M}_p = \bigcup_i \mathcal{E}_i''$ of size $p := (n/k) \cdot (k-2)/2$. Note that $w(\mathcal{M}_{n/2}) + w(\mathcal{M}_p) = w(\mathcal{P}_k^*)$. We have the following bounds.

Lemma 8 (*). $w(\mathcal{C}_k^*) \geq \frac{k-2}{k-1}w(\mathcal{P}_k^*) + \frac{2}{k-1}w(\mathcal{M}_{n/2}).$

Lemma 9 (*). $w(\mathcal{P}_k) \geq \frac{5k-10}{8k}w(\mathcal{P}_k^*) + \frac{2k+3}{4k}w(\mathcal{M}_{n/2}).$

Next, we propose an algorithm, denoted by Algorithm 5, to obtain another k-path packing $\mathcal{P}_k'$, which can be used to make a trade-off with $\mathcal{P}_k$. The framework of Algorithm 5 is similar to Algorithm 3 in Sect. 3.3. Let $\mathcal{M}_p^*$ denote the maximum weight matching of size $p = (n/k) \cdot (k-2)/2$, which can be computed in polynomial time [10,20]. Note that $w(\mathcal{M}_p^*) \geq w(\mathcal{M}_p)$. There are $2n/k$ *isolated vertices* not covered by $\mathcal{M}_p^*$. Next, we construct a k-path packing using $\mathcal{M}_p^*$ with isolated vertices. Algorithm 5 mainly contains three steps.

Step 1. Arbitrarily partition the p edges of $\mathcal{M}_p^*$ into n/k sets with the same size, denoted by $\mathcal{S}_1, \mathcal{S}_2, \ldots, \mathcal{S}_{n/k}$. Note that each edge set contains $m :=$ $(k-2)/2$ edges. For each of the n/k edge sets, arbitrarily assign two isolated vertices.

Step 2. Consider an arbitrary edge set $\mathcal{S}_i = \{e_1, e_2, \ldots, e_m\}$ with the two isolated vertices u and v. Assume w.o.l.g. that $w(e_1) \geq w(e_m) \geq w(e_i)$ for $2 \leq i < m$, i.e., $w(e_1) + w(e_m) \geq (2/m)w(\mathcal{S}_i)$. Orient each edge e_i uniformly at random from the two choices. Let t_i (resp., h_i) denote the tail (resp., head) vertex of e_i. Construct a k-path P_i' such that $P_i' = ut_1h_1t_2h_2\cdots t_mh_mv$.

Step 3. Get a k-path packing $\mathcal{P}_k'$ by packing the k-paths from the edge sets and the isolated vertices.

Algorithm 5 can also be derandomized by conditional expectations.

Next, we analyze the expected weight of $P_i' = ut_1h_1t_2h_2\cdots t_mh_mv$, obtained from the edge set $\mathcal{S}_i$ and the two isolated vertices u and v.

Lemma 10 (*). It holds that $\mathbb{E}[w(u, t_1)] \geq \frac{1}{2}w(e_1)$, $\mathbb{E}[w(v, h_m)] \geq \frac{1}{2}w(e_m)$, and $\mathbb{E}[w(h_i, t_{i+1})] \geq \frac{1}{4}(w(e_i) + w(e_{i+1}))$ for $1 \leq i < m$.

Lemma 11 (*). It holds that $\mathbb{E}[w(\mathcal{P}_k')] \geq \frac{3k-4}{2k-4}w(\mathcal{M}_p^*)$.

Lemma 12 (*). $w(\mathcal{P}_k') \geq \frac{3k-4}{2k-4}w(\mathcal{M}_p)$.

Theorem 5 (*). There is a $\frac{27k^2-48k+16}{32k^2-36k-24}$-approximation algorithm for metric kPP with even k.

The approximation ratio in Theorem 5 is better than $7/8 \cdot (1 - 1/k)$ for even $10 \geq k \geq 6$. For $k = 4$, the ratio is even worse than the ratio $3/4$ in [11]. But, in the next section, we show an improved $14/17 \approx 0.823$-approximation algorithm.

5 Approximation Algorithms for the Case of $k = 4$

In this section, we study the case of $k = 4$ for metric/general kCP and kPP. For metric 4CP, we improve the best-known ratio from $3/4$ [21] to $5/6$. For general 4CP, we improve the best-known ratio from $2/3$ [21] to $3/4$. For metric 4PP, we improve the best-known ratio from $3/4$ [11] to $14/17$.

5.1 General 4CP

Zhao and Xiao [30] observed some structural properties of the minimum weight 4-cycle packing and the minimum weight matching of size $n/2$. In fact, these properties even hold for the maximum weight 4-cycle packing $\mathcal{C}_4^*$ and the maximum weight matching $\mathcal{M}^*$ of size $n/2$.

Lemma 13 ([30]). Given $\mathcal{C}_4^*$ and $\mathcal{M}^*$, there is a way to color edges in $\mathcal{C}_4^*$ with red and blue such that

(1) the blue (resp., red) edges form a matching of size $n/2$ $\mathcal{M}_b$ (resp., $\mathcal{M}_r$);
(2) $\mathcal{C}_4^* = \mathcal{M}_b \cup \mathcal{M}_r$;
(3) $\mathcal{M}_b \cup \mathcal{M}^*$ is a cycle packing and the length of every cycle is divisible by 4.

An alternative proof of Lemma 13 could be found in [23]. Next, we describe the approximation algorithm for general 4CP, denoted by Algorithm 6.

Step 1. Find a maximum weight matching $\mathcal{M}^*$ of size $n/2$.
Step 2. Construct a multi-graph $G/\mathcal{M}^*$ such that there are $n/2$ super-vertices one-to-one corresponding to the $n/2$ edges in $\mathcal{M}^*$, i.e., there is a function f, and for two super-vertices $f(e_i), f(e_j)$ such that $e_i, e_j \in \mathcal{M}^*$, there are four super-edges $f(e_i)f(e_j)$ between them, corresponding to the four edges uv with a weight of $w(u,v)$ $(u \in e_i, v \in e_j)$.
Step 3. Find a maximum weight matching $\mathcal{M}_{n/4}^{**}$ of size $n/4$ in graph $G/\mathcal{M}^*$. Note that $\mathcal{M}^* \cup \mathcal{M}_{n/4}^{**}$ corresponds to a 4-path packing $\mathcal{P}_4$ in graph G.
Step 4. Obtain a 4-cycle packing $\mathcal{C}_4$ by completing the 4-path packing $\mathcal{P}_4$.

Note that $w(\mathcal{C}_4) \geq w(\mathcal{P}_4) = w(\mathcal{M}^*) + w(\mathcal{M}_{n/4}^{**})$.

Lemma 14 (*). $w(\mathcal{M}_{n/4}^{**}) \geq \frac{1}{2}w(\mathcal{M}_b)$.

Lemma 15 (*). $w(\mathcal{P}_4) \geq \frac{1}{2}w(\mathcal{M}^*) + \frac{1}{2}w(\mathcal{C}_4^*)$.

Theorem 6 (*). *Algorithm 6 is a 3/4-approximation algorithm for general 4CP.*

5.2   Metric 4CP

Li and Yu [21] proved an almost trivial approximation ratio of $3/4$. We show that their algorithm, denoted by Algorithm 7, actually achieves an approximation ratio of $5/6$.

Step 1. Find a maximum weight matching $\mathcal{M}^*$ of size $n/2$.
Step 2. Construct a multi-graph $G/\mathcal{M}^*$ such that there are $n/2$ super-vertices one-to-one corresponding to the $n/2$ edges in $\mathcal{M}^*$, i.e., there is a function f, and for two super-vertices $f(e_i), f(e_j)$ such that $e_i, e_j \in \mathcal{M}^*$, there are two super-edges $f(e_i)f(e_j)$ between them, corresponding to the edge sets $\{uz, xy\}$ and $\{uy, xz\}$ with a weight of $w(u,z)+w(x,y)$ and $w(u,y)+w(x,z)$ $(ux \in e_i, yz \in e_j)$.
Step 3. Find a maximum weight matching $\mathcal{M}_{n/4}^{**}$ of size $n/4$ in graph $G/\mathcal{M}^*$. Note that $\mathcal{M}^* \cup \mathcal{M}_{n/4}^{**}$ corresponds to a 4-cycle packing $\mathcal{C}_4$ in graph G if we decompose each super-edge of $\mathcal{M}_{n/4}^{**}$ into two normal edges.
Step 4. Return $\mathcal{C}_4$.

Note that $\mathcal{C}_4$ is the maximum weight 4-cycle packing containing the edges of $\mathcal{M}^*$ by the optimality of $\mathcal{M}_{n/4}^{**}$. Recall that we can get a 4-path packing $\mathcal{P}_4$ such that $w(\mathcal{P}_4) \geq \frac{1}{2}w(\mathcal{M}^*) + \frac{1}{2}w(\mathcal{C}_4^*)$ by Lemma 15. Moreover, if $\mathcal{P}_4 = \{u_i x_i y_i z_i\}_{i=1}^{n/4}$, $\mathcal{M}^*$ represents the edge set $\{u_i x_i, y_i z_i\}_{i=1}^{n/4}$. Let $\overline{\mathcal{P}_4}$ denote the edge set $\{u_i z_i\}_{i=1}^{n/4}$.

Lemma 16 (*). $w(\mathcal{C}_4) \geq \frac{3}{4}w(\mathcal{C}_4^*) + w(\overline{\mathcal{P}_4})$.

Lemma 17 (*). $w(\mathcal{C}_4) \geq w(\mathcal{C}_4^*) - 2w(\overline{\mathcal{P}_4})$.

Theorem 7 (*). *Algorithm 7 is a 5/6-approximation algorithm for metric 4CP.*

On $\{1, 2\}$-weighted graphs we may obtain a better approximation ratio.

Theorem 8 (*). *On $\{1,2\}$-weighted graphs, Algorithm 7 is a 7/8-approximation algorithm for metric 4CP.*

5.3   Metric 4PP

At last, we will consider metric 4PP. Recall that we can get a 4-path packing $\mathcal{P}_4$ such that $w(\mathcal{P}_4) \geq \frac{1}{2}w(\mathcal{M}^*) + \frac{1}{2}w(\mathcal{C}_4^*)$ by Lemma 15. For metric 4PP, we will construct another 4-path packing $\mathcal{P}_4'$. The algorithm, denoted by Algorithm 8, is shown as follows.

Step 1. Obtain a 4-path packing $\mathcal{P}_4$ such that $w(\mathcal{P}_4) \geq \frac{1}{2}w(\mathcal{M}^*) + \frac{1}{2}w(\mathcal{C}_4^*)$ using Algorithm 6.
Step 2. Obtain a maximum weight matching $\mathcal{M}_{n/4}^{**}$ of size $n/4$ in graph G. Note that there are also $n/2$ isolated vertices not covered by $\mathcal{M}_{n/4}^{**}$.
Step 3. Arbitrarily assign two isolated vertices u_i, z_i for each edge $x_i y_i \in \mathcal{M}_{n/4}^{**}$. Assume w.l.o.g. that $w(u_i, x_i) + w(y_i, z_i) \geq w(z_i, x_i) + w(y_i, u_i)$.
Step 4. Obtain another 4-path packing $\mathcal{P}_4'$ by taking a 4-path $u_i x_i y_i z_i$ for every edge $x_i y_i \in \mathcal{M}_{n/4}^{**}$ with the two isolated vertices u_i, z_i.

Let $\mathcal{C}_4$ be the 4-cycle packing obtained by completing the maximum weight 4-path packing $\mathcal{P}_4^*$, i.e., for every 4-path $P_i = u_i x_i y_i z_i \in \mathcal{P}_4$, we obtain a 4-cycle $C_i = u_i x_i y_i z_i u_i$. Then, let $\mathcal{C}_4 = \mathcal{P}_4^* \cup \overline{\mathcal{P}_4^*}$. Moreover, let $\mathcal{C}_4 = \mathcal{M}_1 \cup \mathcal{M}_2$ such that $\mathcal{M}_1$ and $\mathcal{M}_2$ are two matchings of size $n/2$, and $\mathcal{M}_1 \cap \overline{\mathcal{P}_4^*} = \emptyset$. Obtain another 4-cycle packing $\mathcal{C}_4'$ such that for every 4-path $P_i = u_i x_i y_i z_i \in \mathcal{P}_4$ there is a 4-cycle $C_i' = u_i x_i z_i y_i u_i$ in $\mathcal{C}_4'$.

Lemma 18 (*). $w(\mathcal{P}_4) \geq \max\{\frac{1}{2}w(\mathcal{M}_1) + \frac{1}{2}w(\mathcal{P}_4^*) + \frac{1}{2}w(\overline{\mathcal{P}_4^*}), \frac{3}{2}w(\mathcal{M}_1) - w(\overline{\mathcal{P}_4^*})\}$.

Lemma 19 (*). $w(\mathcal{P}_4') \geq 2w(\mathcal{P}_4^*) - 2w(\mathcal{M}_1)$.

Theorem 9 (*). *There is a 14/17-approximation algorithm for metric 4PP.*

6   Conclusion

In this paper, we consider approximation algorithms for metric/general kCP and kPP. Most of our results are based on simple algorithms but with deep analysis. In the future, it would be interesting to improve these approximation ratios, even on $\{0, 1\}$-weighted or $\{1, 2\}$-weighted graphs. In particular, one challenging direction is to design better algorithms for metric/general 3CP and 3PP.

Acknowledgments. The work is supported by the National Natural Science Foundation of China, under grants 62372095 and 61972070.

References

1. Arkin, E.M., Hassin, R.: On local search for weighted k-set packing. Math. Oper. Res. **23**(3), 640–648 (1998)
2. Bar-Noy, A., Peleg, D., Rabanca, G., Vigan, I.: Improved approximation algorithms for weighted 2-path partitions. Discret. Appl. Math. **239**, 15–37 (2018)
3. Berman, P.: A d/2 approximation for maximum weight independent set in d-claw free graphs. Nord. J. Comput. **7**(3), 178–184 (2000)
4. Berman, P., Karpinski, M.: 8/7-approximation algorithm for (1, 2)-TSP. In: SODA 2006, pp. 641–648. ACM Press (2006)
5. Chen, Y., Chen, Z., Lin, G., Wang, L., Zhang, A.: A randomized approximation algorithm for metric triangle packing. J. Comb. Optim. **41**(1), 12–27 (2021)
6. Chen, Z., Nagoya, T.: Improved approximation algorithms for metric MaxTSP. J. Comb. Optim. **13**(4), 321–336 (2007)
7. Chen, Z., Tanahashi, R., Wang, L.: An improved randomized approximation algorithm for maximum triangle packing. Discret. Appl. Math. **157**(7), 1640–1646 (2009)
8. Chen, Z., Tanahashi, R., Wang, L.: Erratum to "an improved randomized approximation algorithm for maximum triangle packing" [Discrete Appl. Math. 157 (2009) 1640–1646]. Discret. Appl. Math. **158**(9), 1045–1047 (2010)
9. Dudycz, S., Marcinkowski, J., Paluch, K., Rybicki, B.: A 4/5 - approximation algorithm for the maximum traveling salesman problem. In: Eisenbrand, F., Koenemann, J. (eds.) IPCO 2017. LNCS, vol. 10328, pp. 173–185. Springer, Cham (2017). https://doi.org/10.1007/978-3-319-59250-3_15
10. Gabow, H.N.: Implementation of algorithms for maximum matching on nonbipartite graphs. Ph.D. thesis, Stanford University (1974)
11. Hassin, R., Rubinstein, S.: An approximation algorithm for maximum packing of 3-edge paths. Inf. Process. Lett. **63**(2), 63–67 (1997)
12. Hassin, R., Rubinstein, S.: A 7/8-approximation algorithm for metric max TSP. Inf. Process. Lett. **81**(5), 247–251 (2002)
13. Hassin, R., Rubinstein, S.: An approximation algorithm for maximum triangle packing. Discret. Appl. Math. **154**(6), 971–979 (2006)
14. Hassin, R., Rubinstein, S.: Erratum to "an approximation algorithm for maximum triangle packing": [discrete applied mathematics 154 (2006) 971–979]. Discret. Appl. Math. **154**(18), 2620 (2006)
15. Hassin, R., Rubinstein, S., Tamir, A.: Approximation algorithms for maximum dispersion. Oper. Res. Lett. **21**(3), 133–137 (1997)
16. Hassin, R., Schneider, O.: A local search algorithm for binary maximum 2-path partitioning. Discret. Optim. **10**(4), 333–360 (2013)
17. Kirkpatrick, D.G., Hell, P.: On the completeness of a generalized matching problem. In: STOC 1978, pp. 240–245. ACM (1978)
18. Kostochka, A., Serdyukov, A.: Polynomial algorithms with the estimates 3/4 and 5/6 for the traveling salesman problem of the maximum. Upravliaemie Syst. **26**, 55–59 (1985)
19. Kowalik, L., Mucha, M.: Deterministic 7/8-approximation for the metric maximum TSP. Theor. Comput. Sci. **410**(47–49), 5000–5009 (2009)
20. Lawler, E.: Combinatorial Optimization: Networks and Matroids. Holt, Rinehart and Winston (1976)
21. Li, S., Yu, W.: Approximation algorithms for the maximum-weight cycle/path packing problems. Asia-Pac. J. Oper. Res. **40**(04), 2340003 (2023)

22. Manthey, B.: On approximating restricted cycle covers. SIAM J. Comput. **38**(1), 181–206 (2008)
23. Monnot, J., Toulouse, S.: Approximation results for the weighted p_4 partition problem. J. Discrete Algorithms **6**(2), 299–312 (2008)
24. Neuwohner, M.: An improved approximation algorithm for the maximum weight independent set problem in d-claw free graphs. In: STACS 2021, LIPIcs, vol. 187, pp. 53:1–53:20. Schloss Dagstuhl - Leibniz-Zentrum für Informatik (2021)
25. Neuwohner, M.: The limits of local search for weighted k-set packing. In: Aardal, K., Sanitá, L. (eds.) IPCO 2022. LNCS, vol. 13265, pp. 415–428. Springer, Cham (2022). https://doi.org/10.1007/978-3-031-06901-7_31
26. Neuwohner, M.: Passing the limits of pure local search for weighted k-set packing. In: SODA 2023, pp. 1090–1137. SIAM (2023)
27. Tanahashi, R., Chen, Z.Z.: A deterministic approximation algorithm for maximum 2-path packing. IEICE Trans. Inf. Syst. **93**(2), 241–249 (2010)
28. Thiery, T., Ward, J.: An improved approximation for maximum weighted k-set packing. In: SODA 2023, pp. 1138–1162. SIAM (2023)
29. Williamson, D.P., Shmoys, D.B.: The Design of Approximation Algorithms. Cambridge University Press, Cambridge (2011)
30. Zhao, J., Xiao, M.: Improved approximation algorithms for capacitated vehicle routing with fixed capacity. CoRR abs/2210.16534 (2022)
31. Zhao, J., Xiao, M.: Improved approximation algorithms for cycle and path packings. CoRR abs/2311.11332 (2023)
32. van Zuylen, A.: Deterministic approximation algorithms for the maximum traveling salesman and maximum triangle packing problems. Discret. Appl. Math. **161**(13–14), 2142–2157 (2013)

Exact and Approximate Hierarchical Hub Labeling

Ruoying Li$^{(\boxtimes)}$ (ID) and Sabine Storandt (ID)

University of Konstanz, Konstanz, Germany
{ruoying.li,sabine.storandt}@uni-konstanz.de

Abstract. Hub Labeling (HL) is a state-of-the-art technique for accelerating shortest path computation in road networks. By utilizing precomputed node labels, it can answer distance queries in microseconds on continent-sized networks. The optimization goal is to get correct query results with a minimum number of labels. There is an $\mathcal{O}(\log n)$ approximation algorithm for the size of an HL with a running time of $\mathcal{O}(n^3 \log n)$. However, existing practical implementations rely mostly on heuristics for a special type of HL, so called Hierarchical HL (HHL). Deciding whether a graph admits a labeling of size at most k is NP-hard for both HL and HHL. For HHL, an $\mathcal{O}(\sqrt{n} \log n)$ approximation algorithm (called w-HHL) is known, as well as a parametrized upper bound of $\mathcal{O}(t \log n)$ on the average label size, where t denotes the treewidth of the network. In this paper, we devise an exact HHL algorithm for general graphs. Furthermore, we improve the parametrized approximation factor to $\mathcal{O}(b)$, where b denotes the balanced separator number with $b - 1 \leq t$. We also show that w-HHL provides a constant factor approximation on trees, and investigate for the first time the practical performance of existing HHL approximation algorithms. Our theoretical results offer some explanatory power for the good performance of HHL on road networks, while our experimental results show that commonly used methods for HHL are noticeably outperformed by w-HHL on general graphs as well as trees.

Keywords: Shortest Path · Hub Labeling · Separator · Approximation

1 Introduction

The concept of Hub Labeling (HL) was introduced by Cohen et al. [8] to quickly answer reachability and distance queries on networks with the help of precomputed auxiliary data. Given a weighted graph $G(V, E, c)$ with $c : E \rightarrow \mathbb{R}_+$, an HL assigns to each node a label $L : V \rightarrow 2^V$ such that the so called *cover property* is fulfilled, that is, for all connected pairs $(s, t) \in V$, $L(s) \cap L(t) \cap \pi(s, t) \neq \emptyset$ where $\pi(s, t)$ denotes the set of nodes on the shortest path from s to t in G. If all nodes $v \in V$ store the shortest path distance $c_v(w)$ to each node $w \in L(v)$, the cover property allows for computing the shortest path distance between nodes s and t via the formula: $\min_{v \in L(s) \cap L(t)}(c_s(v) + c_t(v))$. To compute the set $L(s) \cap L(t)$ efficiently, nodes in $L(v)$ are presorted by node ID. Then, the intersection of

R. Uehara et al. (Eds.): WALCOM 2024, LNCS 14549, pp. 194–211, 2024.
https://doi.org/10.1007/978-981-97-0566-5_15

any two such sets can be computed in time linear in their sizes by a merge-like procedure.

Thus, the performance of HL solely depends on the size of the respective labels. Let $L_{\text{avg}} := \frac{1}{n}\sum_{v \in V}|L(v)|$ denote the average label size. Then the space needed to store an HL is in $\mathcal{O}(nL_{\text{avg}})$ and the average query time over all node pairs is upper bounded by $2L_{\text{avg}}$. Accordingly, the natural optimization goal is to compute an HL with the average node label size as small as possible. Minimizing L_{avg} is an NP-hard problem [5]. Hence efficient heuristics that produce moderate label sizes are used in practice. In particular, heuristics for a special type of HL, a so called Hierarchical HL, are most prominent.

Definition 1 (Hierarchical Hub Labeling (HHL)). *A hub labeling L is a hierarchical hub labeling if there exists a bijective ordering (rank) $r : V \to \{1,\ldots,n\}$ for $n = |V|$, s.t., $\forall w, v \in V : w \in L(v) \to r(w) \geq r(v)$.*

So nodes have ranks and the label of a node is only allowed to contain nodes of higher rank. One of the main advantages of HHL is that it suffices to fix a ranking function and then the smallest possible HHL that respects this ranking can be computed in polynomial time. This labeling is called a canonical HHL.

Definition 2 (Canonical HHL). *Given a node ordering r, the node $w \in V$ is contained in the label set $L(v)$ of $v \in V$, if and only if w has the highest ranking on some shortest paths originating from node v.*

It is easy to prove that the canonical HHL is both necessary and sufficient to fulfill the cover property. Determining the ranking function that minimizes L_{avg} is still NP-hard [6]. A heuristic is to rank the nodes by degree [9]. While this works very fast, label sizes might become huge. More sophisticated approaches use other greedy rankings or are related to Contraction Hierarchy (CH) construction [2], which is another rank-based preprocessing method for faster shortest path computation [12]. However, these do not come with provable guarantees. There do exist approximation algorithms, but with quite large approximation factors of $\mathcal{O}(\sqrt{n}\log n)$ [6]. Using a nested dissection approach, a parametrized approximation factor of $\mathcal{O}(t\log n)$ for label sizes was shown [7], where t denotes the treewidth of the network. This parameter is known to be small in road networks [16]. We will improve the parametrized approximation factor and provide further novel insights into exact and approximate HHL computation.

1.1 Related Work

HHL was proposed by Abraham et al. [2] and shown to perform well on road networks using heuristics inspired by contraction hierarchies (CH). It was shown that HHL outperforms CH significantly with respect to query time. In [10], $\mathcal{O}(\log n)$ approximate solutions for HL were compared to heuristically obtained HHL solutions, showing that HHL labels are 20–50% larger but can also be computed much faster. The gap in label size is also far from the worst-case gap of $\Theta(\sqrt{n})$ shown for specifically constructed graphs [13]. Several scalable HHL heuristics have been investigated in the meantime, offering different trade-offs between preprocessing time and label size [3,15]. The best known approximation algorithms for HHL, called w-HHL and g-HHL come with large approximation factors of $\mathcal{O}(\sqrt{n}\log n)$ [6]. It was also shown that the analysis is tight for g-HHL. For w-HHL, there exist example instances on which the produced labeling is larger than the optimum by a factor of $\Theta(\sqrt[3]{n})$.

There are several parametrized upper bounds for maximum (and thus average) label sizes in HL and HHL. For HL, an expected upper bound of $\mathcal{O}(\kappa\log n)$ was proven in [14], where κ denotes the skeleton dimension of the network. This approach relies on a randomized preprocessing scheme. For HHL, labels in $\mathcal{O}(h\log D)$ exist where h is the highway dimension and D the network diameter. Using polynomial-time preprocessing, the bound increases to $\mathcal{O}(h\log D\log n)$ [1]. However, the preprocessing is still too demanding to be used in practice even on small networks. A more practical approach that also comes with bounded label sizes was proposed in [7]. It is called nested dissection and was originally designed for CH computation. The approach relies on recursively dividing the graph using balanced node separators. The method produces HHL with label sizes in $\mathcal{O}(t\log n)$ where t denotes the treewidth. Allowing only polynomial construction time, the label sizes are in $\mathcal{O}(t\log^2 n)$. On trees, where $t = 1$, the same strategy yields a 2-approximation for L_{avg} [17]. There is also a PTAS on trees [5]. However, the complexity of minimizing label sizes of HHL on trees is still an open problem.

1.2   Contribution

We investigate HHL algorithms with provable guarantees on general graphs and on trees. The following are our main contributions:

- We present an exact algorithm for HHL with a running time of $\mathcal{O}(n^2 2^n)$ that avoids to iterate over all $n!$ possible ranking functions. While the algorithm does not scale to large networks, it can be used to assess the quality of heuristics on small instances.
- We improve the parametrized approximation factor of $\mathcal{O}(t\log n)$ to $\mathcal{O}(b)$ for HHL where b is the balanced separator number of the input graph and $b-1 \leq t$.
- We prove that two known $\mathcal{O}(\sqrt{n}\log n)$ approximation algorithms for HHL on general graphs (namely w-HHL and g-HHL) produce constant-factor approximations on trees. By generalizing an LP-based approach from [5], we prove the more general result that nested dissection based on α-balanced separators for $\alpha \in [1/2, 1)$ yields an approximation factor of $\frac{2\alpha}{1-\alpha}$. For w-HHL and g-HHL we show that $\alpha = 1/\sqrt{2}$.
- We implement and evaluate the performance of the approximation algorithms (g-HHL and w-HHL) on a diverse set of benchmarks. We assess their quality with respect to lower bounds and compare the outcomes to those of the commonly used HHL heuristics. Despite the large theoretical approximation factors, it turns out that the approximation algorithms outperform the heuristics drastically.

2   Preliminaries

Throughout the paper, we assume to be given an undirected connected graph $G(V, E)$ with non-negative edge weights $c : E \rightarrow \mathbb{R}^+$ and unique shortest paths between all pairs of nodes. The latter property is a common assumption that might also be enforced via symbolic perturbation of the edge weights. We refer to the shortest path between two nodes $s, t \in V$ as $\pi(s, t) = s, \ldots, t$ and to its cost as $c(\pi(s, t))$ or $c_s(t) = c_t(s)$ for short.

3   An Exact Algorithm for HHL

There are $n!$ many node permutations, each of them defining a valid HHL. A naive algorithm for computing the optimal HHL with respect to maximum or average label size is to iterate over all these permutations, to compute the canonical HHL for each, and to keep track of the one with smallest average label size. The respective running time amounts to $\mathcal{O}(n^3 n!)$. As the label $L(v)$ of a node v depends on the exact ranking of all the nodes of rank higher than v, it appears to be difficult to improve that running time. However, to get the average label size in a canonical HHL, we can also sum over the inverse label sizes $L^{-1}(v)$ instead, where $L^{-1}(v) := \{w \in V \mid v \in L(w)\}$. Clearly, it yields $\sum_{v \in V} |L(v)| = \sum_{v \in V} |L^{-1}(v)|$. We show next that considering the inverse label size has a crucial advantage in the design of an exact algorithm, as $L^{-1}(v)$ only depends on the set of nodes with a rank higher than v but *not* on their particular order.

Lemma 1. *For a node v, given $R(v) := \{w \in V \mid r(w) > r(v)\}$, the inverse label $L^{-1}(v)$ can be determined in $\mathcal{O}(n^2)$.*

Proof. The inverse label of node v is the set of nodes w for which v is the node of maximum rank on the shortest path from w to v. Thus, $L^{-1}(v)$ can be determined by running reverse Dijkstra from v to retrieve the reverse shortest path tree, and then cutting off all subtrees rooted in a node $u \in R(v)$.   □

The lemma can be leveraged to develop a DP that returns the HHL with smallest possible average label size as follows: We allocate a table with n rows and 2^n columns. Row i corresponds to node i. For each node subset, there is a corresponding column. The columns are sorted increasingly by the subset sizes. The entries in the first column, which corresponds to the empty set, are all filled with zeros. The other columns are filled in order using the following rules: Let the column set be S and the row node v. If $v \notin S$, we fill the respective cell with ∞. Otherwise, we want to enter the smallest summed label size of the nodes in S assuming that the contained nodes have the $|S|$ highest ranks among all nodes and v has the lowest rank among the nodes in S. To compute this value, we consider the smallest entry in column $S \setminus v$ and add to it the size of $L^{-1}(v)$ for $R(v) = S \setminus \{v\}$. Once the whole table is filled, the smallest entry in the last column (which corresponds to the whole node set) divided by n indicates the optimal average label size. The respective ranking can be deduced via backtracking.

Theorem 1. *The DP algorithm computes the minimum average label size of an HHL in $\mathcal{O}(n^2 2^n)$ time using $\mathcal{O}(n 2^n)$ space.*

Proof. According to Lemma 1, computing $L^{-1}(v)$ takes $\mathcal{O}(n^2)$. Using this method to fill each of the $\mathcal{O}(n 2^n)$ cells in the table would result in an overall running time of $\mathcal{O}(n^3 2^n)$. However, we reduce this running time by precomputing all shortest path trees in $\mathcal{O}(n^3)$, using $\mathcal{O}(n^2)$ space. Then, whenever we have to compute the inverse label size of a node v with given set $R(v)$ of higher ranked nodes, we simply traverse the precomputed shortest path tree from v and cap it at nodes in $R(v)$ instead of computing the whole tree from scratch. As the tree has a size of $\mathcal{O}(n)$, the traversal takes only linear time. Therefore, the overall running time is reduced to $\mathcal{O}(n^2 2^n + n^3) = \mathcal{O}(n^2 2^n)$. The space consumption is dominated by the number of table cells.

To show correctness, we prove by induction that the following loop invariant upholds: Once a table column is completely processed, the minimum value in said column equals the minimum summed inverse label size of the nodes in the column set S, given that those have rank higher than the nodes in $V \setminus S$. For columns that refer to single node, that is, $S = \{v\}$, we compute $L^{-1}(v)$ with an empty set $R(v)$ and thus always get $|L^{-1}(v)| = n$. Adding that to the zero obtained from the first column, which encodes $\emptyset$, the value of n does not change. Clearly, this is the smallest possible label size for any choice of a highest rank node. For the induction step, we assume that all columns with set sizes up to k are correctly processed and now consider a column with set S of size $k+1$. Let v^* be the lowest ranked node in S in the optimal ordering of S. When processing the row v^*, the node will assume that role and $L^{-1}(v^*)$ will be computed accordingly. Based on the induction hypothesis that the minimum summed inverse node labels for the nodes in $S \setminus \{v^*\}$ can be read from the respective table column, adding $|L^{-1}(v^*)|$ to said value produces the optimal entry. As in the end the minimum over all row entries is returned, the correct value produced by row v^* is retrieved. □

4 Parametrized Approximation

In [7], it was proven that the maximum (and thus also the average) label size in an HHL can be upper bounded by $\mathcal{O}(t \log n)$ where t denotes the treewidth of the input graph. The rank assignment works via nested graph dissection: First, a balanced node separator S is computed in the input network. The nodes in the separator get assigned the highest available ranks (arbitrarily). The algorithm then recurses on the components that remain after removing S from G. This results in a decomposition tree T, in which each vertex encodes a separator S in G and its children are the separators of the connected components $C_1, \ldots, C_d$ of $G[V \setminus S]$. For a node $v \in V$, its label can be constructed by considering the vertex in T in which it is contained, computing the path from this vertex to the root of T and then accumulating all nodes that are contained in the respective separators. As all S are balanced separators, the depth of T is in $\mathcal{O}(\log n)$. Thus the label size is upper bounded by the size of the largest separator in the decomposition tree multiplied with $\mathcal{O}(\log n)$. The maximum separator size in such a decomposition tree can always be upper bounded by $t + 1$.

However, we observe that it can also be upper bounded by the balanced separator number b, which denotes the smallest integer such that every induced subgraph of G with n' nodes can be separated by a node set of size at most b such that the resulting components have size at most $(n' - b)/2$. This parameter is more useful as $b \leq t + 1$ holds in any graph and thus the bound is stronger when using b. We will now proceed to show that the average label size of an HHL can be even approximated within a factor of $\mathcal{O}(b)$ instead of $\mathcal{O}(b \log n)$, thus shaving a log factor. As node labels cannot be empty, any upper bound on the label size immediately provides a matching approximation factor for the resulting labeling.

To show the improved bound, we define $\mathrm{SOL}_G(G')$ to be the average label size of the nested dissection algorithm on a subgraph G' of G that respects the shortest path structure of G. That means, that only shortest paths between nodes in G' that are completely contained in G' are relevant for the labeling. Shortest paths that leave G' are then taken care of via the inclusion of the separator nodes in the respective labels.

Lemma 2. *Let $G(V, E)$ be a graph, S a node separator of G with $|S| = b$, and $C_1, \ldots, C_d$ the connected components of $G[V \setminus S]$. Then we have following upper bound on the average HHL label size created by the nested dissection algorithm: $\mathrm{SOL}_G(G) \leq \frac{1}{n} \sum_{i=1}^{d} |C_i| \cdot \mathrm{SOL}_G(C_i) + b$.*

Proof. The solution of nested dissection algorithm $\mathrm{SOL}_G(G)$ is determined by declaring the nodes in S to have highest rank among all nodes in G and to then recurse on the connected components $C_1, \ldots C_d$ that result from removing the nodes in S from G. Thus, given the solutions computed on $C_1, \ldots C_d$, the final label of each node consists of the respective component labels plus at most all the nodes in S. The nodes in S can only have other nodes from S in their label based on the ranking property. Summing up all label sizes, we get $|C_i| \cdot \mathrm{SOL}_G(C_i)$ labels per component plus b additional labels per node. Accordingly, we have $\mathrm{SOL}_G(G) \leq \frac{1}{n} \sum_{i=1}^{d} |C_i| \cdot \mathrm{SOL}_G(C_i) + b$. □

Similar to $\mathrm{SOL}_G(G')$ we next define $OPT_G(G')$ as the optimal average label size for nodes in subgraph G' to G, considering only shortest paths from G that are fully contained in G'.

Lemma 3. *Let $G(V, E)$ be a graph, S a balanced node separator of G, and $C_1, \ldots, C_d$ the connected components of $G[V \setminus S]$, then $OPT_G(G) \geq \frac{1}{n} \sum_{i=1}^{d} |C_i| \cdot OPT_G(C_i) + \frac{1}{2}$.*

Proof. Considering a globally optimal HHL L and a component C_i, we have $\sum_{v \in C_i} |L(v) \cap C_i| \geq \sum_{v \in C_i} OPT_G(C_i)$. That means a global solution restricted to C_i can not be smaller than a locally optimal solution. Thus, we established $\mathrm{OPT}_G(G) \geq \frac{1}{n} \sum_{i=1}^{d} |C_i| \cdot \mathrm{OPT}_G(C_i)$. To show the additional $+\frac{1}{2}$ term, we consider shortest paths between different components C_i and C_j, $i \neq j$, as well as paths emerging from separator nodes. These paths are clearly not covered by the local component labels. The nodes in S all need to adhere to the trivial lower bound of having at least one node in their label. Now, we distinguish two cases:

Case I: All nodes in $\biguplus_{i=1}^{d} C_i$ have at least one label outside of their component. As $|\biguplus_{i=1}^{d} C_i| = n - |S|$ and we have established prior that the nodes in S also have non-empty labels, this gives us $n - |S| + |S| = n$ additional labels in total, and 1 additional label on average.

Case II: There exists a component C_i and a node $v \in C_i$ with $L(v) \subset C_i$, that is, v has only labels inside of C_i. To fulfill the cover property for all shortest paths $\pi(v, w)$ with $w \in C_j, j \neq i$ we hence need w to have a label inside of C_i as well. Let C^* be the largest among the d components, and nodes in C^* have only labels inside of C^*. Then minimum additional labels are at least $|\biguplus_{i=1}^{d} C_i \setminus C^*| = n - |S| - |C^*|$. As S is a balanced node separator, we have $|C^*| \leq \frac{n}{2}$. Combining this with the non-empty labels of the separator nodes, we now get $n - |S| - \frac{n}{2} + |S| = \frac{n}{2}$ additional labels in total and $\frac{1}{2}$ additional label on average.
□

Theorem 2. *Using nested dissection, the average label size of the resulting HHL is within a factor of 2b of the optimum.*

Proof. We prove the theorem by induction on n using $\mathrm{SOL}_G(G) \leq 2b\mathrm{OPT}_G(G)$ as induction hypothesis. If $n = 1$, $\mathrm{SOL}_G(G) = \mathrm{OPT}_G(G) = 1$ and $b = 1$. For $n \geq 2$, by virtue of Lemma 2 and the induction hypothesis, we get

$$n \cdot \mathrm{SOL}_G(G) \leq \sum_{i=1}^{d} |C_i| \cdot \mathrm{SOL}_G(C_i) + nb \tag{1}$$

$$\leq b(2 \sum_{i=1}^{d} n_i \cdot \mathrm{OPT}_G(C_i) + n). \tag{2}$$

Plugging in Lemma 3 that says that $2 \sum_{i=1}^{d} |C_i| \cdot \mathrm{OPT}_G(C_i) + n \leq 2n\mathrm{OPT}_G(G)$, we deduce $n\mathrm{SOL}_G(G) \leq b2n\mathrm{OPT}_G(G)$. Dividing by n produces the desired inequality.   □

Our result generalizes the observation that on trees the nested dissection approach yields a constant-factor approximation by showing that indeed the approximation factor solely depends on the separator size but not on the size of the network as a whole. On trees, balanced separators can easily be determined in polytime. For planar graphs, constant-factor approximations exist [4]. Thus, Theorem 2 shows an $\mathcal{O}(b)$ approximation for this graph class. For general graphs, computing the smallest balanced separator constitutes an NP-hard problem in itself. However, pseudo-approximation algorithms can be used to get separators with slightly worse balance guarantee and an increase in separator size by $\mathcal{O}(\log n)$. The resulting label size is then in $\mathcal{O}(b \log n)$. Note that this is still an improvement by a log factor over the previously known result, where demanding polytime preprocessing results in labels of size $\mathcal{O}(b \log^2 n)$.

5   HHL Approximation on Trees

Now, we turn our focus to tree graphs $T(V, E)$. No exact polytime algorithm is known for computing the HHL with minimum average label size on trees. In this section, we show that g-HHL and w-HHL admit constant-factor approximations on trees while for general graphs the approximation factor is in $\mathcal{O}(\sqrt{n} \log n)$.

5.1   Upper Bound for α-Balanced Separator Algorithm

The algorithm proposed by Peleg [17] uses nested dissection on trees as described in the last section. It relies on using a balanced separator on each level of the decomposition tree. We will now generalize this approach to using α-balanced separators for $\alpha \in (0, 1)$. Such a separator S demands that all components in $T[V \setminus S]$ have size at most αn.

It was proven in [5] that the optimal HL on trees is hierarchical. They provide a primal and dual formulation as shown in Table 1 of HL as linear programs to show that applying Peleg's algorithm based on balanced separators with $\alpha = 1/2$ yields a solution that is a 2-approximation compared to the optimal HL on trees. First, we recall their approach and then extend their proof to cater for general α-balanced separators.

Table 1. Primal and Dual LPs for HL on trees [5]

(PRIMAL-LP)

$$\min : \sum_{u \in V} \sum_{v \in V} x_{uv}$$

$$\text{s.t.} : \sum_{w \in P_{uv}} y_{uvw} \geq 1, \quad \forall \{u, v\} \in I$$

$$x_{uw} \geq y_{uvw}, \quad \forall \{u, v\} \in I, \forall w \in P_{uv}$$

$$x_{vw} \geq y_{vuw}, \quad \forall \{u, v\} \in I, \forall w \in P_{uv}$$

$$x_{uv} \geq 0, \quad \forall \{u, v\} \in V \times V$$

$$y_{uvw} \geq 0, \quad \forall \{u, v\} \in I, \forall w \in P_{uv}$$

(DUAL-LP)

variables α_{uv} and b_{uvw} for $w \in P_{uv}$

$$\max : \sum_{\{u,v\} \in I} a_{uv}$$

$$\text{s.t.} : a_{uv} \leq b_{vuw} + b_{uvw}, \quad \forall \{u, v\} \in I, u \neq v$$
$$\forall w \in P_{uv}$$

$$a_{uu} \leq b_{uuu}, \quad \forall u \in V$$

$$\sum_{v : w \in P_{uv}} b_{uvw} \leq 1, \quad \forall (u, w) \in V \times V$$

$$a_{uv} \geq 0, \quad \forall \{u, v\} \in I$$

$$b_{uvw} \geq 0, \quad \forall \{u, v\} \in I, \forall w \in P_{uv}$$

$$b_{vuw} \geq 0, \quad \forall \{u, v\} \in I, \forall w \in P_{uv}$$

In the formulation of the primal linear program, I is the set of all (unordered) pairs of vertices. Note that for all vertices u, the pair $\{u, u\}$ is included as well. The variables x_{uv} indicate whether v is contained in the label set H_u of u. For every node w on the shortest path P_{uv} from u to v, the variable y_{uvw} is lower or equal to both x_{uw} and x_{vw}. For all pairs $\{u, v\} \in I$, the sum over all y_{uvw} for all nodes w on P_{uv} must be at least one. This implies that any shortest path must be covered by at least one node. The goal of the primal linear program is to determine the assignments for the variables x and y, minimizing the sum over the x variables. This is exactly equivalent to minimizing the hub label size.

In the dual linear program, a_{uv} is a variable of the unordered pair $\{u, v\} \in I$, however variables b_{uvw} and b_{vuw} are different variables, where $(u, v) \in V \times V$ is an ordered pair.

The authors in [5] derive the constant approximation factor of Peleg's algorithm by comparing the contribution from each iteration of the heuristic algorithm to the value of DUAL-LP. We adjust the variables assignment in the DUAL-LP associated with parameter α such that the constraints are still fulfilled, then apply the analogous upper bound analysis.

Theorem 3. *The α-balanced separator algorithm is a $\frac{2\alpha}{1-\alpha}$-approximation algorithm on trees for $\alpha \in [1/2, 1)$.*

Proof. We apply the α-balanced separator algorithm and construct a fractional solution for the dual linear program. Let T' be the current tree of size n' in one iteration of the algorithm, and let r be a α-balanced separator. However, since trees are free of cycles, one vertex is sufficient to cut a tree into subtrees of sizes at most $\alpha n'$. Furthermore, such a node r exists for every tree. Let $A = 1/\alpha n'$ and $B = 1/2\alpha n'$. Let $T_1, \ldots, T_a$ be the connected components from T' after removal of node r.

We assign values for the variables $a_{u,v}$, b_{uvw} and b_{vuw} for node pairs $\{u, v\}$, where r is contained on the (unique) path from u to v. In the previous iterations, nodes u and v were in the same subtree. Hence, no previous α-balanced separator is contained in their shortest path. After removing r, they lie in different connected components. Hence, the variables a_{uv}, b_{uvw} and b_{vuw} are assigned exactly once during execution of the algorithm.

The rules of assignment are as follows:

$$
\begin{aligned}
&a_{uv} = A, && \forall u \in T_i, v \in T_j \wedge i \neq j \\
&a_{ur} = A, && \forall u \in T' \setminus \{r\} \\
&a_{rr} = b_{rrr} = B \\
&b_{uvw} = A \text{ and } b_{vuw} = 0, && \forall w \in P_{rv} \setminus \{r\} : \forall u \in T_i, v \in T_j \wedge i \neq j \\
&b_{uvw} = 0 \text{ and } b_{vuw} = A, && \forall w \in P_{ur} \setminus \{r\} : \forall u \in T_i, v \in T_j \wedge i \neq j \\
&b_{uvr} = b_{vur} = B, && \forall u \in T_i, v \in T_j \wedge i \neq j \\
&b_{urr} = b_{rur} = B, && \forall u \in T \setminus \{r\}
\end{aligned}
$$

We verify that these assignments are feasible solutions for the dual linear program. As shown in Fig. 1, for all nodes pairs $(u, v) \in V \times V \setminus \{(r, r)\}$ it follows $a_{uv} = A = b_{uvw} + b_{vuw} = 2B$ and $a_{rr} = b_{rrr}$, so both the first and the second constraints from DUAL-LP are satisfied.

For the third constraint, consider nodes $u \in T_i$, $w \in T_j$ and $v \notin T'$, where $i \neq j$. If w is not on the shortest u-v path, then b_{uvw} will not be assigned. If w lies on the shortest u-v path, then there must be a balanced separator r' in some previous iteration of the algorithm such that $w \in P_{ur'}$. In this case, the variable b_{uvw} receives the value zero. Therefore, for nodes $u \in T_i$, $w \in T_j$, $i \neq j$,

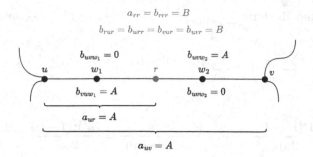

Fig. 1. DUAL-LP Assignments

all nodes v whose b_{uvw} are non-zero lie in T_j. For $u \in T_i \cup \{r\}$ and $w \in T_j$, where $i \neq j$, it follows

$$\sum_{v|w \in P_{uv}} b_{uvw} \leq |T_j| \cdot A \leq \alpha n' \cdot \frac{1}{\alpha n'} = 1. \tag{3}$$

For $u \in T_i \cup \{r\}$ and $w = r$, it follows,

$$\sum_{v|w \in P_{uv}} b_{uvw} = \sum_{v|r \in P_{ur}} b_{uvr} = \sum_{v:r \in P_{uv}} B \leq n' \cdot \frac{1}{2\alpha n'} \leq 1. \tag{4}$$

Hence, the assignments for a and b lead to a feasible solution of DUAL-LP. Furthermore, we show that the assignments lead to a resulting value that is at least $\frac{1-\alpha}{2\alpha}$ of the maximal value. This proves that the algorithm is a $\frac{2\alpha}{1-\alpha}$-approximation algorithm.

During one iteration, we add r into the hub labels of all nodes in T'. This increases the cost of hub labeling by n'. Now consider the contribution C of the assignment in this iteration, which is the sum of all a_{uv} such that r lies on the shortest u-v path, i.e., u and v are in different subtrees after removal of r. Let $t_1, \ldots, t_s$ be the subtree sizes. It follows that

$$C = \sum_{i=1}^{s-1} \sum_{j=i+1}^{s} \left(\sum_{u \in T_i, v \in T_j} a_{uv} \right) + \sum_{i=1}^{s} \sum_{u \in T_i} a_{ur} + a_{rr} \tag{5}$$

$$= A \sum_{i=1}^{s-1} \sum_{j=i+1}^{s} t_i t_j + A(n'-1) + B \tag{6}$$

$$= \frac{A}{2} \sum_{i=1}^{s} t_i \left(\sum_{j=1, i \neq j}^{s} t_j \right) + A(n'-1) + B. \tag{7}$$

Considering that the term $\sum_{j=1,i\neq j}^{s} t_j = n' - 1 - t_i \geq n' - 1 - \alpha n'$, it follows

$$\sum_{i=1}^{s} t_i \left(\sum_{j=1,i\neq j}^{s} t_j \right) \geq \sum_{i=1}^{s} t_i(n' - 1 - \alpha n') = (n' - 1)(n' - 1 - \alpha n'). \quad (8)$$

Moreover,

$$C \geq \frac{(n' - 1)(n' - 1 - \alpha n')}{2\alpha n'} + \frac{n' - 1}{\alpha n'} + \frac{1}{2\alpha n'} = \frac{(n')^2(1 - \alpha)}{2\alpha n'} = \frac{1 - \alpha}{2\alpha}n'. \quad (9)$$

Hence, the α-balanced separator algorithm is a $\frac{2\alpha}{1-\alpha}$-approximation algorithm on trees. $\qquad\square$

5.2 Properties of G-HHL and W-HHL on Trees

Both g-HHL and w-HHL are greedy algorithms for HHL which assign the node ranks from highest to lowest. g-HHL always selects next the node that is contained in the largest number of shortest paths that do not already contain a node of higher rank. w-HHL also uses that shortest path count but divides it by the inverse label size.

To show that g-HHL and w-HHL are $\frac{2}{\sqrt{2}-1}$-approximations on trees, we first define the center graph on a undirected graph as follows (analogous to [6]): A center graph $G_v = (V, E_v)$ of node v in graph $G = (V, E)$ contains all the vertices in G, and there is an edge $\{u, w\} \in E_v$ if and only if there is a shortest path between u and w via v. According to the definition, g-HHL selects the node whose center graph has the most edges in each iteration and assigns it the next highest rank; instead, w-HHL selects the node whose center graph has the highest density (number of edges divided by the number of non-isolated nodes).

It is trivial to show that g-HHL and w-HHL return the same label sizes on trees. Since for any node $v \in T$, there is exactly one shortest path from v to any other node in T. Therefore, the center graph of each node in T does not contain isolated vertices. Let $v \in T$ be the node whose center graph has the highest number of edges. It follows immediately, that the center graph of v has also the highest density among all center graphs.

For the same tree, both g-HHL and w-HHL assign v the next highest rank, every node in T contains v in its label set. After the removal of node v, T decomposes into subtrees of smaller size. The shortest path pairs covered by v are separated into different subtrees. Hence, recursively g-HHL and w-HHL select the same node in each subtree. Hence, the resulting label sizes of both algorithms are the same.

To proof that g-HHL and w-HHL have a constant approximation factor on trees, we show that a node $v \in V$ maximizing the number of edges in its center graph is actually an α-balanced separator of the tree T with $\alpha \in [1/2, 1)$. Then, the assertion readily follows from Theorem 3.

Lemma 4. *Let $T = (V, E)$ be a tree with n nodes. The vertex $v \in V$ whose center graph G_v of T maximizes the number of edges is a $\frac{1}{\sqrt{2}}$-balanced separator.*

Proof. Let v be the node whose center graph G_v maximizes the number of edges. Let T_m be the largest subtree rooted at a neighbor of v. Our goal is to find an upper bound on the size t of T_m. In the worst-case scenario, T_m has as many nodes as possible, and the center graph G_v still has the most edges. This is achieved if the number of edges contributed from $T\backslash T_m$ is as large as possible, i.e., if the nodes in $T\backslash T_m$ form a clique in G_v. Therefore, $T\backslash T_m$ forms a star in T whose center is v. In this case, these nodes contribute $\frac{(n-t)(n-t-1)}{2}$ edges in G_v. If T_m is too large, then there exists some node u on T_m whose center graph G_u has more edges than G_v. In order to make the center graph G_u have as few edges as possible (so that T_m can be as large as possible), T_m must be a path. Hence, u is the balanced separator of T whose center graph has the most edges among the nodes on T_m. Since G_v has at least so many edges as in G_u, it follows that $\frac{(n-t-1)(n-t-2)}{2} + t(n-t-1) + n - 1 \geq \left(\frac{n-1}{2}\right)^2 + n - 1$. Accordingly, the balance ratio of separator v is $\frac{t}{n} \leq \frac{1}{\sqrt{2}}$. □

Based on Lemma 4 and Theorem 3, we obtain the following theorem.

Theorem 4. *g-HHL is a $\frac{2}{\sqrt{2}-1}$-approximation algorithm for HL on trees.*

6 Experimental Evaluation

This section evaluates the performance of g-HHL, w-HHL, CH-based HHL and HHL based on balanced separators. Furthermore, we apply the exact algorithm for HHL on trees and investigate lower and upper bounds.

6.1 On General Graphs

Methods for Contraction Hierarchies (CH) construct node rankings that can be used for HHL [13]. We always compute the canonical HHL associated with the CH node ranking.

The algorithm based on balanced separators computes a preferably small 1/2-balanced separator S in each connected component and assigns them the next highest rankings. We use KaFFPa in our implementation (a tool included in KaHIP [18]), which can find small balanced separators in large graphs efficiently in a heuristic fashion. We again compute the canonical HHL with respect to the obtained node ranking.

To evaluate the resulting label size quality of the heuristics, we compare the results with the label size lower bound from [8]. The authors introduce the efficiency of a pair of nodes and show that it can be used to derive a general lower bound for HL on arbitrary graphs. For $u, v \in V$, let $h_u(v)$ be the number of vertices x whose shortest path from u to x contains v, and let P_{uv} be the set of shortest paths from u to v. Then the efficiency of each node pair (u, v) is defined as follows: $\text{eff}(u, v) = \max\limits_{p \in P_{uv}} \max\limits_{w \in p} \min\{h_u(w), h_v(w)\}$. For the optimal average hub labeling size L_{avg} for the graph $G = (V, E)$ holds that

$$n \cdot L_{\text{avg}} \geq \sum_{(s,t):P_{st} \neq \emptyset} \frac{1}{\text{eff}(s, t)}. \qquad (10)$$

Note that this bound works well for dense graphs, however, it is less tight on sparser graphs. For example, on paths or binary trees, the lower bound above states $n \cdot L_{\text{avg}} \in \Omega(n)$ (where n is the number of nodes), though the optimal hub labeling on these graphs is in $\Omega(n \log n)$.

We implemented all algorithms in C++ and and executed on a single core of an AMD Ryzen 7 3700X processor (clocked at 3.6 GHz) with 128 GB main memory, which had an L3 cache size of 16384K. The operating system was ubuntu 18.04. KaHIP 2.10 was compiled using Clang 11.0.0 with OpenMPI 1.4.1.

Benchmark Sets. The first benchmark set contains the instances from the PACE challenge 2020[1], which were provided for the tree depth decomposition challenge. The second benchmark set contains the instances from the PACE challenge 2019 [11], which were provided for the vertex cover challenge. The last benchmark set consists of road networks extracted from OpenStreetMap[2] (OSM). Table 2 presents an overview over all benchmark sets. Benchmark sets from the PACE challenge contain a variety of graph densities, while the OSM benchmark set is comparably sparse.

Table 2. Number of nodes, edges and average degree of the graphs in the used benchmark sets.

Benchmark Sets	#Instances	$|V|$	$|E|$	$\bar{d}$
PACE 2020 exact	100	$[10, 491]$	$[15, 4.100]$	$[2, 65]$
PACE 2020 heuristic	40	$[105, 7.813]$	$[143, 366.239]$	$[2, 208]$
PACE 2019	30	$[153, 4.579]$	$[625, 126.163]$	$[3, 164]$
OSM	15	$[100, 4000]$	$[99, 4336]$	2

The average label sizes L_{avg} of the different approaches, compared to the lower bound l_{avg} given by efficiency, are shown in Fig. 2. We can observe that g-HHL and w-HHL compute results of comparable quality. No algorithm could find a label size smaller than four times the efficiency lower bound on the road networks. Recall that the efficiency lower bound is less accurate on sparse graphs. In all PACE instances, the label sizes from both algorithms are within the factor of 7 from the efficiency lower bound, while the average value of $L_{\text{avg}}/l_{\text{avg}}$ is around three on the PACE instances and still lower than 5 for the OSM instances. The algorithm based on CH almost always computes worse results than the previously mentioned algorithms. It performs poorly on the OSM road networks. In particular, the average ratio to the lower bound is around twelve, and even the best-case instances have a larger label set than the worst-case instances of g-HHL and w-HHL.

---

[1] PACE challenge.
[2] OSM.

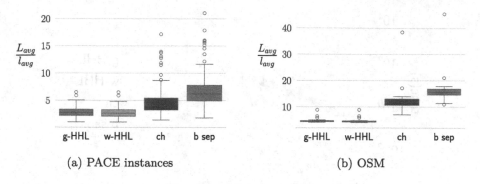

Fig. 2. Comparison of the resulting average label sizes L_{avg} and the efficiency lower bound l_{avg}

The balanced separator based algorithm computes the worst results on average in all instances. On the PACE instances, its average ratio is around six compared to the lower bound, while all other algorithms find smaller hub labels on average. On the road networks, the balanced separator based algorithm could, in the best-performed instances, find a solution within the lower bound by a factor of around 13. It is slightly worse than the worst results of g-HHL and w-HHL. The average ratio of the solution quality of the balanced separator based algorithm compared to the lower bound is around 15, about 3 times as large as g-HHL and w-HHL.

To summarize, we observe that g-HHL and w-HHL compute hub labels of similar size on all inputs. On average, the CH-based algorithm computes solutions smaller than the balanced separator based algorithm but slightly worse hub labels than the previous algorithms. The last algorithm based on balanced separators computes the worst solutions in all instances.

The running times of all algorithms are shown in Fig. 3. Both g-HHL and w-HHL have very similar running times. In almost all instances, the CH-based approach is the fastest algorithm. However, on some of the dense graphs, the running times of CH are even the longest. This results from the fact that CH is originally designed to work on large and sparse road networks. The last algorithm based on balanced separators is time-consuming in small instances. This is due to the computation of a balanced separator involving the execution of KaHIP, which takes considerable time. However, this approach can be used on large instances since its asymptotical growth is slower than that of all other algorithms.

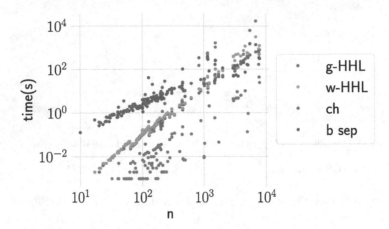

Fig. 3. Running time comparison

6.2 On Trees

To evaluate the performance of Peleg's algorithm and g-HHL on trees, we generated two benchmark sets. The first benchmark set includes extracted shortest path trees from the PACE 2020 heuristic benchmark set. The second one comprises our self-generated random trees, which are constructed from a single node by iteratively adding a new node to a random previous node.

Since both g-HHL and w-HHL behave the same on trees followed from Sect. 5.2, it is sufficient to only compare Peleg's algorithm to g-HHL. We applied Peleg's algorithm and g-HHL to our two benchmark sets and obtained the resulting label sizes L_p and L_g. Furthermore, we also measured the balance ratio $\alpha(v, T')$ of each node v to which g-HHL assigned the highest rank in the subtree T' during the execution. Recall the definition of the balance ratio of a node v in a tree T' with n nodes. The balance ratio of v is the largest connected component size after removing the node v and its adjacent edges divided by n.

Table 3 presents an overview over the benchmark sets and the maximum balance ratios. Figure 4 illustrates the resulting label sizes of g-HHL L_g compared to the label size of Peleg's algorithm L_p.

Table 3. Number of instances, number of nodes, and the maximum balance ratio during the execution of g-HHL.

Benchmark Set	#Instances	$|V|$	$\max(\alpha(v, T'))$
PACE 2020 Heur	64	[100, 1.319.677]	0.66
Random Trees	66	[100, 448.000]	0.64

The maximum balance ratio $\alpha(v, T')$ in the first benchmark set is 0.64 and 0.66 in the second benchmark set, both were significantly lower than $1/\sqrt{2} \approx 0.707$

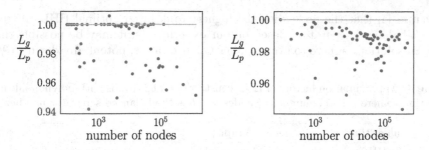

Fig. 4. Label size comparison between g-HHL and Peleg's algorithm. Left: Label size comparison on PACE2020 heuristic public benchmark set. Right: Label size comparison on random tree benchmark set.

from the worst case shown in Lemma 4. Furthermore, there are only two instances in which L_g is larger than L_p, one from each benchmark set. In both instances, the resulting label sizes of g-HHL exceeded the results from Peleg's algorithm by less than 1%. The approximation factor from Theorem 3 depends on the balance ratio. Since g-HHL and w-HHL do not aim to cover the shortest paths by the node that separates the tree in the most balanced way, the approximation factors via Theorem 3 are larger than Peleg's algorithm. However, our results demonstrate that the label sizes of g-HHL are in most cases lower or equal to L_p, and are significantly closer to the optimum than the factor of $\frac{2}{\sqrt{2}-1}$ from Theorem 4.

Furthermore, we implemented the exact algorithm for hierarchical hub labeling and applied it on 20000 randomly generated small trees with node sizes lower or equal to 20. Comparing the hub label sizes resulting from g-HHL and Peleg's algorithm to the optimal solution in the same instances, we observed a lower bound of 1.005 and 1.023. The maximum balance ratio $\alpha(v, T')$ of the exact algorithm never exceed $2/3$. Based on this observation, we implemented an algorithm that tries to assign every $2/3$-balanced separator with the next highest rank in each iteration and returns the HHL with the smallest label size, which has a running time in $\mathcal{O}(n^{\log n})$. After applying this algorithm to randomly generated trees with node sizes up to 2000, we observed a slightly larger lower bound of 1.034 for g-HHL.

7   Conclusions and Future Work

We have shown novel approximation bounds for HHL, which complement and improve existing bounds. Table 4 provides an overview of the currently best known approximation factors for general graphs and trees. Our evaluation incidates that g-HHL and w-HHL perform remarkably well on trees as well as on general graphs. Improving their scalability to make them applicable to larger inputs may be an interesting direction for future work. Also, the approximation factor on trees could potentially be shown to be even smaller than $\frac{2}{\sqrt{2}-1}$. It was proven in [5] that the node with the highest rank in an optimal HHL on trees is a $5/6$-balanced separator. Based on our experiments, it may be possible that optimal solutions adhere to even smaller balance ratios, potentially down to $2/3$.

Table 4. Approximation factors of HHL construction algorithms with polynomial running time, where n is the number of nodes and b is the balanced separator number

algorithm	general graph		tree	
w-HHL	$\mathcal{O}(\sqrt{n}\log n)$	[6]	$\frac{2}{\sqrt{2}-1}$	[Theorem 4]
g-HHL	$\mathcal{O}(\sqrt{n}\log n)$	[6]	$\frac{2}{\sqrt{2}-1}$	[Theorem 4]
balanced separator	$\mathcal{O}(b\log n)$	[Theorem 2]	2	[5]

References

1. Abraham, I., Delling, D., Fiat, A., Goldberg, A.V., Werneck, R.F.: Highway dimension and provably efficient shortest path algorithms. J. ACM (JACM) **63**(5), 1–26 (2016)
2. Abraham, I., Delling, D., Goldberg, A.V., Werneck, R.F.: Hierarchical hub labelings for shortest paths. In: Epstein, L., Ferragina, P. (eds.) ESA 2012. LNCS, vol. 7501, pp. 24–35. Springer, Heidelberg (2012). https://doi.org/10.1007/978-3-642-33090-2_4
3. Akiba, T., Iwata, Y., Kawarabayashi, K.I., Kawata, Y.: Fast shortest-path distance queries on road networks by pruned highway labeling. In: 2014 Proceedings of the Sixteenth Workshop on Algorithm Engineering and Experiments (ALENEX), pp. 147–154. SIAM (2014)
4. Amir, E., Krauthgamer, R., Rao, S.: Constant factor approximation of vertex-cuts in planar graphs. In: Proceedings of the Thirty-Fifth Annual ACM Symposium on Theory of Computing, pp. 90–99 (2003)
5. Angelidakis, H., Makarychev, Y., Oparin, V.: Algorithmic and hardness results for the hub labeling problem. In: Proceedings of the Twenty-Eighth Annual ACM-SIAM Symposium on Discrete Algorithms, pp. 1442–1461. SIAM, Society for Industrial and Applied Mathematics (2017)
6. Babenko, M., Goldberg, A.V., Kaplan, H., Savchenko, R., Weller, M.: On the complexity of hub labeling (extended abstract). In: Italiano, G.F., Pighizzini, G., Sannella, D.T. (eds.) MFCS 2015, Part II. LNCS, vol. 9235, pp. 62–74. Springer, Heidelberg (2015). https://doi.org/10.1007/978-3-662-48054-0_6
7. Bauer, R., Columbus, T., Rutter, I., Wagner, D.: Search-space size in contraction hierarchies. Theoret. Comput. Sci. **645**, 112–127 (2016)
8. Cohen, E., Halperin, E., Kaplan, H., Zwick, U.: Reachability and distance queries via 2-hop labels. SIAM J. Comput. **32**(5), 1338–1355 (2003)
9. Delling, D., Goldberg, A.V., Pajor, T., Werneck, R.F.: Robust exact distance queries on massive networks. Microsoft Research, USA, Technical report 2 (2014)
10. Delling, D., Goldberg, A.V., Savchenko, R., Werneck, R.F.: Hub labels: theory and practice. In: Gudmundsson, J., Katajainen, J. (eds.) SEA 2014. LNCS, vol. 8504, pp. 259–270. Springer, Cham (2014). https://doi.org/10.1007/978-3-319-07959-2_22

11. Dzulfikar, M.A., Fichte, J.K., Hecher, M.: PACE 2019: track 1 - vertex cover instances (2019). https://doi.org/10.5281/zenodo.3368306
12. Geisberger, R., Sanders, P., Schultes, D., Vetter, C.: Exact routing in large road networks using contraction hierarchies. Transp. Sci. **46**(3), 388–404 (2012)
13. Goldberg, A.V., Razenshteyn, I., Savchenko, R.: Separating hierarchical and general hub labelings. In: Chatterjee, K., Sgall, J. (eds.) MFCS 2013. LNCS, vol. 8087, pp. 469–479. Springer, Heidelberg (2013). https://doi.org/10.1007/978-3-642-40313-2_42
14. Kosowski, A., Viennot, L.: Beyond highway dimension: small distance labels using tree skeletons. In: Proceedings of the Twenty-Eighth Annual ACM-SIAM Symposium on Discrete Algorithms, pp. 1462–1478. SIAM (2017)
15. Lakhotia, K., Dong, Q., Kannan, R., Prasanna, V.: Planting trees for scalable and efficient canonical hub labeling. arXiv preprint arXiv:1907.00140 (2019)
16. Maniu, S., Senellart, P., Jog, S.: An experimental study of the treewidth of real-world graph data. In: 22nd International Conference on Database Theory (ICDT 2019). Schloss Dagstuhl-Leibniz-Zentrum fuer Informatik (2019)
17. Peleg, D.: Proximity-preserving labeling schemes. J. Graph Theory **33**(3), 167–176 (2000)
18. Sanders, P., Schulz, C.: Think locally, act globally: highly balanced graph partitioning. In: Bonifaci, V., Demetrescu, C., Marchetti-Spaccamela, A. (eds.) SEA 2013. LNCS, vol. 7933, pp. 164–175. Springer, Heidelberg (2013). https://doi.org/10.1007/978-3-642-38527-8_16

Reconfiguration of Multisets
with Applications to Bin Packing

Jeffrey Kam[1], Shahin Kamali[2], Avery Miller[3(✉)], and Naomi Nishimura[1]

[1] University of Waterloo, Waterloo, ON, Canada
{jeffrey.kam,nishi}@uwaterloo.ca
[2] York University, Toronto, ON, Canada
kamalis@yorku.ca
[3] University of Manitoba, Winnipeg, MB, Canada
avery.miller@umanitoba.ca

Abstract. We use the reconfiguration framework to analyze problems that involve the rearrangement of items among groups. In various applications, a group of items could correspond to the files or jobs assigned to a particular machine, and the goal of rearrangement could be improving efficiency or increasing locality.

To cover problems arising in a wide range of application areas, we define the general REPACKING problem as the rearrangement of multisets of multisets. We present hardness results for the general case and algorithms for various classes of instances that arise in real-life scenarios. By limiting the total size of items in each multiset, our results can be viewed as an offline approach to BIN PACKING, in which each bin is represented as a multiset.

In addition to providing the first results on reconfiguration of multisets, our contributions open up several research avenues: the interplay between reconfiguration and online algorithms and parallel algorithms; the use of the tools of linear programming in reconfiguration; and, in the longer term, a focus on resources in reconfiguration.

1 Introduction

We consider the problem of rearranging items in multisets, from a given source arrangement to a specified target arrangement. Although our techniques draw on the problem of BIN PACKING and the area of reconfiguration, each one is considered in a non-traditional way: we view the rearrangement as an offline problem, not an online one, we view bins and items of the same size as indistinguishable, and we focus on the feasibility of reconfiguration under various conditions. In doing so, we set the stage for the exploration of *resource-focused reconfiguration*, in which the goal is to determine which extra resources, if any, are needed to make reconfiguration possible (Sect. 7).

Research supported by the Natural Sciences and Engineering Research Council of Canada (NSERC).

R. Uehara et al. (Eds.): WALCOM 2024, LNCS 14549, pp. 212–226, 2024.
https://doi.org/10.1007/978-981-97-0566-5_16

The reconfiguration framework [9] has been used to consider the step-by-step modification among configurations, which may encode information such as solutions to the instance of a problem or the state of a geometric object, game, or puzzle. Common types of questions framed using reconfiguration include structural properties of the *reconfiguration graph*, formed by adding edges between a configuration and any other configuration resulting from the execution of a single *reconfiguration step*, as well as the reachability of one configuration from another [6,12]. In our context, we wish to transform a source configuration into a target configuration by a sequence of reconfiguration steps (or *reconfiguration sequence*); of particular import is the fact that each intermediate configuration in the sequence conform to the same constraints as the source and target.

Ito et al. studied a related problem [7,8], where the objective is to reconfigure one "feasible packing" of an instance of the KNAPSACK problem to another, where a feasible packing is defined as a subset of items summing to a value in a given range. Under the assumption that the intermediate packings must be feasible, the authors present hardness results for the decision problem and a polynomial-time approximation scheme (PTAS) for the optimization problem. In contrast, in our work, each configuration consists of multiple bins, not a single bin, and every item must be packed.

We define a problem, REPACKING, where, as in BIN PACKING, the goal is to group items to form a packing. Packings are naturally modeled as multisets, since neither the ordering of multisets nor the ordering of items within a multiset are important. At a high level, we are considering the reconfiguration of items in unlabeled, and hence indistinguishable, bins. To avoid confusion with traditional bin packing (where the bins are distinguishable containers), we refer to each group of items as a *bunch* instead of a bin. For full generality, there are no constraints on the numbers or types of items, composition of bunches, or allowable packings.

Due to the hardness of the most general form of the problem (Sect. 3), we turn to situations that might naturally arise in real-life uses of repacking. In many settings, such as virtual machine placement, the maximum sum of sizes of items in a multiset can be set to a fixed constant (or capacity). To mimic the capacity of a container, we define a *capacity* as the upper bound on the sum of sizes of items in a bunch. From a practical standpoint, it may be reasonable to consider scenarios in which the sizes of items and capacities are constant with respect to the number of multisets under consideration. For example, a typical data center contains thousands of homogeneous Physical Machines (PMs), each having a bounded capacity, in terms of bandwidth, CPU, memory, and so on, and hosting a number of Virtual Machines (VMs). Virtual Machines ("items") have various loads, and the total load of VMs assigned to a PM (a "bunch") must not exceed the uniform capacity of the PMs. It is desirable to reconfigure a given assignment of VMs to PMs for reasons such as performing maintenance, balancing loads, collocating VMs that communicate with each other, and moving VMs apart to minimize the fault domain [2,11]. Such reconfiguration must occur in steps that involve migrating a VM from one PM to another while respecting the bounded capacity of the PMs at each step.

Our paper is structured as follows. In Sect. 2, we formally define the REPACK-ING problem. We prove that REPACKING is NP-hard in Sect. 3. Then, we provide algorithms for instances relevant to real-life scenarios. In Sect. 4, we consider instances in which all item sizes are bounded above by a constant fraction of the capacity, and give an algorithm that reconfigures any instance in which the unoccupied space is sufficiently large. In Sect. 5, we consider a setting in which all item sizes and the capacity are powers of 2, fully characterize when reconfiguration is possible in such instances, and give a reconfiguration algorithm for all reconfigurable instances. Motivated by the possibility of solving reconfiguration problems in parallel, Sect. 6 gives an algorithm that determines, for any given instance, whether or not reconfiguration is possible by partitioning the source configuration into smaller parts and only moving an item within its assigned part. In Sect. 7, we present directions for future work. Proofs that have been omitted due to space constraints will appear in the full version of the paper.

2   Preliminaries

Our goal is to determine whether it is possible to rearrange items from one configuration (multiset of multisets of items) to another. Stated in the most general terms, items can have a variety of attributes, such as distinct IDs, and multisets can have not only various attributes but also different capacities. In the remainder of the paper, we simplify the discussion and notation by assuming that *items* are positive integers representing sizes (where items of the same size are indistinguishable) and *bunches* are multisets of items (where bunches corresponding to the same multiset are indistinguishable). To facilitate explanation of algorithms or analysis, at times we might assign names to specific items or bunches for greater clarity.

To avoid confusion among items, bunches (multisets of items), and multisets of bunches (multisets of multisets of items), in our choice of notation, we use the convention that a lower-case letter refers to an item or a number (where at times Greek letters will also be used for numbers), an upper-case letter refers to a set or multiset, and a calligraphic letter refers to a multiset of multisets.

To give an example of rearrangement, suppose we wish to transform the two bunches, $\{1, 1, 2, 6\}$ and $\{2, 3, 5\}$, (the first bunch and second bunch, respectively) into the two bunches $\{1, 3, 6\}$ and $\{1, 2, 2, 5\}$. In the absence of constraints, we can achieve our goal by moving items 1 and 2 from the first bunch to the second bunch and item 3 from the second bunch to the first bunch. The task proves more challenging when there is an upper bound on the sum of items in any bunch, and rendered impossible without extra storage space if the upper bound is 10.

Although in general the sum of all items in a bunch may be unconstrained, in this paper we view each bunch as having a positive integer *capacity*, typically denoted by κ, as an upper bound. We use $vol(B)$, the *volume of B*, to denote the sum of all the items in bunch B, and require that $vol(B) \le \kappa$. The term *slack*, denoted $slack(B)$, is used to refer to the amount of space in B that is not occupied by items, where $slack(B) = \kappa - vol(B)$. By definition, the slack of a bunch is non-negative. A bunch is *empty* if it contains no items; in such a case, the slack will equal the capacity.

Formalizing our terminology, we define a *configuration* to be a multiset of bunches, and a *legal configuration for capacity* κ to be a configuration in which the volume of each bunch is at most κ. The *underlying set of a configuration* $\mathcal{C}$, denoted $U(\mathcal{C})$, is the multiset union of items in all bunches in a configuration; U is used without a parameter to denote the underlying set of the source and target configurations.

We transform one configuration to another by a series of steps, each corresponding to the move of a single item. More formally, we move a single item u from one bunch (the *donor bunch*, B_d) to another bunch (the *recipient bunch*, B_r), where we reuse the names B_d and B_r to describe two different bunches, and two different points in time (before and after the move). Between the two points of time, B_d is changed only by the removal of u, and B_r is changed only by the addition of u. The pairs of bunches (B_d and B_r at the first point in time and B_d and B_r at the second point in time) are said to correspond to the *move of a single item*.

We consider legal configurations $\mathcal{C}$ and $\mathcal{D}$ for capacity κ to be *adjacent* if it is possible to form $\mathcal{D}$ from $\mathcal{C}$ by the move of a single item. More formally, $\mathcal{C} \setminus \mathcal{D}$ consists of two *old bunches*, $\mathcal{D} \setminus \mathcal{C}$ consists of two *new bunches*, and the pair of old bunches and the pair of new bunches correspond to the move of a single item. We say that we can get from $\mathcal{C}$ to $\mathcal{D}$ in a single *reconfiguration step* if the two configurations are adjacent. A *reconfiguration sequence for capacity* κ from a source configuration $\mathcal{C}_S$ to $\mathcal{C}_T$ consists of a sequence of legal configurations for capacity κ, $\mathcal{C}_S = \mathcal{C}_0, \mathcal{C}_1, \ldots, \mathcal{C}_T$ such that each pair of consecutive configurations $\mathcal{C}_i$ and $\mathcal{C}_{i+1}$ are adjacent.

We define the following problem REPACKING, with instance $(\mathcal{C}_S, \mathcal{C}_T, \kappa)$, as defined below:

Input: Source and target legal configurations $\mathcal{C}_S$ and $\mathcal{C}_T$ for capacity κ.
Question: Is there a reconfiguration sequence for capacity κ from $\mathcal{C}_S$ to $\mathcal{C}_T$?

Since, as we will show in Theorem 1, REPACKING is NP-hard in general, it is natural to consider extensions and variants of the problem. As defined, the number of bunches is preserved at each step in the sequence, and hence the number of bunches available for reconfiguration is the number of bunches in the source (and hence the target) configuration, denoted $|\mathcal{C}_S|$. More generally, we consider the interplay among various properties of an instance and the relationship between the sum of the items and the total capacity of all bunches.

3   Complexity of Repacking

We use a reduction from MULTIWAY NUMBER PARTITIONING [5] to establish the NP-hardness of REPACKING. The MULTIWAY NUMBER PARTITIONING problem, known to be NP-hard [4], asks whether a given multiset of integers can be partitioned into m multisets such that the numbers in each multiset sum to at most α. Given an instance P of the MULTIWAY NUMBER PARTITIONING problem, formed by n positive integers, a positive integer m, and a positive integer α, we define an instance $(\mathcal{C}_S, \mathcal{C}_T, \kappa)$ of REPACKING as follows. We may assume $m \geq 1$, $n \geq 2m + 2$, and α is at least as large as the largest of the n integers in P.

We set the capacity to be $\kappa = 2(n - m)\alpha$ and create configurations $\mathcal{C}_S$ and $\mathcal{C}_T$, each containing $n + 2$ bunches, n of which are identical in the two configurations (the *matching bunches*), and two of which will differ in the two configurations (the *non-matching bunches*). Each of the n matching bunches consists of an item of size equal to one of the n integers in P and an item of size $\kappa - \alpha$. The two non-matching bunches store the following items: two items a, b, each of size $\kappa/2$, an item c of size $\kappa/2 - 1$, and $n - m$ items each of size α (and hence total volume $\kappa/2$). In the source, a and b are in the same bunch, and in the target, a and c are in the same bunch.

We prove that P is a yes-instance of MULTIWAY NUMBER PARTITIONING if and only if $(\mathcal{C}_S, \mathcal{C}_T, \kappa)$ is a yes-instance of REPACKING. In particular, we show that reconfiguration is possible if and only if all of the $n - m$ items of size α can be moved from the non-matching bunches to the matching bunches. This is indeed possible if and only if we can rearrange the data items in the n matching bunches so that $n - m$ bunches each have slack α and the remaining m bunches contain the n items with sizes corresponding to integers from P, which is equivalent to having a yes-instance of MULTIWAY NUMBER PARTITIONING.

Theorem 1. REPACKING *is NP-hard.*

4   Special Case: Instances with Small Items

Intuitively, if items are relatively small and there is enough slack in the source and target configurations, it seems feasible to reconfigure them. In this section, we formalize and prove this intuition. In particular, we consider a setting where all items are of size at most κ/α, for some integer $\alpha > 1$. We define the *average slack* of bunches in a configuration $\mathcal{C}$ by $\frac{1}{|\mathcal{C}|} \sum_{B \in \mathcal{C}} slack(B)$, and we assume that the average slack of $\mathcal{C}_S$ is at least $\frac{\kappa}{\alpha+1} + \frac{3\alpha\kappa}{(\alpha+1)|\mathcal{C}_S|}$. Since $|\mathcal{C}_S| = |\mathcal{C}_T|$ and the total volume of all bunches is the same in $\mathcal{C}_S$ and $\mathcal{C}_T$, this assumption implies that the average slack of bunches of $\mathcal{C}_T$ is at least $\frac{\kappa}{\alpha+1} + \frac{3\alpha\kappa}{(\alpha+1)|\mathcal{C}_T|}$. For larger inputs, this means that the average slack of bunches converges to a value at least $\frac{\kappa}{\alpha+1}$. We will describe an algorithm that, under these assumptions, reconfigures $\mathcal{C}_S$ to $\mathcal{C}_T$ (as formally stated in Theorem 3).

Consider a configuration resulting from sorting items in the underlying set U in non-increasing order of their sizes and placing them, one by one, in the first bunch with enough slack, and placing them in a new bunch if no such bunch exists. This configuration, which we call the *First-Fit-Decreasing (FFD) configuration*, and denote by $\mathcal{C}_{FFD}$, is a canonical configuration for our reconfiguration algorithm. That is, we show how to reconfigure both $\mathcal{C}_S$ and $\mathcal{C}_T$ to $\mathcal{C}_{FFD}$ and therefore to each other.

In what follows, we describe how a configuration $\mathcal{C}$ with bunches of average slack at least $\frac{\kappa}{\alpha+1} + \frac{3\alpha\kappa}{(\alpha+1)|\mathcal{C}|}$ can be reconfigured to $\mathcal{C}_{FFD}$. This process is applied to reconfigure $\mathcal{C}_S$ and $\mathcal{C}_T$ to $\mathcal{C}_{FFD}$. The algorithm works in stages: each stage starts with a "compression phase" and then an "FFD-retrieval phase" follows. As we will show, the compression phase always results in at least one empty bunch, which is subsequently packed according to the FFD rule. Next, we formally describe this process.

At the beginning of stage i, we have two types of bunches: FFD bunches and general bunches. At the beginning of the reconfiguration process, all bunches are general bunches. As mentioned earlier, items in FFD bunches are packed according to the FFD rule.

In the compression phase, we will process general bunches one by one in an arbitrary order. Items within the bunch being processed are also processed one by one in an arbitrary order. For each item x, we check the previously processed bunches and move x to the first bunch (with respect to the ordering) with slack at least equal to x; if no such bunch exists, x remains in its bunch. The following lemma shows that the compression process always results in empty bunches, which are subsequently used to increase the number of FFD bunches in the FFD-retrieval phase. For the proof, we will show that the average slack of the general bunches at the beginning of stage i, for any $i \geq 1$, is at least $\frac{\kappa}{\alpha+1} + \frac{3\alpha\kappa}{(\alpha+1)\beta_i}$, where β_i is the number of general bunches at stage i. At stage 1, this invariant follows from the assumed lower bound for the average slack of bunches in $\mathcal{C}_S$. For other stages, we prove in Lemma 2 that there is at least one empty bunch at the end of the compression stage.

Lemma 2. *If all items in the underlying set U of the configuration $\mathcal{C}$ are of size at most κ/α for some integer $\alpha > 1$ and the average slack of all bunches in $\mathcal{C}$ is at least $\frac{\kappa}{\alpha+1} + \frac{3\alpha\kappa}{(\alpha+1)|\mathcal{C}|}$, then at the end of the compression phase of any stage $i \geq 1$, there is at least one empty general bunch.*

In the FFD-retrieval phase of stage i, we will declare any empty general bunch to be an FFD bunch. By Lemma 2, at least one such bunch exists. We pack such bunches according to the FFD rule: we process items in the general bunches in non-increasing order of their sizes and place each item into the first FFD bunch with enough slack to host it. This process continues until no more items are in the general bunches or no FFD bunch has enough slack for the largest remaining item in the general bunches. In the former case, the canonical configuration $\mathcal{C}_{FFD}$ is reached and the process ends. In the latter case, stage i ends and the process continues with the compression phase of the new stage $i+1$.

Theorem 3. *Suppose the underlying set U of $\mathcal{C}_S$ and $\mathcal{C}_T$ is formed by items of size at most κ/α for some integer $\alpha > 1$ and the average slack of bunches in $\mathcal{C}_S$ and $\mathcal{C}_T$ is at least $\frac{\kappa}{\alpha+1} + \frac{3\alpha\kappa}{(\alpha+1)|\mathcal{C}_S|}$. Then, it is always possible to reconfigure both $\mathcal{C}_S$ and $\mathcal{C}_T$ to the FFD configuration and, hence, to each other.*

Proof. We use the reconfiguration process described above to reconfigure $\mathcal{C}_S$ to $\mathcal{C}_{FFD}$ (and then $\mathcal{C}_{FFD}$ to $\mathcal{C}_T$). Given the presence of at least one empty bunch at the end of the compression phase (Lemma 2), the number of FFD bunches increases at the end of each stage. Eventually, at the end of some stage, all bunches become FFD bunches, and we obtain the canonical configuration $\mathcal{C}_{FFD}$.   □

5   Special Case: Items and Capacities Are Powers of 2

Consider a setting where the capacity κ and all item sizes are powers of 2. We characterize input instances in which reconfiguration is possible. In particular, we show that, as long as the total slack across all bunches is at least equal to the size of the largest item that must be moved, it is possible to reconfigure $\mathcal{C}_S$ to $\mathcal{C}_T$. Since this is the minimal requirement for a feasible reconfiguration, in essence our results show that powers of 2 provide an ideal setting for reconfiguration.

We present an algorithm that processes items in stages by non-increasing order of size, such that once items in the current stage have been "settled", they remain in place during all subsequent stages. Each stage entails repeatedly identifying two bunches, B_s and B_d, with a surplus and a deficit, respectively, of items of the current size with respect to the target configuration, and then moving an item from B_s to B_d. Each move takes place in two phases: in the *compression phase*, slack equal to the size of the item is accumulated in a bunch B_{temp}, and in the *transfer phase*, B_{temp} is used as temporary storage in the move of an item from B_s to B_d. The remainder of this section formalizes what it means for an item to be "settled", and provides the details of the compression and transfer phases.

Settling. Our results depend on the ability to look at an item size a in a configuration $\mathcal{C}$ and determine: is there a reconfiguration sequence that reconfigures $\mathcal{C}$ to $\mathcal{C}_T$ such that no step involves moving an item with size at least a? If we ask this question about the item sizes in $\mathcal{C}$ from largest to smallest, the first item size for which the answer is 'no' is called *the size of the largest item in $\mathcal{C}$ that must be moved* and denoted by $\ell(\mathcal{C}, \mathcal{C}_T)$. For example, if $\mathcal{C} = \{\!\{\,\{\!\{32, 16\}\!\}, \{\!\{4, 4, 2\}\!\}\,\}\!\}$ and $\mathcal{C}_T = \{\!\{\,\{\!\{32, 4, 4, 2\}\!\}, \{\!\{16\}\!\}\,\}\!\}$, then $\ell(\mathcal{C}, \mathcal{C}_T) = 16$.

To enable us to find $\ell(\mathcal{C}, \mathcal{C}_T)$, we introduce the idea of an item size being "settled". Informally, an item of size a is settled in its current bunch B if there is a bunch in the target configuration that contains the same multiset of items of size at least a as B does, i.e., none of these items need to be moved from their current positions. The largest item in B whose size is not settled would be "out of place" if all larger items in B stay where they are for the remainder of the reconfiguration sequence, i.e., its size is a candidate for the size of the largest item in the configuration that must be moved. Formally, for any configuration $\mathcal{D}$, any bunch $B \in \mathcal{D}$, and any item size $a \in B$, we define $AtLeast(a, B, \mathcal{D})$ to be the multiset of items in B that have size at least a. For any a, we say that item size a is *settled* if there is a bijection φ between the multiset of bunches containing an item of size a in $\mathcal{C}$ and the multiset of bunches containing an item of size a in $\mathcal{C}_T$, and $AtLeast(a, B, \mathcal{C}) = AtLeast(a, \varphi(B), \mathcal{C}_T)$ for each bunch B in $\mathcal{C}$ that contains an item of size a.

Practically speaking, one can check if an item size a is settled by iterating through the multiset of bunches in $\mathcal{C}$ that contain an item of size a, pairing each such bunch B in $\mathcal{C}$ with any unpaired bunch B_T in $\mathcal{C}_T$ such that $AtLeast(a, B, \mathcal{C}) = AtLeast(a, B_T, \mathcal{C}_T)$, and concluding that a is settled if and only if this process terminates with each bunch in $\mathcal{C}$ containing an item of size a and each bunch in $\mathcal{C}_T$ containing an item of size a belonging to exactly one pair. The largest item size in $\mathcal{C}$ that is not settled is the size of the largest item in $\mathcal{C}$ that must be moved, that is, $\ell(\mathcal{C}, \mathcal{C}_T)$. In what follows, we assume that the total slack across all bunches is at least $\ell(\mathcal{C}_S, \mathcal{C}_T)$.

Our algorithm works in stages. For an arbitrary $i \geq 1$, let $\mathcal{C}_i$ denote the configuration at the start of the ith stage, and denote by u_i the ith largest item size in the underlying set U. During the ith stage, the algorithm proceeds by moving items of size at most u_i between bunches to ensure that, at the end of the stage, all items of size u_i are settled. After a finite number of stages, all item sizes are settled, which means that the current configuration is $\mathcal{C}_T$.

Throughout its execution, the algorithm maintains a bijection φ between a subset of the bunches of $\mathcal{C}_i$ and those of $\mathcal{C}_T$. Initially, the domain of this bijection is empty. Intuitively, for any bunch $B \in \mathcal{C}_i$, $\varphi(B)$ indicates the bunch in $\mathcal{C}_T$ that B must reconfigure to. At the end, the domain of φ contains all bunches of $\mathcal{C}_i$, and for any $B \in \mathcal{C}_i$, bunches B and $\varphi(B)$ contain the same multiset of items, i.e., $\mathcal{C}_i$ is $\mathcal{C}_T$.

To extend φ, we form a one-to-one mapping between multisets of bunches $\mathcal{C}'_i$ and $\mathcal{C}'_T$ selected from $\mathcal{C}_i$ (resp., $\mathcal{C}_T$) that were not in the domain (resp., co-domain) of φ at the start of the ith stage. To form $\mathcal{C}'_i$ and $\mathcal{C}'_T$ of equal size, we first select any bunch that contains an item of size u_i but no item larger than u_i; if no such bunch exists, then we can proceed to the next step, as all bunches containing u_i were mapped in φ during previous stages. If the numbers of bunches in the multisets are not equal, we add to the smaller of $\mathcal{C}'_i$ and $\mathcal{C}'_T$ arbitrary bunches with all items smaller than u_i. Such bunches must exist as φ is a bijection, and hence $\mathcal{C}_i$ and $\mathcal{C}_T$ contain the same number of multisets unmapped by φ.

Example 1. Suppose $\mathcal{C}_2 = \{\{32, 8\}, \{8, 8, 4, 4\}, \{8, 8, 4, 2, 2\}, \{8, 8\}, \{4, 4, 1\}\}$ and $\mathcal{C}_T = \{\{32, 8, 1\}, \{8, 8, 8, 8\}, \{8, 8, 4, 2\}, \{4, 4, 2\}, \{4, 4\}\}$. At this point, we are at the start of the second stage of the algorithm, only 32 is settled, and φ maps $\{32, 8\}$ to $\{32, 8, 1\}$. Therefore, $u_2 = 8$, $\mathcal{C}'_2 = \{\{8, 8, 4, 4\}, \{8, 8, 4, 2, 2\}, \{8, 8\}\}$, and $\mathcal{C}'_T = \{\{8, 8, 8, 8\}, \{8, 8, 4, 2\}, \{4, 4, 2\}\}$ (the last bunch is included in $\mathcal{C}'_T$ to ensure $|\mathcal{C}'_2| = |\mathcal{C}'_T|$). An arbitrary one-to-one mapping between bunches in $\mathcal{C}'_2$ to bunches in $\mathcal{C}'_T$ extends the domain of φ to bunches that contain 8. For example, $\{8, 8, 4, 4\} \leftrightarrow \{8, 8, 8, 8\}$, $\{8, 8, 4, 2, 2\} \leftrightarrow \{8, 8, 4, 2\}$, and $\{8, 8\} \leftrightarrow \{4, 4, 2\}$.

To settle the items of size u_i, we consider two bunches in $\mathcal{C}_i$: a surplus bunch B_s containing more items of size u_i than in $\varphi(B_s)$ and a deficit bunch B_d containing fewer items of size u_i than in $\varphi(B_d)$. Since $\mathcal{C}_i$ and $\mathcal{C}_T$ have the same number of items of size u_i, if not all items of size u_i have been settled, two such bunches must exist. In the example above, $B_s = \{\!\{8, 8\}\!\}$ and $B_d = \{\!\{8, 8, 4, 4\}\!\}$. We now describe a procedure for moving an item of size u_i from B_s to B_d. Repeatedly applying this procedure, first for all items in B_s and B_d and then for all other such pairs of bunches, results in a state where φ bijectively maps each bunch B in $\mathcal{C}_i$ containing at least one item of size u_i to a bunch in $\mathcal{C}_T$ containing at least one item of size u_i such that B and $\varphi(B)$ contain the same number of items of size u_i. At that point, all items of size u_i are settled, and the current stage ends, yielding Lemma 4.

Lemma 4. *For any $i \geq 2$, and any bunch $B \in \mathcal{C}_i$ in the domain of φ, at the start of ith stage, B and $\varphi(B)$ contain the same number of items of size u_j for all $j < i$.*

In the ith stage, suppose that there is at least one item of size u_i that is not settled (which implies that $u_i = \ell(\mathcal{C}_i, \mathcal{C}_T) \leq \ell(\mathcal{C}_S, \mathcal{C}_T)$). We move an item of size u_i from B_s to B_d, in two phases: in the *compression phase*, items of size at most u_i are moved between bunches to accumulate enough slack in at least one bunch B_{temp} to host an item of size u_i, and in the *transfer phase*, an item of size u_i is moved from B_s to B_d by using B_{temp} as a temporary host for the item. We now provide terminology and a procedure to aid in formalizing the two phases.

We partition the slack of each bunch B into *slack items* that are maximal powers of 2; for example, when $slack(B) = 14$, B contains slack items of sizes $2, 4$, and 8. For clarity, we refer to items in the packing as "actual items", using $All(B)$ to denote the multiset formed by the actual and slack items in B, all powers of 2, which sum to κ. We use the term *bundle* to refer to any multiset of items from $All(B)$, the sum of which (actual items and slack items) will be called its *bsum*.

Procedure $Bundles(B, p)$. Given a bunch B and an integer $p < \log_2 \kappa$ such that $All(B)$ contains at least one element of size at most 2^p, the procedure returns two disjoint bundles, each with bsum equal to 2^p. Starting with bundles consisting of single actual or slack items, $Bundles(B, p)$ repeatedly merges any two bundles with bsum 2^i to form a new bundle with bsum 2^{i+1} until there is no pair of bundles each with bsum 2^i for any $i < p$. Finally, two bundles with bsum 2^p are chosen arbitrarily and returned as output.

Example 2. Suppose $\kappa = 64$, and $B = \{32, 4, 4, 4, 4, 2\}$. Then $slack(B) = 14$, and slack items are $\underline{8}, \underline{4}, \underline{2}$ (underbars represent slack items). Therefore, $All(B) = \{32, \underline{8}, \underline{4}, 4, 4, 4, 4, \underline{2}, 2\}$. Suppose $p = 3$, and note that there are elements less than or equal to 2^p in $All(B)$. $Bundles(B, 3)$ repeatedly merges bundles of $All(B)$ that have bsum less than 8. After all merges, the bundles will be $\{32\}, \{\underline{8}\}, \{\underline{4}, 4\}, \{4, 4\}, \{4, \underline{2}, 2\}$. Note that there are four bundles with bsum equal to 8. $Bundles(B, 3)$ returns any of these bundles, say the first two, $\{\underline{8}\}$ and $\{\underline{4}, 4\}$, as its output.

$Bundles(B, p)$ terminates after a finite number of iterations; an item in $All(B)$ with initial size 2^j can be involved in at most $p - j$ merges, since more merges would require the merging of two bundles with bsum 2^p. Lemma 5 proves the correctness of the procedure. The proof follows from the bunch capacity κ, the actual item sizes, and the slack item sizes all being powers of 2.

Lemma 5. *Let p be an integer such that $p < \log_2 \kappa$ and there is at least one element of size at most 2^p in a bunch B. Bundles(B, p) returns two disjoint bundles consisting of elements of $All(B)$, each with bsum equal to 2^p.*

The Compression Phase. This phase reduces the number of slack items in the configuration by repeatedly merging pairs of slack items of the same size, resulting in enough space to make a transfer from B_s to B_d possible (Lemma 6). Merges take place on any two bunches that contain slack items of equal size less than u_i. Since each merge reduces the total number of slack items by one, after a finite number of merge operations, the process ends in a state where, in the current configuration, there is at most one slack item of size 2^i for each $i < \log_2(u_i)$.

More formally, the procedure *Merge-slack* takes two bunches B_1 and B_2, each having a slack item of size 2^q with $q < \log_2(u_i)$, and results in one of the bunches having a slack item of size 2^{q+1}. Merge-slack applies Bundles(B_1, q) to obtain two disjoint bundles consisting of elements of $All(B_1)$, each with bsum 2^q. At least one of these bundles, *Move*, does not contain y, the slack item of B_1 that has size 2^q. Since slack item sizes are defined to be maximal powers of 2, and each bundle in the output of Bundles(B_1, q) has bsum 2^q (Lemma 5), *Move* cannot be entirely formed by slack items (otherwise, its slack items, together with y, should have formed a larger slack item of size 2^{q+1}). Merge-slack moves the actual items of the *Move* bundle from B_1 to B_2. This is possible because the sum of the actual items in such a bundle is at most 2^q, which is no more than the available slack in B_2 since B_2 is assumed to have a slack item of size 2^q. As a result, there will be two slack items of size 2^q in B_1, which are subsequently combined into one slack item of size 2^{q+1}.

Example 3. Suppose $\kappa = 64$ and let $B_1 = \{32, 4, 4, 4, 4, 2\}$ (with slack items $\{\underline{8}, \underline{4}, \underline{2}\}$) and $B_2 = \{32, 16, 8\}$ (with slack item $\{\underline{8}\}$). Each bunch contains a slack item of size 8; thus $q = 3$. Merge-slack calls Bundles$(B_1, 3)$ which, as discussed in Example 2, returns two multisets $\{\underline{8}\}$ and $\{\underline{4}, 4\}$. The actual item, of size 4, in the second multiset is moved to B_2. After the merge, bunches become $B_1 = \{32, 4, 4, 4, 2\}$ and $B_2 = \{32, 16, 8, 4\}$. Note that the set of slack items of B_1 is changed to $\{\underline{16}, \underline{2}\}$.

Lemma 6. *Suppose that, at the start of the ith stage, there is at least one item with size u_i that is not settled. By the end of the compression phase of the ith stage, there is a bunch B_{temp} with slack at least u_i.*

The Transfer Phase. We move an item of size u_i from B_s to B_d in C_i in three steps: (1) moving an item size u_i from B_s to B_{temp}, (2) ensuring there is slack of size at least u_i in B_d, and then (3) moving an item of size u_i from B_{temp} to B_d. The third step is possible due to the second step, and the first step is possible due to the slack of B_{temp} being at least equal to u_i at the end of compression phase (Lemma 6). It thus suffices to consider the second step.

First, consider the case where B_d does not contain an item of size less than u_i. Then, by Lemma 4, for each $j < i$, the bunch $\varphi(B_d)$ has the same number of items of size u_j as the bunch B_d. Moreover, by the definition of B_d, the bunch B_d has fewer items of size u_i than the bunch $\varphi(B_d)$, and by assumption, B_d has no items of size less than u_i. Since B and $\varphi(B_d)$ have the same capacity, it follows that B_d must have slack of size at least u_i, as desired. Next, consider the case where B_d contains at least one item of size less than u_i. Then, we apply Bundles$(B_d, \log_2(u_i))$ to obtain two bundles from $All(B_d)$, each with bsum equal to u_i. We move the actual items from one of the two subsets returned by Bundles$(B_d, \log_2(u_i))$ to B_s, which is always possible since, in step (1), we moved an item of size u_i from B_s to B_{temp}. After moving the items to B_s, there is slack of size at least u_i in B_d, as desired. From the discussion above, we can conclude the following theorem:

Theorem 7. *Suppose that all items in the underlying set U of the configuration are powers of 2, and assume that bunch capacity κ is also a power of 2. Let $\ell(\mathcal{C}_S, \mathcal{C}_T)$ denote the size of the largest item that must be moved. It is possible to reconfigure $\mathcal{C}_S$ to $\mathcal{C}_T$ if and only if the total slack across all bunches is at least $\ell(\mathcal{C}_S, \mathcal{C}_T)$.*

6   Special Case: Reconfiguration Within Partitions

In order to efficiently parallelize the process of repacking, we wish to determine whether the source and target configurations can be each split into smaller pieces, such that each piece can be reconfigured independently. Determining the existence of such splits can be formulated as a variant of a transshipment problem [1,10], that is, a network flow problem that in turn can be solved using integer linear programming (ILP) in linear time [3]. To represent an instance of repacking as a directed graph for PARTITION ILP (defined below), we create a vertex for each way of assigning the items in each piece to a bounded number of bunches, and add edges between vertices that capture the movement of a single item from one bunch to another. We will show that if the reconfiguration is possible, we can decompose the total flow into a set of paths, *path flows*, such that each corresponds to the reconfiguration of a single piece.

Below, we define the repacking problem β-REPACKING-κ (Sect. 6.1) and PARTITION ILP (Sect. 6.2). We solve β-REPACKING-κ (Theorem 9) by proving that it is equivalent to PARTITION ILP (Lemma 8).

Lemma 8. PARTITION ILP *has a feasible solution if and only if the corresponding instance $(\mathcal{C}_S, \mathcal{C}_T)$ for β-REPACKING-κ is reconfigurable.*

Theorem 9. *β-REPACKING-κ can be solved in linear time.*

6.1   Defining Repacking with Partitions

To decompose our source and target (and hence, all intermediate) configurations into pieces, we split the underlying set U into "underlying subsets" and allocate a specific number of bunches to each "underlying subset". Then, each configuration of U can be viewed as the disjoint union of "pieces", where each piece is a configuration of an "underlying subset". By restricting moves to occur within pieces, each piece can be reconfigured independently. Below, we formalize these definitions, and extend notations of adjacency and reconfiguration sequences to capture the idea of repacking with partitions.

To formalize "underlying subsets", we define a *partition* $\mathbf{P} = (P_1, P_2, \ldots, P_{|\mathbf{P}|})$ of underlying set U to be a sequence of parts, where a *part* is a pair $P_i = (U_i, \beta_i)$, U_i is a multiset of items, β_i is a positive integer number of bunches, and the multiset union of all U_i's is equal to U; it is a β-bounded partition if $\beta_i \leq \beta$ for all values of i.

We define the *assignment of a partition for capacity* κ as a sequence of multisets of bunches $\mathbf{A} = (\mathcal{A}_1, \mathcal{A}_2, \ldots, \mathcal{A}_{|\mathbf{A}|})$, such that for each part $P_i = (U_i, \beta_i)$ in $\mathbf{P}$, the multiset union of items in $\mathcal{A}_i$ is equal to U_i, the number of bunches in $\mathcal{A}_i$ is β_i, and each bunch in $\mathcal{A}_i$ has volume at most κ. The multiset $\mathcal{A}_i$ in $\mathbf{A}$ is called the *ith portion*, sometimes written as $\text{PORTION}(\mathbf{A}, i)$. An assignment $\mathbf{A}$ is *consistent with a configuration* $\mathcal{C}$ if the disjoint multiset union of all $\mathcal{A}_i$'s is equal to $\mathcal{C}$; in this case, the underlying set of $\mathcal{C}$ will be the union of the U_i's.

In a reconfiguration sequence that ensures that each move occurs within a part, each configuration can be expressed as an assignment, where two consecutive assignments differ only in a single portion. Two assignments $\mathbf{A}$ and $\mathbf{A}'$ of a partition $\mathbf{P}$ for capacity κ are *adjacent* if there exists exactly one index j such that $\text{PORTION}(\mathbf{A}, j)$ and $\text{PORTION}(\mathbf{A}', j)$ are adjacent (viewed as configurations with underlying set U_j) and for all $i \neq j$, $\text{PORTION}(\mathbf{A}, i) = \text{PORTION}(\mathbf{A}', i)$. Accordingly, a $\mathbf{P}$-*conforming reconfiguration sequence* for capacity κ from a source configuration $\mathcal{C}_S$ to a target configuration $\mathcal{C}_T$, where $\mathbf{P}$ is a partition for the underlying set of $\mathcal{C}_S$ (and $\mathcal{C}_T$), is a sequence of assignments of $\mathbf{P}$ for capacity κ, such that the first and last assignments are consistent with $\mathcal{C}_S$ and $\mathcal{C}_T$, respectively, and each consecutive pair of assignments are adjacent.

For any positive integer constants β, κ, we define the following problem as β-REPACKING-κ, with instance $(\mathcal{C}_S, \mathcal{C}_T)$:

Input: Source and target configurations $\mathcal{C}_S$ and $\mathcal{C}_T$ for capacity κ.
Question: Is there a β-bounded partition $\mathbf{P}$ of the underlying set of $\mathcal{C}_S$ such that there exists a $\mathbf{P}$-conforming reconfiguration sequence for capacity κ from $\mathcal{C}_S$ to $\mathcal{C}_T$?

6.2   Defining Partition ILP

In order to represent reconfiguration as a directed graph, we view it as a transshipment problem in which instead of transporting goods from supply to demand nodes through intermediate nodes (with negative, positive, and zero weights, or *demands*, respectively), the network represents the reconfiguration of a packing. Moreover, since our goal is to determine feasibility, not cost, the objective function and edge costs are all set to zero, and demands are constrained to ensure that we use exactly the bunches in the source configuration $\mathcal{C}_S$ and target configuration $\mathcal{C}_T$: the demands of the supply (resp. demand) nodes induce an assignment $\mathbf{A}_S$ (resp. $\mathbf{A}_T$) that is consistent with the source (resp. target) configuration.

To determine whether it is possible to partition the source and target configurations, we try all possibilities. A configuration with underlying set $U_i \subseteq U$ is a *subconfiguration* of a configuration $\mathcal{C}$ with underlying set U, namely a multiset of bunches that is formed by deleting zero or more items and removing zero or more empty bunches from $\mathcal{C}$. Given an underlying set U and a capacity κ, we define β-SUBS(U, κ) to be the set of non-empty subconfigurations using subsets of items in U and at most β bunches of capacity κ (or β-SUBS when U and κ are clear from context).

To form our network, we create three nodes for each subconfiguration $\mathcal{I}$ in β-SUBS: an input node $x_{\mathcal{I}}$ with non-positive demand, an intermediate node $y_{\mathcal{I}}$ with zero demand, and an output node $z_{\mathcal{I}}$ with non-negative demand. Using X, Y, and Z to denote the sets of input, intermediate, and output nodes, respectively, the graph G is defined as $V(G) = X \cup Y \cup Z$ and $E(G) = \{x_{\mathcal{I}}y_{\mathcal{I}} : \mathcal{I} \in \beta\text{-SUBS}\} \cup \{y_{\mathcal{I}}y_{\mathcal{J}} : \mathcal{I}, \mathcal{J} \in \beta\text{-SUBS}, \mathcal{I} \text{ and } \mathcal{J} \text{ are adjacent}\} \cup \{y_{\mathcal{I}}z_{\mathcal{I}} : \mathcal{I} \in \beta\text{-SUBS}\}$.

We design Partition ILP (Definition 1) in such a way that, intuitively, given a solution $[f \; d]$, we can think of f as the flow vector and d as the demand vector for each node. We wish to ensure that the total flow out of a particular node is equal to the demands for the node (Constraints 1, 2, 3), and that supply, intermediate, and demand nodes have demands in the appropriate ranges (Constraint 7). We want each unit of flow to correspond to a single reconfiguration step, so we require that a feasible flow be non-negative and integral (Constraint 6). Constraints 4 and 5 relate the assigned demands for each node in X and Z, respectively, to the multiplicity of bunches in the source and target configurations, respectively. In particular, Constraint 4 ensures that the demands d from a feasible solution $[f \; d]$ can be directly related to a β-bounded partition $\mathbf{P}$ of the underlying set U and an assignment $\mathbf{A}_S$ of $\mathbf{P}$ that is consistent with $\mathcal{C}_S$: each unit of demand on each $x_{\mathcal{I}}$ corresponds to a portion in $\mathbf{A}_S$ that is equivalent to $\mathcal{I}$, and, corresponds to a part in $\mathbf{P}$ with the same items and number of bunches as $\mathcal{I}$. In a similar way, Constraint 5 ensures that the demands d can be directly related to an assignment $\mathbf{A}_T$ that is consistent with $\mathcal{C}_T$.

In the formal definition below, $d : V(G) \to \mathbb{Z}$ and $f : E(G) \to \mathbb{R}_{\geq 0}$ are functions with f extended to arbitrary subsets E of $E(G)$ by defining $f(E) = \sum_{uv \in E} f(uv)$. For any vertex $v \in V(G)$, we use $E^+(v)$ and $E^-(v)$ to denote the sets of in-edges to and out-edges from v. We use $mult(B, \mathcal{C})$ to denote the number of bunches that are equal to B in the configuration $\mathcal{C}$, namely, the *multiplicity* of B in $\mathcal{C}$, and denote as $mult(\mathcal{I}, \mathbf{A})$ the number of elements of the assignment $\mathbf{A}$ that are equal to the subconfiguration $\mathcal{I}$. We use BUNCHTYPES to denote the set of all possible bunches formed of items taken from U such that the volume of the bunch is at most κ.

Definition 1. [PARTITION ILP]

$$minimize\ 0^T \begin{bmatrix} f \\ d \end{bmatrix}$$

$$-f(E^-(x_{\mathcal{I}})) - d(x_{\mathcal{I}}) = 0 \qquad\qquad \forall \mathcal{I} \in \beta\text{-SUBS} \qquad (1)$$

$$f(E^+(y_{\mathcal{I}})) - f(E^-(y_{\mathcal{I}})) = 0 \qquad\qquad \forall \mathcal{I} \in \beta\text{-SUBS} \qquad (2)$$

$$f(E^+(z_{\mathcal{I}})) - d(z_{\mathcal{I}}) = 0 \qquad\qquad \forall \mathcal{I} \in \beta\text{-SUBS} \qquad (3)$$

$$\sum_{\mathcal{I} \in \beta\text{-SUBS}} mult(B, \mathcal{I}) \cdot d(x_{\mathcal{I}}) = -mult(B, \mathcal{C}_S) \qquad \forall B \in \text{BUNCHTYPES} \qquad (4)$$

$$\sum_{\mathcal{I} \in \beta\text{-SUBS}} mult(B, \mathcal{I}) \cdot d(z_{\mathcal{I}}) = mult(B, \mathcal{C}_T) \qquad \forall B \in \text{BUNCHTYPES} \qquad (5)$$

$$f(uv) \in \mathbb{Z}_{\geq 0} \qquad\qquad \forall uv \in E(G) \qquad (6)$$

$$d(x_{\mathcal{I}}) \in \mathbb{Z}_{\leq 0},\ d(z_{\mathcal{I}}) \in \mathbb{Z}_{\geq 0} \qquad\qquad \forall \mathcal{I} \in \beta\text{-SUBS} \qquad (7)$$

7   Conclusions and Future Work

We introduced the area of the reconfiguration of multisets, demonstrated the hardness of the general problem, and provided algorithms for situations in which items are of bounded size, item and bunch capacities are powers of two, and items can be partitioned into smaller groups of bunches. Our results are applicable to a variety of application areas, as well as to the problem of BIN PACKING.

In this paper, we have restricted our attention to instances in which all bunches have the same capacity and in which items and bunches are indistinguishable. Future directions of research include the non-uniform case, in which bunches have different capacities; variants could include restrictions on the minimum total size, restrictions on both the minimum and maximum total size, or other specifications for particular sets of bunches. The instances could be further generalized by allowing us to distinguish among items or bunches, even those of the same size or capacity. Another direction for future work is to optimize the time complexities of the algorithms studied in this paper: while our focus has been to establish the feasibility of reconfiguration, it is natural to ask how quickly one can reconfigure one configuration to another.

Our work demonstrates the use of linear programming to make use of parallel computation in reconfiguration. Further investigations are required to determine whether similar techniques may be applicable more generally to other types of reconfiguration problems, and whether other techniques are amenable to parallelization of other reconfiguration problems.

Each of the algorithms presented in the paper is restricted to using the amount of space provided in the source configuration. A natural extension is to ask whether there are no-instances for our problems that can be rearranged using extra space, and if so, how much space would be needed. In shifting the focus of reconfiguration from length of reconfiguration sequences to the impact of extra space, we open up a new type of reconfiguration problem.

We anticipate that *resource-focused reconfiguration* will find widespread use in practical settings for a host of various problems. As examples, we consider two commonly-studied reconfiguration problems, namely POWER SUPPLY RECONFIGURATION [9] and INDEPENDENT SET RECONFIGURATION [9]. In the former problem, the goal is to reassign customers (each with a specified requirement) to power stations (each with a specified capacity) in such a way that aside from the customer being moved, all other customers have an uninterrupted flow of power. In a resource-focused setting, our goal would be determine how many generators might be needed to temporarily provide power in order to make reconfiguration possible. For the latter problem, a possible extra resource could be isolated vertices to which tokens could be temporarily moved in order to allow reconfiguration to occur. Depending on the problem, there may be more than one resource that can be measured in the reconfiguration process, yielding further new research directions as the impacts of resources are considered individually and in concert.

References

1. Cook, W.J., Cunningham, W.H., Pulleyblank, W.R., Schrijver, A.: Minimum-cost flow problems. In: Combinatorial Optimization, chap. 4, pp. 91–126. Wiley (1997). https://doi.org/10.1002/9781118033142.ch4
2. VMWare Docs: Migrating virtual machines (2021). https://docs.vmware.com/en/VMware-vSphere/7.0/com.vmware.vsphere.vcenterhost.doc/GUID-FE2B516E-7366-4978-B75C-64BF0AC676EB.html. Accessed 28 Sept 2023
3. Eisenbrand, F.: Fast integer programming in fixed dimension. In: Di Battista, G., Zwick, U. (eds.) ESA 2003. LNCS, vol. 2832, pp. 196–207. Springer, Heidelberg (2003). https://doi.org/10.1007/978-3-540-39658-1_20
4. Garey, M.R., Johnson, D.S.: Computers and Intractability: A Guide to the Theory of NP-Completeness. W. H. Freeman (1979)
5. Graham, R.L.: Bounds on multiprocessing timing anomalies. J. SIAM Appl. Math. **17**(2), 416–429 (1969)
6. van den Heuvel, J.: The complexity of change. In: Surveys in Combinatorics 2013. London Mathematical Society Lecture Note Series, pp. 127–160. Cambridge University Press (2013). https://doi.org/10.1017/CBO9781139506748.005
7. Ito, T., Demaine, E.D.: Erratum to "approximability of the subset sum reconfiguration problem". http://www.dais.is.tohoku.ac.jp/take/erratum_subsetsumreconf.pdf. Accessed 15 Nov 2023
8. Ito, T., Demaine, E.D.: Approximability of the subset sum reconfiguration problem. J. Comb. Optim. **28**(3), 639–654 (2014)
9. Ito, T., et al.: On the complexity of reconfiguration problems. Theor. Comput. Sci. **412**(12–14), 1054–1065 (2011). https://doi.org/10.1016/j.tcs.2010.12.005
10. Korte, B., Vygen, J.: Minimum cost flows. In: Korte, B., Vygen, J. (eds.) Combinatorial Optimization. AC, vol. 21, pp. 215–244. Springer, Heidelberg (2018). https://doi.org/10.1007/978-3-662-56039-6_9
11. Medina, V., García, J.M.: A survey of migration mechanisms of virtual machines. ACM Comput. Surv. **46**(3), 30:1–30:33 (2014)
12. Nishimura, N.: Introduction to reconfiguration. Algorithms **11**(4) (2018). https://www.mdpi.com/1999-4893/11/4/52

The Shortest Path Reconfiguration Problem Based on Relaxation of Reconfiguration Rules

Naoki Domon$^{(\boxtimes)}$, Akira Suzuki, Yuma Tamura, and Xiao Zhou

Graduate School of Information Sciences, Tohoku University, Sendai, Japan
naoki.domon.r7@dc.tohoku.ac.jp,
{akira,tamura,zhou}@tohoku.ac.jp

Abstract. The shortest path problem is the most classical and fundamental problem in the field of graph algorithm. Recently, its reconfiguration variant, namely the SHORTEST PATH RECONFIGURATION problem, has received a lot of attention. In this paper, we study the complexity of k-SPR, which generalizes the SHORTEST PATH RECONFIGURATION problem, with respect to k. In k-SPR, we are allowed to replace at most k consecutive vertices of the current shortest path at a time. We first show that, for any fixed rational numbers c and ε such that $c > 0$ and $0 < \varepsilon \leq 1$, k-SPR with $k = cn^{1-\varepsilon}$ is polynomially solvable if $\varepsilon = 1$ and $c < 1$; otherwise, PSPACE-complete. This intractability holds even when given graphs are restricted to bipartite graphs and r-th power graphs, where r is any positive integer. Furthermore, when we restrict $0 < \varepsilon < 1$, the PSPACE-completeness holds for graphs with maximum degree 3. Then, we design an FPT algorithm parameterized by $\mu = n/2 - k \geq 0$ that runs in $O(m + 6.730^\mu \mu^4 n)$ time. Finally, we show that, for any k, k-SPR can be solved in linear time for $K_{2,3}$-minor-free graphs.

Keywords: Reconfiguration problem · Graph algorithm · Parameterized complexity · Shortest path

1 Introduction

There are frequent situations where one needs to know the shortest route between two points in the real-world road network. Such a situation can be modeled as the *shortest path problem* on a graph. It is well known that the shortest path problem is solvable in polynomial time by using classic algorithms such as Dijkstra's algorithm.

Often the shortest path is used when selecting routes in scheduled bus or delivery planning. However, one may encounter a situation where the obtained initial shortest path needs to be changed to another target shortest path, for

A. Suzuki—Partially supported by JSPS KAKENHI Grant Number JP20K11666, Japan.

Y. Tamura—Partially supported by JSPS KAKENHI Grant Number JP21K21278, Japan.

© The Author(s), under exclusive license to Springer Nature Singapore Pte Ltd. 2024
R. Uehara et al. (Eds.): WALCOM 2024, LNCS 14549, pp. 227–241, 2024.
https://doi.org/10.1007/978-981-97-0566-5_17

Table 1. The complexity of k-SPR on several graph classes and values of k.

	$k < 2$	any const. $k \geq 2$	$k = cn^{1-\varepsilon}$	$k = n/2 - \mu$
general	PSPACE-comp.			FPT parameterized by μ [Theorem 4]
bipartite	PSPACE-comp. [3]	PSPACE-comp. [Theorem 1]		
r-th power	PSPACE-comp. [5]	PSPACE-comp. [Theorem 2]		
claw-free	P [3]	PSPACE-comp.	open	
line		[5]		
max. degree 3	open		PSPACE-comp. [Theorem 3]	
planar	P [4]	open		
$K_{2,3}$-minor-free	P [Theorem 6]			

example, due to avoiding road construction or traffic congestion. The simplest approach is to change directly the former path to the latter path, but such a drastic change may not be desired in the real-world road network. We want to change the initial shortest path to the target shortest path in a step-by-step manner.

The SHORTEST PATH RECONFIGURATION problem (SPR for short), which is introduced by Kamiński et al. [8], is one way to model such a situation. In this problem, we are given two s-t shortest paths on a graph G. Then we are asked to determine whether or not we can transform one into the other by iteratively replacing a single vertex of the current s-t shortest path at a time, so that intermediate results remain s-t shortest paths on G. This problem models "dynamic" transformations of paths in the real-world road network, while the conventional shortest path problem deals with a "static" path.

Since SPR was introduced by Kamiński [8], computational complexities based on various graph classes have been studied. It is known that SPR is PSPACE-complete for bipartite graphs [3], graphs with bounded bandwidth [13], and r-th power graphs, where r is any positive integer [5]. On the other hand, SPR can be solved in polynomial time for:

- chordal graphs and claw-free graphs [3];
- planar graphs and graphs where every vertex has at most two neighbors closer to s and at most two neighbors closer to t [4];
- grid graphs [2]; and
- permutation graphs, circle graphs, Boolean hypercube, circular-arc graphs, bridged graphs and constant diameter graphs [5].

As mentioned above, various studies have been conducted on SPR. However, when considering real-world applications, the constraint of SPR is too strict that only a single vertex can be replaced at a time. To overcome this situation, k-SPR was introduced by Gajjar et al. [5]. For a positive rational number[1] k, roughly

---

[1] In [5], k was defined as a positive integer. In this paper, however, we allow k to be a positive rational number in order to accurately state our results.

speaking, we are allowed to replace at most k consecutive vertices of a path. Note that 1-SPR is equivalent to SPR, and hence k-SPR is a generalization of SPR. Gajjar et al. showed that k-SPR is PSPACE-complete for all integer constants $k \geq 2$, even when input graphs are restricted to line graphs [5]. This is in contrast to the fact that 1-SPR is solvable for line graphs because all line graphs are claw-free. On the other hand, k-SPR can be solved in polynomial time when $k \geq n/2$ [5], where n is the number of vertices in a graph.

SPR and k-SPR are included in the framework of *combinatorial reconfiguration* proposed by Ito et al. [7]. In the last decade, the framework is applied to many "static" (conventional) problems, and their "dynamic" variants have been extensively studied in the field of theoretical computer science. (See, e.g., the surveys of van den Heuvel [6] and Nishimura [11].) In most cases, a "static" problem and its "dynamic" variant are both intractable. SPR and k-SPR are rare examples that the problems are intractable while their "static" problem (the shortest path problem) is tractable.

1.1 Our Contribution

Let n and m denote the number of vertices and edges in a graph, respectively. The results of Gajjar et al. [5] inform us that k-SPR has a tractability transition depending on k, that is, k-SPR is intractable if $k = \Theta(1)$, whereas k-SPR is tractable if $k \geq n/2$. The question we ask is whether or not k-SPR is tractable when k lies between $\Theta(1)$ and $n/2$, such as $k = \Theta(\sqrt{n})$.

In this paper, we first show that, for any fixed rational numbers c and ε such that $c > 0$ and $0 < \varepsilon \leq 1$, k-SPR with $k = cn^{1-\varepsilon}$ is polynomially solvable if $\varepsilon = 1$ and $c < 1$; otherwise, PSPACE-complete[2]. Thus, the problem is intractable for a wider range of k than Gajjar et al. indicated. This intractability holds even when given graphs are restricted to bipartite graphs and r-th power graphs, where r is any positive integer. Furthermore, when we restrict $0 < \varepsilon < 1$, the PSPACE-completeness holds for graphs with maximum degree 3. It is worth noting that the complexity of SPR on graphs with maximum degree 3 remains open.

We then focus on tractable cases. Suppose that $k = n/2 - \mu$ for a non-negative rational number $\mu \leq n/2$. Since k-SPR is solvable in polynomial time when $k \geq n/2$, the parameter μ represents how small k is from the upper bound that guarantees the solvability of k-SPR. Such a parameter, so-called a *(below) guaranteed value*, was introduced by Mahajan and Raman [9] and has been studied in the literature, for example, [1,10]. We design an FPT algorithm parameterized by μ that runs in $O(m + 6.730^\mu \mu^4 n)$ time via a dynamic programming algorithm that generalizes the algorithm proposed by Bonsma [4]. This result means that k-SPR is polynomially solvable even when $k = n/2 - O(\log n)$. Finally, we show that for any positive rational number k, k-SPR can be solved in linear time for $K_{2,3}$-minor-free graphs (See Table 1).

Due to the space limitation, proofs for the claims marked with ($\spadesuit$) are omitted from this extended abstract.

---

[2] In general, k can be an irrational number even if c and ε are rational numbers, but as we will see later, we guarantee in our proof that k is a rational number.

2   Preliminaries

Let $G = (V, E)$ be a graph: we also denote by $V(G)$ and $E(G)$ the vertex set and the edge set of G, respectively. All the graphs considered in this paper are finite and undirected. For a vertex v of G, we denote by $N_G(v)$ the *neighborhood* of v in G, that is, $N_G(v) = \{w \in V \mid vw \in E\}$. The *degree* $\deg_G(v)$ of v is the size of $N_G(v)$, that is, $\deg_G(v) = |N_G(v)|$. For a vertex subset $S \subseteq V(G)$, we denote by $G[S]$ the subgraph induced by S. For two positive integers i and j such that $i \leq j$, we denote $[i, j] = \{i, i+1, \ldots, j\}$. We use the shorthand $[j] = [1, j]$. A *path* P of G is a sequence $\langle v_0, v_1, \ldots, v_\ell \rangle$ of distinct vertices such that $v_{i-1}v_i \in E$ for each $i \in [\ell]$, where the integer ℓ is called the *length* of P. We denote by $V(P)$ the set of vertices in P. For two vertices s, t of G, P is called an *s-t path* of G if $v_0 = s$ and $v_\ell = t$. An *s-t shortest path* P of G is an *s-t* path such that the length of P is minimized. For vertices v and w of G, the *distance* between v and w, denoted by $\mathsf{dist}_G(v, w)$, is defined as the length of a *v-w* shortest path. A graph G is said to be *connected* if there is a path between any two distinct vertices of G. A maximal connected subgraph of G is called a *connected component* of G. An *independent set* I of G is a vertex subset of G such that there are no edges between any two vertices in I.

Let k be a rational number, and let $P = \langle p_0, p_1, \ldots, p_\ell \rangle$ and $Q = \langle q_0, q_1, \ldots, q_\ell \rangle$ be distinct two *s-t* shortest paths of a graph G. We say that P and Q are *k-adjacent* if and only if there are two integers i and j with $i \leq j$ such that $p_h \neq q_h$ for each $h \in [i, j]$, $p_h = q_h$ for each $h \in [0, \ell] \setminus [i, j]$, and $j - i \leq k - 1$. In other words, Q is obtained by replacing the subpath $\langle p_i, p_{i+1}, \ldots, p_j \rangle$ of P with $\langle q_i, q_{i+1}, \ldots, q_j \rangle$ with $p_h \neq q_h$ for every $h \in [i, j]$. Such a replacement of a subpath with at most k vertices is called a *reconfiguration step*. The *reconfiguration graph* $\mathcal{R}_k(G, s, t)$ is a graph such that its vertex set consists of all *s-t* shortest paths of G, and two vertices are adjacent if and only if the corresponding *s-t* paths are *k-adjacent*. To avoid confusion, we refer vertices and edge of $\mathcal{R}_k(G, s, t)$ as *nodes* and *links*, respectively. Given a graph G, distinct two vertices $s, t \in V(G)$, and two *s-t* shortest paths P, Q of G, the *k-SPR* problem asks whether there exists a path of $\mathcal{R}_k(G, s, t)$ between the nodes P and Q, that is, Q is obtained from P by iteratively replacing subpaths with at most k vertices[3]. The path of $\mathcal{R}_k(G, s, t)$ between the nodes P and Q is called a *solution* of the given instance. When $k < 1$, any subpath of P cannot be replaced and hence *k-SPR* is trivially solvable. In the remainder of this paper, we assume that $k \geq 1$. Observe that 1-SPR is exactly SPR.

We introduce here the key notion that will be used throughout this paper. Let s and t be two vertices of a graph G, and let $\ell = \mathsf{dist}_G(s, t)$. For each integer $0 \leq i \leq \ell$, the *i-th layer* $L_i(G)$ is a vertex subset of G defined as follows:

$$L_i(G) = \{v \in V(G) : (\mathsf{dist}_G(s, v) + \mathsf{dist}_G(v, t) = \mathsf{dist}_G(s, t)) \wedge (\mathsf{dist}_G(s, v) = i)\}.$$

---

[3] We will also consider the case where k depends on an input graph, such as $k = cn^{1-\varepsilon}$ and $k = n/2 - \mu$. When using such notation, we implicitly assume that k is part of the input.

Note that $L_0(G) = \{s\}$ and $L_\ell(G) = \{t\}$, and that all the layers of G can be obtained in linear time by breadth-first search from s and t.

We assume that every vertex in G is contained in some layer. If there is a vertex v that is not contained in any layer, we consider a graph G' that is obtained by removing v from G. This graph reduction is safe because $\mathcal{R}_k(G', s, t) = \mathcal{R}_k(G, s, t)$. For the same reason, we may remove edges between vertices in the same layer. For a vertex v in $L_i(G)$, let denote $N_G^+(v) = N_G(v) \cap L_{i+1}(G)$ and $N_G^-(v) = N_G(v) \cap L_{i-1}(G)$. We say that $L_{i-1}(G)$ and $L_{i+1}(G)$ are the *previous* and *next* layer of $v \in L_i(G)$, respectively.

3 PSPACE-Completeness

3.1 Bipartite Graphs

We first show the PSPACE-completeness of k-SPR on bipartite graphs. A graph $G = (V, E)$ is *bipartite* if V can be partitioned into two independent sets.

Theorem 1. *Let G be a graph with n vertices, and let c and ε be any fixed rational numbers such that $c > 0$ and $0 < \varepsilon \leq 1$. Unless $\varepsilon = 1$ and $c < 1$, k-SPR with $k = cn^{1-\varepsilon} \geq 1$ is PSPACE-complete even for bipartite graphs.*

Note that, if $\varepsilon = 1$ and $c < 1$, then $k < 1$ holds and hence k-SPR is trivially solvable as mentioned before. It follows from Savitch's Theorem that k-SPR is in PSPACE [12]. To show that k-SPR is PSPACE-hard, we provide a polynomial-time reduction from SPR on bipartite graphs to k-SPR. SPR is known to be PSPACE-complete for bipartite graphs [3].

Since c and ε are positive rational numbers, there exist positive integers a and b such that ca and εb are integers, and consider minimum such integers a and b. Note that $b - \varepsilon b$ is a non-negative integer because $\varepsilon \leq 1$. One can see that k-SPR is equivalent to $\lfloor k \rfloor$-SPR from the definition. This implies that it suffices to consider the following two cases: $b - \varepsilon b \geq 1$; and $b - \varepsilon b = 0$ (hence $\varepsilon = 1$) and $a = 1$.

We denote by (G_0, s, t, P, Q) an instance of SPR and n_0 the number of vertices in G_0. We first construct an intermediate instance (G, s, t, P, Q) of SPR from (G_0, s, t, P, Q) to adjust the number of vertices in a graph for k-SPR. Let R be a graph consisting of a path with $r = (ca)^{\varepsilon b - 1} a^{\varepsilon b + 1} n_0^{\varepsilon b} - n_0$ vertices. Since ca, a, and εb are fixed positive integers, the path is well-defined and constructed in polynomial time of n_0. We add R into G_0 and then connect an endpoint of R and s in G_0 by an edge. The constructed graph is defined as G. Let n denote the number of vertices in G, and then we have $n = c^{\varepsilon b - 1} a^{2\varepsilon b} n_0^{\varepsilon b}$. Clearly, no s-t shortest path of G contains vertices of R. This immediately leads to the following proposition.

Proposition 1. *(G_0, s, t, P, Q) has a solution for SPR if and only if (G, s, t, P, Q) has a solution for SPR.*

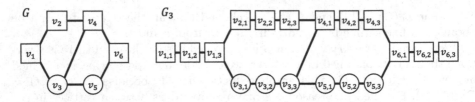

Fig. 1. The construction of G_3 from G. For a path $P = \langle v_1, v_2, v_4, v_6 \rangle$ of G and its corresponding path P_3 of G_3, their vertices are illustrated by squares.

We next construct an instance $(G_k, s_k, t_k, P_k, Q_k)$ of k-SPR from (G, s, t, P, Q). We set

$$k = c^{1/\varepsilon} n^{(1-\varepsilon)/\varepsilon}. \tag{1}$$

It should be noted that k is an integer and a polynomial of n_0. We have

$$
\begin{aligned}
k &= c^{\frac{1}{\varepsilon}} \left(c^{\varepsilon b - 1} a^{2\varepsilon b} n_0^{\varepsilon b} \right)^{\frac{1-\varepsilon}{\varepsilon}} \\
&= c^{\frac{1+(\varepsilon b-1)(1-\varepsilon)}{\varepsilon}} a^{2b(1-\varepsilon)} n_0^{b-\varepsilon b} \\
&= c^{b-\varepsilon b+1} a^{2b-2\varepsilon b} n_0^{b-\varepsilon b} \\
&= (ca)^{b-\varepsilon b+1} a^{b-\varepsilon b-1} n_0^{b-\varepsilon b}.
\end{aligned}
$$

Recall that ca, a, and $b - \varepsilon b$ are fixed integers, and we have assumed that $b - \varepsilon b \geq 1$; or $\varepsilon = 1$ and $a = 1$. In both cases, k is an integer and a polynomial of n_0 from the above equation.

We then partition vertices of G into layers by breadth-first search. We construct a graph G_k from G as follows: label vertices of G with $v_1, v_2, \ldots, v_n$, where $s = v_1$ and $t = v_n$; replace v_i with a path $Z_i = \langle v_{i,1}, v_{i,2}, \ldots, v_{i,k} \rangle$; and add an edge between $v_{i,k}$ and $v_{j,1}$ of G_k for any vertices v_i and v_j with $v_i v_j \in E(G)$ and $\mathrm{dist}_G(s, v_i) < \mathrm{dist}_G(s, v_j)$. We set $s_k = v_{1,1}$ and $t_k = v_{n,k}$. Moreover, we construct a path P_k of G_k from $P = \langle v_{i_0}, v_{i_1}, \ldots, v_{i_\ell} \rangle$ by replacing v_{i_j} with the path Z_{i_j}, where ℓ is the length of P and $i_j \in [n]$ for each $j \in [0, \ell]$ (See also Fig. 1). A path Q_k of G_k is obtained in the same way. This completes the construction of the instance $(G_k, s_k, t_k, P_k, Q_k)$ of k-SPR. Since k is a polynomial of n_0, the construction can be done in polynomial time.

We prove the correctness of our reduction. The following three lemmas complete the proof of Theorem 1.

Lemma 1. G_k is a bipartite graph.

Proof. We show that G_k can be partitioned into two independent sets X_k and Y_k. Recall that the original graph G is bipartite and hence it can be partitioned into two independent sets X and Y. If k is even, let

$$X_k = \{v_{i,j} : v_i \in V(G) \text{ and } j \text{ is even}\}.$$

If k is odd, let

$$X_k = \{v_{i,j} : v_i \in X \text{ and } j \text{ is even}\} \cup \{v_{i,j} : v_i \in Y \text{ and } j \text{ is odd}\}.$$

In both cases, we define $Y_k = V(G_k) \setminus X_k$. From the construction of G_k, observe that X_k and Y_k are independent sets of G_k. □

Lemma 2. Let $n_k = |V(G_k)|$. Then, $k = cn_k^{1-\varepsilon}$.

Proof. Observe that $n_k = kn = c^{1/\varepsilon}n^{1/\varepsilon}$. By Eq. (1), we have

$$k = c^{1/\varepsilon}n^{(1-\varepsilon)/\varepsilon}$$
$$= c \cdot c^{(1-\varepsilon)/\varepsilon}n^{(1-\varepsilon)/\varepsilon}$$
$$= c\left(c^{1/\varepsilon}n^{1/\varepsilon}\right)^{1-\varepsilon}$$
$$= cn_k^{1-\varepsilon}.$$

□

Lemma 3. (G, s, t, P, Q) *has a solution of SPR if and only if* $(G_k, s_k, t_k, P_k, Q_k)$ *has a solution of k-SPR.*

Proof. First, from the construction of G_k, observe that there exists a bijection from s-t shortest paths $R = \langle v_{i_0}, v_{i_1}, \ldots, v_{i_\ell} \rangle$ of G to s_k-t_k shortest paths R_k of G_k consisting of subpaths $Z_{i_0}, Z_{i_1}, \ldots, Z_{i_\ell}$. In particular, P and Q correspond to P_k and Q_k, respectively.

(Only-if direction.) Suppose that an s-t shortest path R' of G is obtained by replacing a vertex $v_{j_1} \in L_i(G)$ in another s-t shortest path R of G with a vertex $v_{j_2} \in L_i(G)$ for some $i \in [\ell - 1]$. Let R_k and R'_k be s_k-t_k shortest paths of G_k corresponding to R and R', respectively. Observe that there exists two vertices v_{j_3} and v_{j_4} of G such that $v_{j_3} \in L_{i-1}(G)$, $v_{j_4} \in L_{i+1}(G)$, and there exist edges $v_{j_3}v_{j_1}, v_{j_1}v_{j_4}, v_{j_3}v_{j_2}$, and $v_{j_2}v_{j_4}$ in G. From the construction of G_k, there exist edges $v_{j_3,k}v_{j_1,1}, v_{j_1,k}v_{j_4,1}, v_{j_3,k}v_{j_2,1}$, and $v_{j_2,k}v_{j_4,1}$ in G_k. Thus, we obtain an s_k-t_k shortest path R'_k by replacing a subpath $Z_{j_1} = \langle v_{j_1,1}, v_{j_1,2}, \ldots, v_{j_1,k} \rangle$ in R_k with a subpath $Z_{j_2} = \langle v_{j_2,1}, v_{j_2,2}, \ldots, v_{j_2,k} \rangle$, where Z_{j_1} and Z_{j_2} contain k vertices, respectively. Therefore, if Q is obtained from P in G by iteratively applying reconfiguration steps, then Q_k is obtained from P_k in G_k by iteratively applying the corresponding reconfiguration steps.

(If direction.) Consider that an s_k-t_k shortest path R'_k of G_k is obtained by replacing a subpath S_k of an s_k-t_k shortest path R_k with a subpath S'_k. Note that each of S_k and S'_k contains at most k vertices. We denote by x and x' the two vertices closest to s_k of these subpaths, and denote by y and y' the two vertices closest to t_k of these subpaths. It is clear that x and x' are contained in the same layer $L_{i_x}(G_k)$, and y and y' are contained in the same layer $L_{i_y}(G_k)$. Then, there exists a vertex $u \in L_{i_x-1}(G_k)$ adjacent to x and x', and there exists a vertex $z \in L_{i_y+1}(G_k)$ adjacent to y and y'. Recall that, from the construction of G_k, only vertices contained in $\bigcup_{i\in[\ell]} L_{ik-1}(G_k)$ can have at least two neighbors in the next layers. Similarly, only vertices contained in $\bigcup_{i \in [\ell]} L_{ik}(G_k)$ can have at least two neighbors in the previous layers. Thus, $u \in L_{ik-1}(G_k)$ and $z \in L_{(i+1)k}(G_k)$ for some $i \in [\ell - 1]$; otherwise, S_k contains more than k vertices. This means that S_k and S'_k are actually subpaths Z_{j_1} and Z_{j_2} of G_k for some $j_1, j_2 \in [n-1]$ such that v_{j_1} and v_{j_2} are contained in the same layer $L_i(G)$. Moreover, it follows from the existence of u and z in G_k that v_{j_1} and v_{j_2} have common neighbors in both the previous and next layers of G. Let R and R' be s-t shortest paths of G corresponding to R_k and R'_k, respectively. We conclude that an s-t shortest path R' is obtained by replacing a vertex v_{j_1} in R with a vertex v_{j_2}. Therefore, if Q_k is obtained from P_k in G_k by iteratively applying reconfiguration steps, Q is obtained from P in G by iteratively applying the corresponding reconfiguration steps.   □

3.2 r-th Power Graphs

In this section, we show the PSPACE-completeness of k-SPR for r-th power graphs.

Definition 1. *Let r be a positive integer. The r-th power G^r of a graph G is a graph such that $V(G^r) = V(G)$ and two vertices $u, v \in V(G^r)$ are adjacent if and only if $\mathrm{dist}_G(u, v) \leq r$. A graph G is r-th power graph if there exists a graph H such that $G = H^r$.*

Theorem 2 (♠). *Let G be a graph with n vertices, and let c and ε be any fixed rational numbers such that $c > 0$ and $0 < \varepsilon \leq 1$. Unless $\varepsilon = 1$ and $c < 1$, k-SPR with $k = cn^{1-\varepsilon} \geq 1$ is PSPACE-complete even for r-th power graphs, where r is any positive integer.*

3.3 Graphs with Maximum Degree 3

Theorem 3 (♠). *Let G be a graph with n vertices, and let c and ε be any fixed rational numbers such that $c > 0$ and $0 < \varepsilon < 1$. When $k = cn^{1-\varepsilon} \geq 1$, k-SPR is PSPACE-complete even for graphs with maximum degree 3.*

4 Fixed-Parameter Tractability

Gajjar et al. showed that k-SPR is solvable in polynomial time if $k \geq n/2$ [5]. It is natural to expect that the problem remains tractable when k is slightly smaller than $n/2$. In this section, we show that the expectation is correct.

Theorem 4. *Let k and μ be non-negative rational numbers, and let n be the number of vertices in a graph G. When $k = n/2 - \mu$, there exists an algorithm that solves k-SPR in $O(m + 6.730^{\mu}\mu^4 n)$ time.*

To prove Theorem 4, we extend a dynamic programming algorithm for SPR developed by Bonsma [4]. We here give an overview of our algorithm. Let G_i denote the graph obtained as follows: remove all vertices in layers $L_{i+1}(G) \cup L_{i+2}(G) \cup \cdots \cup L_{\ell-1}(G)$ from a given graph G, and then add edges between t and all vertices in $L_i(G)$. Observe that $G_\ell = G$. In the algorithm, we construct a new graph H_i for each $i \in [0, \ell]$, called the encoding graph, which is obtained by compressing a reconfiguration graph $\mathcal{R}_k(G_i, s, t)$. The encoding graph H_i preserves necessary information contained in $\mathcal{R}_k(G_i, s, t)$ to decide whether a given instance (G, s, t, P, Q) has a solution. Note that, in general, our algorithm runs in exponential time because the size of H_i becomes super-polynomial. Nevertheless, we will show that the size of H_i is bounded by a single exponential function of μ when $k = n/2 - \mu$.

First of all, we define a *link-weight function* w for the reconfiguration graph $\mathcal{R}_k(G, s, t)$. For each link PQ of $\mathcal{R}_k(G, s, t)$, we set $w(PQ) = i$ for an integer $1 \le i \le k$ if two s-t shortest paths P and Q of G differ by exactly i consecutive vertices. Moreover, for each node P in $\mathcal{R}_k(G, s, t)$, we add a self-loop PP with $w(PP) = 0$. The self-loops do not affect on the solution of k-SPR. We will see that they are useful for constructing our dynamic programming algorithm.

We next introduce the notion of the encoding graph. Suppose that $\mathrm{dist}_G(s, t) = \ell$. For an integer $i \in [0, \ell]$, let $L_i(G)$ be the i-th layer. For an s-t shortest path R of G, let R_i denote the s-t shortest path of G_i obtained by removing vertices in $L_{i+1} \cup L_{i+2} \cup \cdots \cup L_{\ell-1}$ from R. For two s-t shortest paths R_i and R'_i, we write $R_i \sim_i R'_i$ if the node R'_i is reachable from the node R_i on $\mathcal{R}_k(G_i, s, t)$ without changing a vertex $v \in L_i(G)$ in R_i. It is not hard to see that $\sim_i$ is an equivalence relation. Let $V(\mathcal{R}_k(G_i, s, t))/\sim_i$ denote the family of equivalence classes for $V(\mathcal{R}_k(G_i, s, t))$ with respect to $\sim_i$. Observe that a subgraph of $\mathcal{R}_k(G_i, s, t)$ induced by the nodes of an equivalence class in $V(\mathcal{R}_k(G_i, s, t))/\sim_i$ is connected. The *encoding graph* H_i for a triple (G_i, P, Q), where P and Q are s-t shortest paths of G, is a graph obtained from $\mathcal{R}_k(G_i, s, t)$ as follows:

1. remove all connected components of $\mathcal{R}_k(G_i, s, t)$ not containing a node P_i;
2. contract the nodes of each equivalence class in $V(\mathcal{R}_k(G_i, s, t))/\sim_i$ into a single node x, where we denote by S_x the set of nodes in $\mathcal{R}_k(G_i, s, t)$ contracted into x; and
3. add a self-loop to each node.

We will construct the encoding graph H_i for each $i \in [0, \ell]$ by means of dynamic programming (without making the reconfiguration graph $\mathcal{R}_k(G_i, s, t)$). To this end, we provide a link-weight function η_i for H_i defined as follows:

$$\eta_i(xy) = \min\{w(XY) : X \in S_x, Y \in S_y, XY \in E(\mathcal{R}_k(G_i, s, t))\},$$

where xy is a link of H_i. The function η_i means that, in order to obtain an s-t shortest path in S_y from an s-t shortest path in S_x, a subpath with at least $\eta_i(xy)$ vertices containing a vertex in $L_i(G)$ must be replaced. We also define functions c_i, p_i, and q_i used in our algorithm as follows: for each node x in H_i,

- if every node in S_x is an s-v shortest path of G_i for a vertex $v \in L_i(G)$, then $c_i(x) = v$;
- $p_i(x) = 1$ if $P_i \in S_x$, otherwise $p_i(x) = 0$; and
- $q_i(x) = 1$ if $Q_i \in S_x$, otherwise $q_i(x) = 0$.

4.1 A Dynamic Programming Algorithm

As the base case, consider $i = 0$. Then, a graph G_0 consists of vertices s, t and an edge st. Therefore, H_0 has a unique node x and $P_i = Q_i = \langle s, t \rangle$. We set $\eta_0(xx) = 0$, $c_0(x) = s$, $p_0(x) = q_0(x) = 1$.

Consider $i > 0$. Assume that H_{i-1}, η_{i-1}, c_{i-1}, p_{i-1}, and q_{i-1} have already been computed. For a vertex $v \in L_i(G)$, denote $X_v = \{a \in V(H_{i-1}) : c_{i-1}(a) \in N_G^-(v)\}$. We construct H_i as follows:

1. initialize H_i to a graph with no nodes;
2. for each $v \in L_i(G)$, consider the subgraph H_{i-1}^v of H_{i-1} induced by the nodes in X_v;
3. for each connected component of H_{i-1}^v, add a node x with $c_i(x) = v$ into H_i, where we denote by C_x the connected component corresponding to x;
4. for two distinct nodes x and y of H_i, add an edge between x and y if and only if there exist nodes $a \in V(C_x)$ and $b \in V(C_y)$ such that $ab \in E(H_{i-1})$ and $\eta_{i-1}(ab) \leq k - 1$;
5. for each link xy of H_i, let

$$\eta_i(xy) = \min\{\eta_{i-1}(ab) + 1 : a \in V(C_x), b \in V(C_y)\};$$

6. for each node x of H_i, add a self-loop xx with $\eta_i(xx) = 0$;
7. for each node x of H_i, set $p_i(x) = 1$ if and only if $c_i(x)$ is a vertex in P_i and C_x has a node a with $p_{i-1}(a) = 1$. Similarly, set $q_i(x) = 1$ if and only if $c_i(x)$ is a vertex in Q_i and C_x has a node a with $q_{i-1}(a) = 1$; and
8. remove all connected components of H_i not containing a node x with $p_i(x) = 1$.

We conclude that a given instance (G, s, t, P, Q) has a solution if H_ℓ has a node x with $q_\ell(x) = 1$; otherwise, (G, s, t, P, Q) has no solution.

4.2 Correctness of Our Algorithm

We show that our algorithm correctly computes the encoding graph H_i for (G_i, P, Q) and functions η_i, c_i, p_i, and q_i. Recall that our algorithm generalizes the algorithm for SPR developed by Bonsma [4]. In particular, the definitions of c_i, p_i, and q_i are the same as the ones in the original algorithm. Therefore, c_i, p_i, and q_i are correctly computed. What we have to do here is to show the correctness of H_i and η_i. Obviously, they are correctly computed when $i = 0$. Assume that $i > 0$, and H_{i-1} and η_{i-1} have been obtained.

For an s-t shortest path $R = \langle s = r_0, r_1, \ldots, r_{i-1}, t \rangle$ of G_{i-1} and a vertex $v \in L_i(G)$, we denote by $R + v$ the new sequence of vertices obtained by inserting

v between r_{i-1} and t. For a vertex $v \in L_i(G)$ and a node $a \in X_v$, let $S_a \oplus v = \{R + v : R \in S_a\}$. Note that, since $c_{i-1}(a) \in N_{G_i}^-(v)$ for every $a \in X_v$, every sequence in $S_a \oplus v$ is indeed an s-t shortest path of G_i that contains v.

Let $\widehat{\mathcal{R}}_k(G_i, s, t)$ be the connected component of $\mathcal{R}_k(G_i, s, t)$ that contains the node P_i. We first show that there is a bijection from $V(H_i)$ to $V(\widehat{\mathcal{R}}_k(G_i, s, t))/\sim_i$ by the following lemmas. In fact, the lemmas and their proof are completely the same as the ones shown by Bonsma [4]. For this reason, we here omit their proofs.

Lemma 4 ([4]). *Let H_{i-1} be the encoding graph for (G_{i-1}, P, Q), and $v \in L_i(G)$. Let C be a connected component of H_{i-1}^v. Then, $\bigcup_{a \in V(C)}(S_a \oplus v)$ is a set of s-t shortest paths of G_i that is an equivalence class of $\sim_i$.*

Lemma 5 ([4]). *Let H_{i-1} and H_i be encoding graphs for (G_{i-1}, P, Q) and (G_i, P, Q), respectively. For every node $x \in V(H_i)$, there exists a connected component C_x of $H_{i-1}^{c_i(x)}$ such that $S_x = \bigcup_{a \in V(C_x)}(S_a \oplus c_i(x))$.*

The following lemma shows the correctness of links in H_i and a link-weight function η_i.

Lemma 6 (♠). *Let H_i be the encoding graph and let $x, y \in V(H_i)$. Then, $xy \in E(H_i)$ and $\eta_i(xy) = j$ with $1 \leq j \leq k$ if and only if there exist nodes a and b in H_{i-1} such that $a \in V(C_x)$, $b \in V(C_y)$, $ab \in E(H_{i-1})$, and $\eta_{i-1}(ab) = j - 1$.*

4.3   Running Time

Theorem 5 (♠). *Let (G, s, t, P, Q) be an instance of k-SPR such that G has n vertices and m edges and $\mathrm{dist}_G(s, t) = \ell$. We denote $z = \max_{0 \leq i \leq \ell} |L_i(G)|$ and $N = \max_{0 \leq i \leq \ell} |V(H_i)|$, where H_i is the encoding graph for (G_i, P, Q). Then, our dynamic programming algorithm solves k-SPR in $O(n + m + \ell z^2 N^2)$ time.*

4.4   Proof of Theorem 4

Suppose that an instance (G, s, t, P, Q) with $\mathrm{dist}_G(s, t) = \ell$ of k-SPR is given. We first provide two simple observations.

Proposition 2. *When $k \geq \ell - 1$, any instance (G, s, t, P, Q) has a solution.*

Proposition 3. *Let i be any integer in $[\ell - 1]$, and suppose that there exists a vertex $v \in V(G)$ such that $L_i(G) = \{v\}$. Let P^1 and P^2 be the subpaths of P that consist of vertices in $\bigcup_{0 \leq j \leq i} L_j(G)$ and $\bigcup_{i \leq j \leq \ell} L_j(G)$, respectively. The subpaths Q^1 and Q^2 of Q are similarly defined. Then, (G, s, t, P, Q) has a solution if and only if both (G, s, v, P^1, Q^1) and (G, v, t, P^2, Q^2) have solutions.*

Applying Proposition 3 iteratively, assume that every layer $L_i(G)$ of G except for $i = 0$ and $i = \ell$ has at least two vertices. For each $i \in [\ell]$, we say that the layer $L_i(G)$ is *narrow* if $|L_{i-1}(G)| = |L_i(G)| = 2$.

Let $z = \max_{0 \leq i \leq \ell} |L_i(G)|$. Lemmas 7 and 8 bound the size of the encoding graph H_i for each $i \in [0, \ell]$.

Lemma 7 (♠). *For an integer $i \in [0, \ell]$, let H_i be the encoding graph for (G_i, P, Q). If $0 \leq i \leq k + 1$, then $|V(H_i)| \leq z$.*

Lemma 8 (♠). *For an integer $i \in [0, \ell]$, let H_i be the encoding graph for (G_i, P, Q). If $L_i(G)$ with $i \in [\ell]$ is narrow, then $|V(H_i)| \leq \max\{|V(H_{i-1})|, 2\}$.*

We here define an integer z_i for $i \in [0, \ell]$ as follows:

$$z_i = \begin{cases} 1 & \text{if } i \leq k + 1 \text{ or the layer } L_i(G) \text{ is narrow;} \\ |L_i(G)| & \text{otherwise.} \end{cases}$$

We bound the size of $V(H_i)$ with these values.

Lemma 9 (♠). *Let $N = \max_{0 \leq i \leq \ell} |V(H_i)|$. Then, $N \leq z \prod_{0 \leq j \leq \ell} z_j$.*

We say that the layer is *large* if it has at least three vertices. Let α be the number of large layers after $\lfloor k + 1 \rfloor$-th layer of G. We denote by $p_1, p_2, \ldots, p_\alpha$ integers such that $k + 1 < p_j \leq \ell - 1$ and $L_{p_j}(G)$ is large for every $j \in [\alpha]$. In addition, we denote by $q_1, q_2, \ldots, q_\beta$ integers such that $k + 1 < q_j \leq \ell - 1$ and $L_{q_j}(G)$ is neither narrow nor large for every $j \in [\beta]$. Let $\gamma = \sum_{j \in [\alpha]} |L_{p_j}(G)|$. The following lemma is the final piece of the puzzle in proving Theorem 4.

Lemma 10. *If $k < \ell - 1$, then $3\alpha + 2\beta < 2\mu - 2$, $\gamma < 2\mu - 2$, and $z < 2\mu$.*

Proof. Recall that $k = n/2 - \mu$. The Layers $L_0(G)$ and $L_\ell(G)$ have exactly one vertex, respectively, and $L_j(G)$ for each integer j with $1 \leq j \leq k + 1$ has at least two vertices. Moreover, $|L_{p_j}(G)| \geq 3$ for each $j \in [\alpha]$ and $|L_{q_j}(G)| = 2$ for each $j \in [\beta]$. Thus, we have

$$\begin{aligned} n &\geq 2 + 2\lfloor k + 1 \rfloor + 3\alpha + 2\beta \\ &> 2 + 2k + 3\alpha + 2\beta \\ &= 2 + n - 2\mu + 3\alpha + 2\beta. \end{aligned}$$

This immediately implies that $3\alpha + 2\beta < 2\mu - 2$. Similarly, we have $\gamma < 2\mu - 2$ because

$$\begin{aligned} n &\geq 2 + 2\lfloor k + 1 \rfloor + \gamma \\ &> 2 + 2k + \gamma \\ &= 2 + n - 2\mu + \gamma, \end{aligned}$$

and $z < 2\mu$ because each of the $(\ell - 2)$ layers, except for $L_0(G)$, $L_\ell(G)$, and the layer with z vertices, has at least two vertices and hence we have

$$\begin{aligned} n &\geq 2 + 2(\ell - 2) + z \\ &> 2 + 2(k - 1) + z \\ &= n - 2\mu + z. \end{aligned}$$

$\square$

We are now ready to prove Theorem 4. Let (G, s, t, P, Q) be an instance of k-SPR after applying Proposition 3 iteratively. In the remainder of this proof, assume that every layer of G except for $L_0(G)$ and $L_\ell(G)$ has at least two vertices. If $k \geq \ell - 1$, then we immediately conclude that (G, s, t, P, Q) has a solution by Proposition 2.

Suppose that Proposition 2 is not applicable, that is, $k < \ell - 1$. From Lemma 10, we have $3\alpha + 2\beta < 2\mu - 2$ and $\gamma < 2\mu - 2$. Since $\gamma = \sum_{j \in [\alpha]} |L_{p_j}(G)|$ and $L_{p_j}(G)$ is large for every $j \in [\alpha]$, it follows from the definition of z_i that

$$\sum_{j \in [\alpha]} z_{p_j} < 2\mu - 2. \tag{2}$$

We next bound the size of $N = \max_{0 \leq i \leq \ell} |V(H_i)|$. Recall that $z_i = 1$ holds if $i \leq k+1$ or $L_i(G)$ is narrow. In addition, observe that $z_{q_j} = 2$ for every $j \in [\beta]$. Therefore, from Lemma 9 and $z_0 = z_\ell = 1$, we have

$$N \leq z \prod_{0 \leq j \leq \ell} z_j$$

$$= z \prod_{k+1 < j \leq \ell-1} z_j$$

$$= z \left(\prod_{j \in [\beta]} z_{q_j} \right) \cdot \left(\prod_{j \in [\alpha]} z_{p_j} \right)$$

$$= z 2^\beta \cdot \prod_{j \in [\alpha]} z_{p_j}. \tag{3}$$

By combining Eqs. (2), (3), $3\alpha + 2\beta < 2\mu - 2$, and the inequality of arithmetic and geometric means, we have

$$N < z 2^{\mu-1-3\alpha/2} \cdot \left(\frac{\sum_{j \in [\alpha]} z_{p_j}}{\alpha} \right)^\alpha$$

$$< z 2^{\mu-1-3\alpha/2} \cdot \left(\frac{2\mu - 2}{\alpha} \right)^\alpha$$

$$= z 2^{\mu-1} \left(\frac{\mu - 1}{\sqrt{2}\alpha} \right)^\alpha,$$

where $0 \leq \alpha < (2\mu - 2)/3$. The function $f(\alpha) = ((\mu - 1)/(\sqrt{2}\alpha))^\alpha$ for $0 \leq \alpha < (2\mu - 2)/3$ is maximized when $\alpha = (\mu - 1)/(\sqrt{2}e)$, where e is Euler's number. Therefore, we have $N < z 2^{\mu-1} f((\mu - 1)/(\sqrt{2}e)) = z(2e^{1/(\sqrt{2}e)})^{\mu-1}$. Recall that $z < 2\mu$ by Lemma 10. We conclude that, by Theorem 5, our algorithm runs in time

$$O(n + m + \ell z^2 N^2) = O(n + m + \ell z^4 (4e^{\sqrt{2}/e})^{\mu-1})$$

$$= O(m + 6.730^\mu \mu^4 n).$$

5   Polynomial-Time Algorithms

In Sect. 4, we showed that k-SPR is efficiently solvable if k is close to $n/2$. In this section, we consider a case where k-SPR is solvable for any positive rational number k.

The *contraction* of an edge uv in a graph G is to remove the vertices u and v together with their incident edges and then add a vertex w that is adjacent to all vertices in $N_G(u) \cup N_G(v) \setminus \{u, v\}$. A graph G contains a graph H as a *minor* if there is a graph G' that is obtained from G by iteratively applying contractions of edges and H is a subgraph of G'. A graph G is said to be H-*minor-free* if G does not contain H as a minor. We denote by $K_{2,3}$ the graph whose vertex set can be partitioned into two independent sets X of size 2 and Y of size 3 such that $xy \in E(K_{2,3})$ for any two vertices $x \in X$ and $y \in Y$. It is well known that all outerplanar graphs, which are a famous subclass of planar graphs, are $K_{2,3}$-minor-free.

Theorem 6 (♠). *For any positive integer k, k-SPR can be solved in linear time for $K_{2,3}$-minor-free graphs. Moreover, if a given instance (G, s, t, P, Q) has a solution, then the distance from P to Q in $\mathcal{R}_k(G, s, t)$ is at most $|V(P) \setminus V(Q)| \leq \ell - 1$.*

6   Conclusion

In this paper, we investigated the complexity of k-SPR depending on k. We first showed that, for any fixed rational numbers c and ε such that $c > 0$ and $0 < \varepsilon \leq 1$, unless $\varepsilon = 1$ and $c < 1$, k-SPR with $k = cn^{1-\varepsilon} \geq 1$ is PSPACE-complete even for bipartite graphs and r-th power graphs, where r is any positive integer. On the other hand, we designed a single exponential FPT algorithm for k-SPR parameterized by $\mu = n/2 - k \geq 0$. This implies that k-SPR with $k = n/2 - O(\log n)$ is solvable in polynomial time. Future research should narrow the gap between tractability and intractability with respect to k. We also obtained the similar intractability of k-SPR on graphs with maximum degree 3. However, our approach does not work when $k = \Theta(1)$. It would be interesting to settle the complexity of k-SPR with $k = \Theta(1)$ for graphs with maximum degree 3. Finally, we designed a linear-time algorithm for $K_{2,3}$-minor-free graphs. This result also means the tractability of k-SPR on outerplanar graphs. Surprisingly, Bonsma showed that 1-SPR is solvable even for planar graphs via a dynamic programming algorithm [4]. Our dynamic programming algorithm may also be the key to solving k-SPR on planar graphs when $k \geq 2$.

Acknowledgements. We are grateful to Shunta Suzumura for valuable discussions with him. We thank anonymous referees for their valuable comments and suggestions which greatly helped to improve the presentation of this paper.

References

1. Alon, N., Gutin, G.Z., Kim, E.J., Szeider, S., Yeo, A.: Solving MAX-r-SAT above a tight lower bound. Algorithmica **61**(3), 638–655 (2011). https://doi.org/10.1007/s00453-010-9428-7
2. Asplund, J., Edoh, K.D., Haas, R., Hristova, Y., Novick, B., Werner, B.: Reconfiguration graphs of shortest paths. Discret. Math. **341**(10), 2938–2948 (2018). https://doi.org/10.1016/j.disc.2018.07.007
3. Bonsma, P.S.: The complexity of rerouting shortest paths. Theoret. Comput. Sci. **510**, 1–12 (2013). https://doi.org/10.1016/j.tcs.2013.09.012
4. Bonsma, P.S.: Rerouting shortest paths in planar graphs. Discret. Appl. Math. **231**, 95–112 (2017). https://doi.org/10.1016/j.dam.2016.05.024
5. Gajjar, K., Jha, A.V., Kumar, M., Lahiri, A.: Reconfiguring shortest paths in graphs. In: Thirty-Sixth AAAI Conference on Artificial Intelligence, AAAI 2022, Thirty-Fourth Conference on Innovative Applications of Artificial Intelligence, IAAI 2022, The Twelveth Symposium on Educational Advances in Artificial Intelligence, EAAI 2022 Virtual Event, 22 February–1 March 2022, pp. 9758–9766. AAAI Press (2022). https://ojs.aaai.org/index.php/AAAI/article/view/21211
6. van den Heuvel, J.: The complexity of change. In: Blackburn, S.R., Gerke, S., Wildon, M. (eds.) Surveys in Combinatorics. London Mathematical Society Lecture Note Series, vol. 409, pp. 127–160. Cambridge University Press, Cambridge (2013). https://doi.org/10.1017/CBO9781139506748.005
7. Ito, T., et al.: On the complexity of reconfiguration problems. Theoret. Comput. Sci. **412**(12), 1054–1065 (2011). https://doi.org/10.1016/j.tcs.2010.12.005
8. Kamiński, M., Medvedev, P., Milanic, M.: Shortest paths between shortest paths. Theoret. Comput. Sci. **412**(39), 5205–5210 (2011). https://doi.org/10.1016/j.tcs.2011.05.021
9. Mahajan, M., Raman, V.: Parameterizing above guaranteed values: MaxSat and MaxCut. J. Algorithms **31**(2), 335–354 (1999). https://doi.org/10.1006/jagm.1998.0996
10. Mahajan, M., Raman, V., Sikdar, S.: Parameterizing above or below guaranteed values. J. Comput. Syst. Sci. **75**(2), 137–153 (2009). https://doi.org/10.1016/j.jcss.2008.08.004
11. Nishimura, N.: Introduction to reconfiguration. Algorithms **11**(4), 52 (2018). https://doi.org/10.3390/a11040052
12. Savitch, W.J.: Relationships between nondeterministic and deterministic tape complexities. J. Comput. Syst. Sci. **4**(2), 177–192 (1970). https://doi.org/10.1016/S0022-0000(70)80006-X
13. Wrochna, M.: Reconfiguration in bounded bandwidth and tree-depth. J. Comput. Syst. Sci. **93**, 1–10 (2018). https://doi.org/10.1016/j.jcss.2017.11.003

Combinatorial Reconfiguration with Answer Set Programming: Algorithms, Encodings, and Empirical Analysis

Yuya Yamada[1], Mutsunori Banbara[1]([✉])[iD], Katsumi Inoue[2][iD],
Torsten Schaub[3][iD], and Ryuhei Uehara[4][iD]

[1] Nagoya University, Nagoya, Japan
{yuya.yamada,banbara}@nagoya-u.jp
[2] National Institute of Informatics, Tokyo, Japan
inoue@nii.ac.jp
[3] Universität Potsdam, Potsdam, Germany
torsten@cs.uni-potsdam.de
[4] Japan Advanced Institute of Science and Technology, Nomi, Japan
uehara@jaist.ac.jp

Abstract. We propose an approach called *bounded combinatorial reconfiguration* for solving combinatorial reconfiguration problems based on Answer Set Programming (ASP). The general task is to study the solution spaces of combinatorial problems and to decide whether or not there are sequences of feasible solutions that have special properties. The resulting *recongo* solver covers all metrics of the solver track in the most recent international competition on combinatorial reconfiguration (CoRe Challenge 2022). *recongo* ranked first in the shortest metric of the single-engine solvers track. In this paper, we present the design and algorithm of bounded combinatorial reconfiguration, and also present ASP encodings of the independent set reconfiguration problem under the token jumping rule that is one of the most studied combinatorial reconfiguration problems. Finally, we present empirical analysis considering all instances of CoRe Challenge 2022.

Keywords: Combinatorial Reconfiguration · Independent Set Reconfiguration · Answer Set Programming

1 Introduction

Combinatorial reconfiguration [14,16,24] aims at analyzing the structure and properties (e.g., connectivity and reachability) of the solution spaces of combinatorial problems. Each solution space has a graph structure in which each node represents an individual feasible solution, and the edges are defined by a certain adjacency relation. *Combinatorial Reconfiguration Problems* (CRPs) are defined in general as the task of deciding, for a given combinatorial problem and two among its feasible solutions, whether or not one is reachable from another via a sequence of adjacent feasible solutions. A CRP is *reachable* if there exists

ⓒ The Author(s), under exclusive license to Springer Nature Singapore Pte Ltd. 2024
R. Uehara et al. (Eds.): WALCOM 2024, LNCS 14549, pp. 242–256, 2024.
https://doi.org/10.1007/978-981-97-0566-5_18

such a sequence, otherwise it is *unreachable*. We refer to the original problem as *source problem* in contrast to its reconfiguration problem. *CRP solvers* are programs solving combinatorial reconfiguration problems. The solvers output a reconfiguration sequence as a solution if reachable.

A great effort has been made to investigate the theoretical aspects of CRPs in the field of theoretical computer science over the last decade. For many NP-complete source problems, their reconfigurations have been shown to be PSPACE-complete, including SAT reconfiguration [12,21], independent set reconfiguration [16,19], dominating set reconfiguration [13,26], graph coloring reconfiguration [3–5], clique reconfiguration [18], Hamiltonian cycle reconfiguration [15,27], and set covering reconfiguration [16].

However, little attention has been paid so far to its practical aspects. To stimulate research and development on practical CRP solving, the first international combinatorial reconfiguration competition (CoRe Challenge 2022; [25]) has been held in 2022. Eight solvers participated in the solver track of CoRe Challenge 2022, including a planning-based solver [6], a ZDD-based solver [17], etc. The competition used the independent set reconfiguration problem under the token jumping rule, which is one of the most studied combinatorial reconfiguration problems.

In this paper, we describe an approach for solving combinatorial reconfiguration problems based on Answer Set Programming (ASP; [1,11,23]). The resulting *recongo* solver reads a CRP instance and converts it into ASP facts. In turn, these facts are combined with an ASP encoding for CRP solving, which are afterward solved by efficient ASP solvers, in our case *clingo* [8]. To show the effectiveness of our approach, we conduct experiments on the benchmark set of CoRe Challenge 2022.

ASP is a declarative programming paradigm for knowledge representation and reasoning in artificial intelligence. The declarative approach of ASP has distinct advantages. First, ASP provides an expressive language and is well suited for modeling combinatorial (optimization) problems in artificial intelligence and computer science [7]. Second, the extension to their reconfiguration problems can be easily done. Finally, recent advances in multi-shot ASP solving [10] allow for efficient reachability checking of combinatorial reconfiguration problems.

The main contributions of our paper are as follows:

(1) We present the design and algorithm of *bounded combinatorial reconfiguration* for solving combinatorial reconfiguration problems based on ASP. Our declarative approach is inspired by Bounded Model Checking (BMC; [2]), which is widely used in formal verification of finite state transition systems.

(2) We develop an ASP-based CRP solver *recongo*[1] using *clingo*'s multi-shot ASP solving [10]. *recongo* ranked first at the shortest metric of the single-engine solvers track in the CoRe Challenge 2022, and ranked second or third in the other four metrics.

---

[1] An overview of *recongo* is given in a short paper [28]. The present paper gives more detailed algorithms, encodings, and empirical analysis of bounded combinatorial reconfiguration.

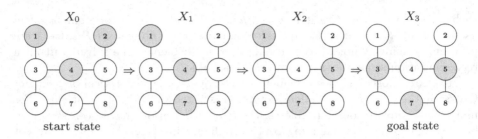

Fig. 1. An ISRP example

(3) We present an ASP encoding for solving the independent set reconfiguration problem under the token jumping rule. In algorithm theory, this problem has been shown to be W[1]-hard when parameterized by $k + \ell$, where k is the size of independent sets and ℓ is the length of reconfiguration sequences [22].

(4) Our empirical analysis considers all 369 instances publicly available from the CoRe Challenge 2022 website.[2] We address the competitiveness of our declarative approach by contrasting it to other approaches.

Overall, the proposed declarative approach can make a significant contribution to the state-of-the-art of CRP solving as well as ASP application to combinatorial reconfiguration.

Our encodings are given in the language of *clingo* [8]. Although we provide a brief introduction to ASP and its basic language constructs in Sect. 2.2, a comprehensive introduction to ASP can be found in [9].

2   Background

2.1   Combinatorial Reconfiguration

The combinatorial reconfiguration problem (CRP) is defined as the task of deciding, for a given source problem and two of its feasible solutions X_s and X_g, whether or not there are sequences of transitions:

$$X_s = X_0 \rightarrow X_1 \rightarrow X_2 \rightarrow \cdots \rightarrow X_\ell = X_g. \tag{1}$$

Each state X_i represents a feasible solution of the source problem. We refer to X_s and X_g as the start and the goal states, respectively. We write $X \rightarrow X'$ if state X at step t can be followed by state X' at step $t + 1$ subject to a certain *adjacency relation*. We refer to the sequence (1) as a *reconfiguration sequence* from X_s to X_g. The *length* of the reconfiguration sequence, denoted by ℓ, is the number of transitions. Regarding the reconfiguration sequences, combinatorial reconfiguration problems can be classified into three categories: *existent*, *shortest*, and *longest*. The existent-CRP is to decide whether or not there are reconfiguration sequences. The shortest-CRP is to find a shortest reconfiguration sequence. The

---

[2] https://core-challenge.github.io/2022/.

longest-CRP is to find a longest reconfiguration sequence that cannot include any loop.

Let us consider the independent set reconfiguration problem (ISRP). Its source is *the independent set problem*, that is, to decide whether or not there is an independent set in G of size k, for a given graph $G = (V, E)$ and an integer k. A subset $V' \subseteq V$ is called an *independent set* in G of size k if $(u, v) \notin E$ for all $u, v \in V'$ and $|V'| = k$. In the ISRP, each state X in (1) represents an independent set. Regarding adjacency relations, we focus on one of the most studied relations called *token jumping* [19]. Suppose that a *token* is placed on each node in an independent set. The token jumping meaning of $X \to X'$ is that a single token "jumps" from the single node in $X \setminus X'$ to the one in $X' \setminus X$.

Figure 1 shows an example of ISRP. The example consists of a graph having 8 nodes and 8 edges, and the size of independent sets is $k = 3$. The independent sets (tokens) are highlighted in yellow. We can observe that the goal state can be reachable from the start state with length $\ell = 3$. For instance, in the transition from X_0 to X_1, a token jumps from node 2 in X_0 to node 7 in X_1.

2.2 Answer Set Programming

Answer Set Programming (ASP) is a declarative programming paradigm widely used in artificial intelligence. In ASP, problems are represented as *logic programs*, which are finite sets of *rules* of the form:

$$a_0 \;\text{:-}\; a_1, \ldots, a_m, \texttt{not}\; a_{m+1}, \ldots, \texttt{not}\; a_n.$$

Each a_i is a propositional *atom*. The connectives ':-', ',', and 'not' denote 'if', 'conjunction', and 'default negation', respectively. Each rule is terminated by a period '.'. A *literal* is an atom $\texttt{a}$ or $\texttt{not}\;\texttt{a}$. Intuitively, the rule means that a_0 must be true if $a_1, \ldots, a_m$ are true and $a_{m+1}, \ldots, a_n$ are false. Semantically, a logic program induces a collection of so-called *answer sets*, which are distinguished models of the program based on stable model semantics [11]. ASP rules have two special cases. One is a *fact* of the form a_0. A fact is always true and belongs to every answer set. Another is an *integrity constraint* of the form $\text{:-}\; a_1, \ldots, a_m, \texttt{not}\; a_{m+1}, \ldots, \texttt{not}\; a_n$. An integrity constraint means that the conjunction of literals must not hold, and can be used to filter solution candidates. For instance, $\text{:-}\; a_1, a_2$. means a_1 and a_2 must not hold simultaneously.

Several extensions have been made to facilitate the use of ASP in practice. First of all, rules with first-order variables are viewed as shorthand for the set of their *ground* instances (i.e., variable-free rules). Further language constructs include *conditional literals* and *cardinality constraints*. The former are of the form $\ell_0 : \ell_1, \ldots, \ell_m$ and the latter can be written as $\ell b\; \{c_1; \ldots; c_n\}\; ub$, where all ℓ_i are literals, and all c_j are conditional literals; ℓb and ub indicate lower and upper bounds on the number of satisfied literals in the cardinality constraint. For instance, a conditional literal like $\texttt{a(X):b(X)}$ in a rule's antecedent expands to the conjunction of all instances of $\texttt{a(X)}$ for which the corresponding instance of $\texttt{b(X)}$ holds. Similarly, $1\; \{\; \texttt{a(X):b(X)}\; \}\; 1$ is true whenever the exact one instance of $\texttt{a(X)}$ (subject to $\texttt{b(X)}$) is true.

3   The *recongo* Approach

Basic Design. Combinatorial reconfiguration problems can be readily expressed as satisfiability problems. Let $x = \{x_1, x_2, \ldots, x_n\}$ and $C(x)$ be the variables and the constraints of a source problem, respectively. For its reconfiguration problem, each state X at step t can be represented by a set of variables $x^t = \{x_1^t, x_2^t, \ldots, x_n^t\}$. Each adjacent relation can be represented by a set of constraints $T(x^{t-1}, x^t)$ that must be satisfied. Optionally, additional constraints $S(x^0)$ and $G(x^\ell)$ can be added to specify conditions on the start state X_s and/or the goal state X_g, respectively, as well as any other constraints that we want to enforce. Then, the existence of a reconfiguration sequence (1) of bounded length ℓ is equivalent to the following satisfiability problem

$$\varphi_\ell = S(x^0) \wedge \bigwedge_{t=0}^{\ell} C(x^t) \wedge \bigwedge_{t=1}^{\ell} T(x^{t-1}, x^t) \wedge G(x^\ell). \tag{2}$$

We use φ_ℓ to check properties of a reconfiguration relation (a transition relation) between the possible feasible solutions of the source problem. We call this general framework "bounded combinatorial reconfiguration", because we consider only reconfiguration sequences that have a bounded length ℓ. For reachability checking, if φ_ℓ is satisfiable, there is a reconfiguration sequence of length ℓ. Otherwise, we keep on reconstructing a successor (e.g., $\varphi_{\ell+1}$) and checking its satisfiability until a reconfiguration sequence is found. Bounded combinatorial reconfiguration is an incomplete method, because it can find reconfiguration sequences if they exist, but cannot prove unreachability in general. However, it can be a complete method if the diameters of solution spaces are given. Any off-the-shelf satisfiability solvers, such as SAT solvers and CSP solvers, can be used as back-end. In this paper, we make use of ASP solvers, in our case *clingo*.

Algorithm. We present an algorithm of bounded combinatorial reconfiguration. Obviously it is inefficient to fully reconstruct φ_ℓ in each transition because of the expensive grounding, which transforms first-order logic programs to variable-free ones. Instead, we incrementally construct φ_ℓ from its predecessor $\varphi_{\ell-1}$ by adding $C(x^\ell)$, $T(x^{\ell-1}, x^\ell)$, and $G(x^\ell)$ and deactivating $G(x^{\ell-1})$. This can be easily done by utilizing *clingo*'s multi-shot ASP solving [10]. The multi-shot ASP solving allows for incremental grounding and solving for logic programs and is well suited to achieve bounded combinatorial reconfiguration.

The pseudo code of a grounding-conscious algorithm for bounded combinatorial reconfiguration is shown in Algorithm 1. We use five variables at the loop in Lines 9–24 to control the successive grounding and solving of an input logic program. The input consists of a problem instance P of ASP fact format and four subprograms $S(x^0)$, $C(x^t)$, $T(x^{t-1}, x^t)$, and $G(x^t)$. The values of I^{min} and I^{max} respectively indicate the minimum number of steps (1 by default) and the maximum number of steps (*none* by default). A variable I^{stop} is used to specify a termination criterion (SAT by default or UNSAT). The value of i indicates each step, and a variable ret is used to store the solving result. In addition, I^{search} is

---

Algorithm 1. Grounding-conscious bounded combinatorial reconfiguration

---

Input P: problem instance of ASP fact format
Input $S(x^0)$, $C(x^t)$, $T(x^{t-1}, x^t)$, $G(x^t)$: logic program
Parameter I^{min}: the minimum number of steps [1]
Parameter I^{max}: the maximum number of steps [none]
Parameter I^{stop}: termination criterion [SAT]
Parameter I^{search}: path search [shortest]

1: $ctl \leftarrow$ create an object of ASP solver
2: $i \leftarrow 0$
3: $ret \leftarrow none$
4: $model \leftarrow none$
5: **if** $I^{search} =$ longest **then**
6:     $I^{min} \leftarrow I^{max}$
7: **end if**
8: add a statement "#external query(t)." to $G(x^t)$
9: **while** $(I^{max} = none$ **or** $i < I^{max})$ **and**
      $(I^{min} = none$ **or** $i < I^{min}$ **or** $ret = none$ **or** $ret \neq I^{stop})$ **do**
10:     $parts \leftarrow$ an empty list
11:     $parts.append(C(x^i))$
12:     $parts.append(G(x^i))$
13:     **if** $i > 0$ **then**
14:         $parts.append(T(x^{i-1}, x^i))$
15:         $ctl.release\_external(query(i-1))$ {deactivating $G(x^{i-1})$}
16:     **else**
17:         $parts.append(P)$
18:         $parts.append(S(x^0))$
19:     **end if**
20:     $ctl.ground(parts)$
21:     $ctl.assign\_external(query(i), \textbf{true})$ {activating $G(x^i)$}
22:     $(ret, model) = ctl.solve()$
23:     $i \leftarrow i + 1$
24: **end while**
25: $ctl.close()$
26: **if** $model \neq none$ **then**
27:     print REACHABLE
28: **else if** $I^{max} \neq none$ **and** $i \geq I^{max}$ **then**
29:     print UNREACHABLE
30: **else**
31:     print UNKNOWN
32: **end if**

---

used to switch a search type (shortest by default, longest, or existent), and $model$ is used to store the answer sets. We note that the external atom query(t) in Line 8 is used to deactivate $G(x^t)$ as well as to activate it.

In each step of the loop, the subprograms stored in the list $parts$ are grounded and solved. For instance, in step $i = 0$, the $ctl.ground(parts)$ method in Line 20 grounds $P \land S(x^0) \land C(x^0) \land G(x^0)$ in which query(0) is set to true in Line 21 in order to activate $G(x^0)$. Then, the method $ctl.solve()$ in Line 22 checks its satisfiability, that is, whether the goal state is reachable from the start state. Finally, the value of step i is incremented by 1. This process iterates until the termination criterion is met. Each following step checks the satisfiability of the logic program $P \land S(x^0) \land \bigwedge_{t=0}^{i} C(x^t) \land \bigwedge_{t=1}^{i} T(x^{i-1}, x^i) \land G(x^i)$, in which the current external atom query(i) is set to true, but the previous query($i-1$) is permanently set to false in Line 15. The algorithm searches shortest reconfiguration sequences in a default setting. For searching longest ones, all we have to do is just assigning the value of I^{max} to I^{min} in Line 6.

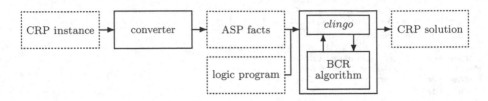

Fig. 2. The architecture of *recongo*

Implementation. Bounded combinatorial reconfiguration (BCR) in Algorithm 1 can be easily implemented using *clingo*'s Python API. The resulting *recongo* system is a general-purpose CRP solver. The architecture of *recongo* is shown in Fig. 2. *recongo* reads an CRP instance and converts it into ASP facts. In turn, these facts are combined with an ASP encoding for CRP solving, which are afterward solved by the BCR algorithm powered by *clingo*. *recongo* covers all metrics of the solver track in the most recent international competition on combinatorial reconfiguration (CoRe Challenge 2022).

4   ASP Encoding for Independent Set Reconfiguration

We present an ASP encoding for solving the independent set reconfiguration problem (ISRP).

Fact Format. The input of ISRP is an independent set problem, a start state, and a goal state. Listing 1.1 shows an ASP fact format of the ISRP instance in Fig. 1. The predicate k/1 represents the size of independent sets. The predicates node/1 and edge/2 represent the nodes and edges, respectively. The predicates start/1 and goal/1 represent the independent sets of the start and goal states, respectively. For instance, the atom start(4) means that node 4 is in an independent set at the start state.

First Order Encoding. Listing 1.2 shows an ASP encoding for ISRP solving. The encoding consists of three parts: base, step(t), and check(t). The parameter t represents each step in a reconfiguration sequence. The atom in(X,t) is intended to represent that the node X is in an independent set at step t. The base part specifies the constraints on the start state $S(x^0)$. The rule in Line 3 enforces that in(X,0) holds for each node X in the start state. The step(t) part specifies the constraints that must be satisfied at each step t. The rules in Lines 7–8 represent the constraints of independent set $C(x^t)$. The rule in Line 7 generates a candidate independent set with size K. The rule in Line 8 enforces that no two nodes connected by an edge are in an independent set. The rules in Lines 11–12 represent the adjacency relation $T(x^{t-1}, x^t)$. The auxiliary atom moved_from(X,t) in Line 11 represents that a token jumps from node X to any other node, from step t-1 to t. The rule in Line 12 enforces that exactly one token jumps at each step t. The check(t) part specifies the termination condition that must be satisfied at the goal state $G(x^t)$. The rule in Line 16

```
k(3).
node(1).    node(2).    node(3).    node(4).
node(5).    node(6).    node(7).    node(8).
edge(1,3). edge(2,5). edge(3,4). edge(3,6).
edge(4,5). edge(5,8). edge(6,7). edge(7,8).
start(1).   start(2).   start(4).
goal(3).    goal(5).    goal(7).
```

Listing 1.1. ASP fact format of ISRP instance in Fig. 1

```
1  #program base.
2  % start state
3  :- not in(X,0), start(X).
4
5  #program step(t).
6  % independent set constraints
7  K { in(X,t): node(X) } K :- k(K).
8  :- in(X,t), in(Y,t), edge(X,Y).
9
10 % adjacency relation: token jumping
11 moved_from(X,t) :- in(X,t-1), not in(X,t), t > 0.
12 :- not 1 { moved_from(X,t) } 1, t > 0.
13
14 #program check(t).
15 % goal state
16 :- not in(X,t), goal(X), query(t).
```

Listing 1.2. ASP encoding for ISRP solving

enforces that in(X,t) holds for each node X in the goal state. The volatility of this rule is handled by a truth assignment to the external atom query(t), as explained in Algorithm 1. In addition, Listing 1.3 shows a simple encoding that ensure there is no loop in reconfiguration sequences. This constraint is essential for longest-ISRP solving.

Hint Constraints. We present a search heuristics and four hint constraints to accelerate ISRP solving. Their ASP encodings are shown in Listing 1.4. The hints d1 and d2 are constraints on the bound of distance from the start and goal states, respectively, in reconfiguration sequences. The hint d1 in Line 3 enforces that, for each step t, there are at most t nodes that are in the start state but not in step t. Similarly, the hint d2 in Line 6 enforces that, for each step t and $T \in \{0 \ldots t-1\}$, there are at most t-T nodes that are in the goal state but not in step T. An example of invalid reconfiguration sequences forbidden by d2 is shown in Fig. 3. In the sequence, the lower bound of distance between $X_{\ell-2}$ and X_ℓ is 3, but it is greater than the possible number of transitions, namely 2. The hints t1 and t2 are constraints that forbid redundant token moves. The hint t1 in Line 9 enforces that, in two consecutive transitions, no token jumps

```
#program step(t).
:- in(X,t):in(X,T); not in(X,t):not in(X,T),node(X); T = 0..t-1; t > 0.
```

Listing 1.3. No loop constraints for ISRP solving

```
1   #program step(t).
2   % distance constraint (d1)
3   :- not { not in(X,t): start(X) } t.
4
5   % distance constraint (d2)
6   :- not { not in(X,T): goal(X) } t-T, T = 0..t-1, query(t).
7
8   % token constraint (t1)
9   :- moved_from(X,t-1), in(X,t), t > 1.
10
11  % token constraint (t2)
12  moved_to(X,t) :- not in(X,t-1), in(X,t), t > 0.
13  :- moved_to(X,t-1), not in(X,t), t > 1.
14
15  % maximal independent set heuristic (h)
16  #heuristic in(Y,t): edge(X,Y), not in(X,t). [level_max-t,true]
17  #heuristic in(Y,t): edge(Y,X), not in(X,t). [level_max-t,true]
```

Listing 1.4. Hint constraints for ISRP solving

back to a node X at step t from which a token jumped before. Similarly, the hint t2 in Lines 12–13 enforces that no token jumps from a node X at step t to which a token jumped before. An example of invalid reconfiguration sequences forbidden by t1 is shown in Fig. 4. The sequence is obviously redundant since X_ℓ can be reachable from $X_{\ell-2}$ in one transition. The heuristics h is a domain-specific heuristics that attempts to make each state to be a maximal independent set. This can be easily done by using *clingo*'s #heuristic statements [8]. They allow for modifying the search heuristic of *clingo* from within logic programs. In *clingo*'s heuristic programming, each atom has a level, and its default value is 0. For each step t and for each edge(X,Y), the statement in Line 16 gives a higher level to the atom in(Y,t) if its adjacent node X is not in an independent set at step t. The statement in Line 17 works in the same way. The distance constraints d1 and d2 are domain-independent and can be applied to many CRPs. In contrast, the others are domain-specific constraints for ISRP. We note that the token hints t1 and t2 cannot be used for longest-ISRP solving.

5   Experiments

To evaluate the *recongo* approach in Sect. 3 and the *recongo* encoding in Sect. 4, we conduct experiments on the benchmark set of the most recent international competition on combinatorial reconfiguration (CoRe Challenge 2022).

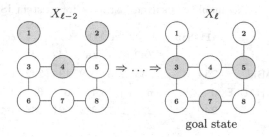

Fig. 3. An invalid reconfiguration sequence forbidden by d2

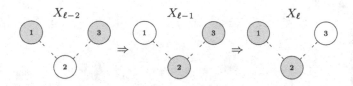

Fig. 4. An invalid reconfiguration sequence forbidden by t1

Our empirical analysis considers all ISRP instances, namely 369 in a total. They are publicly available from CoRe Challenge 2022 website.[3] This ISRP benchmark set is classified into seven families. The family color04 consists of 202 instances, grid of 49, handcraft of 6, power of 17, queen of 48, sp of 30, and square of 17. The number of nodes of input graphs ranges from 7 to 40,000. We use the proposed encoding in Listing 1.2 and the hint constraints in Listing 1.4. The no loop constraints in Listing 1.3 are used only for longest-ISRP solving. We use *recongo* 0.2 powered by *clingo* 5. *recongo* 0.2 searches shortest reconfiguration sequences when I^{search} is existent as well as shortest. For 23 instances, we assign the total number of all feasible solutions to I^{max} (*none* by default), which can be easily computed by *clingo* in our preliminary experiments. We ran our experiments on a Mac OS machine equipped with Intel Xeon W 12-core 3.3 GHz processors and 96 GB RAM. We imposed a time limit of 1 h for each instance.

Table 1 shows the number of solved instances for existent-ISRP. The columns display in order reachability and the number of solved instances for each encoding. The best results are highlighted in bold. The no-hint means the *recongo* encoding without any hints. The all-hints means the *recongo* encoding with all hints. The all-hints/X means the all-hints except $X \in \{d1, d2, t1, t2, h\}$. The all-hints solved the most, namely 240 instances out of 369. The hints are highly effective since the all-hints was able to solve 48 instances more than the no-hint. Regarding single hint, the distance hint d2 is the most effective since the difference of all-hints/d2 from the all-hints is the largest, namely $240 - 201 = 39$. It is followed by $240 - 233 = 7$ of the heuristics h based on the idea of the maximal independent set. Figure 5 shows a cactus plot where the

---

[3] https://core-challenge.github.io/2022/.

Table 1. The number of solved instances for existent-ISRP

	all-hints	all-hints					no-hint
		/t2	/t1	/h	/d2	/d1	
REACHABLE	**234**	233	233	229	195	232	188
UNREACHABLE	**6**	6	4	4	6	6	4
Total	**240**	239	237	233	201	238	192

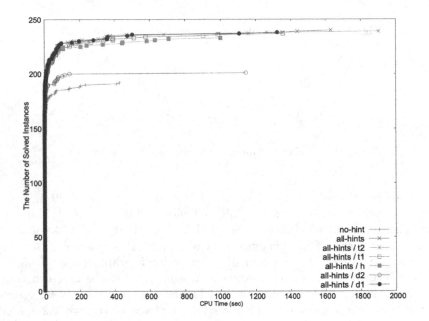

Fig. 5. Cactus plot of existent-ISRP

horizontal axis (x-axis) indicates CPU times in seconds, and the vertical axis (y-axis) indicates the number of solved instances. The cactus plot can clearly show the contrast in performance discussed above between `all-hints` and `all-hints/d2`, as well as between `all-hints` and `no-hint`.

Second, we examine the longest-ISRP that is to find longest reconfiguration sequences without any loop. We here consider two evaluation criteria since finding optimal sequences is quite a hard task. One criterion is the sum of steps gained from the length of shortest reconfiguration sequences for all the solved instances. For instance, the instance `hc-square-02_01` has a shortest sequence of length 30 and a longest sequence of length 74. Thus, the number of steps gained is $74 - 30 = 44$. Another criterion is the number of instances for which at least one sequence was found. The comparison results between encodings are shown in Table 2. The best results are highlighted in bold. The first row displays all possible combinations of three hints `d1`, `d2`, and `h`. It is noted that the token hints `t1` and `t2` cannot be applied to longest-ISRP. For the former criterion, the

Table 2. The number of gained steps and solved instances for longest-ISRP

	no-hint	d1d2h	d1d2	d1h	d2h	d1	d2	h
#steps gained	4,995	**21,693**	8,058	12,456	20,321	6,656	7,318	12,044
#solved instances	189	221	**224**	194	213	213	223	191

Table 3. The result of the single-engine solver track in CoRe Challenge 2022

metric		1st	2nd	3rd
existent	solver name	*PARIS single*	*recongo*	@toda5603
	method	Planning	ASP	Greedy & BMC
	score	299 (275/24)	244 (238/6)	207 (207/0)
shortest	solver name	*recongo*	@tigrisg	*PARIS single*
	method	ASP	Brute force & SARSA	Planning
	score	238	232	213
longest	solver name	*PARIS single*	*recongo*	*ReconfAIGERation*
	method	Planning	ASP	SAT & BMC
	score	144	115	54

recongo encoding with the hints d1d2h has gain the most, namely 21,693 steps in a total sum. We can observe that the combination of d2 and h can drastically improve the performance for longest-ISRP solving. For the latter criterion, the *recongo* encoding with the hints d1d2 solved the most, but less steps gained.

CoRe Challenge 2022. Finally, we discuss the competitiveness of our approach by empirically contrasting it to the top-ranked solvers of the CoRe Challenge 2022 [25]. The competition consists of two tracks: solver track and graph track. The solver track is divided into three metrics. The *existent* metric is to decide the reachability of ISRP. Its evaluation index is the number of instances that contestants can solve. The *shortest* metric is to find reconfiguration sequences as short as possible of ISRP. Its evaluation index is the number of instances that contestants can find the shortest sequence among all contestants. The *longest* metric is to find reconfiguration sequences of as long as possible of ISRP. Its evaluation index is the number of instances that contestants can find the longest sequence among all contestants. Each metric is evaluated by two indices: single-engine solvers and overall solvers. The former index can be applied only to sequential solvers. The latter index can be applied to all solvers, including portfolio solvers as well as sequential solvers. The benchmark instances are the same as ones used in our experiments above. There are no restrictions on solvers used, time limits, or execution environments. Eight solvers (from seven groups) participated in the solver track of CoRe Challenge 2022.

Table 3 shows the results of the top-ranked solvers of single-engine solvers track. The columns display in order the metric and the winner (name, method, and score) for each rank. Our *recongo* solver ranked first in the shortest metric, ranked second both in the existent and longest metrics. Furthermore, *recongo* ranked second in the longest metric and ranked third in the shortest metric of overall solvers track. Overall, our declarative approach can be highly competitive in performance. On the other hand, we can observe that many top-ranked solvers are based on the techniques of BMC [2] and classical planning [20]. Besides them, ZDD-based solver [17] participated in the solver track.

Discussion. We discuss some more details of the results from a practical point of view. *recongo* showed good performance for the `color04` and `queen` families for all metrics. In particular, *recongo* was able to find the shortest reconfiguration sequences for all instances of `color04`. The `color04` family contains many instances that have relatively short reconfiguration sequences. In contrast, *recongo* is less effective for the `grid` family since most instances are unreachable. To resolve this issue, we will investigate the possibility of incorporating *the numeric abstraction* used in the *PARIS* solver [6] to our declarative approach.

6   Conclusion

We proposed an approach called bounded combinatorial reconfiguration to solving combinatorial reconfiguration problems. We presented the design and algorithm of bounded combinatorial reconfiguration based on ASP. We also presented an ASP encoding of the independent set reconfiguration problem. Our empirical analysis considered all 369 instances publicly available from the CoRe Challenge 2022 site. We showed the competitiveness of our approach by empirically contrasting it to other approaches. The resulting *recongo* system is an ASP-based CRP solver, which is available from https://github.com/banbaralab/recongo. In addition to independent set reconfiguration, *recongo* has been recently applied to Hamiltonian cycle reconfiguration [28].

The most relevant related fields are bounded model checking [2] and classical planning [20], in the sense of transforming a given state to another state. Bounded model checking in general is to study properties (e.g., safety and liveness) of finite state transition systems and to decide whether there is no sequence $X_s = X_0 \rightarrow X_1 \rightarrow X_2 \rightarrow \cdots \rightarrow X_\ell = X_g$, for which X_s is a start state and X_g is an error state expressed by rich temporal logic. Classical planning is to develop action plans for more practical applications and to decide whether there are sequences of actions for which X_s is a start state and X_g is a goal state.

In contrast, combinatorial reconfiguration is to study the structure and properties of solution spaces for given combinatorial problems and has several variations, such as *reachability, connectivity,* and *diameter.* This paper focused on the reachability problem that decides whether there are reconfiguration sequences for given two of feasible solutions X_s and X_g. When X_s and X_g are not given, the connectivity problem is to check whether the solution space is connected for any feasible solution. The diameter problem is to find the diameter of the solution space, that is, the maximum length of shortest reconfiguration sequences for any two of feasible solutions. From a broader perspective, combinatorial reconfiguration can involve the task of constructing problem instances that have the maximum length of shortest reconfiguration sequences. Such a distinctive task has been used at the graph track of CoRe Challenge 2022. On the other hand, combinatorial reconfiguration is a relatively new research field. Therefore, the relationship between those fields has not been well investigated, both from theoretical and practical points of view. We will investigate the relationship and will explore the possibility of synergy between techniques independently developed in those closely related research fields.

References

1. Baral, C.: Knowledge Representation, Reasoning and Declarative Problem Solving. Cambridge University Press, Cambridge (2003)
2. Biere, A.: Bounded model checking. In: Handbook of Satisfiability, pp. 457–481. IOS Press (2009)
3. Bonsma, P.S., Cereceda, L.: Finding paths between graph colourings: PSPACE-completeness and superpolynomial distances. Theoret. Comput. Sci. **410**(50), 5215–5226 (2009)
4. Brewster, R.C., McGuinness, S., Moore, B.R., Noel, J.A.: A dichotomy theorem for circular colouring reconfiguration. Theoret. Comput. Sci. **639**, 1–13 (2016)
5. Cereceda, L., van den Heuvel, J., Johnson, M.: Finding paths between 3-colorings. J. Graph Theory **67**(1), 69–82 (2011)
6. Christen, R., et al.: PARIS: planning algorithms for reconfiguring independent sets. In: Gal, K., Nowé, A., Nalepa, G.J., Fairstein, R., Radulescu, R. (eds.) Proceedings of the 26th European Conference on Artificial Intelligence (ECAI 2023). Frontiers in Artificial Intelligence and Applications, vol. 372, pp. 453–460. IOS Press (2023)
7. Erdem, E., Gelfond, M., Leone, N.: Applications of ASP. AI Mag. **37**(3), 53–68 (2016)
8. Gebser, M., et al.: Potassco User Guide, 2nd edn. University of Potsdam (2015). http://potassco.org
9. Gebser, M., Kaminski, R., Kaufmann, B., Schaub, T.: Answer Set Solving in Practice. Morgan and Claypool Publishers, San Rafael (2012)
10. Gebser, M., Kaminski, R., Kaufmann, B., Schaub, T.: Multi-shot ASP solving with clingo. Theory Pract. Logic Program. **19**(1), 27–82 (2019)
11. Gelfond, M., Lifschitz, V.: The stable model semantics for logic programming. In: Kowalski, R., Bowen, K. (eds.) Proceedings of the Fifth International Conference and Symposium of Logic Programming (ICLP 1988), pp. 1070–1080. MIT Press (1988)
12. Gopalan, P., Kolaitis, P.G., Maneva, E.N., Papadimitriou, C.H.: The connectivity of Boolean satisfiability: Computational and structural dichotomies. SIAM J. Comput. **38**(6), 2330–2355 (2009)
13. Haddadan, A., et al.: The complexity of dominating set reconfiguration. Theoret. Comput. Sci. **651**, 37–49 (2016)
14. van den Heuvel, J.: The complexity of change. In: Blackburn, S.R., Gerke, S., Wildon, M. (eds.) Surveys in Combinatorics 2013, London Mathematical Society Lecture Note Series, vol. 409, pp. 127–160. Cambridge University Press (2013)
15. Hirate, T., et al.: Hamiltonian cycle reconfiguration with answer set programming. In: Gaggl, S.A., Martinez, M.V., Ortiz, M. (eds.) JELIA 2023. LNCS, vol. 14281, pp. 262–277. Springer, Cham (2023). https://doi.org/10.1007/978-3-031-43619-2_19

16. Ito, T., et al.: On the complexity of reconfiguration problems. Theoret. Comput. Sci. **412**(12–14), 1054–1065 (2011)
17. Ito, T., et al.: ZDD-based algorithmic framework for solving shortest reconfiguration problems. In: Ciré, A.A. (ed.) CPAIOR 2023. LNCS, vol. 13884, pp. 167–183. Springer, Cham (2023). https://doi.org/10.1007/978-3-031-33271-5_12
18. Ito, T., Ono, H., Otachi, Y.: Reconfiguration of cliques in a graph. In: Jain, R., Jain, S., Stephan, F. (eds.) TAMC 2015. LNCS, vol. 9076, pp. 212–223. Springer, Cham (2015). https://doi.org/10.1007/978-3-319-17142-5_19
19. Kaminski, M., Medvedev, P., Milanic, M.: Complexity of independent set reconfigurability problems. Theoret. Comput. Sci. **439**, 9–15 (2012)
20. Kautz, H.A., Selman, B.: Planning as satisfiability. In: Proceedings of the 10th European Conference on Artificial Intelligence (ECAI 1992), pp. 359–363 (1992)
21. Mouawad, A.E., Nishimura, N., Pathak, V., Raman, V.: Shortest reconfiguration paths in the solution space of Boolean formulas. SIAM J. Discret. Math. **31**(3), 2185–2200 (2017)
22. Mouawad, A.E., Nishimura, N., Raman, V., Simjour, N., Suzuki, A.: On the parameterized complexity of reconfiguration problems. Algorithmica **78**(1), 274–297 (2017)
23. Niemelä, I.: Logic programs with stable model semantics as a constraint programming paradigm. Ann. Math. Artif. Intell. **25**(3–4), 241–273 (1999)
24. Nishimura, N.: Introduction to reconfiguration. Algorithms **11**(4), 52 (2018)
25. Soh, T., Okamoto, Y., Ito, T.: Core challenge 2022: Solver and graph descriptions. CoRR abs/2208.02495 (2022)
26. Suzuki, A., Mouawad, A.E., Nishimura, N.: Reconfiguration of dominating sets. J. Comb. Optim. **32**(4), 1182–1195 (2016)
27. Takaoka, A.: Complexity of hamiltonian cycle reconfiguration. Algorithms **11**(9), 140 (2018)
28. Yamada, Y., Banbara, M., Inoue, K., Schaub, T.: Recongo: bounded combinatorial reconfiguration with answer set programming. In: Gaggl, S.A., Martinez, M.V., Ortiz, M. (eds.) JELIA 2023. LNCS, vol. 14281, pp. 278–286. Springer, Cham (2023). https://doi.org/10.1007/978-3-031-43619-2_20

A Bisection Approach to Subcubic Maximum Induced Matching

Gordon Hoi[1(✉)], Sanjay Jain[2], Ammar Fathin Sabili[2], and Frank Stephan[2,3]

[1] School of Informatics and IT, Temasek Polytechnic, 21 Tampines Ave 1, Singapore 529757, Republic of Singapore
gordon_hoi@tp.edu.sg
[2] School of Computing, National University of Singapore, 13 Computing Drive, Singapore 117417, Republic of Singapore
{sanjay,fstephan}@comp.nus.edu.sg, athin2008@gmail.com
[3] Department of Mathematics, National University of Singapore, 10 Lower Kent Ridge Road, Singapore 119076, Republic of Singapore

Abstract. In this paper, we present a faster exact algorithm which solves the Maximum Induced Matching problem for subcubic graphs. Here let n be the overall number of vertices and k be the number of those vertices of degree 3 where all neighbours have also at least degree 2. Then the runtime is at most $O(1.2335^k) \cdot Poly(n)$, giving an FPT bound for the time used by the algorithm; the algorithm uses the result of Monien and Preis combined with a bound obtained by applying the measure and conquer technique where the number k replaces n as the measure used; note that $k \leq n$.

Keywords: Branch and Bound · Exponential Time Algorithms · Graph Theory · Measure and Conquer · Fixed Parameter Tractable Problems

1 Introduction

In this paper we will be dealing with only undirected graphs. Given a graph $G = (V, E)$, where V is its set of vertices and E is its set of edges. We let n denote the number of vertices in the graph G. For $U \subseteq V$, we let N(U) denote the set of neighbours of U in the graph G, that is $\{e \in V : (\exists e' \in U)[(e, e') \in E]\}$. We let $E(U)$ denote the set of edges in E with both endpoints in U.

For a graph $G = (V, E)$ and a subset $S \subseteq V$, the induced graph $G[S] = (S, E(S))$, has as set of edges all those members $(e, e') \in E$ where e, e' are both in S. S is said to be induced matching if every vertex in S has exactly one neighbour in S. The Maximum Induced Matching problem asks for the largest possible S such that S is an induced matching. This problem has a number of

G. Hoi (as RF), S. Jain (as Co-PI) and F. Stephan (as PI) were supported in part by the Singapore Ministry of Education Academic Research Fund Tier 2 grant MOE2019-T2-2-121/R146-000-304-112. Additionally S. Jain was supported by NUS grant E-252-00-0021-01.

R. Uehara et al. (Eds.): WALCOM 2024, LNCS 14549, pp. 257–272, 2024.
https://doi.org/10.1007/978-981-97-0566-5_19

practical applications, such as in VLSI design, network flow problems [9] and risk-free marriages [18]. The Maximum Induced Matching (MIM) problem is much harder than the similarly defined Maximum Matching problem where one asks only for a subset $F \subseteq E$ such that F is a matching (i.e., no vertex is an endpoint of two edges in F). While the MIM problem is **NP**-hard, the second problem can be solved in polynomial time [21]. Gupta, Raman and Saurabh [11] provided two subsequent algorithms to solve the MIM problem in time $O(1.6957^n)$ and $O(1.4786^n)$, respectively. Chang, Chen and Hung [2] improved the running time to $O(1.4658^n)$ and subsequently they improved it to $O(1.4321^n)$. Finally, Xiao and Tan [20] provided further improvements, running in time $O(1.4231^n)$ and $O(1.3752^n)$ where the last algorithm uses exponential space in contrast to the former ones which use polynomial space. The value $O(1.4231^n)$ is also the state of the art for the special case of graphs with maximum degree 4 [20, Lemma 11] with polynomial space; however, Xiao and Tan [20] did not comment on the complexity of solving the MIM problem for graphs with maximum degree 3 for polynomial space.

In this paper, we study a special case of the MIM problem that is limited to subcubic graphs, where each vertex has at most three neighbours. Prior algorithms that were used to solve the MIM problem on subcubic graphs were relying on applying the fastest Maximum Independent Set algorithm [19] on the line graph of G^2 ($L(G^2)$) to obtain a running time of $O(1.3139^n)$, using polynomial space.

We present a faster exact algorithm to solve the MIM problem for subcubic graphs in polynomial space and $O(1.2335^n)$ time; this paper supersedes the technical report [12] by solving the MIM problem for subcubic graphs faster than the previous $O(1.2630^n)$ and with a simpler algorithm. To achieve this, we design a branch and bound algorithm in conjunction with the Measure and Conquer technique [4] and the bisection result of Monien and Preis (stated below); though Monien and Preis also provided bounds for graphs of degree 4, the bisection width of that subcase is so large that the approach chosen here cannot be adjusted to higher degrees in a competitive way.

If we allow for the use of exponential space to solve this problem, then the general pathwidth bounds of Fomin and Høie [5] imply that the algorithm runs in time $O(r^n)$ for every $r > 3^{1/6}$ and thus the problem can be solved in time $O(1.2010^n)$; Kumar and Kumar [16, Theorems 2 and 4] provide their own way to this and other results. Our algorithm will not match this exponential space bound but provides with time $O(1.2335^n)$ the best polynomial space algorithm so far.

The interested reader will find more information on exponential time algorithms in the textbooks of Gaspers [7] and of Fomin and Kratsch [6]. Families of graphs for which maximum induced matching problem is **NP**-hard has been studied in great detail by the following references [1, 3, 9, 10, 13, 14].

2   Definitions and Preliminaries

Now we proceed more formally to introduce the notations used in this paper.

Definition 1. A subcubic graph is a graph where each vertex has at most three neighbours. For any subcubic graph, we call a vertex W_1-vertex if it has degree three and all its neighbours have degree at least two. Otherwise, the vertex is called W_0-vertex.

For the naming of vertices in the present work, we use the following conventions: vertices are usually denoted by a, b, c, d, e where a is of degree one, b of degree two, c of degree three, d of degree two or three and e of degree zero to three. Furthermore, subscripts and superscripts can be used to have more names.

Definition 2. Given a subcubic graph (V, E), we also denote by W_1 (respectively W_0) the set of all the W_1-vertices (respectively W_0-vertices) in the graph. We say that there is a *meta-edge* between two W_1-vertices c_1 and c_2 if there is a sequence (possibly empty) of W_0-vertices $d_1, d_2, \ldots, d_t$ such that the graph has the edges $(c_1, d_1), (d_t, c_2)$ and (d_i, d_{i+1}) for i with $1 \le i < t$ (in case t is 0, then there is an edge between c_1 and c_2). The meta-edges between c_1 and c_2 are denoted by $[c_1, c_2]$, $[c_1, c_2]'$ and $[c_1, c_2]''$ where the primed versions are only used if there are several such meta-edges and one needs to distinguish them.

The (multi) graph of all W_1-vertices and meta-edges between these vertices is called the meta-graph associated to (V, E). Note that the meta-graph might be a "multi-graph", as there maybe multiple meta-edges between two vertices in the meta-graph.

Sometimes, we also consider the meta-edge (as in above) as the sequence of edges $(c_1, d_1), (d_1, d_2), \ldots, (d_t, c_2)$ and denote/write it as $c_1 - d_1 - d_2 \ldots d_t - c_2$. We also sometimes say that the meta-edge $[c_1, c_2]$ contains/consists of the edges $(c_1, d_1), (d_1, d_2), \ldots, (d_t, c_2)$ or contains the vertices $c_1, d_1, \ldots, d_t, c_2$.

If they exist, $[c_1, c_2]$, $[c_1, c_2]'$ and $[c_1, c_2]''$ have only the end vertices c_1 and c_2 in common. We note that there are at most three meta-edges between c_1, c_2 and if there are three, then c_1 and c_2 are the only W_1-vertices in their component of the graph.

If S is a subset of V then S defines an induced matching of G if and only if every vertex in S has exactly one neighbour in S. For the ease of notations, we also say that an edge is in S if and only if it is in E and both its endpoints are in S. As the matching is induced, the endpoints of distinct edges in S are not neighbours in (V, E), that is, not connected by an edge in E.

Note that given a graph G, we can find its W_1-vertices and the meta-edges between them in polynomial time; meta-edges are also called paths and, more precisely, they are paths whose endpoints are W_1-vertices and where no inner node of the path is a W_1-vertex. As we can easily detect a degree one neighbour of a degree three W_0-vertex, one can easily find all the meta-edges starting from any W_1-vertex (to some W_1-vertex, including possibly itself). There are at most three meta-edges between two W_1-vertices and the latter happens if and only if these W_1-vertices are only connected to each other through meta-edges.

Furthermore, Monien and Preis showed the following.

Theorem 3 (Monien and Preis [17]**).** *For every $\varepsilon > 0$, there is a number k_ε such that for all subcubic graphs with at least $n \geq k_\varepsilon$ vertices, a polynomial time algorithm finds a partition of the graph into two halves, with each half having at least $n/2 - 1/2$ vertices, such that there are at most $(1/6 + \varepsilon) \cdot n$ edges between the two halves. This set of edges is called the "bisection" of the graph. For graphs with less than k_ε vertices, the algorithm returns the optimal bisection by using table look-up.*

In our earlier technical report [12, Corollary 2], we used this fact to give the corresponding algorithm for meta-graphs where they originally run the algorithm for possible double-meta-edges between two W_1-vertices to be considered as single meta-edges and then adjust the result such that for each case of original double-meta-edges between two W_1-vertices they choose one of the neighbouring single meta-edges to replace it. Thus one W_1-vertex may be moved from one half of the bisection into the other. The W_1-vertex chosen to be moved is the one which makes the partition more balanced, that is, which keeps the difference in the number of W_1-vertices between the two sides of the bisection bounded by two. Thus by relaxing the permitted size difference to two instead of one (as in the original result of Monien and Preis), one can get that the bisection of the meta-graph does not have double-meta-edges.

The following proposition is straightforward.

Proposition 4 (Subset Principle). *Suppose $G = (V, E)$ is a graph whose MIM we need to find and that we are considering two alternatives for MIM,*

(a) *An induced matching S union with MIM of induced subgraph from $U \subseteq V$, where $(S \cup N(S)) \cap U = \emptyset$ and*

(b) *An induced matching S' union with MIM of induced subgraph from $U' \subseteq V$, where $(S' \cup N(S')) \cap U' = \emptyset$.*

Suppose further that above U, U', S, S' satisfy $U \subseteq U'$ and $|S| \leq |S'|$.

Then, one can ignore the alternative MIM computed by (a) above as it is bounded in size by the MIM obtained in (b).

Note that we use the above subset principle to cut down on some alternative expansions S, S' of a given preliminary matching (with associated remaining graph U, U' respectively).

Lemma 5. *Suppose we are considering the MIM problem for a (sub) graph $(U, E(U))$. Suppose $Y \subseteq X \subseteq U$ such that there are no edges of the form (e, e'), with $e \in X - Y$ and $e' \in U - X$. Suppose M is an MIM of U.*

Suppose $S \subseteq X$ such that each member of S has exactly one neighbour in S in the graph $(U, E(U))$. Suppose that,

(i) *$(Y - M) \cap S = \emptyset$, and*

(ii) *$|E(S)| - |E(M \cap X)| \geq$ number of vertices in $M \cap Y$ which have edges in $E(M)$ to vertices outside X.*

Then S is a part of some MIM in $(U, E(U))$.

Proof. Suppose M' be the set of vertices in $M - X$ which have an edge in $E(M)$ to a vertex in X. Let $S' = M' \cup (M \cap X)$. Note that $|S| \geq |S'|$. Also, $U - X - N(S)$ is a superset of $U - X - N(M \cup X \cup M')$. Lemma follows using the subset principle (Proposition 4). □

Intuitively, in the above Lemma, think of Y as "boundary of X", i.e., the only vertices in Y which have edges to vertices outside X.

Definition 6 (Simplification Rule SR). Consider a graph (U, E) in which we want to find an MIM S. Let e be a vertex with a neighbour d of degree at least two such that all other neighbours of d are of degree one (there may be one or two such neighbours, say a and perhaps a'). Then using the subset principle, it can be assumed without loss of generality that $d, a \in S$. Thus we can remove the vertices e, d, a, a' from further consideration for finding remaining part of MIM.

The above situation is known in the literature. It is very similar to the pending edge elimination of Xiao and Tan [20, Lemma 5] and implied by the Simplification Rule S3 of our earlier technical report [12]. One can adjust the notions of Xiao and Tan to cover this case fully by saying that d has pending edges if and only if d has neighbours of degree 1 and furthermore all neighbours of d of degree 2 or more form a clique – note that a single neighbour of degree at least 2 forms always a clique of size 1. In this situation, we can assume by arguments of Xiao and Tan [20, Lemma 5] that there is a maximum induced matching of the graph which contains an edge between d and one of its degree 1 neighbours. For completeness we give a proof below for soundness of the simplification rule.

Proposition 7. *Simplification Rule SR is sound.*

Proof. Suppose that S is an MIM. The case that $e \notin S$ is trivial. We now consider the case that e is in S. Then either d must be in S or some other vertex adjacent to e must be in S, along with e. Without loss of generality, we can assume that $\{e, d\} \subseteq S$ as the proof for either case is similar. Let $U = V - S - N(S) - \{a, a'\}$, since after the removal of S' and $N(S')$ from V, the vertices a and a' have 0 degree, and can thus be removed from consideration.

Now, consider $S' = S \cup \{d, a\} - \{e\}$. Let $U' = V - S' - N(S') - \{a'\}$, as the vertex a' becomes a 0 degree vertex after the removal of S and $N(S)$, and can thus be removed from consideration.

We have that $U \subseteq U'$ and $|S| \leq |S'|$, and thus from the Subset Principle given earlier, we can assume without loss of generality that $\{d, a\}$ is in MIM. □

Proposition 8. *(a) If a subcubic graph (component) is acyclic, then an MIM of the graph can be found in polynomial time in the number of vertices.*

(b) If a subcubic graph consists only of W_0-vertices, then an MIM of the graph can be found in time polynomial in the number of vertices.

(c) If a subcubic graph consists of at most a constant κ number of W_1-vertices, then an MIM of the graph can be found in time polynomial in the number of vertices (though it may be exponential in κ).

Proof. These three items are consequences of the fact that a graph with at most κ W_1-vertices has path width at most $(1/6 + \varepsilon)\kappa + O(1)$ for any $\varepsilon > 0$ and therefore the MIM problem can be solved in polynomial time for fixed κ where the usage of both the computation time and the space is exponential in κ [5,16]. For the sake of completeness, we give here the full proofs.

(a) Note that repeated use of Simplification Rule SR will leave the graph to be having only disconnected edges. Thus, we can find an MIM in polynomial time in the number of vertices.

(b) Without loss of generality assume that the graph is connected (otherwise consider each component separately). If the graph is acyclic, then part (a) gives us the result. If the graph has a cycle, then one can pick an edge (d, d') from it and branch the three cases that both d, d' are in MIM, or d is not in MIM or d' is not in MIM (both d, d' not being in MIM is covered by both of the last two cases). This gives at most three cases and after that one can use part (a) to solve the MIM problem in polynomial time for the graph.

(c) As the number of W_1-vertices is bounded by a constant κ, we can consider for each W_1-vertex d one of its neighbours d' and branch as in (b) for the cases that both d, d' are in MIM or d is not in MIM or d' is not in MIM. This gives a branching algorithm of complexity $O(3^\kappa) \cdot Poly(n)$. We assume that $\kappa \geq 2$ in order to also handle the pathological case that there are two W_1-vertices connected by three meta-edges – $\kappa \geq 2$ is sufficient as the graph component containing these two W_1-vertices is then isolated and does neither contain a further W_1-vertex nor a further meta-edge, see Definition 2.

This case-distinction completes the proof. $\square$

Note that for any δ with $0 < \delta < 1$, one can find $\varepsilon > 0$ and corresponding κ such that for all $m \geq \kappa$, $[\frac{m}{2} - 1]/[m(1/6 + \varepsilon)] \geq 3 - \delta$. We will take δ to be small enough value.

3   Overview of the Algorithm

Intuitively, we will be constructing a branching tree, where the root of the tree represents the original graph G with n vertices whose MIM we need to find. Nodes in the branching tree are of the form (G, U, S, B), where U is a subset of the vertices of G (the node then represents finding an MIM for the induced subgraph of G on the vertices U), S denotes a matching in G (not necessarily maximum) such that the vertices in S and their neighbours in G do not belong to U. B is a set of meta-edges between two partitions (initially obtained using Monien Pries algorithm) which are remaining to process when we are in the branching phase (more on this below). We implicitly assume that the two partitions associated with the bisection edges B are also kept (we omit mentioning them explicitly for ease of notation).

At a particular node (G, U, S, B) of the branching tree, the algorithm will first do some simplification of the problem (using the Simplification Rule SR mentioned above), and then, if needed, do branching. The branching would only be in the "branching phases", where initially a bisection is obtained using the Monien Preis method for bisecting and then the bisection meta-edges are removed one by one for each branching as we go down the branching tree (some of the meta-edges may get automatically removed due to the simplification rule).

It will always be the case that the MIM problem for the subgraph of G at a node (G, U, S, B) of the branching tree can be solved in polynomial time (in n) using the solutions of the MIM problem for the subgraphs at its children. Furthermore, the children of the nodes in the branching tree can be obtained in polynomial time (in n) from the subgraph represented by the node. Thus the overall complexity of finding an MIM of the graph G is within a polynomial factor of the number of leaves in the branching tree. Thus, we wish to bound the number of leaves in the branching tree.

Let μ denote our measure of complexity for a graph/subgraph (see more details in the next section). Then we let $T(\mu)$ denote the maximum number of leaf nodes generated by the algorithm when we have μ as the parameter for the input problem. Since the search tree is only generated by applying a branching rule, it suffices to consider the number of leaf nodes generated by these rules (as simplification rules take time only polynomial in n, the number of vertices in the original graph). To do this, we employ techniques in [15]. Suppose a branching rule has $r \geq 2$ children, with $t_1, t_2, \ldots, t_r$ reduction in the complexity measure for these children (compared to the measure of the subgraph at the parent). Then, any function $T(\mu)$ which satisfies $T(\mu) \geq T(\mu - t_1) + T(\mu - t_2) + \ldots T(\mu - t_r)$, with appropriate base cases, would satisfy the bounds for the branching rule. To solve the above linear recurrence, one can model this as $x^{-t_1} + x^{-t_2} + \ldots + x^{-t_r} = 1$. Let β be the root of this recurrence, where $\beta \geq 1$. Then any $T(\mu) = (\beta')^\mu$, with $\beta' \geq \beta$ would satisfy the recurrence for this branching rule. In addition, we denote the branching factor $\tau(t_1, t_2, \ldots, t_r)$ as β. Tuple $(t_1, t_2, \ldots, t_r)$ is also known as the branching vector [6]. If there are k branching rules, with branching factors $\beta_1, \ldots, \beta_k$, in the branch and bound algorithm, then the overall complexity of the algorithm can be seen as the largest branching factor among all k branching rules; i.e. $z = max\{\beta_1, \beta_2, \ldots, \beta_k\}$, and therefore the time complexity of the algorithm is bounded above by $O(z^\mu) \cdot Poly(n)$.

4   Measure and Conquer Algorithm

Assign weight 0 to each of the vertices in W_0, and weight 1 to each of the vertices in W_1. The measure μ for a graph is defined as the sum of the weights of all the vertices in the graph. In other words, μ for a graph is the number of vertices in W_1. By definition, $\mu \leq n$. Therefore, any algorithm solving in time $O(z^\mu)$ also runs in time $O(z^n)$ since $O(z^\mu) \subseteq O(z^n)$, for any $z > 1$.

We now state the main result of this work. Note that the proof implies the easier formula that the MIM of a subcubic graph with n vertices can be solved in $O(1.2335^n)$ time; the $Poly(n)$ is absorbed by the uprounding of the branching factors to 1.2335.

Theorem 9. *The MIM problem of a subcubic graph with n vertices out of which k are W_1-vertices can be solved in $O(1.2335^k) \cdot Poly(n)$ steps.*

Here also the constant κ from above is absorbed as a constant into the O-expression and κ is a constant independent of k. Next we give the algorithm.

Algorithm $MIM(G, U, S, B)$

Invariant: $G = (V, E)$ is the original graph. S consists of a matching in G such that in the original graph G, none of the vertices in S and none of their neighbours are in $U \subseteq V$. Thus, if S is a part of some maximum induced matching of G, then the union of S and any maximum induced matching of $(U, E(U))$ will give a maximum induced matching of G.

B if non-empty, denotes that the algorithm is in a branching phase, where the set of meta-edges in B bisects the W_1-vertices of U into two (nearly equal) parts. We will be branching and removing the meta-edges in the bisection B one by one.

1. Simplify the graph by setting $(G, U, S, B) = \text{Simplify}(G, U, S, B)$.
    The procedure Simplify() is given below this algorithm.
2. If $B = \emptyset$ and $(U, E(U))$ is the disjoint union of nonempty graphs (U_1, E_1) and (U_2, E_2) with no edge between U_1 and U_2 in (V, E) then solve both subgraphs independently by using $MIM(G, U_1, S, \emptyset)$ and $MIM(G, U_2, S, \emptyset)$ and return the union of the two matchings as the answer.
3. If $B = \emptyset$ and the induced subgraph over U has at most κ number of W_1-vertices, then compute an MIM $S1$ for the induced subgraph over U and return $S \cup S1$.
4. If $B = \emptyset$ and $(U, E(U))$ has more than κ W_1-vertices, then do a bisection based on the Monien Preis algorithm obtaining a bisection meta-edge set B' and return $MIM(G, U, S, B')$. This starts a branching phase.
5. If $B \neq \emptyset$ (we are in a branching phase), then pick one meta-edge $[c_1, c_2]$ in B where c_1, c_2 denote its endpoints.
    5.1. Suppose the bisection partitions are P_1 and P_2. If the component in U connected to $[c_1, c_2]$ on one side, say P_1 without loss of generality, consists of at most two W_1-vertices, then:
        Transfer this component in P_1 connected to $[c_1, c_2]$ to P_2. This move causes some meta-edges $B' \subseteq B$ to be from P_2 to P_2. Then return $(G, U, S, B - B')$ (i.e., the meta-edges in B' are removed). Note that this gives only one child in the branching tree. Note that at least one meta-edge $[c_1, c_2]$ was deleted.
    5.2. If the meta-edge $[c_1, c_2]$ contains a W_0-vertex with degree three, then pick one such vertex c, and let a be its neighbouring vertex of degree one. Then branch based on whether c will be in the MIM or not:
        – $MIM(G, U - \{c, a\}, S, B - \{[c_1, c_2]\})$ and
        – $MIM(G, U - \{c, a\} - \text{N}(c), S \cup \{c, a\}, B - \{[c_1, c_2]\})$.

The algorithm then returns the better answer among these two branches.

Note that, if in the second item above one considers that the neighbour of c used in the matching is not a but some other neighbour d of c, then not only c and $N(c)$ but also all $N(d)$ have to be removed from U, and while $N(a) = \{c\}$, $N(\{c,d\})$ is a proper superset of $N(\{c,a\})$. Thus by the subset principle only the two branchings above need to be considered and the above two branches are exhaustive for the current case in which the meta-edge $[c_1, c_2]$ contains a W_0-vertex of degree three.

Note that the simplification process (step 2) in the two children will then remove all the vertices in the meta-edge (and maybe more), and thus the meta-edge is deleted; similar comment applies for each of the cases below.

5.3.   If the meta-edge $[c_1, c_2]$ contains no W_0-vertices with degree three, then all the W_0-vertices in the meta-edge are of degree two. Suppose the meta-edge is c_1-b_1-b_2-...-b_t-c_2 where c_1, c_2 are the two bordering W_1-vertices and all vertices $b_1, \ldots, b_t$ have degree 2. For the ease of notation, we use b_0, b_{t+1} as aliases for c_1, c_2 in formulas for S_0, S_1, S_2, in 5.3.3 below, though these vertices do not have degree 2.

5.3.1 If $t = 0$, then branch based on whether both c_1, c_2 are in the MIM, or c_1 is not in the MIM or c_2 is not in the MIM (there is some overlap, when both c_1 and c_2 are not in the MIM but this is ok, as we are taking the best of the cases).

That is consider three children,

- $MIM(G, U - \{c_1, c_2\} - N(c_1) - N(c_2),\ S \cup \{c_1, c_2\},\ B - \{[c_1, c_2]\})$,
- $MIM(G, U - \{c_1\},\ S,\ B - \{[c_1, c_2]\})$,
- $MIM(G, U - \{c_2\},\ S,\ B - \{[c_1, c_2]\})$

and return the best of the three answers.

5.3.2 If $t = 1$, then branch based on whether (c_1, b_1) is an edge in the MIM, or (b_1, c_2) is an edge in the MIM, or b_1 is not in the MIM.

That is, consider three children:

- $MIM(G, U-\{c_1,b_1\}-N(c_1)-N(b_1), S\cup\{c_1,b_1\}, B-\{[c_1,c_2]\})$,
- $MIM(G, U-\{b_1,c_2\}-N(b_1)-N(c_2), S\cup\{b_1,c_2\}, B-\{[c_1,c_2]\})$,
- $MIM(G, U - \{b_1\}, S, B - \{[c_1, c_2]\})$,

and return the best of the three answers.

5.3.3 If $t \geq 2$, for $\ell \in \{0, 1, 2\}$, let $S_\ell = \{b_{3h+\ell}, b_{3h+\ell+1} : h \geq 0$ and $3h + \ell \leq t\}$, where we take $b_0 = c_1$ and $b_{t+1} = c_2$. Branch based on putting S_0 or S_1 or S_2 in S, and appropriately removing the corresponding vertices and their neighbours.

That is, for $\ell \in \{0, 1, 2\}$, let
$$U_\ell = \bigcup \{\{b_{3h+\ell}, b_{3h+\ell+1}\} \cup N(b_{3h+\ell}) \cup N(b_{3h+\ell+1}) : 3h + \ell \leq t, \ell \in \{0, 1, 2\}, h \geq 0\}$$
and consider the three children:

$MIM(G, U - U_\ell, S \cup S_\ell, B - \{[c_1, c_2]\})$ for $\ell \in \{0, 1, 2\}$.

The algorithm then returns the best answer among these three branches.

End Algorithm MIM

Function Simplify(G, U, S, B).

While at least one of the following three if-conditions applies to the graph
$(U, E(U))$ do the following three steps.

1. If there is a W_0-vertex c_1 in U such that (a) c_1 was a W_1-vertex before the
    previous step of the algorithm and was part of a meta-edge in B, and
    (b) there is a meta-edge $[c_2, c_3]$ in U which has c_1 as a vertex in it, and
    c_2, c_3 are in different partitions of the bisection B. Then, place $[c_2, c_3]$ in
    B. Intuitively, above replaces an old meta-edge in B, if c_2, c_3 above are in
    the different partitions of the bisection.
2. If there is a meta-edge $[c_1, c_2]$ in B such that some of the vertices on this
    meta-edge are not in U then remove this meta-edge from B.
3. If there is a vertex d of degree at least two which has a neighbour e of
    arbitrary degree and all other neighbours of d (say a and perhaps a') have
    degree 1 in U then update $S = S \cup \{a, d\}$ and $U = U - \{d, e, a, a'\}$. (Here
    $a' = a$ in the case that the degree of d is 2.)

End While

Return the updated (G, U, S, B).

End Function Simplify

5   Verification of the Properties of the Algorithm

Now we consider the analysis of the algorithm. We shall prove that the cases
are exhaustive, show that the algorithm preserves the optimality of the solution,
the correctness of the algorithm and, finally, establish the overall runtime of the
algorithm.

Proposition 10. *The case-distinction of the algorithm is exhaustive.*

Proof. In Step 1 of the algorithm, we do some cleaning via the simplify algorithm.
The aim is to (i) update the meta-edges in case some of the vertices in previous
step have been converted from degree 3 to degree 2 vertices or some of the
meta-edges no longer exist (see steps 1 and 2 of procedure Simplify), and (ii)
handle the case when there is a vertex d of degree at least two with exactly one
neighbour e of degree 2 or more (in which case we can without loss of generality,
using the Simplification Rule SR, assume that the MIM has the edge from d to
one of the other neighbours of d).

In Step 2, we deal with disconnected components of the graph. Hence, after
this line, we can assume that our graph is a connected graph.

In Step 3, if there are at most κ number of W_1-vertices, then we solve the
problem directly. After which, we can assume that any given instance has more
than κ number of W_1-vertices.

In Step 4, we apply the Monien Preis algorithm to obtain a bisection of meta-edges B and begin the branching phase.

In Step 5, we deal with the different cases of branching based on a selected meta-edge. Step 5.1 deals with the case that some component in one the partitions P_1 and P_2 has no more than three W_1-vertices. If the selected meta-edge contains a degree three W_0-vertex, then this case is handled by Step 5.2. Otherwise, the case is handled by Step 5.3, where we consider the cases of the number $t + 1$ of edges in the meta-edge being $t = 0$ (Step 5.3.1), $t = 1$ (Step 5.3.2) and $t \geq 2$ (Step 5.3.3).

Therefore, we have covered all cases. $\square$

Next, we show that algorithm preserves the optimality of the solution.

Proposition 11. *The algorithm generates a maximum induced matching.*

Proof. Note that the Simplification Rule SR clearly preserves optimality of the solution. Step 2 clearly preserves optimality as it works on different components of the graph separately. Step 3 explicitly computes optimal answer. Step 4 and Step 5.1 preserve optimality as the subgraph does not change. Step 5.2 preserves optimality as if c is in an MIM, then by the subset principle we can assume that (c, a) is in the MIM. Step 5.3.1 preserves optimality as we consider all possible cases of c_1, c_2 being in the MIM, where if both c_1, c_2 are in the MIM, then it must be via edge (c_1, c_2) being in the matching. Step 5.3.2 preserves optimality as we have considered all possible cases for (c_1, b_1) and (b_1, c_2) being in the matching. Now we consider Step 5.3.3.

The edges in the meta-edge $[c_1, c_2]$ are $(c_1, b_1), (b_1, b_2), \ldots, (b_t, c_2)$. Consider some MIM M of the subgraph U. For the case analysis below, we will use Lemma 5 with $X = \{c_1, b_1, \ldots, b_t, c_2\}$, and $Y = \{c_1, c_2\}$ (except for case 3.1.2 where X, Y are explicitly defined differently), to claim that one of S_1, S_2, S_3 is contained in some MIM for U, and thus optimality is preserved in step 5.3.3.

We now consider the following cases.

Case 1: Both c_1 and c_2 are not in M.

Case 1.1: $t+1 = 3h$. In this case by Lemma 5, there is an MIM of U containing S_1 as $E(S_1)$ gives maximum number of edges from $[c_1, c_2]$ without using c_1, c_2.

Case 1.2: $t + 1 = 3h + 1$. In this case $E(M)$ can have at most h edges from $[c_1, c_2]$, and both $E(S_1)$ and $E(S_2)$ achieve this without using either of c_1, c_2. Thus, by Lemma 5, there is an MIM of U containing S_1 or S_2.

Case 1.3: If $t+1 = 3h+2$, then $E(M)$ has at most h edges from $[c_1, c_2]$. $E(S_2)$ achieves h edges from $[c_1, c_2]$ without using either of c_1, c_2. Thus, by Lemma 5, there is an MIM of U containing S_2.

Case 2: Only one of c_1, c_2 is in M. Assume without loss of generality that c_1 is in M. Then, (b_1, b_2) and (b_t, c_2) are not in $E(M)$.

Case 2.1.1: $t + 1 = 3h$ for some h and $(c_1, b_1) \in E(M)$. Now, $E(S_0)$ achieves maximum number of edges in $[c_1, c_2]$ without using c_2. Thus, by Lemma 5, there is an MIM of U containing S_2.

Case 2.1.2: $t+1 = 3h$ for some h and $E(M)$ does not contain (c_1, b_1). In this case $E(M)$ can have at most $h - 1$ edges from $[c_1, c_2]$. Now, $E(S_1)$ has h edges from $[c_1, c_2]$ and does not contain c_2. Thus, by Lemma 5, there is an MIM of U containing S_1 (as the edge in M from c_1 can be dropped due to extra edge in $E(S_1)$).

Case 2.2: $t + 1 = 3h + 1$, for some h. In this case $E(M)$ can have at most h edges from $[c_1, c_2]$. $E(S_2)$ has h edges, and S_2 does not contain c_1, c_2, b_1. Thus, there is an MIM of U containing S_2 as we can replace the edges of $E(M)$ having at least one end point in $b_1, \ldots, b_t$ by edges of $E(S_2)$.

Case 2.3.1: $t + 1 = 3h + 2$ for some h and $(c_1, b_1) \in E(M)$. In this case $E(S_0)$ has at least the same number of edges from $[c_1, c_2]$ as $E(M)$ has from $[c_1, c_2]$. Also, S_0 does not contain c_2. By Lemma 5, there is an MIM of U containing S_0.

Case 2.3.2: $t + 1 = 3h + 2$ for some h and $(c_1, b_1) \notin E(M)$. In this case $E(M)$ can have at most h of the edges from $[c_1, c_2]$. $E(S_2)$ has h edges from $[c_1, c_2]$ and S_2 does not contain c_1, c_2, b_1, b_t. Thus, there is an MIM of U containing S_2 as we can replace the edges of $E(M)$ having at least one endpoint in $b_1, \ldots, b_t$ by edges of $E(S_2)$.

Case 3: Both c_1, c_2 are in M.

In this case (b_1, b_2) and (b_{t-1}, b_t) are not in $E(M)$.

Case 3.1.1: $t + 1 = 3h$ for some h and (c_1, b_1) is in $E(M)$. In this case $E(S_0)$ contains at least the same number of edges from $[c_1, c_2]$ as $E(M)$, and $c_2, b_t \notin S_0$. By Lemma 5, there is an MIM of U containing S_0.

Case 3.1.2: $t + 1 = 3h$ and (c_1, b_1) is not in $E(M)$. In this case if M has h edges from $[c_1, c_2]$, then it could be only be by M containing S_2. Otherwise, $E(M)$ has at most $h - 1$ edges from $[c_1, c_2]$. $E(S_2)$ has h edges from $[c_1, c_2]$ and it does not contain c_1, b_1. Thus, using Lemma 5 with $X = \{b_1, b_2, \ldots, b_t, c_2\}$ and $Y = \{b_1, c_2\}$, we have that there is an MIM containing S_2.

Case 3.2.1: $t + 1 = 3h + 1$ for some h and $E(M)$ has at least one of (c_1, b_1) or (b_t, c_2). In case $E(M)$ has $h+1$ edges from $[c_1, c_2]$, then S_0 is contained in $E(M)$. In case $E(M)$ has at most h edges from $[c_1, c_2]$, then as $E(S_0)$ has $h + 1$ edges and S_0 contains both c_1, c_2, by Lemma 5, we have that some MIM contains S_0.

Case 3.2.2: $t + 1 = 3h + 1$ for some h and $E(M)$ does not contain any of (c_1, b_1) or (b_t, c_2). In this case $E(M)$ has at most $h - 1$ edges from $[c_1, c_2]$. As $E(S_0)$ has $h + 1$ edges, using Lemma 5, we have that some MIM contains S_0.

Case 3.3.1: $t + 1 = 3h + 2$ for some h and $E(M)$ has at most h edges from $[c_1, c_2]$. Then, $E(S_2)$ has h edges from $[c_1, c_2]$ and S_2 does not contain c_1, c_2, b_1, b_t. Thus, there is an MIM of U containing S_2 as we can replace the edges of $E(M)$ having at least one endpoint in $b_1, \ldots, b_t$ by edges of $E(S_2)$.

Case 3.3.2: $t + 1 = 3h + 2$ for some h and $E(M)$ has $h + 1$ edges from $[c_1, c_2]$. Then, $E(M)$ must contain at least one of (c_1, b_1) or (b_t, c_2). In these cases, consider $E(S_0)$ and $E(S_1)$ respectively, which both contain $h + 1$ edges from $[c_1, c_2]$ and do not contain c_2 and c_1 respectively. Thus, using Lemma 5, we have that some MIM contains S_0 or S_1 respectively in this case.

Thus, step 5.3.3 preserves optimality.   □

Note that it follows from Propositions 10 and 11 that the algorithm is correct.

Proposition 12. *The running time of the algorithm is $O(1.2335^k) \cdot Poly(n)$.*

Proof. Note that the general idea of the algorithm works by repeatedly applying the Monien Preis algorithm to get a bisection and we branch on the meta-edges until the component has small enough number of W_1-vertices for us to apply Step 3 directly. Recall that the leaves of the branching tree are small enough sub-graphs of the graph G. We bound the overall runtime of the algorithm by bounding the number of leaves in the branching tree. For this, we bound the number of leaves $R(\mu, e)$, of the branching tree whose associated subgraph has a fixed vertex e of the graph G (here μ denotes the complexity measure, number of W_1-vertices in the graph). Thus, $\sum_{e \in V} R(\mu, e)$ will bound the total number of leaves. As mentioned earlier, the complexity of the algorithm is bounded by $Poly(n)$ times the number of leaves in the branching tree. As the number of vertices in V can be absorbed in $Poly(n)$, it is thus sufficient to bound $R(\mu, e)$, for each vertex e.

Now we explain the details. As mentioned above, branching happens only in phases, where initially using the Monien Preis method, the graph is partitioned in two parts (nearly equal), where the number of bisection edges is at most $m(1/6+\varepsilon)$, where m is the number of W_1-vertices in the corresponding subgraph.

For bounding $R(\mu, e)$, for arbitrary vertex e, we will consider only those branches of the computation tree whose subgraph contains e. In particular, if there are two disjoint components of the graph, only one will produce a sub-computation tree which contributes to $R(\mu, e)$. This can be used to calculate the branching factor of rules removing a bisection meta-edge. Only one side of the bisection will eventually contain e and therefore all W_1-vertices on the other side of the bisection can be counted as removed; these vertices will be counted in an amortised way per cutting of an meta-edge in the bisection.

Thus, in the analysis below, we only need to worry about the W_1-vertices from the partition corresponding to the vertex e as above (plus some of the W_1-vertices which were moved over from the other partion due to step 5.1 above). As there are at most $m(1/6 + \varepsilon)$ number of bisection edges and each partition consists initially of at least $m/2-1$ number of W_1-vertices, we can allocate credit of $(m/2-1)/(m(1/6+\varepsilon)) \geq 3-\delta$ for each deletion of bisection meta-edge, i.e., for each of the branching steps in step 5 of the algorithm. We call this AuxCredit in the analysis below. Note that this also means that for the other reductions in the number of W_1-vertices, we should consider only reductions due to W_1-vertices in the same partition side as e, since the reductions due to other partition has already been taken into account due to AuxCredit, and thus cannot be counted twice.

Now our aim is to inductively bound $R(s, e)$, the number of leaves of the subtree starting with s W_1-vertices which contain the vertex e among others.

We will show that $R(s, e) \leq p^s$, where $p = 1.2335$ is the above chosen strict upper bound of all the "branching factors" that are obtained in the cases below. Here the choice of 1.2335 as the upper bound follows the conventions, as usually only the first four digits after the decimal dot are considered, but any other upper bound would also do.

Suppose that at any node (with s W_1-vertices in the subgraph associated with it) in the branching tree we have reduction in number of W_1-vertices for its r children respectively as $\alpha_1, \alpha_2, \ldots, \alpha_r$ (where each α_i is positive) after the simplification process (step 1) in that child; we do this "after the simplification" for ease of our calculations. Note that this is fine as the simplification process does not do any branching. Note also that for bounding the value $R(s, e)$ when computing the reductions in $\alpha_1, \alpha_2, \ldots$, we need to consider only the reduction from "the same part in the bisection partition to which vertex e belongs", as we have already considered the vertices from the other side in AuxCredit; the subtrees produced by other branches will be counted in other terms $R(n, e')$.

Let $\gamma > 1$ be such that

$$\gamma^s = \sum_{i:1 \leq i \leq r} \gamma^{s - \alpha_i}$$

where this γ is represented as $\tau(\alpha_1, \alpha_2, \ldots, \alpha_r)$. Then, using any $\beta \geq \gamma$ would give a valid bound for $R(s, e)$ with respect to "this branching" in the branching tree.

Now, step 5.1 above has only one branch and is therefore a reduction rule. This will only take polynomial time in n.

In step 5.2, the reduction on each side of the partition reduces at least one W_1-vertex after the simplification process (the W_1-vertices at the end of the meta-edge $[c_1, c_2]$) and thus the reduction in the number of W_1-vertices is at least $4 - \delta$ (taking into account AuxCredit mentioned above) on each side, for each of the two cases. Thus, the branching factor is at least $\tau(4 - \delta, 4 - \delta)$.

In step 5.3, first note that following:

If the vertex c_1 (respectively c_2) is removed from the children subgraph, then either it causes another bisection edge to be deleted, or it causes at least 2 other W_1-vertices to be removed or become W_0-vertices in the same partition side as c_1 (respectively c_2) after the simplification process in the child nodes. To see this, consider the following exhaustive cases. If the two meta-edges (different from $[c_1, c_2]$) originating from c_1 both lead to a W_1-vertex c, then there must be a meta-edge from c to another W_1-vertex c'. After simplification process, as c_1 is removed, c is also removed and c' either gets removed or becomes a W_0-vertex. If the two meta-edges (different from $[c_1, c_2]$) originating from c_1 lead to two different W_1-vertices c and c', then after simplification process both c and c' either get removed or become W_0-vertices. The only remaining case is when the meta-edge from c_1 leads to itself, which is already considered in step 5.1. Thus, in each of the above possibilities, when c_1 (respectively c_2) gets removed from child subgraph, we get at least two additional decrease of W_1-vertices on the same side of the partition or a further reduction $3 - \delta$ due to AuxCredit for the additional deletion of bisection edge (in addition to c_1 being removed). Thus, besides the deletion of the bisection edge from c_1, there is at least an additional reduction of three W_1-vertices in this case.

Now we consider each of the substeps in step 5.3.

In step 5.3.1, clearly, c_1 (respectively c_2) is removed from at least 2 of the three children nodes. Thus, the branching factor is at least $\tau(4 - \delta, 6 - \delta, 6 -$

δ). Here $3 - \delta$ in each child is due to AuxCredit for the bisection edge to c_1 (respectively c_2), 1 in each child is due to c_1, c_2 becoming W_0-vertices or being removed from children, furthermore 2 in at least two of the children is due to two additional W_1-vertices being removed or becoming W_0-vertices as mentioned above or additional $3 - \delta$ AuxCredit due to another additional bisection edge being deleted.

In step 5.3.2, similarly, c_1 (respectively c_2) is removed in at least two of the children as both are neighbours of b_1. Thus, branching factor is at least $\tau(4 - \delta, 6 - \delta, 6 - \delta)$ as in the case of step 5.3.1.

In step 5.3.3, note that c_1 is removed from at least two of the children sub-graph as c_1 is a neighbour of b_1 which is in S_0 and S_1. Similarly for c_2 (though which of S_0, S_1, S_2 is used depends on $t \mod 3$). Thus, branching factor is at least $\tau(4 - \delta, 6 - \delta, 6 - \delta)$ as in the case of step 5.3.1.

Now $\tau(4, 4) < 1.1893$ and $\tau(4, 6, 6) < 1.2335$. Thus, as the worst branching factor is strictly below $p = 1.2335$, the value δ can be chosen to be small enough due to the gap above in the inequality ($\delta = 10^{-6}$ works).   □

6   Conclusion

The present work investigates the complexity of the MIM problem on subcubic graphs. For graphs of degree at most four, the best known polynomial space algorithm has the same bound as the overall best polynomial space algorithm using time $O(1.4231^n)$ [20] and any improvement to the special case of degree 4 would result in an improvement of the algorithms for the general problem as well. However, the case of graphs with maximum degree 3 has in general only been solved by invoking the maximum independent set of line graphs, a method which gives a $O(1.3139^n)$ time and polynomial space algorithm. The current work gives an improved use of the bisection method of Monien and Preis and obtains a polynomial space algorithm running in time $O(1.2335^n)$. Here, by adding a polynomial multiplicative factor $Poly(n)$, 1.2335^n can be replaced by 1.2335^k, where k is the number of those degree three vertices in the graph whose neighbours all have at least degree two.

References

1. Cameron, K.: Induced matchings. Discret. Appl. Math. **24**(1–3), 97–102 (1989)
2. Chang, M.-S., Chen, L.-H., Hung, L.-J.: Moderately exponential time algorithms for the maximum induced matching problem. Optim. Lett. **9**(5), 981–998 (2015)
3. Duckworth, W., Manlove, D.F., Zito, M.: On the approximability of the maximum induced matching problem. J. Discret. Algorithms **3**(1), 79–91 (2005)
4. Fomin, F.V., Grandoni, F., Kratsch, D.: A measure & conquer approach for the analysis of exact algorithms. J. ACM (JACM) **56**(5), 1–32 (2009)
5. Fomin, F.V., Høie, K.: Pathwidth of cubic graphs and exact algorithms. Inf. Process. Lett. **97**(5), 191–196 (2006)

6. Fomin, F.V., Kaski, P.: Exact Exponential Algorithms. In: Texts in Theoretical Computer Science (TTCS). An EATCS Series. Springer, Berlin, Heidelberg (2010). https://doi.org/10.1007/978-3-642-16533-7

7. Gaspers, S.: Exponential Time Algorithms: Structures, Measures, and Bounds, p. 216, VDM Verlag Dr. Müller (2010)

8. Gaspers, S., Sorkin, G.B.: Separate, measure and conquer: faster polynomial-space algorithms for Max 2-CSP and counting dominating sets. ACM Trans. Algorithms (TALG) **13**(4), 44:1–36 (2017)

9. Golumbic, M.C., Laskar, R.C.: Irredundancy in circular arc graphs. Discret. Appl. Math. **44**(1–3), 79–89 (1993)

10. Golumbic, M.C., Lewenstein, M.: New results on induced matchings. Discret. Appl. Math. **101**(1–3), 157–165 (2000)

11. Gupta, S., Raman, V., Saurabh, S.: Maximum r-regular induced subgraph problem: fast exponential algorithms and combinatorial bounds. SIAM J. Discret. Math. **26**(4), 1758–1780 (2012)

12. Hoi, G., Sabili, A.F., Stephan, F.: An algorithm for finding maximum induced matching in subcubic graphs. Technical report on http://www.arxiv.org/abs/2201.03220 (2022)

13. Ko, C.W., Shepherd, F.B.: Bipartite domination and simultaneous matroid covers. SIAM J. Discret. Math. **16**(4), 517–523 (2003)

14. Kobler, D., Rotics, U.: Finding maximum induced matchings in subclasses of claw-free and P 5-free graphs, and in graphs with matching and induced matching of equal maximum size. Algorithmica **37**(4), 327–346 (2003)

15. Kullmann, O.: New methods for 3-SAT decision and worst-case analysis. Theor. Comput. Sci. **223**(1–2), 1–72 (1999)

16. Kumar, A., Kumar, M.: Deletion to Induced Matching. Technical Report on https://arxiv.org/abs/2008.09660 (2020)

17. Monien, B., Preis, R.: Upper bounds on the bisection width of 3- and 4-regular graphs. J. Discret. Algorithms **4**(3), 475–498 (2006)

18. Stockmeyer, L.J., Vazirani, V.V.: NP-completeness of some generalizations of the maximum matching problem. Inf. Process. Lett. **15**(1), 14–19 (1982)

19. Xiao, M., Nagamochi, H.: Exact algorithms for maximum independent set. Inf. Comput. **255**, 126–146 (2017)

20. Xiao, M., Tan, H.: Exact algorithms for maximum induced matching. Inf. Comput. **256**, 196–211 (2017)

21. Micali, S., Vazirani, V.V.: An $O(V^{1/2}E)$ algoithm for finding maximum matching in general graphs. In: 21st Annual Symposium on Foundations of Computer Science (SFCS 1980), pp. 17–27. IEEE (1980)

Stable and Dynamic Minimum Cuts

Mark de Berg, Andrés López Martínez$^{(\boxtimes)}$, and Frits Spieksma

Department of Mathematics and Computer Science, TU Eindhoven,
Eindhoven, The Netherlands
{m.t.d.berg,a.lopez.martinez,f.c.r.spieksma}@tue.nl

Abstract. We consider the problems of maintaining exact *minimum cuts* and *ρ-approximate cuts* in dynamic graphs under the vertex-arrival model. We investigate the trade-off between the stability of a solution—the minimum number of *vertex flips* required to transform an induced bipartition into another when a new vertex arrives—and its quality. Trivially, in a graph with n vertices any cut can be maintained with $n/2$ vertex flips upon a vertex arrival. For the two problems, in general graphs as well as in planar graphs, we obtain that this trivial stability bound is tight up to constant factors, even for a clairvoyant algorithm—one that knows the entire vertex-arrival sequence in advance. When $ρ$ is relaxed more than certain thresholds, we show that there are simple and stable algorithms for maintaining a $ρ$-approximate cut in both general and planar graphs. In view of the negative results, we also investigate the quality-stability trade-off in the amortized sense. For maintaining exact minimum cuts, we show that the trivial $O(n)$ amortized stability bound is also tight up to constant factors. However, for maintaining a $ρ$-approximate cut, we show a lower bound of $\Omega(\frac{n}{ρ^2})$ average vertex flips, and give a (clairvoyant) algorithm with amortized stability $O\left(\frac{n \log n}{ρ \log ρ}\right)$.

Keywords: Dynamic Minimum Cut · Stability · Approximation

1 Introduction

Given an undirected graph $G = (V, E)$, a *cut* $(S, \overline{S})$ is a partition of V into two non-empty sets S and $\overline{S}$. The *size* or *value* of the cut, denoted by $w(S, \overline{S})$, is the total number of edges connecting a node in S with a node in $\overline{S}$. A *minimum cut*, or *min-cut*, is a cut with the smallest size. Finding such a cut is a classic combinatorial optimization problem and has numerous practical and theoretical applications [1]. Throughout, we call this problem MINIMUM CUT. For a positive integer k, graph G is said to be *k-edge connected* if every cut in G has size at least k. Let $λ(G)$ denote the value of a min-cut in G. For a parameter $ρ \geq 1$, a $ρ$-approximate cut $(X, \overline{X})$ is a cut with value at most $ρλ(G)$; that is, $w(X, \overline{X}) \leq ρλ(G)$.

This research was supported by the European Union's Horizon 2020 research and innovation programme under the Marie Skłodowska-Curie grant agreement no. 945045, and by the NWO Gravitation project NETWORKS under grant no. 024.002.003.

We consider the problem of maintaining an exact or approximate min-cut in the *vertex-arrival model*, where the graph G is subject to changes over time due to new vertices being inserted into G. Starting with the empty graph G_0, new vertices arrive *one by one* together with *all* their incident edges to previously arrived vertices, thus producing a sequence of graph instances $(G_0, G_1, \ldots, G_n)$ with n the number of vertices in G_n. We further assume that each graph in the sequence is connected. Traditionally, maintaining near-optimal solutions is the main objective in a dynamic setting. In practice, however, it may also be costly to implement the necessary changes to go from a valid solution at time i to a valid solution at time $i + 1$. As a result, we are also interested in the *stability* of the maintained solutions or how *different* consecutive solutions are from each other. Following the framework by De Berg *et al.*[2], we say that a dynamic algorithm is a *γ-stable ρ-approximation algorithm* if, upon each vertex arrival, at most γ changes are required to transform the currently maintained solution into a solution in the augmented graph, and each solution is a ρ-approximation.

To define the difference between consecutive solutions in dynamic graph cuts we use the notion of *vertex flips*. Let $(X, \overline{X})$ be a cut in a graph G, a *vertex flip* is the operation of a vertex v switching sides from X to $\overline{X}$ (or vice versa). Consider two consecutive graphs $G_i = (V, E)$ and $G_{i+1} = (V \cup \{v\}, E')$, and let $S_i = (X, \overline{X})$ and $S_{i+1} = (Y \cup \{v\}, \overline{Y} \setminus \{v\})$ be cuts in G_i and G_{i+1}, respectively. We say that the *difference* between S_i and S_{i+1} is the minimum number of vertex flips required to transform one cut into the other, and denote it by $D(S_i, S_{i+1}) = \min(\delta(S_i, S_{i+1}), |V| - \delta(S_i, S_{i+1}))$, where $\delta(S_i, S_{i+1}) = |X \cup Y| - |X \cap Y|$ is the cardinality of the symmetric difference $X \triangle Y$. (Equivalently, we may write $D(S_i, S_{i+1}) = \min(|X \triangle Y|, |X \triangle \overline{Y}|)$.) We remark that the newly arrived vertex v has no contribution to the calculation.

Related Work. In general, the challenge to maintain a (high-quality) solution to a dynamic problem while aiming to minimize changes to the solution, is known as optimization with bounded recourse. Here, the phrase "recourse" refers to the changes one is allowed to make to a solution. For various problems, results are known; we mention Gupta et al. [9] and Bernstein et al. [4] for work on maintaining matchings and flows, Imase and Waxman [13], Megow *et al.*[14] and Gu *et al.*[8] for work on maintaining (Steiner) trees and Hamiltonian cycles, and Feldkord *et al.*[5] and Han and Makino [10] for work on bin packing and knapsack. In many cases, the computational time spent in an iteration (the update time) is a relevant aspect of these works; in particular, for the min-cut problem results along these lines can be found in [6,7,12,15,16]. There is also work that focuses on the "difference" between two consecutive solutions, while not taking explicitly computational time into account; we mention Wasim and King [17] for work on MAX-CUT, and De Berg et al. [3] for work on independent and dominating set. We follow this latter line of work, i.e., given the definition of difference between two min-cuts as formulated above, we establish trade-offs between the stability of a solution and its quality.

Our Results. We study stable approximation algorithms for MINIMUM CUT in the vertex-arrival model. More precisely, we obtain lower and upper bounds on the stability of dynamic min-cuts on general graphs as well as planar graphs. The results are summarized in Table 1.

Table 1. Summary of results on γ-stability for MINIMUM CUT.

Graph class	Exact		ρ-Approximation	
	Lower bound	Upper bound	Lower bound	Upper bound
General	$\frac{n-1}{2}$	$\frac{n-1}{2}$	$\frac{n-2}{2}$ (for $\rho < \frac{n-2}{2} - 2$)	$\frac{n-1}{2}$ (for $\rho < \frac{n-2}{2} - 2$)
			0 (for $\rho > \frac{n-1}{2}$)	2 (for $\rho > \frac{n-1}{2}$)
Planar	$\frac{n-1}{2}$	$\frac{n-1}{2}$	$\frac{n-2}{2}$ (for $\rho < 5$)	$\frac{n-1}{2}$ (for $\rho < 5$)
			0 (for $\rho \geq 5$)	2 (for $\rho \geq 5$)
General (amortized)	$\frac{n}{16}$	$\frac{n-1}{4}$	$\Omega(n/\rho^2)$	$O\left(\frac{n \log n}{\rho \log \rho}\right)$

For general graphs, we show that an algorithm maintaining an exact minimum cut may need $\frac{n-1}{2}$ vertex flips in each iteration. This result is tight (as one can always change from one cut to another one using at most $\frac{n-1}{2}$ vertex flips), and applies to both the *oblivious setting*—when the algorithm has no knowledge of the vertex-arrival sequence other than the previously arrived elements—and the *clairvoyant setting*—when the algorithm is allowed to see the entire sequence of vertex arrivals in advance. Similar results apply to the special case of planar graphs. In contrast, the problem becomes trivial in trees (where starting from a tree consisting of a single edge, we can always keep the partition of the vertex set induced by that edge as the cut) and in complete graphs (where we can always keep the same vertex as one of the parts of the cut). For general graphs in the amortized case, we show that in order to maintain an exact minimum cut, $\Theta(n)$ vertex flips are needed.

We now turn to the case of maintaining a ρ-approximate cut. For general graphs, we show that similar to the exact case, an algorithm may need $\frac{n-2}{2}$ vertex flips in each iteration, but only when $\rho < \frac{n-2}{2} - 2$. In contrast, when $\rho > \frac{n-1}{2}$, we show that two vertex flips per iteration suffices to maintain a ρ-approximate cut. Both results are tight up to constant terms. Similar results apply for planar graphs when $\rho < 5$ and $\rho \geq 5$, respectively. Finally, for general graphs in the amortized case, we show that to maintain a ρ-approximate cut at least $\Omega(n/\rho^2)$ vertex flips are needed. We accompany this result by giving a clairvoyant algorithm with amortized stability $O\left(\frac{n \log n}{\rho \log \rho}\right)$.

Roadmap. Like Table 1, the presentation of the results is split into two parts. First, in Sect. 2, we present the results about maintaining exact minimum cuts. Then, in Sect. 3, we discuss the results on maintaining ρ-approximate cuts. We conclude with some general remarks in Sect. 4.

2   Maintaining Exact Minimum Cuts

We start with the *oblivious* setting, in which an algorithm has no knowledge of the input sequence, other than the previously arrived vertices. We use $deg(v)$ to denote the degree of a vertex v.

Theorem 1. *There is no exact γ-stable algorithm for* MINIMUM CUT *in general graphs of size $n \geq 9$ such that $\gamma < \lfloor \frac{n-1}{2} \rfloor$.*

Proof. For every $n \geq 9$, we present a sequence of graph instances $(G_1, \ldots, G_n)$ (see Fig. 1) for which an exact stable algorithm requires at least $\lfloor \frac{n-1}{2} \rfloor$ vertex flips to maintain an exact minimum cut. Let G_{n-1} be the graph consisting of two cliques A and B, where $|A| = \lceil \frac{n-1}{2} \rceil$ and $|B| = \lfloor \frac{n-1}{2} \rfloor$, connected to each other by means of two edges (a_1, b_1) and (a_2, b_2) for arbitrary $a_1, a_2 \in A$ and $b_1, b_2 \in B$. The graph G_n has one more vertex u, which is connected by a single edge to an arbitrary vertex in clique A.

Note that for $n - 1 \geq 8$ the cliques A and B have at least four vertices, and so the only minimum cut for G_{n-1} is (A, B) which has value 2. When the vertex u arrives, $(\{u\}, A \cup B)$ is the unique minimum cut. Hence, any algorithm maintaining a minimum cut must move all $\lfloor \frac{n-1}{2} \rfloor$ vertices of B into the part of the cut containing A.   □

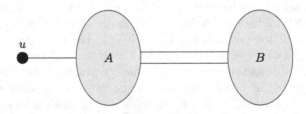

Fig. 1. Graph G_n for the proof of the lower bound in Theorem 1. Cliques A and B (gray) are connected by means of two edges between arbitrary vertices. Vertex u (black) is connected to an arbitrary vertex in clique A. After its arrival, any algorithm must perform $\frac{n-1}{2}$ vertex flips to maintain an exact minimum cut.

At first glance, it might appear that an algorithm that can react to an incoming update only with past information is too restrictive when trying to obtain a good quality-stability trade-off. However, even in a setting where the algorithm has access to the entire vertex-arrival sequence in advance—what we call, the *clairvoyant* setting—Theorem 1 still holds. Simply observe that, in the proof of Theorem 1, the algorithm must find solutions $S_{n-1} = (A, B)$ and $S_n = (\{u\}, A \cup B)$ at timesteps $n - 1$ and n respectively since these are the only available min-cuts in G_{n-1} and G_n, respectively.

Corollary 1. *There is no clairvoyant exact γ-stable algorithm for* MINIMUM CUT *in general graphs of size $n \geq 9$ such that $\gamma < \lfloor \frac{n-1}{2} \rfloor$.*

In search of better quality-stability trade-offs, we turn our attention to planar graphs. But like the general case, we obtain that $\Omega(n)$ vertex flips may be needed. This is tight with respect to the trivial upper bound of $\frac{n-1}{2}$ vertex flips.

Corollary 2. *There is no exact γ-stable algorithm for* MINIMUM CUT *in planar graphs such that $\gamma < \lfloor \frac{n-1}{2} \rfloor$, even in the clairvoyant setting.*

Proof. Similar to the proof of Theorem 1, but replacing the cliques A and B by two planar graphs of connectivity at least 3 (e.g., each a maximal planar graph on $\frac{n-1}{2}$ vertices). $\qquad\square$

Amortized Analysis. We saw in Theorem 1 that there are vertex arrival sequences where at least one iteration requires $\frac{n-1}{2}$ vertex flips to maintain an exact min-cut. It is natural to ask whether this behavior is only limited to a handful of iterations. If true, we could design an algorithm that, on average, requires only a few vertex flips per iteration. However, as the following shows, the average stability of maintaining an exact min-cut is not much better than the worst case: there exists a sequence of vertex arrivals such that each arrival induces $\Omega(n)$ many vertex flips.

Theorem 2. *There is no stable and exact algorithm for* MINIMUM CUT *in general graphs with amortized stability $\overline{\gamma} < \frac{n}{16} - 2$.*

Proof. For every $n \geq 32$, we present a sequence of graph instances $(G_1, \ldots, G_n)$ for which an exact stable algorithm requires at least $\frac{n}{16} - 2$ vertex flips *on average* (hence $\Omega(n^2)$ flips *in total*) to maintain an exact minimum cut. The graph G_n consists of three cliques A, B, and C of equal size $\frac{n+5}{4}$, connected to each other by the edges (a_1, b_1), (a_2, c_1), and (b_2, c_2), for arbitrary $a_1, a_2 \in A$, $b_1, b_2 \in B$, and $c_1, c_2 \in C$. Additionally, there is one more edge (a_3, c_3) for arbitrary $a_3 \in A$ and $c_3 \in C$, and two more edges (b_3, c_4), (b_4, c_5) for arbitrary $b_3, b_4 \in B$ and $c_4, c_5 \in C$. Graph G_n has $\frac{n-15}{4}$ additional vertices, denoted by set D, each of which shares an edge with every vertex in clique C. Moreover, D is partitioned into two disjoint sets D_1 and D_2 of equal size $|D|/2$, where each vertex in D_1 (resp. D_2) is connected to exactly two arbitrary vertices in clique A (resp. B). There are no more edges in G_n. See Fig. 2 for an illustration.

Consider the vertex arrival sequence $\sigma = (v_1, v_2, \ldots, v_n)$ where the vertices in $A \cup B \cup C$ all arrive in the prefix subsequence $\sigma_1 = (v_1, v_2, \ldots, v_\ell)$, with $\ell = \frac{3(n+5)}{4}$, and vertices in D arrive according to subsequence $\sigma_2 = (v_{\ell+1}, \ldots, v_n)$. Let σ_2 be a permutation of vertices in D such that $v_{\ell+i} \in D_1$ if i is odd, and $v_{\ell+i} \in D_2$ if i is even. We prove the main claim by showing that each vertex arrival in σ_2 induces $\frac{n+5}{4}$ vertex flips. The idea is to have the minimum cut *oscillate* between cuts $(A, \overline{A})$ and $(B, \overline{B})$ as vertices in D_1 and D_2 arrive alternately.

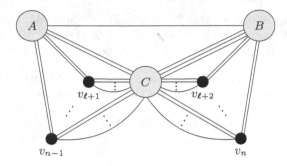

Fig. 2. Graph G_n for the proof of the lower bound in Theorem 2. Cliques A, B, and C are highlighted in gray. The vertices in set D are highlighted in black. Of these vertices, those to the left (resp. right) of clique C belong to the set D_1 (resp. D_2). The alternated arrival of vertices from D_1 and D_2 induce $\Omega(n)$ vertex flips per iteration.

First, we notice that after the ℓ-th vertex has arrived, cuts $(A, \overline{A})$, $(B, \overline{B})$ and $(C, \overline{C})$ have values 3, 4 and 5, respectively. Any other cut in the subgraph $G[A \cup B \cup C]$ must cross a clique and thus have value at least $\frac{n+1}{4}$. Thus, the min-cut at timestep ℓ is $S_\ell = (A, \overline{A})$. Next, after vertex $v_{\ell+1} \in D_1$ arrives, the value of cuts $(A, \overline{A})$ and $(C, \overline{C})$ increase by 2 and $\frac{n+5}{4}$ units, respectively; while the value of cut $(B, \overline{B})$ remains unchanged. Therefore, the min-cut at timestep $\ell+1$ becomes $S_{\ell+1} = (B, \overline{B})$. Similarly, after vertex $v_{\ell+2} \in D_2$ arrives, the value of cut $(B, \overline{B})$ increases by 2 units while cut $(A, \overline{A})$ remains unchanged, thus making $S_{\ell+2} = (A, \overline{A})$ the min-cut again; and so on for the remaining vertex arrivals in σ_2. The key observations are (i) that the min-cut at every timestep i is unique and has value less than $\frac{n+5}{4} - 1$, and (ii) $|w(A, \overline{A}) - w(B, \overline{B})| = 1$ is an invariant throughout the arrival sequence σ_2. From observation (ii), it follows that every vertex arrival in σ_2 increases the connectivity of the graph in a single unit. Now, by definition of G_n, we know that $\delta_{A,B} = D((A, \overline{A}), (B, \overline{B})) = \frac{n+5}{4}$. So, the total number of vertex flips performed for sequence σ_2 is $|\sigma_2| \cdot \delta_{A,B} = \frac{n-15}{4} \cdot \frac{n+5}{4}$. And averaged over the entire sequence σ, we obtain an amortized stability of at least[1] $\frac{n}{16} - 2$; which proves the theorem. ☐

Like Theorem 1, the lower bound of Theorem 2 is tight up to constant factors. To see this, simply consider a vertex arrival sequence where each update induces the maximum number of vertex flips at each iteration. Clearly, the amortized stability in this case is $\frac{n-1}{4}$.

3   Maintaining Approximate Cuts

We now consider the stability of maintaining approximate cuts. Theorem 1 shows that maintaining an exact solution is very expensive in terms of stability. Perhaps

---

[1] Because $\frac{n}{16} - 2 < \frac{n-15}{4} \cdot \frac{n+5}{4} \cdot \frac{1}{n} = \frac{n-10}{16} - \frac{75}{16 \cdot n}$ for any $n > 5$.

surprisingly, the following result shows that no better trade-off can be achieved for approximate solutions.

Theorem 3. *There is no γ-stable ρ-approximation algorithm for* MINIMUM CUT *in general graphs of size $n \geq 10$ such that $\rho < \lfloor \frac{n-2}{2} \rfloor - 2$ and $\gamma < \lfloor \frac{n-2}{2} \rfloor$.*

Proof. For every $n \geq 10$, we present a sequence of graph instances $(G_1, \ldots, G_n)$ for which a stable approximation algorithm requires at least $\lfloor \frac{n-2}{2} \rfloor$ vertex flips to obtain an approximation ratio less than ℓ where $1 < \ell \leq \lfloor \frac{n-2}{2} \rfloor - 2$ (when $\ell = \lfloor \frac{n-2}{2} \rfloor - 2$ the main claim follows)[2]. The graph G_n has two cliques A and B of roughly equal size such that $|A| \geq \lfloor \frac{n-2}{2} \rfloor$ and $|B| \geq \lfloor \frac{n-2}{2} \rfloor$, connected to each other by means of a single edge (a, b), for arbitrary $a \in A$ and $b \in B$. In addition, G_n has two more vertices u and w. Vertex w has $deg(w) = 2(\ell - 1)$ and shares half of its edges with arbitrary vertices from clique A and the other half with arbitrary vertices from clique B. Vertex u has $deg(u) = 1$ and is connected to an arbitrary vertex in clique A. See Fig. 3 for an illustration of graph G_n.

Consider any dynamic algorithm for maintaining a ρ-approximate cut and let S_i denote the cut maintained by the algorithm after the first i vertices have arrived. Consider the graph defined above, where the vertices in $A \cup B$ arrive in the first $n - 2$ timesteps, followed by vertex u at timestep $n - 1$ and w at timestep n. First, we show that at timestep $n - 2$—that is, right after the vertices in $A \cup B$ have arrived—the algorithm must maintain the cut (A, B) as the solution; *i.e.*, $S_{n-2} = (A, B)$. This follows from the fact that the graph G_{n-2} has a single min-cut of value 1—namely, the cut (A, B)—and any other cut in G_{n-2} has value at least $\lfloor \frac{n-2}{2} \rfloor - 1 > \ell$. Hence, only the cut (A, B) has approximation ratio less than ℓ.

We now show that at timestep n—after vertices u and w arrive—our algorithm will have performed $\lfloor \frac{n-2}{2} \rfloor$-many vertex flips. First, we observe that at timestep $n - 1$ (after vertex u arrives) the graph G_{n-1} presents only three ℓ-approximate cuts: the two min-cuts $(A \cup \{u\}, B)$ and $(\{u\}, A \cup B)$, and the 2-approximate cut $(A, B \cup \{u\})$. These are the only ℓ-approximate cuts because any other cut partitions clique A and/or clique B into two non-empty sets, thus cutting at least $\lfloor \frac{n-2}{2} \rfloor - 1$ edges. Thus, at timestep $n - 1$, our algorithm must pick one of these cuts as S_{n-1}). Next, after the final vertex w arrives at timestep n, only the (unique) min-cut $(\{u\}, A \cup B \cup \{w\})$ is a valid ρ-approximate cut[3]. Therefore, at timestep n, our algorithm must find $S_n = (\{u\}, A \cup B \cup \{w\})$. Now we show that no matter the choice for S_{n-1}, there is a timestep where the difference between two consecutive solutions is $\lfloor \frac{n-2}{2} \rfloor$.

Case 1. Let $S_{n-1} = (A \cup \{u\}, B)$. (The case for $S_{n-1} = (A, B \cup \{u\})$ is similar and is thus omitted.)

---

[2] Solving $1 < \lfloor \frac{n-2}{2} \rfloor - 2$ for integer n results in our stated bound of $n \geq 10$.

[3] Because any clique-crossing cut has value at least $\lfloor \frac{n-2}{2} \rfloor - 1 > \ell$ since $deg(v) \geq \lfloor \frac{n-2}{2} \rfloor - 1 \, \forall v \in A \cup B$. And any non-clique-crossing cut (except the min-cut) must cut at least ℓ edges: one edge shared by cliques and $\ell - 1$ edges shared by w with one of the cliques.

As mentioned above, the only valid solution at timestep n is $S_n = (\{u\}, A \cup B \cup \{w\})$, but $D(S_{n-1}, S_n) \geq \lfloor \frac{n-2}{2} \rfloor$; that is, cut S_n is at least $\lfloor \frac{n-2}{2} \rfloor$ vertex flips away from S_{n-1}. Therefore, at least $\lfloor \frac{n-2}{2} \rfloor$ vertex flips are needed at timestep n.

Case 2. Let $S_{n-1} = (\{u\}, A \cup B)$. In contrast to the previous case, the difference between consecutive solutions S_{n-1} and S_n here is $D(S_{n-1}, S_n) = 0$. However, the difference between S_{n-2} and S_{n-1} is $D(S_{n-2}, S_{n-1}) \geq \lfloor \frac{n-2}{2} \rfloor$, because $S_{n-2} = (A, B)$. Therefore, at least $\lfloor \frac{n-2}{2} \rfloor$ vertex flips are performed at timestep $n - 1$.

This proves that any algorithm on $(G_1, \ldots, G_n)$ requires at least $\lfloor \frac{n-2}{2} \rfloor$ vertex flips to find an ρ-approximate cut such that $\rho < \ell$. Since $\ell = \lfloor \frac{n-2}{2} \rfloor - 2$ in the worst case, the claim follows. □

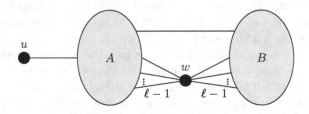

Fig. 3. Graph G_n for the proof of the lower bound in Theorem 3. Cliques A and B (gray) are connected by a single edge between arbitrary vertices. Vertex u (black) is connected to an arbitrary vertex in clique A, and vertex w (black) to $\ell - 1$ arbitrary vertices in A and B, respectively. After the arrival of both u and w, any dynamic algorithm must perform $\frac{n-2}{2}$ vertex flips to maintain a ρ-approximate cut.

Similar to the case of maintaining an exact minimum cut, a clairvoyant algorithm fares no better than an oblivious one.

Corollary 3. *There is no clairvoyant γ-stable ρ-approximation algorithm for* MINIMUM CUT *in general graphs of size $n \geq 10$ such that $\rho < \lfloor \frac{n-2}{2} \rfloor - 2$ and $\gamma < \lfloor \frac{n-2}{2} \rfloor$.*

Proof. This follows directly from the proof of Theorem 3. Observe that at timesteps $n - 2$ and n, respectively, the space of valid ρ-approximate cuts contains a single solution. Hence, even a clairvoyant algorithm is required to find solutions $S_{n-2} = (A, B)$ and $S_n = (\{u\}, A \cup B)$ at timesteps $n - 2$ and n respectively. The only freedom that the algorithm can exert is at timestep $n-1$, where the space of valid ρ-approximate cuts contains three possible solutions. But as we have proved, any of the three possibilities for S_{n-1} still lead the algorithm to make $\lfloor \frac{n-2}{2} \rfloor$ vertex flips in some timestep. □

Similarly, we have the following for planar graphs.

Corollary 4. *There is no γ-stable ρ-approximation algorithm for* MINIMUM CUT *in planar graphs such that $\rho < 5$ and $\gamma < \frac{n-2}{2}$, even in the clairvoyant setting.*

Proof. For every $n \geq 24$ divisible by 12, there is a planar graph H on n vertices with edge-connectivity five, and with at least six vertices incident to the *outer face* (see e.g. [11, Fig. 1]). Then, we can use the proof of Theorem 3 by replacing the clique clusters A and B with two copies of H on $\frac{n-2}{2}$ vertices (assuming that $\frac{n-2}{2}$ is divisible by 12) and setting $1 < \ell \leq 4$. Notice that, since the planar graph H has more than four vertices incident to the outer face, the arrived vertex w can indeed share at most one edge with each of these vertices while the overall graph remains planar. □

Amortized Analysis. Using a similar construction as in the proof of Theorem 2 it is not hard to obtain an $\Omega(\log n / \log \rho)$ lower bound on the amortized stability of maintaining a ρ-approximate cut in a graph. In the following, however, we derive an even better bound.

Theorem 4. *Any dynamic ρ-approximation algorithm for* MINIMUM CUT *in general graphs has average stability $\Omega(n/\rho^2)$, even in the clairvoyant setting.*

Proof. We present a sequence of graph instances $(G_1, \ldots, G_n)$ for which maintaining a ρ-approximate cut requires $\Omega(n/\rho^2)$ vertex flips on average. We assume that $\rho = o(\sqrt{n})$ since, otherwise, the theorem is trivial. In the following, we use $V(t)$ to denote the vertex set of graph instance G_t.

Consider the vertex arrival sequence $\sigma = (v_1, \ldots, v_n)$ where, at time $t = 2n/3$, the graph G_t consists of two cliques A and B of equal size $n/3$, with $\rho + 1$ edges between them. Notice that at this time, the cut $X(t) = (A, B)$ is a minimum cut, and is in fact the only ρ-approximate cut available. We partition the rest of the sequence $(v_{t+1}, \ldots, v_n)$ into $\frac{n}{3(1+(\rho+1)^2)}$ batches b_i of size $\ell = 1 + (\rho + 1)^2$. We will argue that for each batch, there is a sequence of vertex arrivals such that any algorithm must perform $\Omega(n)$ vertex flips.

Let $b_i = (v_{p(i)}, \ldots, v_{q(i)})$ denote the vertex arrival sequence of the i-th batch, with $p(i) = (t+1) + \ell \cdot (i-1)$ and $q(i) = p(i) + \ell$. The vertices arrive as follows. First, vertex $v_{p(i)}$ arrives with an edge to an arbitrary vertex in $V(p(i) - 1) \setminus A$ and no other incident vertices. At this point in time, the cut $X(p(i)) = (\{v_{p(i)}\}, V(p(i) - 1))$ is a minimum cut, and the only available ρ-approximate cut. Now, each new vertex arriving at time $j \in [p(i) + 1, q(i)]$ has edges to all other vertices in $V(j-1) \setminus A$. Notice that at the end of the batch—that is, at time $q(i)$—the cut $X(q(i)) = (A, V(q(i)) \setminus A)$ will be the only available ρ-approximate cut since any other cut must partition either the set $V(q(i)) \setminus A$, the set A, or both and thus has value at least $\rho + 1$. See Fig. 4 for an illustration of the graph after the arrival of vertex $v_{q(1)}$.

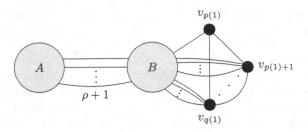

Fig. 4. Graph G_t for the proof of the lower bound in Theorem 4 at time $t = 2n/3 + \ell + 1$; that is, after the arrival of the vertices in the first batch $b_1 = (v_{p(1)}, \ldots, v_{q(1)})$. Cliques A and B are highlighted in gray, while vertices in b_1 are highlighted in black.

The claim that $\Omega(n)$-many vertex flips are required in a batch b_i follows from the fact that any algorithm must maintain cuts $X(p(i))$ and $X(q(i))$ at times $p(i)$ and $q(i)$, respectively. Then, by definition of difference between two cuts, we have that: (i) for any batch b_i we have $D(X(p(i)), X(q(i))) = |A|$, and (ii) for any two consecutive batches b_i and $b+i+1$ we have $D(X(q(i)), X(p(i+1))) = |A|$. (Notice that, at the start of the first batch b_1, we also have $D(X(t), X(p(1))) = |A|$, with $t = 2n/3$). In other words, in every batch, all the vertices in the set A must be flipped twice.

Now, since there are $\frac{n}{3(1+(\rho+1)^2)}$ batches and each one performs $\Omega(n)$-many vertex flips, the main claim follows. □

3.1 Improved Upper Bounds

Theorem 3 is tight with respect to the trivial upper bound of $\frac{n-1}{2}$ vertex flips. However, as Theorem 5 below shows, when the approximation factor of the maintained cut is large, very simple and stable algorithms exist. First, we prove the following lemma. We say that a cut $(X, \overline{X})$ is a *singleton* cut if one of X or $\overline{X}$ consists of a single vertex.

Lemma 1. *Any graph $G = (V, E)$ has an $\frac{n-1}{2}$-approximate cut that is a singleton.*

Proof. Consider the singleton cut induced by the vertex of minimum degree, and let $d_{\min}$ be its degree. Now consider an optimal cut (A, B). Define $m := |A|$ and assume without loss of generality that $m \leq n/2$. Since $d_{\min} \leq n - 1$, we are done when the value of a minimum cut is at least 2, so assume that it is 1. Note that A has at most $\binom{m}{2}$ internal edges, accounting for a total degree of $2 \cdot \binom{m}{2}$. Since each vertex in A has degree at least $d_{\min}$, we thus have at least $m \cdot d_{\min} - m(m-1)$ edges crossing the cut. But we just observe that the minimum cut value is 1, so $m \cdot d_{\min} - m(m-1) \leq 1$, which implies that $d_{\min} \leq (m-1) - \frac{1}{m} < \frac{n-1}{2}$. □

Lemma 1 immediately implies that there is a 2-stable $\frac{n-1}{2}$-approximation algorithm, namely the algorithm that maintains a singleton cut that gives a

$\frac{n-1}{2}$-approximation. (Note that switching between two singleton cuts requires at most two vertex flips.) Thus we obtain the following theorem.

Theorem 5. *There is a 2-stable $\frac{n-1}{2}$-approximation algorithm for* MINIMUM CUT *in general graphs.*

For planar graphs, a similar situation occurs when $\rho \geq 5$, since we know that any planar graph has a vertex of degree at most 5. A singleton cut consisting of such a vertex is thus a 5-approximate cut, and maintaining one such cut requires at most two vertex flips per iteration.

Theorem 6. *There is a 2-stable 5-approximation algorithm for* MINIMUM CUT *in planar graphs.*

Amortized Analysis. Contrary to the case of maintaining an exact minimum cut, the $\Omega(n/\rho^2)$ amortized lower bound of Theorem 4 is not tight with respect to the trivial $O(n)$ upper bound. We now reduce this gap by showing a new upper bound for maintaining a ρ-approximate cut in the clairvoyant setting. '

Theorem 7. *There exists a clairvoyant ρ-approximation algorithm for* MINI- MUM CUT *with amortized stability* $O\left(\frac{n \log n}{\rho \log \rho}\right)$.

For the sake of clarity, we introduce a slightly different notation. We use $v(t)$ to denote the vertex arriving at time t and let $G(t) := (V(t), E(t))$ represent the graph obtained after the arrival of vertex $v(t)$. We let OPT(t) denote the value of a minimum cut in $G(t)$, and let ALG(t) denote the value of the cut maintained by our algorithm at time t. To identify a cut in $G(t)$, we specify only the bipartition set $X(t) \subset V(t)$ that contains vertex $v(1)$, and use cost$(X(t))$ to denote its value in $G(t)$. We use $D(X(t), X(t+1))$ to denote the difference—as defined in the introduction—between cuts $X(t)$ and $X(t+1)$.

Next, we state two simple results that form the basis of our algorithm.

Lemma 2. *Let $X(t)$ be any cut in $G(t)$ and let $t' \leq t$. If $V(t') \not\subseteq X(t)$ then the set $Y(t') = V(t') \cap X(t)$ is a feasible cut in $G(t')$ with* cost$(Y(t')) \leq$ cost$(X(t))$.

Proof. Note that $V(t') \cap X(t) \neq \emptyset$ since $v(1) \in V(t') \cap X(t)$. Moreover, $V(t') \cap \overline{X}(t) \neq \emptyset$ since $V(t') \not\subseteq X(t)$. Hence, $Y(t')$ is a feasible cut. Further, the edges crossing the cut $Y(t')$ must be a subset of the edges crossing the cut $X(t)$, hence the value of the cut $Y(t')$ cannot be greater than that of $X(t)$. □

Lemma 3. *If* OPT$(t+1) <$ OPT(t)*, then the cut $X(t+1) = V(t)$ is the unique minimum cut in $G(t+1)$.*

Proof. For the sake of contradiction, suppose there is a cut $Y(t+1) \subset V(t)$ with cost$(Y(t+1)) \leq$ OPT$(t+1)$. But then the cut $Y(t) = Y(t+1) \setminus \{v(t+1)\}$ has cost$(Y(t)) <$ OPT$(t+1) \leq$ OPT(t), contradicting that OPT(t) is minimum. □

The Algorithm. Consider the sequence $OPT(2), \ldots, OPT(n)$ of minimum cut values at times $t = 2, \ldots, n$. We can partition the time interval $[2, n]$ into sub-intervals, or *phases*, I_i such that $OPT(t)$ is non-decreasing for all $t \in I_i$. Notice that, by Lemma 3, the minimum cut at the start of a phase is always a singleton cut. Now, let s be a parameter. With the aid of clairvoyance, the algorithm can distinguish between two types of phases: a *short phase*—when $|I_i| \leq s$—and a *long phase*—when $|I_i| > s$.

In a short phase $I_{\text{short}} = [t_{\text{start}}, t_{\text{end}}]$, the algorithm adopts the following simple strategy: For all $t \in I_{\text{short}}$, maintain the cut $X_{\text{alg}}(t) = V(t) \setminus \{v(t_{\text{start}})\}$.

Lemma 4. *For any short phase* $I_{\text{short}} = [t_{\text{start}}, t_{\text{end}}]$, *we have:*

1. $D(X_{\text{alg}}(t), X_{\text{alg}}(t + 1)) = 0$ *for all* $t \in [t_{\text{start}}, t_{\text{end}} - 1]$, *and*
2. $\text{ALG}(t) \leq (s - 1) \cdot \text{OPT}(t)$ *for all* $t \in I_{\text{short}}$.

Proof. The first part of the lemma trivially follows from the definition of cut difference and the fact that $X_{\text{alg}}(t + 1) \cap X_{\text{alg}}(t) = \{v(t)\}$. The second part follows from the fact that by Lemma 3, the starting cut $X_{\text{alg}}(t_{\text{start}})$ is minimum, and from observing that each vertex arrival after t_{start} can only increase the degree of $v(t_{\text{start}})$ in one unit. Putting this together with the fact that $\text{OPT}(t) \geq \text{OPT}(t_{\text{start}})$ for all $t \in I_{\text{start}}$ implies the result. $\square$

Now, let I_{long} be a long phase, and let ρ be the approximation guarantee we want to achieve with the algorithm. We define a *sub-phase* $I_{\text{sub}} = [t_{\text{start}}, t_{\text{end}}]$ of phase I_{long} as a maximal time interval such that $\text{OPT}(t_{\text{end}}) \leq \rho \cdot \text{OPT}(t_{\text{start}})$, with $t_{\text{start}}, t_{\text{end}} \in I_{\text{long}}$. (With a slight abuse of notation, we are re-using the notation t_{start} and t_{end} here, to also denote the start and end of a sub-phase.) Notice that there can be up to $O(\log n / \log \rho)$ sub-phases in a long phase. Ideally, we would like our algorithm to identify sub-phases in a long phase and for each sub-phase adopt the following strategy: For all $t \in I_{\text{sub}}$, maintain the cut $X_{\text{alg}}(t) = V(t) \cap X_{\text{opt}}(t_{\text{end}})$, where $X_{\text{opt}}(t_{\text{end}})$ is a minimum cut in $G(t_{\text{end}})$. This has the potential to grant us similar results to Lemma 4. The strategy, however, is flawed: the cut $X_{\text{alg}}(t)$ might be infeasible since there can be some time $t' \in I_{\text{sub}}$ for which $V(t') \subseteq X_{\text{opt}}(t_{\text{end}})$.

To refine this strategy, we further partition a sub-phase $I_{\text{sub}} = [t_{\text{start}}, t_{\text{end}}]$ into sub-intervals $I_{\text{sub}}^i = [t_i, t_{i+1})$ as follows. First, we let $t_0 := t_{\text{start}}$. Then, given t_i, we define t_{i+1} as the time immediately after the "furthest" time $t \in I_{\text{sub}}$ such that $X_{\text{opt}}(t)$—a minimum cut in $G(t)$—induces a feasible cut in $G(t_i)$. More formally, $t_{i+1} = 1 + \max\{t \mid t \leq t_{\text{end}} \text{ and } V(t_i) \not\subseteq X_{\text{opt}}(t)\}$. Now, as our new strategy, for each sub-interval $I_{\text{sub}}^i = [t_i, t_{i+1})$ let the algorithm perform the following: For all $t \in I_{\text{sub}}^i$, maintain the cut $X_{\text{alg}}(t) = V(t) \cap X_{\text{opt}}(t_{i+1} - 1)$.

Lemma 5. *For any sub-phase* $I_{\text{sub}} = [t_{\text{start}}, t_{\text{end}}]$ *of a long phase we have:*

1. $\sum_t D(X_{\text{alg}}(t), X_{\text{alg}}(t + 1)) = O(n)$ *for all* $t \in [t_{\text{start}}, t_{\text{end}} - 1]$, *and*
2. $\text{ALG}(t) \leq \rho \cdot \text{OPT}(t)$ *for all* $t \in I_{\text{sub}}$.

Proof. We start with the second part of the lemma. First, observe that $X_{\text{alg}}(t)$ is feasible throughout I_{sub}, since on each sub-interval $[t_i, t_{i+1})$ of I_{sub} we have that $X_{\text{opt}}(t_{i+1} - 1)$ induces a feasible cut on $G(t_i)$, hence also on $G(t)$ for all $t \in [t_i, t_{i+1})$. Now, for each sub-interval I_{sub}^i of I_{sub}, by Lemma 2 we have $\text{ALG}(t) \leq \text{OPT}(t_{i+1} - 1)$ for all $t \in I_{\text{sub}}^i$. But $\text{OPT}(t_{i+1} - 1) \leq \rho \cdot \text{OPT}(t_{\text{start}})$ for every sub-interval of I_{sub}. Hence, $\text{ALG}(t) \leq \rho \cdot \text{OPT}(t_{\text{start}})$ for all $t \in I_{\text{sub}}$.

Now we prove the first part of the lemma. Recall that $v(1) \in X_{\text{alg}}(t)$ for all $t \in I_{\text{sub}}$. First, observe that any vertex is placed into $X_{\text{alg}}(t)$ at most once during any given sub-interval. (It is simply assigned to the maintained set $X_{\text{alg}}(t)$ of the bipartition or its complement.) Now, let $I_{\text{sub}}^i = [t_i, t_{i+1})$ be a sub-interval of I_{sub}. We claim that a vertex $v \in X_{\text{alg}}(t_i)$ cannot be flipped out of $X_{\text{alg}}(t)$ in any t such that $t_{i+1} \leq t \leq t_{\text{end}}$. This follows because, otherwise, there would be a time $t' \geq t_{i+1} - 1$ such that $X_{\text{opt}}(t')$ induces a feasible cut in $G(t_i)$, which violates the condition that $t_{i+1} - 1$ was maximal. Therefore, a vertex can be flipped in the sub-phase I_{sub} at most once. Accounting for all vertices then gives the result. □

We are now ready to prove Theorem 7.

Proof of Theorem 7. The approximation ratios of short and long phases are $(s - 1)$ and ρ, respectively. Hence, the approximation ratio of the algorithm is $\max(s - 1, \rho)$. Now we analyze the stability of the algorithm. First observe that, by Lemma 4, there are no vertex flips performed in short phases. As for long phases, we know that each can have at most $O(\frac{\log n}{\log \rho})$ sub-phases and, by Lemma 5, each sub-phase performs at most $O(n)$ vertex flips in total. There are at most $\frac{n}{s}$ long phases, hence the total number of vertex flips performed by long phases is $\frac{n}{s} \cdot O(n \cdot \frac{\log n}{\log \rho}) = O(\frac{n^2 \log n}{s \log \rho})$.

We only have left to account for the number of vertex flips induced at the start of each phase and sub-phase; namely, when going from one phase (resp. sub-phase) to the next. Notice that, by Lemma 3, going from a short phase to another phase (either short or long) induces a single vertex flip. On the other hand, going from a long phase to a short phase, as well as from a long phase to another long phase, can each induce $O(n)$ vertex flips. Hence, the total number of vertex flips performed at the start of long phases is $O(\frac{n^2}{s})$. Finally, within each long phase, the total number of vertex flips performed when going from the end of one subphase to the beginning of the next subphase is $O(n \cdot \frac{\log n}{\log \rho})$. In total for every long phase then, we have $O(\frac{n^2 \log n}{s \log \rho})$ vertex flips.

Putting all this together, we get that the total number of vertex flips performed by the algorithm is $O(\frac{n^2 \log n}{s \log \rho})$. The result of the theorem then follows by setting $s = \rho$ and dividing the total number of vertex flips by n. □

4   Concluding Remarks

We studied the stability of dynamic algorithms for MINIMUM CUT under the vertex-arrival model. We showed that, for general and planar graphs, the trivial stability bound is tight up to constant factors in both the oblivious and clair-voyant settings. This holds for maintaining both exact and ρ-approximate cuts. When the approximation ratio satisfies $\rho \geq \frac{n-1}{2}$ in general graphs and $\rho \geq 5$ in planar graphs, we show that there are simple 2-stable ρ-approximation algo-rithms for MINIMUM CUT. In the amortized case, we also obtained that the trivial stability bound is tight up to constant factors, but only for the exact case. When maintaining ρ-approximate cuts, we showed that there are better-than-trivial average stability bounds—namely, a lower bound of $\Omega(n/\rho^2)$ and a clairvoyant algorithm with amortized stability $O\left(\frac{n \log n}{\rho \log \rho}\right)$.

The lower bound proofs in this work rely on specific constructions that may never show up in practice. We believe that situations in which a vertex inser-tion induces many vertex flips are rare. As such, the average case analysis of amortized stability of min-cuts seems like an interesting research direction. This is further motivated by the average-case results obtained in this work. Another promising approach toward improved stable approximation algorithms for min-cut is to consider graphs of bounded degree. Finally, we believe that exploring other problems from the viewpoint of stability is an interesting endeavor.

References

1. Ahuja, R., Magnanti, T., Orlin, J.: Network Flows: Theory, Algorithms, and Appli-cations. Prentice Hall, New York (1993)
2. de Berg, M., Sadhukhan, A., Spieksma, F.: Stable approximation algorithms for the dynamic broadcast range-assignment problem. In: 18th Scandinavian Symposium and Workshops on Algorithm Theory, pp. 15:1–15:21 (2022)
3. de Berg, M., Sadhukhan, A., Spieksma, F.: Stable approximation algorithms for dominating set and independent set. In: Approximation, Randomization, and Com-binatorial Optimization. Algorithms and Techniques (APPROX/RANDOM 2023), pp. 27:1–27:19 (2023)
4. Bernstein, A., Dudeja, A.: Online matching with recourse: random edge arrivals. In: 40th IARCS Annual Conference on Foundations of Software Technology and Theoretical Computer Science (FSTTCS 2020), pp. 11:1–11:16 (2020)
5. Feldkord, B., et al.: Fully-dynamic bin packing with little repacking. In: 45th Inter-national Colloquium on Automata, Languages, and Programming (ICALP 2018), pp. 51:1–51:24 (2018)
6. Goranci, G., Henzinger, M., Nanongkai, D., Saranurak, T., Thorup, M., Wulff-Nilsen, C.: Fully dynamic exact edge connectivity in sublinear time. In: Proceedings of the 2023 Annual ACM-SIAM Symposium on Discrete Algorithms (SODA), pp. 70–86. SIAM (2023)
7. Goranci, G., Henzinger, M., Thorup, M.: Incremental exact min-cut in polylog-arithmic amortized update time. ACM Trans. Algorithms (TALG) 14(2), 1–21 (2018)
8. Gu, A., Gupta, A., Kumar, A.: The power of deferral: maintaining a constant-competitive steiner tree online. In: Proceedings of the 45th Annual ACM Sympo-sium on Theory of Computing, pp. 525–534 (2013)
9. Gupta, A., Kumar, A., Stein, C.: Maintaining assignments online: matching, scheduling, and flows. In: Proceedings of the 25th Annual ACM-SIAM Sympo-sium on Discrete Algorithms, pp. 468–479. SIAM (2014)

10. Han, X., Makino, K.: Online minimization knapsack problem. In: Bampis, E., Jansen, K. (eds.) Approximation and Online Algorithms. WAOA 2009. LNCS, vol. 5893, pp. 182–193. Springer, Berlin, Heidelberg (2010). https://doi.org/10.1007/978-3-642-12450-1_17

11. Hasheminezhad, M., McKay, B.D., Reeves, T.: Recursive generation of simple planar 5-regular graphs and pentangulations. J. Graph Algorithms Appl. **15**(3), 417–436 (2011)

12. Henzinger, M., Noe, A., Schulz, C.: Practical fully dynamic minimum cut algorithms. In: Proceedings of the Symposium on Algorithm Engineering and Experiments (ALENEX), pp. 13–26 (2022)

13. Imase, M., Waxman, B.M.: Dynamic steiner tree problem. SIAM J. Discret. Math. **4**(3), 369–384 (1991)

14. Megow, N., Skutella, M., Verschae, J., Wiese, A.: The power of recourse for online MST and TSP. SIAM J. Comput. **45**(3), 859–880 (2016)

15. Thorup, M.: Fully-dynamic min-cut. Combinatorica **27**(1), 91–127 (2007)

16. Thorup, M., Karger, D.R.: Dynamic graph algorithms with applications. In: Scandinavian Workshop on Algorithm Theory, pp. 1–9 (2000)

17. Wasim, O., King, V.: Fully dynamic sequential and distributed algorithms for maxcut. In: 40th IARCS Annual Conference on Foundations of Software Technology and Theoretical Computer Science (FSTTCS 2020), pp. 33:1–33:19 (2020)

Black Hole Search in Dynamic Cactus Graph

Adri Bhattacharya[1], Giuseppe F. Italiano[2],
and Partha Sarathi Mandal[1]($^{(\boxtimes)}$)

[1] Indian Institute of Technology Guwahati, Guwahati, India
{a.bhattacharya,psm}@iitg.ac.in
[2] Luiss University, Rome, Italy
gitaliano@luiss.it

Abstract. We study the problem of black hole search by a set of mobile agents, where the underlying graph is a dynamic cactus. A black hole is a dangerous vertex in the graph that eliminates any visiting agent without leaving any trace behind. Key parameters that dictate the complexity of finding the black hole include: the number of agents required (termed as *size*), the number of moves performed by the agents in order to determine the black hole location (termed as *move*) and the *time* (or round) taken to terminate. This problem has already been studied where the underlying graph is a dynamic ring [6]. In this paper, we extend the same problem to a dynamic cactus. We introduce two categories of dynamicity, but still, the underlying graph needs to be connected: first, we examine the scenario where, at most, one dynamic edge can disappear or reappear at any round. Secondly, we consider the problem for at most k dynamic edges. In both scenarios, we establish lower and upper bounds for the necessary number of agents, moves and rounds.

Keywords: Black hole search · Dynamic cactus graph · Dynamic networks · Time-varying graphs · Mobile Agents · Distributed Algorithms

1 Introduction

We study the black hole search problem (also termed as BHS) in a dynamic cactus graph, where edges can reappear and disappear, i.e., goes missing over time so the underlying graph remains connected. More precisely, the network is a synchronous cactus graph where one of the vertices (or nodes) is a malicious node that eliminates any visiting agent without leaving any trace of their existence upon arrival on such nodes; that node is termed as *Black Hole* [7]. This scenario

A. Bhattacharya—Supported by CSIR, Govt. of India, Grant Number: 09/731(0178)/2020-EMR-I

P. S. Mandal—This work was done while Partha Sarathi Mandal was in the position of Visiting Professor at Luiss University, Rome, Italy.

frequently arises within networked systems, particularly in situations requiring the safeguarding of agents from potential host attacks. Presently, apart from the research paper concerning ring networks [6], there exists limited knowledge regarding this phenomenon when the network exhibits dynamic characteristics. Therefore, the focus of our study is to expand upon our findings in this context. In our investigation, we consider a collection of mobile agents, all of whom execute the same algorithm synchronously. Initially, these agents are positioned at a node that is confirmed to be free from any black hole threat; these nodes are referred to as 'safe nodes'. The primary objective is to efficiently determine the location of the black hole within the network in the shortest possible time while ensuring at least one agent survives and possesses knowledge of the black hole's whereabouts.

Related Work. The black hole search problem is well-studied in varying underlying topologies such as rings, grids, torus, etc. This problem has been first introduced by Dobrev et al. [7], in which they solved this problem in a static arbitrary topology. They have obtained tight bounds on the number of agents, while also establishing cost of a size-optimal solution. After this seminal paper, there has been a plethora of work done in this domain under different graph classes such as trees [3], rings [8], tori [16] and in graphs of arbitrary topology [2,7]. Now, in all the above literature, the study has been performed when the underlying graph is static.

While most of the study has been done on static networks, very little literature is known about black hole search especially in dynamic graphs. The study of mobile agents in dynamic graphs is a fairly new area of research. Previously, the problem of exploration has been studied in a dynamic rings [4,13], torus [12], cactuses [14] and in general graphs [11]. Di Luna et al. [5] investigated the gathering problem by mobile agents in a dynamic ring. Flocchini et al. [9] studied the black hole search problem in a different class of dynamic graphs, defined as periodically varying graphs, they showed the minimum number of agents required to solve this problem is $\gamma + 1$, where γ is the minimum number of carrier stops at black holes. Di Luna et al. [6], studied the black hole search problem in a dynamic ring, where they have established optimal algorithms in terms of number of agents, moves and rounds in two communication models. In this paper, we aim to solve similar problem, where we want to determine the position of a black hole with the least number of agents, but in our case, we have considered the underlying topology to be a dynamic cactus graph.

Our Contributions. We obtain the following results when the cactus graph has at most one dynamic edge at any round.

- Establish the impossibility of finding black hole in a dynamic cactus with 2 agents.
- With 3 agents we establish lower bound of $\Omega(n^{1.5})$ rounds, $\Omega(n^{1.5})$ moves, and we also establish upper bound of $O(n^2)$ rounds and $O(n^2)$ moves.
- With 4 agent improved lower bounds are $\Omega(n)$ rounds and $\Omega(n)$ moves.

Next, when the cactus graph has at most k $(k > 1)$ dynamic edges at any round.

- Establish the impossibility of finding the black hole with $k + 1$ agents.
- With $k + 2$ agents we establish lower bound of $\Omega((n + 2 - 3k)^{1.5})$ rounds and $\Omega((n + 2 - 3k)^{1.5} + 2k)$ moves.
- With $2k + 3$ agents improved lower bounds are $\Omega(n + 2 - 3k)$ rounds, $\Omega(n + 2 - k)$ moves, and we establish an upper bound of $O(kn)$ rounds and $O(k^2n)$ moves (Table 1).

Table 1. Summary of Results $(k > 1)$, where LB, UB and DE represent lower bound, upper bound and dynamic edge, respectively.

# DE	# Agents	Moves	Rounds	
1	3	$\Omega(n^{1.5})$	$\Omega(n^{1.5})$	LB (Cor 1 & Thm 2)
	3	$O(n^2)$	$O(n^2)$	UB (Thm 9)
	4	$\Omega(n)$	$\Omega(n)$	LB (Thm 3)
k	$k+2$	$\Omega((n + 2 - 3k)^{1.5} + 2k)$	$\Omega((n + 2 - 3k)^{1.5})$	LB (Cor 2 & Thm 6)
	$2k+3$	$\Omega(n + 2 - k)$	$\Omega(n + 2 - 3k)$	LB (Thm 7)
	$2k+3$	$O(k^2n)$	$O(kn)$	UB (Thm 10 & Thm 11)

Organization: Rest of the paper is organized as follows. In Sect. 2, we discuss the model and preliminaries. Section 3, we give the lower bounds. In Sect. 4, we present the algorithm and its correctness for both the single and multiple dynamic edge cases and finally conclude in Sect. 5. Due to the restrictions in the page limit, the pseudo codes of the algorithms, proofs of the theorems and lemmas, as well as a detailed explanation of *cautious* [6], *pendulum* [6] and *pebble* walk are omitted and can be found in the full version of this paper [1].

2   Model and Preliminaries

Dynamic Graph Model: We adapt the synchronous dynamic network model by Kuhn et al. [15] to define our dynamic cactus graph $\mathcal{G}$. The vertices (or nodes) in $\mathcal{G}$ are static, whereas the edges are dynamic, i.e., the edges can disappear (or in other terms go missing) and reappear at any round. The dynamicity of the edges holds as long as the graph is connected. The dynamic cactus graph $\mathcal{G} = (V, \mathcal{E})$ is defined as a collection of undirected cactus graphs $< G_0, G_1, \cdots, G_r, \cdots >$, where $G_r = (V, E_r)$ is the graph at round r, $|V| = n$ and $\mathcal{E} = \cup_{r=0}^{\infty} E_r$, where $|E_r| = m_r$ denotes the number of edges in G_r. The adversary maintains the dynamicity of $\mathcal{G}$, by disappearing or reappearing certain edges at any round r such that the underlying graph is connected. This model of dynamic networks is studied in [15] and is termed as a 1-interval connected network. The degree of a node $u \in \mathcal{G}$ is denoted by $deg(u)$, in other words, $deg(u)$ denotes the degree of the node u in G_0. The maximum degree of the graph $\mathcal{G}$ is denoted as Δ. The vertices (or nodes) in $\mathcal{G}$ are anonymous, i.e., they are unlabelled, although, the edges are labelled, an edge incident to u is labelled via the port numbers $0, \cdots, deg(u) - 1$. The ports are labelled in ascending order along the counter-clockwise direction, where a port with the port number i denotes the i-th incident edge at u in the counter-clockwise direction. Any edge $e = (u, v)$ is labelled by two ports, one among them is incident to u and the other incident to v, they have no relation in common. Any number of agents can pass through an edge concurrently. Each node in $\mathcal{G}$ has a local storage in the form of a whiteboard, where the size of the whiteboard at a node $v \in V$ is $O(deg(v)(\log deg(v) + k \log k))$, where $deg(v)$ is the degree of v and k is the maximum number of dynamic edge. The whiteboard is essential in order to store the list of port numbers attached to the node. Any visiting agent can read and/or write information corresponding to port numbers. Fair mutual exclusion to all incoming agents restricts access to the whiteboard. The network $\mathcal{G}$ contains a malicious node termed as *black hole*, which eliminates any incoming agent without leaving any trace of its existence.

Agent: Let $\mathcal{A} = \{a_1, \cdots, a_m\}$ be a set of $m \leq n$ agents, they are initially co-located at a safe node termed as *home*. Each agent has a distinct Id of size $\lfloor \log m \rfloor$ bits taken from the set $[1, m]$ where each agent is a t-state automata, with local storage of $O(n \log \Delta)$ bits of memory, where $t \geq \alpha n \log \Delta$ and α is any positive integer. The agents visiting a node know the degree of that node and can determine which of the edges has disappeared (or missing) at that particular node, based on the whiteboard information. The agent while moving from u to v along the edge e knows the port along which it left u and the port along which it entered v. Further, the agents can see the Ids of other agents residing at the same node and can communicate with them.

Round: The agents operate in *synchronous* rounds, where each agent gets activated in each round. At any time an agent $a_i \in \mathcal{A}$ gets activated, and they perform the following steps in a round: "Communicate-Compute-Move" (CCM), while it is active. The steps are defined as follows:

- *Communication:* Agents can communicate among themselves when they are at the same node at the same time. They also communicate via whiteboard. In this step, the agents can also observe their own memory.
- *Compute:* An agent based on the gathered information, local snapshot (i.e., information gathered on whether any other agent is present at the current node), internal memory and contents of the whiteboard, either decides to stay or choose the port number in case it decides to move.
- *Move:* The agent moves along the chosen port to a neighboring port, if it decides to move. While it starts to move, the agent writes the information in its memory and also writes on the whiteboard of its current node.

An agent takes one unit of time to move from a node u to another node v following the edge $e = (u, v)$.

Time and Move Complexity: Since the agents operate in synchronous rounds, each agent gets activated at each round to perform one CCM cycle synchronously. So, the time taken by the algorithm is measured in terms of *rounds*. Another parameter is *move* complexity, which counts the total number of moves performed by the agents during the execution of the algorithm.

Configuration: We define C_r to be the *configuration* at round r which holds the following information: the contents in the whiteboard at each node, the contents of the memory of each agent and the locations of the agent at the start of r. So, C_0 in the initial configuration at the start of an algorithm $\mathcal{H}$, whereas C_r is the configuration obtained from C_{r-1} after an execution of $\mathcal{H}$ at round $r-1$.

3   Lower Bound Results

In this section, we first study the lower bound on the number of agents, move and round complexities required to solve the BHS problem when at most one edge is dynamic in a round. Subsequently, we generalize this when at most k dynamic edges can be present in a round such that the underlying graph remains connected.

3.1   Lower Bound Results on Single Dynamic Edge

Here, we present all the results related to a dynamic cactus graph when at most one edge is dynamic at any round.

Theorem 1 (Impossibility for single dynamic edge). *Given a dynamic cactus graph $\mathcal{G}$ of size $n > 3$ with at most one dynamic edge at any round such that the underlying graph is connected. Let the agents know that the black hole is located in any of the three consecutive nodes $S = \{v_1, v_2, v_3\}$ inside a cycle of $\mathcal{G}$. Then it is not possible for two agents to successfully locate the black hole position. The impossibility holds even if the nodes are equipped with whiteboard.*

The above theorem is a consequence of Lemma 1 in [6].

Corollary 1 (Lower bound for single dynamic edge). *In order to locate the black hole in a dynamic cactus graph $\mathcal{G}$ with at most one dynamic edge at any round, any algorithm requires at least 3 agents to solve the black hole search problem in $\mathcal{G}$.*

Lemma 1 ([6]). *If an algorithm solves the black hole search problem in $O(n \cdot f(n))$ moves with 3 agents, then there exists an agent that explores a sequence of at least $\Omega(\frac{n}{f(n)})$ nodes such that:*

- *The agent does not communicate with any other agents while it explores the sequence.*
- *The agent visits at most $\frac{n}{4}$ nodes outside the sequence while exploring.*

In the next theorem, we give a lower bound on the move and round complexity required by any algorithm in order to solve the black hole search problem in a dynamic cactus.

Theorem 2. *Given a dynamic cactus graph $\mathcal{G}$, with at most one dynamic edge, any algorithm $\mathcal{H}$ which solves the black hole search problem with 3 agents requires $\Omega(n^{1.5})$ rounds and $\Omega(n^{1.5})$ moves, when the agents have distinct Ids, they are co-located and each node has a whiteboard.*

The above theorem is a consequence of Theorem 6 in [6]. The next theorem, gives an improved lower bound on the move and round complexity when 4 agents try to locate the black hole instead of 3.

Theorem 3. *Given a dynamic cactus graph $\mathcal{G}$, with at most one dynamic edge at any round. In the presence of whiteboards, any algorithm $\mathcal{H}$ solves the black hole search problem with 4 agents in $\Omega(n)$ rounds and $\Omega(n)$ moves when the agents have distinct Ids and they are co-located.*

The next observation gives a brief idea about the movement of the agents on a cycle inside a dynamic cactus graph. It states that, when a single agent is trying to explore any unexplored cycle, the adversary has the power to confine the agent on any single edge of the cycle. Moreover, in case of multiple agents trying to explore a cycle inside a cactus graph, but their movement is along one direction, i.e., either clockwise or counter-clockwise, then also the adversary has the power to prevent the team of agents from visiting further unexplored nodes.

Observation 4 ([6]). *Given a dynamic cactus graph $\mathcal{G}$, and a cut U (with $|U| > 1$) of its footprint connected by edges e_1 in the clockwise direction and e_2 in the counter-clockwise direction to the nodes in $V \setminus U$. If we assume that all the agents at round r are at U, and there is no agent which tries to cross to $V \setminus U$ along e_1 and an agent tries to cross along e_2, then the adversary may prevent agents to visit nodes outside U.*

The above observation follows from Observation 1 of [6]. The next lemma follows from the structural property of a cactus graph.

Lemma 2. *Consider three consecutive nodes $\{v_0, v, v_1\}$ in a cactus graph $\mathcal{G}$, then any path from u to v in $\mathcal{G}$ must pass through either v_0 or v_1.*

Proof. We prove the above claim by contradiction. Suppose there exists a u to v path which neither passes through v_0 nor v_1, so in order to have an alternate path which does not pass through v_0 or v_1, implies that there must be at least one edge or a path passing from half-1 to half-2 (refer to Fig. 1), where we define half-1 to be the subgraph on and above the horizontal half-line passing through v, whereas half-2 is the subgraph below the horizontal half-line passing through v. Now, the presence of such an edge or path, implies that there is at most one common edge between two cycles, and this violates the characteristic of a cactus graph. Hence, there cannot be any such u to v path which neither passes through v_0 nor v_1. □

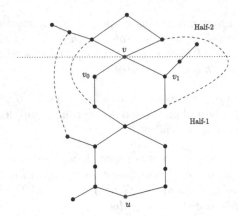

Fig. 1. Represents that any (u, v) path either passes through v_0 or v_1

3.2 Lower Bound Results for Multiple Dynamic Edges

Here, we present all the lower bound results for a dynamic cactus graph G, when at most k edges are dynamic at any round.

Theorem 5 (Impossibility for multiple dynamic edges). *Given, a dynamic cactus graph $\mathcal{G}$ with at most k dynamic edges at any round. It is impossible for $k + 1$ agents, to successfully locate the black hole position, regardless of the knowledge of n, k and the presence of a whiteboard at each node.*

Corollary 2 (Lower bound for k dynamic edges). *In order to locate the black hole in a dynamic cactus graph $\mathcal{G}$ with at most k dynamic edges at any round, any algorithm requires at least $k + 2$ agents.*

The next two theorems give lower bound and improved lower bound on move and round complexity with $k + 2$ and $2k + 3$ agents, respectively.

Theorem 6. *Given a dynamic cactus graph $\mathcal{G}$, with at most k dynamic edges at any round. In the presence of whiteboard, any algorithm $\mathcal{H}$ which solves the black hole search problem with $k + 2$ agents requires $\Omega((n + 2 - 3k)^{1.5})$ rounds and $\Omega((n + 2 - 3k)^{1.5} + 2k)$ moves when the agents have distinct Ids and they are co-located.*

Theorem 7. *Any algorithm $\mathcal{H}$ in the presence of whiteboard which solves the black hole search problem with $2k + 3$ agents in a dynamic cactus graph with at most k dynamic edges at any round, requires $\Omega(n+2-3k)$ rounds and $\Omega(n+2-k)$ moves when each agent have a distinct Id and they are co-located.*

4 Black Hole Search in Dynamic Cactus

In this section, we first present a black hole search algorithm in the presence of at most one dynamic edge, and then we present an algorithm to find the black hole in the presence of at most k dynamic edges. The number of agents required to find the black hole is presented and further, the move and round complexities are analyzed for each algorithm.

4.1   Black Hole Search in Presence of Single Dynamic Edge

In this section, we present two black hole search algorithms, one for the agents and the other for the $LEADER$, in the presence of at most one dynamic edge in a cactus graph. Our algorithms require each node v to have some local storage space, called *whiteboard*. Moreover, it requires 3 agents a_1, a_2 and a_3, respectively. Among these three agents, we consider a_3 to be the $LEADER$. The task of the $LEADER$ is different from the other 2 agents. More precisely, $LEADER$ follows SINGLEEDGEBHSLEADER(), whereas a_1 and a_2 follows SINGLEEDGEBH-SAGENT(). Next, we discuss the contents of the whiteboard.

Whiteboard: For each node $v \in \mathcal{G}$, a whiteboard is maintained with a list of information for each port of v. For each j, where $j \in \{0, \cdots, deg(v) - 1\}$, an ordered tuple $(f(j), Last.LEADER)$ is stored on it, where the function f is defined as follows: $f : \{0, \cdots, deg(v) - 1\} \longrightarrow \{\bot, 0, 1\}^*$,

$$f(j) = \begin{cases} \bot, & \text{if } j \text{ is yet to be explored by any agent} \\ 0 \circ A, & \text{if the set of agents in } A \text{ has visited } j \text{ but cannot} \\ & \text{fully explore the sub-graph originating from } j \\ 1, & \text{if the sub-graph corresponding to } j \text{ is fully explored} \\ & \text{and no agent is stuck} \end{cases}$$

The symbol "$\circ$" refers to the concatenation of two binary strings. We define A to be the set of agents which has visited that particular port. More precisely, if a_1 and a_2 both visits the port j, then we have $A = \{a_1, a_2\}$. We discuss the entries on the whiteboard with the help of the following example. Consider a port j at a node u, along which only a_1 has passed, but is unable to completely explore the sub-graph originating from j. In this case, the function $f(j)$ returns the binary string 001, where the first 0 represents that the sub-graph originating from j is not fully explored, and the next 01 represents the Id of a_1, so we have $A = \{a_1\}$.

The entry $Last.LEADER$ stores the bit 1 if j is the last visited port in v by the $LEADER$, otherwise, it stores 0.

Each agent (i.e., a_1 and a_2) performs a t-INCREASING-DFS [10], where the movement of a_1 and a_2 can be divided in to two categories *explore* and *trace*:

- In *explore*, an agent performs either *cautious* or *pendulum* walk depending on the instruction of the $LEADER$. In this case, an agent visits a node for the first time, i.e., it only chooses a port j, such that $f(j) = \bot$.
- In *trace*, an agent performs *pendulum* walk, where it visits a node that has already been visited by the other agent. In this case, an agent a_1 (or a_2) chooses a port j, where $f(j) = 0 \circ A$ and $A = \{a_2\}$ (or $A = \{a_1\}$).

The task of the *LEADER* can be explained as follows:

- Instructs a_1 or a_2 to perform either *cautious* or *pendulum* walk.
- Maintains the variables $length_{a_i}$ and P_{a_i} ($a_i \in \{1,2\}$), while a_i is performing *pendulum* walk. The *LEADER* increments $length_{a_i}$ by 1, whenever a_i traverses a new node and report this information to *LEADER*. Moreover, P_{a_i} is the sequence port traversed by a_i from the initial node from which it started its *pendulum* walk to the current explored node.
- It also maintains the variables L_{a_i} and PL_{a_i}, where L_{a_i} calculates the length of the path traversed by *LEADER* away from the initial position of a_i and PL_{a_i} stores the sequence of these ports.
- Lastly, terminates the algorithm whenever it knows the black hole position.

An agent a_1 or a_2 *fails to report* to *LEADER*, for one of the following reasons: it has either entered the black hole or it has encountered a missing edge along its forward movement. The algorithm SINGLEEDGEBHSLEADER() assigns only the *LEADER* to communicate with the remaining agents, in order to instruct them regarding their movements, whereas the agents a_1 and a_2 do not communicate among themselves. The agents a_i (where $i \in \{1,2\}$) stores P_{a_i} while performing *pendulum* walk and passes this information, whenever it meets with *LEADER*.

Next, we discuss in detail the description of the algorithm.

Algorithm Description: In this section, we give a high-level description of the algorithms SINGLEEDGEBHSAGENT() and SINGLEEDGEBHSLEADER().

Description of SINGLEEDGEBHSAGENT(): This algorithm is executed by the agents a_1 and a_2, in which they are either instructed to perform *cautious* or *pendulum* walk. The agents perform t-INCREASING-DFS [10] strategy for deciding the next port, further that port is indeed chosen by the agent for its movement based on the whiteboard information. Before commencing the algorithm, each port is labelled as $(\bot, 0)$. Initially, the agents start from *home*, where without loss of generality, a_1 is instructed to perform *cautious* and a_2 is instructed to perform *pendulum* walk. Let us consider a_1 is at a node u, then the decision taken by a_1 based on the contents of the whiteboard is as follows:

- If $\exists$ at least one port with $f()$ value as $\bot$ and i being the minimum among them, then choose that port and move to its adjacent node u' via i from u.
- If there is no such port i at u, with $f(i) = \bot$, then it backtracks accompanying the *LEADER* to a node where $\exists$ such i with $f(i) = \bot$.

Now, suppose a_1 reaches u' through the i-th port and it is safe, then it returns back to u in the subsequent round, on condition that the edge (u, u') remains. Otherwise, it stays at u' until the edge reappears. Now, while it returns back to u, if *LEADER* is found, then it accompanies *LEADER* to u' in the next round. Otherwise, it moves towards *LEADER* following *Last.LEADER* entries at the whiteboard.

Next, suppose a_2, is at a node v, then the decision taken by a_2 based on the whiteboard contents at v is as follows:

- If $\exists$ at least one port with $f()$ value as $\perp$ and i being the minimum among them, then choose that port and move to its adjacent node v' via i from v.
- If there is no such port i with $f(i) = \perp$, but $\exists$ a port j with $f(j) = 0 \circ A$, where $A = \{a_1\}$ then it chooses that port and moves to its adjacent node v'. Otherwise, if $A = \{a_2\}$ or $A = \{a_1, a_2\}$, then it chooses a different available port, or backtracks to a node where there exists an available port.
- If all ports have $f()$ value as 1, then it backtracks to a port where each port value is not 1.

Suppose, a_2 travels to v' using the port i, then it updates $P_{a_2} = P_{a_2} \cup \{i\}$, and after visiting v' it moves towards $LEADER$, based on the stored ports, and if it is unable to find the $LEADER$, then it follows $Last.LEADER$ to meet the $LEADER$. Now, whenever it meets the $LEADER$ it provides the sequence of ports P_{a_i} of the path it has traversed since it started *pendulum* walk to the $LEADER$. Moreover, if at any moment it encounters a missing edge and no other agent is waiting for that missing edge, then it waits until the edge reappears. Now, irrespective of *cautious* or *pendulum* walk, whenever an agent a_1 (or a_2) moves forward along a port i (say), it updates $f(i) = f(i) \circ a_1$ (or $f(i) = f(i) \circ a_2$) in whiteboard with respect to i.

Description of SINGLEEDGEBHSLEADER(): This algorithm is executed by the $LEADER$. It initially instructs a_1 to perform *cautious* and a_2 to perform *pendulum* walk. Whenever an agent, suppose a_2, while performing *pendulum* walk, explores a new node and meets with $LEADER$, it increments $length_{a_2}$ by 1 and stores the sequence of port P_{a_2} from a_2. On the other hand, if the $LEADER$ moves from its current position away from a_2, it increments L_{a_2} by 1 and updates the sequence of ports PL_{a_2} it has taken, after each such movement. Suppose, if the $LEADER$ moves away from a_i from its current node u to a node v along the port i. It does the following things: first, it updates $Last.LEADER$ at u corresponding to i as 1, while the rest to 0. Second, it increments L_{a_i} by 1 and lastly, updates $PL_{a_i} = PL_{a_i} \cup \{i\}$. Further, whenever $LEADER$ finds a missing edge along its path, it stops until the edge reappears. The instructions made by $LEADER$ related to the movement of a_1 and a_2 are as follows:

- If a_2 without loss of generality (w.l.o.g), fails to report while performing *pendulum*, but a_1 is performing *cautious* walk, then $LEADER$ instructs a_1 to perform *pendulum* walk.
- If $LEADER$ is stuck at one end of the missing edge, then it instructs both a_1 and a_2 to perform *pendulum* walk if not already performing the same.
- If $LEADER$ finds a dynamic edge reappear, while both a_1 and a_2 are performing *pendulum* walk, then it instructs either a_1 or a_2 to perform *cautious* walk, based on the fact which among a_1 or a_2 is faster to report to $LEADER$.

Next, we discuss the situations, when $LEADER$ decides that either a_1 or a_2 has entered the black hole and terminates the algorithm.

- If a_1 w.l.o.g, while performing *cautious* walk fails to return, but the edge between $LEADER$ and a_1 remains, then $LEADER$ identifies a_1 to be in black hole.

- If a_1 w.l.o.g, while performing *cautious* walk fails to return, but the edge between $LEADER$ and a_1 is missing, on the other hand a_2 which is performing *pendulum* walk also fails to return, then $LEADER$ identifies a_2 to be in black hole.
- If both a_1 and a_2 is performing *pendulum* walk, and $LEADER$ is stuck at one end of the missing edge. In this situation, if both a_1 and a_2 fails to report, then $LEADER$ moves towards the agent a_1 (or a_2) with the help of P_{a_1} and PL_{a_1} (or P_{a_2} and PL_{a_2}) based on the fact which among them is last to report to the $LEADER$, suppose that agent is a_1, then we have the following cases. If a_1 is found, then the $LEADER$ understands a_2 has entered the black hole. If a_1 cannot be found and there is no edge that is missing, then $LEADER$ identifies a_1 to be in black hole. If a_1 cannot be found, and there is an edge which is missing, then $LEADER$ waits, and if then also a_2 fails to report then $LEADER$ identifies a_2 to be in the black hole.

Correctness and Complexity: In this section, we prove the correctness of our algorithm, as well as give the upper bound results in terms of move and round complexity.

Lemma 3. *Given a dynamic cactus graph $\mathcal{G}$ with at most one dynamic edge at any round r. Our Algorithms* SINGLEEDGEBHSAGENT() *and* SINGLEEDGEBH-SLEADER(), *ensures that at most 2 agents are stuck due to a missing edge.*

Lemma 4. *The* $LEADER$ *following the algorithm* SINGLEEDGEBHS LEADER(), *ensures that among the 2 agents stuck due to a missing edge, eventually, one must be the* $LEADER$.

Lemma 5. *An agent a_1 or a_2 following the* SINGLEEDGEBHSAGENT(), *does not enter an infinite cycle.*

Lemma 6. *Algorithm* SINGLEEDGBHSAGENT(), *ensures that in the worst case, every node in $\mathcal{G}$ is explored by either a_1 or a_2 until it either enters the black hole or detects it.*

Observation 8. *Algorithm* SINGLEEDGEBHSLEADER(), *ensures that* $LEADER$ *does not enter the black hole.*

Lemma 7. *In the worst case, a_1 and a_2 executing algorithm* SINGLEEDGBH-SAGENT(), *enters black hole in $O(n^2)$ rounds.*

Lemma 8. *Let us consider, the agents a_1 or a_2 enter the black hole at round r while executing* SINGLEEDGEBHSAGENT(), *then the* $LEADER$ *following* SIN-GLEEDGEBHSLEADER() *detects the black hole node in $r + O(n^2)$ rounds.*

Lemma 9. *The* $LEADER$ *following the algorithm* SINGLEEDGEBHS LEADER() *correctly locates the black hole position.*

Proof. We discuss all possible scenarios that can occur while executing the algorithm SINGLEEDGEBHSLEADER().

a_1 **enters black hole at** u **during** *cautious* **walk:** In this case, if the edge $e = (u, v)$ between $LEADER$ and a_1 is missing, then $LEADER$ waits at v until e reappears. Meanwhile, a_2 is performing *pendulum* walk, and since there is at most one dynamic edge at any round, hence a_2 will eventually reach u, i.e., enter the black hole from an alternate path. Now, in this case, the $LEADER$ after waiting $2(length_{a_2} + 1 + L_{a_2})$ rounds, concludes that a_2 has entered black hole. The conclusion is indeed correct, as one end of the edge which is missing, is occupied by the $LEADER$ and the other end contains the black hole. Since there is at most one dynamic edge in $\mathcal{G}$ at any round. So, a_2 faces no obstruction while exploring the graph, until it enters the black hole. Hence, within $2(length_{a_2} + 1 + L_{a_2})$ rounds a_2 otherwise will have reported back to $LEADER$.

a_1 **enters black hole during** *pendulum* **walk:** a_1 which is initially performing *cautious walk*, then performs *pendulum* walk for two reasons.

a_2 fails to report: This situation arises when a_1 is initially performing *cautious* walk with $LEADER$, and a_2 which is performing *pendulum* walk, fails to report. In this scenario, the $LEADER$ instructs a_1 to perform *pendulum* walk, while it moves towards a_2 following P_{a_2} and PL_{a_2} for $L_{a_2} + length_{a_2} + 1$ round. Now we have two possibilities: first, a_2 is found and second a_2 is not found. If a_2 is found, then it is instructed to either perform *cautious* or *pendulum* walk, based on the fact that there is a forward missing edge or not. Now, in this situation, since a_1 has entered the black hole and fails to report, then $LEADER$ moves towards a_1 while instructing a_2 to perform *pendulum* walk. If the $LEADER$ does not find the edge missing corresponding to the last visited port of a_1, then it concludes that a_1 has entered the black hole. Otherwise, if $LEADER$ gets stuck at a node v due to a missing edge $e = (u, v)$, then it remains stationary, and while a_2 will either get stuck at u or eventually enter black hole. So, ultimately a_2 also fails to report and $LEADER$ not knowing the reason behind a_2's failure to report, moves towards a_2 leaving the node v. If a_2 is found, i.e., it got stuck at u then it correctly concludes a_1 has entered the black hole, whereas if $LEADER$ encounters another missing edge while moving towards a_2, then after waiting for an additional $2(L_{a_1} + length_{a_2} + 1)$ round, it correctly concludes a_1 has entered the black hole, as otherwise, a_1 will have reported to $LEADER$. Otherwise, if a_2 is not found, and there is no missing edge, then also $LEADER$ correctly concludes that a_2 has entered the black hole, as otherwise, a_2 will have reported back to $LEADER$.

a_1 and a_2 both performs *pendulum* walk : This situation has arised because a_2 initially stopped reporting due to a missing edge, while a_1 is performing *cautious* walk. Now, as $LEADER$ moves towards a_2 it instructs a_1 to change its movement to *pendulum* walk. Moreover, it finds a_2 and the edge has not reappeared, in this situation, a_2 is further instructed to continue *pen-*

dulum walk whereas $LEADER$ remains stationary at one end of the missing edge. If both a_1 and a_2 fail to return, then $LEADER$ identifies at least one among them has entered the black hole. It is because, $LEADER$ holds one end of the missing edge, and at most one agent may be stuck at the other end, whereas the remaining agent must return if it has not entered the black hole. So, $LEADER$ moves towards the last reported agent a_1 (say w.l.o.g). If a_1 is not found and there is no edge that is missing in the forward direction, then $LEADER$ correctly identifies the black hole position. Otherwise, if $LEADER$ finds a missing edge along its path towards a_1 and then it further waits for $2(length_{a_2} + 1 + L_{a_2})$ rounds, within which if a_2 also fails to return, then $LEADER$ concludes a_2 has entered the black hole. Since $LEADER$ has moved towards a_1 and encountered a missing edge, this implies that the earlier missing edge has reappeared. Hence, a_2 has no other obstruction towards $LEADER$, if it is originally stuck due to the earlier missing edge.

The explanation is similar when a_2 enters the black hole while performing either *cautious* or *pendulum* walk. So, we have shown that in each case the $LEADER$ correctly determines the black hole location.   □

Theorem 9. *The agent following algorithms* SINGLEEDGEBHSAGENT() *and* SINGLEEDGEBHSLEADER(), *correctly locates the black hole in a dynamic cactus graph* $\mathcal{G}$ *with at most one dynamic edge at any round with* $O(n^2)$ *moves and in* $O(n^2)$ *rounds.*

4.2   Black Hole Search in Presence of Multiple Dynamic Edges

In this section, we present a black hole search algorithm MULTIEDGEBHS() for the agents, where the underlying cactus graph can have at most k dynamic edges, such that the underlying graph remains connected. A team of $2k+3$ agents, $\mathcal{A} = \{a_1, \cdots, a_{2k+3}\}$ executes MULTIEDGEBHS(), where they start from a safe node also termed as *home*. Next, we define the content of information that can be present on a whiteboard.

Whiteboard: For each node $v \in \mathcal{G}$, a whiteboard is maintained with a list of information for each port of v. For each port j, where $j \in \{0, \cdots, deg(v) - 1\}$, an ordered tuple $(g_1(j), g_2(j))$ is stored on the whiteboard. The function g_1 is defined to be exactly the same as the function f in Sect. 4.1. On the other hand, the function g_2 is defined as follows, $g_2 : \{0, \ldots, deg(v) - 1\} \to \{\bot, 0, 1\}$,

$$g_2(j) = \begin{cases} \bot, & \text{if an agent is yet to visit the port } j \\ 0, & \text{if no agent has returned to the node } v \text{ along } j \\ 1, & \text{otherwise} \end{cases}$$

Each agent performs a t-INCREASING-DFS [10], where the movement of each agent can be divided into two types *explore* and *trace*:

- In *explore*, an agent performs either *cautious* walk or *pebble* walk depending on the situation.
- In *trace*, an agent walks along the safe ports of a node v, i.e., all such port $j \in v$ with $g_2(j) = 1$.

Our algorithm MULTIEDGEBHS() requires no *LEADER*, unlike our previous algorithms in Sect. 4.1. In this case, each agent executes their operations, based on the whiteboard information they gather at each node.

Next, we give a detailed description of the algorithm MULTIEDGEBHS().

Outline of Algorithm: The team of agents $\mathcal{A} = \{a_1, \ldots, a_{2k+3}\}$ are initially located at a safe node, termed as *home*. Initially, the whiteboard entry corresponding to each port at each node in $\mathcal{G}$ is $(\perp, \perp)$. The lowest Id agent present at *home*, i.e., a_1 decides to perform *cautious* walk. At the first round, it chooses the 0-th port and if the edge corresponding to 0-th port exists, then it moves along it to an adjacent node v while updating $(g_1(0), g_2(0))$ at *home* from $(\perp, \perp)$ to $(0 \circ a_1, 0)$. Now, if v is safe, and the edge $(home, v)$ exists then it returns to *home* at the second round, while updating $g_2(0)$ at *home* to 1, i.e., marking the edge $(home, v)$ as safe. Further, if at the third round the edge $(home, v)$ exists, then a_1 accompanies $\mathcal{A}\backslash\{a_1\}$ to v while updating $g_1(0) = g_1(0) \circ A'$, where $A' = \{a_2, \cdots, a_{2k+3}\}$. Otherwise, if the edge disappears at the second round, then a_1 and a_2 remains at v and *home*, respectively until the edge reappears, whereas the remaining agent continues to perform their respective movement.

Consider a scenario where the edge $(home, v)$ disappears at the third round when all the $\mathcal{A}$ agents are at *home*. In this case, a_1 remains at *home* until the edge reappears, whereas the remaining agents continue to perform *cautious* walk along the other available ports. Whenever the edge reappears, suppose at r-th round, then it starts *pebble* walk. The movement of a_1 performing *pebble* walk is as follows: at $r + 1$-th round a_1 moves to v, further the agent moves as follows:

- If there exists a port i at v with $g_2(i) = 0$, and the edge exists, then at the $r + 2$-th round a_1 stays at v. If no agent returns along i-th port at the $r + 2$-th round, a_1 concludes that the node w.r.to the port i is the black hole node. Otherwise, if an agent a_j returns (for some $j > 1$), then both a_1 and a_j start *cautious* walk. Moreover, if the edge does not exist, and there is no other agent at v waiting for the missing i-th port, then a_1 waits until the edge reappears. Otherwise, if there is already an agent waiting, then a_1 decides to move from v to some adjacent node or backtrack based on the available ports at v.
- If there exists a port i with $g_2(i) = \perp$, then at $r + 2$-th round, a_1 chooses that port and moves to the adjacent node (if the edge exists) while updating $(g_1(i), g_2(i))$ to $(0 \circ a_1, 0)$.
- If each port at v is having its $g_2()$ value as 1, then at $r + 2$-th round a_1 backtracks to *home*, if $(home, v)$ exists, otherwise stays at v until the edge reappears.

In general, whenever multiple agents meet at a node, they start *cautious* movement. Moreover, when a single agent waiting for a dynamic edge reappears, then it starts *pebble* walk. In addition, whenever an agent finds a port i at some node in $\mathcal{G}$ with $g_2(i) = 0$ and the edge exists, then after waiting for a round if no agent returns via the port i, then it concludes the adjacent node w.r.to i is the black hole node.

Correctness and Complexity: In this section we analyze the correctness and complexity of our algorithm MULTIEDGEBHS().

Lemma 10. *Given a dynamic cactus graph $\mathcal{G}$ with at most k dynamic edges at any round r. Our algorithm, MULTIEDGEBHS() ensures that at most 2 agents are stuck due to a dynamic edge at any round.*

Lemma 11. *In the worst case at most 2 agents are consumed by a black hole, while the agents are following the algorithm MULTIEDGEBHS().*

Lemma 12. *Our algorithm MULTIEDGEBHS() ensures that no agent enters an infinite cycle.*

Lemma 13. *MULTIEDGEBHS() ensures that any agent that is not stuck due to a missing edge can explore the remaining graph until it either enters the black hole or detects it.*

Theorem 10. *Given a dynamic cactus graph $\mathcal{G}$ with at most k dynamic edges at any round. Our algorithm MULTIEDGEBHS() ensures that it requires at most $2k + 3$ agents to successfully locate the black hole position.*

Theorem 11. *The team of agents $\mathcal{A} = \{a_1, a_2, \cdots, a_{2k+3}\}$ following MULTI-EDGEBHS(), locates the black hole in a dynamic cactus graph $\mathcal{G}$ with at most k dynamic edges at any round with $O(kn)$ rounds and in $O(k^2n)$ moves.*

5   Conclusion

In this paper, we studied the black hole search problem in a dynamic cactus for two types of dynamicity. We propose algorithms and lower bound and upper bound complexities in terms of a number of agents, rounds and moves in each case of dynamicity. First, we studied at most one dynamic edge case, where we showed with 2 agents it is impossible to find the black hole, and designed a black hole search algorithm for 3 agents. Our algorithm is tight in terms of number of agents. Second, we studied the case when at most k edges are dynamic. In this case, also we propose a black hole search algorithm with $2k + 3$ agents. Further, we propose that it is impossible to find the black hole with $k + 1$ agents in this scenario. A future work is to design an algorithm that has a tight bound in terms of a number of agents when the underlying graph has at most k dynamic edges. Further, it will be interesting to find an optimal algorithm in terms of complexity in both cases of dynamicity.

References

1. Bhattacharya, A., Italiano, G.F., Mandal, P.S.: Black hole search in dynamic cactus graph (2023). https://doi.org/10.48550/arXiv.2311.10984
2. Czyzowicz, J., Kowalski, D., Markou, E., Pelc, A.: Complexity of searching for a black hole. Fund. Inform. **71**(2–3), 229–242 (2006)
3. Czyzowicz, J., Kowalski, D., Markou, E., Pelc, A.: Searching for a black hole in synchronous tree networks. Comb. Probab. Comput. **16**(4), 595–619 (2007)
4. Di Luna, G., Dobrev, S., Flocchini, P., Santoro, N.: Distributed exploration of dynamic rings. Distrib. Comput. **33**, 41–67 (2020)
5. Di Luna, G.A., Flocchini, P., Pagli, L., Prencipe, G., Santoro, N., Viglietta, G.: Gathering in dynamic rings. Theor. Comput. Sci. **811**, 79–98 (2020)
6. Di Luna, G.A., Flocchini, P., Prencipe, G., Santoro, N.: Black hole search in dynamic rings. In: 2021 IEEE 41st International Conference on Distributed Computing Systems (ICDCS), pp. 987–997. IEEE (2021)
7. Dobrev, S., Flocchini, P., Prencipe, G., Santoro, N.: Searching for a black hole in arbitrary networks: optimal mobile agents protocols. Distrib. Comput. **19** (2006)
8. Dobrev, S., Santoro, N., Shi, W.: Using scattered mobile agents to locate a black hole in an un-oriented ring with tokens. Int. J. Found. Comput. Sci. **19**(06), 1355–1372 (2008)
9. Flocchini, P., Kellett, M., Mason, P.C., Santoro, N.: Searching for black holes in subways. Theory Comput. Syst. **50**, 158–184 (2012)
10. Fraigniaud, P., Ilcinkas, D., Peer, G., Pelc, A., Peleg, D.: Graph exploration by a finite automaton. Theor. Comput. Sci. **345**(2–3), 331–344 (2005)
11. Gotoh, T., Flocchini, P., Masuzawa, T., Santoro, N.: Exploration of dynamic networks: tight bounds on the number of agents. J. Comput. Syst. Sci. **122**, 1–18 (2021)
12. Gotoh, T., Sudo, Y., Ooshita, F., Kakugawa, H., Masuzawa, T.: Exploration of dynamic tori by multiple agents. Theor. Comput. Sci. **850**, 202–220 (2021)
13. Gotoh, T., Sudo, Y., Ooshita, F., Masuzawa, T.: Dynamic ring exploration with (h, s) view. Algorithms **13**(6), 141 (2020)
14. Ilcinkas, D., Wade, A.M.: Exploration of dynamic cactuses with sub-logarithmic overhead. Theory Comput. Syst. **65**, 257–273 (2021)
15. Kuhn, F., Lynch, N., Oshman, R.: Distributed computation in dynamic networks. In: Proceedings of the Forty-Second ACM Symposium on Theory of Computing, pp. 513–522 (2010)
16. Markou, E., Paquette, M.: Black hole search and exploration in unoriented tori with synchronous scattered finite automata. In: Baldoni, R., Flocchini, P., Binoy, R. (eds.) Principles of Distributed Systems. OPODIS 2012. LNCS, vol. 7702, pp. 239–253. Springer, Berlin, Heidelberg (2012). https://doi.org/10.1007/978-3-642-35476-2_17

Recognition and Isomorphism of Proper H-Graphs for Unicyclic H in FPT-Time

Deniz Ağaoğlu Çağırıcı[1]($^{(\boxtimes)}$) (iD) and Peter Zeman[2] (iD)

[1] Masaryk University, Brno, Czech Republic
agaoglu@mail.muni.cz
[2] University of Neuchâtel, Neuchâtel, Switzerland

Abstract. An *H-graph* is an intersection graph of connected subgraphs of a suitable subdivision of a fixed graph H. Many important classes of graphs can be expressed as H-graphs, and in particular, every graph is an H-graph for a suitable graph H. An H-graph is called *proper* if it has a representation where no subgraph properly contains another. We consider the recognition and isomorphism problems for proper U-graphs where U is a unicyclic graph, i.e. a graph which contains exactly one cycle. We prove that testing whether a graph is a (proper) U-graph, for some U, is *NP*-hard. On the positive side, we give an *FPT*-time recognition algorithm for a fixed U, parameterized by $|U|$. As an application, we obtain an *FPT*-time isomorphism algorithm for proper U-graphs, parameterized by $|U|$. To complement this, we prove that the isomorphism problem for (proper) H-graphs is *GI*-complete for every fixed H which is not unicyclic nor a tree.

Keywords: H-graph · recognition · isomorphism · FPT-time

1 Introduction

The concept of H-graphs was introduced originally by Biró, Hujter and Tuza in 1992 [6]. This notion generalizes and relates to many important classes of graphs that are well-known in the literature, for instance, interval graphs, circular-arc graphs, and chordal graphs.

A graph G has an *H-representation* if G can be represented as an intersection graph of connected subgraphs of the graph H, i.e., if each vertex of G can be assigned a connected subgraph of H such that two subgraphs intersect if and only if the corresponding vertices are adjacent. A *subdivision of H* is a graph obtained by replacing any edge of H by an induced path of arbitrary length. Then, G is an *H-graph* if there is a subdivision H' of H such that G has an

D. Ağaoğlu Çağırıcı—This author was supported by the Czech Science Foundation, project no. 20-04567S.
P. Zeman—This author was supported by Swiss National Science Foundation project PP00P2-202667, and by the Carlsberg Foundation Young Researcher Fellowship CF21-0682 - "Quantum Graph Theory".

R. Uehara et al. (Eds.): WALCOM 2024, LNCS 14549, pp. 304–318, 2024.
https://doi.org/10.1007/978-981-97-0566-5_22

H'-representation. In this language, interval graphs are K_2-graphs, circular-arc graphs are K_3-graphs, and chordal graphs are the union of all T-graphs, where T runs through all trees.

The study of H-graphs was revived in [9] after partially answering the open question posed by Biró, Hujter and Tuza: What is the complexity of testing whether a given graph is an H-graph for a fixed H? In particular, the recognition problem is NP-complete if H contains the diamond graph as a minor, and when H is a tree, XP-time algorithms were given parameterized by the size of H.

There are several variations of the recognition problem that are relevant for H-graphs. Firstly, both the graph G and the graph H can be a part of the input and the question is whether G is an H-graph. This problem is NP-complete even for chordal graphs and for the case when H is a tree [18]. Another variation is when the input is a graph G and the graph H is fixed – this is the question asked by Biró, Hujter, and Tuza. Further, we consider a third variant. To that end, let $\mathcal{H}$ be a class of graphs. By $\mathcal{H}$-graphs, we mean the class of graphs G for which there exists a graph $H \in \mathcal{H}$ such that G is an H-graph. For instance, if $\mathcal{T}$ is the class of all trees, then $\mathcal{T}$-graphs are exactly all chordal graphs [16]. The recognition problem for $\mathcal{H}$-graphs, has a graph G on the input and asks whether there is an $H \in \mathcal{H}$ such that G is an H-graph. In the case of $\mathcal{T}$-graphs, the problem is well-known and solvable in polynomial time [16,23].

Two graphs G and H are called *isomorphic*, denoted by $G \simeq H$, if there exists a bijection f from $V(G)$ to $V(H)$ such that $\{u, v\} \in E(G)$ if and only if $\{f(u), f(v)\} \in E(H)$. The *graph isomorphism problem* is to determine whether two graphs are isomorphic, and a problem is called *GI-complete* if it can be reduced to the graph isomorphism problem in polynomial time. Considering H-graphs, the isomorphism problem is *GI*-complete for (proper) S_d-graphs where S_d is a star with d rays when d is a part of the input. This extends to (proper) H-graphs when H is a tree or a unicyclic graph, and it can be solved in FPT-time for all H-graphs when H contains no cycle, parameterized by the size of H [1–3,5].

Our Results. Lately, there has been a noticeable interest towards H-graphs concerning the recognition and isomorphism problems [1–6,8–10,15,18]. In this paper, we aim to advance this research further considering two special types of H-graphs. Let G be an H-graph and R be an H-representation of G. R is called *proper* if no representing subgraph in R contains another, and *Helly* if the common intersection of all subgraphs in R representing the vertices of each clique in G is non-empty. Then, G is a proper (resp. Helly) H-graph if it has a proper (resp. Helly) H-representation. We focus on the parameterized complexity of the recognition and isomorphism problems on proper U-graphs where U is a unicyclic graph.

- In Sect. 2, we show that the recognition problem is NP-hard on $\mathcal{U}$-graphs and proper $\mathcal{U}$-graphs, where $\mathcal{U}$ is the class of all unicyclic graphs.
- Sections 3 and 4 deal with proper U-graph recognition for a fixed unicyclic graph U. In Sect. 3, we show that chordal proper U-graphs and proper Helly

U-graphs can be recognized in FPT-time. Section 4 extends this result to all proper U-graphs.
- Section 5 and 6 deal with the isomorphism problem. In Sect. 5, we show that the isomorphism of proper U-graphs can be tested in FPT-time. Section 6 complements this by showing that (proper) H-graph isomorphism is GI-complete if H is not unicyclic.

Preliminaries. Let U be the fixed unicyclic graph that contains exactly one cycle, and several paths and trees connected to this cycle, and its order $|U|$ be our parameter. We call the unique cycle and the vertices of U the *"circle"* and *"nodes"* not to be confused with the cycles and vertices of a U-graph, respectively.

An *interval graph* is the intersection graph of a set of intervals on the real line. Both the recognition and isomorphism problems can be solved in linear time on interval graphs [7] and proper interval graphs [11,17].

A *circular-arc graph* is the intersection graph of a set of arcs around a circle. Circular-arc graphs can be recognized in linear time [22] and it was announced that they can be tested for isomorphism in polynomial time [19]. Moreover, both the recognition and isomorphism problems are linear time solvable on proper circular-arc graphs [11,12]. Therefore, we assume that $U \neq K_3$ since K_3-graphs are circular-arc graphs.

Consider a particular U-representation R and a clique C of a U-graph G. Then, C is called *Helly* in R if all vertices of C mutually intersect at some point of U in R, and otherwise, *non-Helly*. By definition, the size of a non-Helly clique is at least 3. A U-graph is *Helly* (resp. *non-Helly*) if it has some (resp. no) U-representation where all its cliques are Helly. Note that non-Helly circular-arc graphs may have exponentially many maximal cliques which do not correspond to the points of the circle [21]. We also call such cliques of proper U-graphs around the unique circle of U *non-Helly*. It is known that Helly circular-arc graphs can be recognized and tested for isomorphism in linear time [11,20].

A graph is called *chordal* if it contains no *hole*, i.e. an induced cycle of length at least 4. They can be recognized in linear time [23] while their isomorphism is *GI-complete* [25], i.e. *"as hard as"* the isomorphism of general graphs that is neither known to be solvable in polynomial time nor NP-complete. They are equivalently defined as the intersection graphs of subtrees of a suitable tree T [16]. With T on the input, T-graph recognition is NP-complete [18] and T-graph isomorphism is GI-complete [25]. However, when T is fixed, T-graph recognition can be solved in XP-time [9] and T-graph isomorphism can be solved in FPT-time [1,10]. On proper T-graphs, both the recognition and the isomorphism problems are solvable in FPT-time [5,8]. Therefore, we also assume that $U \neq T$ for any tree T.

2 NP-Hardness of (Proper) $\mathcal{U}$-Graph Recognition

In this section, we show that the recognition problem is NP-hard for $\mathcal{U}$-graphs and proper $\mathcal{U}$-graphs where $\mathcal{U}$ is the class of unicyclic graphs. A *hypergraph* is a graph where an edge, called a *hyperedge*, can join more than two vertices. If every edge of a hypergraph X joins k vertices, then X is called *k-uniform*. *Hypergraph c-coloring* is the problem of assigning one of c pre-given colors to each vertex such that every hyperedge joins vertices of at least two different colors.

We define the problem HELLYCLIQUESCARC$(G; C)$ as follows: Given a circular-arc graph G and a set C of cliques of G, decide if there exists a circular-arc representation of G such that all cliques in C are Helly. The 2-coloring problem for 3-uniform hypergraphs was shown to be NP-hard [14], and we reduce it to the HELLYCLIQUESCARC$(G; C)$ problem and give the following.

Lemma 1. *The problem* HELLYCLIQUESCARC$(G; C)$ *is* NP-*hard even if we restrict to circular-arc graphs that are the complements of perfect matchings for which every circular-arc representation is proper.*

Using Lemma 1, we get the following.

Corollary 2. *The recognition problem is* NP-*complete for* U-*graphs and proper* U-*graphs where* U *is the class of all unicyclic graphs.*

3   Recognizing Chordal Proper and Proper Helly U-Graphs

In this section, we give two FPT-procedures to decide whether *i*) a given chordal graph G is a proper U-graph, and *ii*) a given graph G is a proper Helly U-graph. We refer to the maximal cliques of G placed on the branching nodes (i.e., the nodes of degree at least 3) of U as the *branching cliques*. In a proper U-representation of G (if G is a proper U-graph), it is known that every branching clique can be extended to a maximal clique [9].

We first consider the case when the input graph G is chordal. It is known that chordal graphs can be recognized in linear time and they have linearly many maximal cliques which can be listed in linear time [23]. Therefore, given a graph that is chordal, we can list its set of maximal cliques efficiently. Among these maximal cliques, our aim is to identify a bounded number of maximal cliques which can exclusively be placed on the branching nodes of the circle in FPT-time parameterized by $|U|$. The following was proven in [5].

Lemma 3 [5, **Lemma 7.3**]. *Given a proper T-graph G for a fixed tree T, one can identify a set of maximal cliques of G that can be used as branching cliques in a proper T-representation of G in* FPT-*time, and its size is* $\mathcal{O}(|T|^2)$.

Building upon Lemma 3, we get the following.

Lemma 4. *Given a proper U-graph G with polynomially many maximal cliques for a fixed unicyclic graph U, one can identify an isomorphism-invariant set C of cliques that contains all maximal cliques placed on the branching nodes of U in some proper U-representation of G in* FPT-*time, and its size is* $\mathcal{O}(|U|^3)$.

We call this set of maximal cliques *rich cliques*. Since their number in proper U-graphs is bounded, we can try all assignments of them to the branching nodes on the circle of U in *FPT*-time. Given a graph G, we assume the following:

A1. G *is not a proper U_1-graph for some subgraph or minor U_1 of U* since the number of subgraphs and minors of U is bounded and we can apply Procedure 5 to test if G is a proper U_1-graph for each such U_1.

A2. G *is connected* due to two facts. Firstly, every component of a disconnected proper U-graph without a rich clique forms an interval graph. Thus, all of them can be placed on an edge incident to a leaf of U. Secondly, the number of rich cliques, thus the components carrying a rich clique is bounded, and we can try all possible assignments of such components to the subgraphs of U.

Procedure 5. Given a connected chordal graph G on n vertices and a fixed unicyclic graph U, we decide whether G is a proper U-graph as follows:

1. Let $\mathcal{B}$ be the branching nodes on the circle of U, and $deg(\mathcal{B})$ be the sum of degrees of nodes of U in $\mathcal{B}$.
2. Find the set $\mathcal{C}$ of rich cliques of G. If $|\mathcal{C}| < |\mathcal{B}|$, or $|\mathcal{C}|$ is not bounded by $\mathcal{O}(|U|^3)$, return that G is not a proper U-graph.
3. Looping through each assignment $f : \mathcal{C} \to \mathcal{B}$:
   (a) Let $\mathcal{C}_f$ denote the branching cliques placed on the circle of U, and $\mathcal{X}$ denote the (connected) components of $G - \mathcal{C}_f$.
   (b) If there are more than $deg(\mathcal{B}) - |\mathcal{B}|$ components in $\mathcal{X}$, move on to another assignment.
   (c) For each branching node $b_i \in \mathcal{B}$:
       i. For every $b_t \in \mathcal{B}$, let $C_t \subseteq \mathcal{C}_f$ denote the clique placed on b_t w.r.t. f.
       ii. Let P denote the maximal induced subgraph of the circle of U containing b_i and every $b_j \in \mathcal{B}$ such that $C_i \cap C_j \neq \emptyset$.
       iii. **If P does not contain at least one branching node b_j:**
           – Let $Y \subsetneq U$ denote the maximal connected subtree of U which contain only P from the circle.
           – Let $\mathcal{Y}^*$ be the union of the branching cliques and the components placed on Y, and additionally the components X_k and X_l placed on the edges next to the ends of P on the circle if they exist.
           – If $G[\mathcal{Y}^*]$ is not a proper Y-graph, move on to another assignment.
       iv. **Else if P is a circle and no component in $\mathcal{X}$ is placed on P:**
           – Let $\mathcal{D} \subseteq \mathcal{C}_f$ denote the set of branching cliques which have minimal intersections with C_i by vertex inclusion.
           – If more than two branching cliques in $\mathcal{D}$ have distinct intersections with C_i, move on to another assignment.
           – If no pair of branching cliques in $\mathcal{D}$ is placed on consecutive branching nodes on P, move on to another assignment.
           – Let $\{b_j, b_{j+1}\} \in \mathcal{B}$ be any pair of consecutive branching nodes such that $C_l \cap C_i \subsetneq C_k \cap C_i$ for all b_k appearing on the path between b_i and b_l not passing through $b_{l'}$ where $l \in \{j, j+1\}$ and $l' = \{j, j+1\} \setminus l$.

  – Let e be the edge between b_j and b_{j+1}, $Y = U - e$ be the maximally connected subtree of U which contains $P - e$ from the circle, and $\mathcal{Y}^* = G - (C_j \cap C_{j+1})$.
  – If $G[\mathcal{Y}^*]$ is a proper Y-graph, return that G is a proper U-graph. Otherwise, move on to another assignment.
  v. **Else, P is a circle and some components in $\mathcal{X}$ are placed on P:**
  – If there exist more than one such component, move on to another assignment.
  – Let $X_k \in \mathcal{X}$ be that unique component placed on the edge e of P between the branching nodes b_j and b_{j+1}.
  – Let $Y = U - e$ denote the maximal connected subtree of U which contains only $P - e$ from the circle.
  – Let $\mathcal{Y}^*$ be the union of the branching cliques and the components placed on Y, and additionally two copies of X_k one with its attachment in C_j and the other with its attachment in C_{j+1}.
  – If $G[\mathcal{Y}^*]$ is a proper Y-graph and $G[C_j \cup X_k \cup C_{j+1}]$ is a proper interval graph, return that G is a proper U-graph. Otherwise, move on to another assignment.
(d) Return that G is a proper U-graph.
4. Return that G is not a proper U-graph.

We first give the following sequence of statements to prove that Procedure 5 correctly decides whether a given chordal graph is a proper U-graph in FPT-time (Fig. 1).

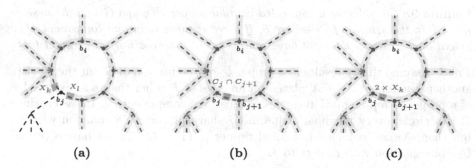

(a)   (b)   (c)

Fig. 1. (a), (b) and (c) illustrate the Step iii, iv and v of Procedure 5, respectively.

Lemma 6. *Let G be a connected proper U-graph and $\mathcal{C}_f$ be the set of maximal cliques placed on the branching nodes on the circle of U in some proper U-representation of G. Then, $G - \mathcal{C}_f$ has at most $\deg(\mathcal{B}) - |\mathcal{B}|$ components.*

Lemma 7. *Let G be a connected chordal proper U-graph. In the setting of Procedure 5:*

1. If the assignment f results in a proper U-representation of G, then every $G[\mathcal{Y}^*]$ in Step iii is a proper Y-graph for the obtained tree Y.
2. If there exists no $G[\mathcal{Y}^*]$ obtained from Step iv or v, then the application of Step iii for distinct branching nodes in $\mathcal{B}$ correctly tests if G has a proper U-representation with respect to f.

Lemma 8. *Consider a connected chordal proper U-graph G in the setting of Procedure 5. Assume that the maximal cliques in $\mathcal{C}_f$ can be placed on the branching nodes of the circle of U in some proper U-representation of G. If P is as defined in Step iv, then the following hold:*

1. *There exist at most two distinct minimal intersections with C_i.*
2. *At least two such maximal cliques are placed on consecutive nodes in $\mathcal{B}$.*

Proof. Assume that $C_i \cap C_j$ and $C_i \cap C_k$ are distinct minimal intersections with C_i by inclusion of the vertices, and also for $C_l \in \mathcal{C}_f$, $C_i \cap C_l$ is another distinct minimal intersection with C_i. Since G is a connected proper U-graph, the induced subgraph of G placed on the circle of U forms a circular-arc graph and its cliques have a circular ordering. However, $C_i \cap C_m \neq \emptyset$ for all $C_m \neq C_i \in \mathcal{C}_f$ in Step iv, thus the existence of at least three distinct minimal intersections with C_i means that the induced subgraph of G placed on the circle of U does not form a circular-arc graph which means that there are at most two maximal cliques $C_j \neq C_k \in \mathcal{C}_f$ which have distinct minimal intersections with C_i. In addition, due to the connectedness of G, and again since the induced subgraph of G placed on the circle of U forms a circular-arc graph, there exists at least one pair of maximal cliques C_j and C_k placed on consecutive nodes of $\mathcal{B}$. $\square$

Lemma 9. *Considering a connected chordal proper U-graph G and the assignment f in the setting of Procedure 5, if there is more than one component of $\mathcal{X}$ placed on P, then G does not have a proper U-representation with respect to f.*

Proof. Assume that f results in a proper U-representation of G but there exists another component X_l of $\mathcal{X}$ placed on an edge of P other than e as every edge of a proper U-representation carries at most one component [5]. However, since P is a circle, every maximal clique in $\mathcal{C}_f$ shares a vertex in common with C_i but then, X_l can not be represented properly. Thus, G can not have a proper U-representation with respect to f. $\square$

Lemma 10. *Considering a connected chordal graph G and the assignment f in the setting of Procedure 5, Steps iv and v correctly test whether G is a proper U-graph with respect to f.*

Theorem 11. *Proper chordal U-graphs can be recognized in* FPT-*time.*

Since all maximal cliques of graphs with polynomially many maximal cliques can be listed in polynomial time [24], we obtain the following.

Theorem 12. *One can decide whether an input graph with polynomially many maximal cliques is a proper U-graph in* FPT-*time parameterized by $|U|$.*

We now focus on the recognition problem for proper Helly U-graphs, i.e. proper U-graphs which have at least one proper U-representation such that all cliques are Helly. It has been shown that every (proper) Helly H-graph G has at most $|H| + |E(H)| \cdot |G|$ maximal cliques which is linear on the order of the input graph G [9]. Recall that all maximal cliques of graphs with polynomially many maximal cliques can be listed in polynomial time [24]. With the same assumptions, we modify Procedure 5 to decide whether a given graph G is a proper Helly U-graphs as follows:

1. Before Step 1, we start listing the maximal cliques of G using the algorithm of [24]. If the listing procedure outputs at least $1 + |U| + |E(U)| \cdot |G|$ maximal cliques, stop the enumeration and return that G is not a proper Helly U-graph.
2. In Step 2, we only consider the assignments f such that $G[\mathcal{C}_f]$ is a Helly circular-arc graph tested using the algorithm of [20].
3. We return that G is a proper Helly U-graph or not analogous to Procedure 5.

Theorem 13. *Proper Helly U-graphs can be recognized in* FPT-*time.*

4   Recognizing Proper U-Graphs in General

In this section, we assume that the input graphs do not have polynomially many maximal cliques since otherwise we can use Procedure 5 by Theorem 12. Therefore, we assume that the input graph is not a proper Helly U-graph and not a chordal graph. Given a graph G, we make use of the following observations:

- If G is a proper U-graph but not Helly, every proper U-representation of G contains a non-Helly clique C. Since T-graphs satisfy the Helly property, C (along with all such cliques) must be placed on the circle in every U-representation. As being non-Helly may depend on a particular U-representation, we aim to identify *the set of non-Helly clique vertices*, i.e. the vertices belonging to the cliques that are non-Helly in all proper U-representations.
- If G is not chordal, it contains at least one "hole", i.e. an induced cycle L of length at least 4. Since T-graphs are chordal, L (along with all holes in G) must be placed on the circle in every U-representation if G is a (proper) U-graph. Hence, our goal is to identify the set of all vertices contained in the holes, referred as *the set of hole vertices*.
- The set of all vertices that are placed completely outside of the circle of U in a proper U-representation forms a (possibly disconnected) chordal graph. Here, the term "completely on" $S \subseteq U$ means that a vertex v is represented with some subgraph of (a subdivision of) S. Conversely, if v is placed "outside of" $S \subseteq U$, it is represented with some subgraph of (a subdivision of) $U \setminus S$.

We first give the following lemma which provides i) an upper bound on the length of every hole L of a proper U-graph which is placed completely on the circle of U, and ii) an upper bound on the length of any hole L.

Lemma 14. *Let G be a proper U-graph that is not Helly nor chordal. In a proper U-representation of G, the length of a hole that is placed completely on the circle of U is at most 6, and the length of any hole is linearly bounded in $|U|$.*

Note that in proper Helly U-graphs, the length of holes is not bounded since they do not contain non-Helly cliques. However, they can be recognized in *FPT*-time by Theorem 11. On the other hand, one can identify all holes of proper U-graphs by brute-force in *XP*-time by testing all d-tuples of vertices for $4 \le d \le |U|$ by Lemma 14 and we will show how to compute them in *FPT*-time.

Recall that a dominating edge of a circular-arc graph is an edge covering the whole circle and having two disconnected intersections. Considering a proper U-graph G, we focus on the dominating edges around the circle of U which may not form dominating edges in G. We refer to the union of end-vertices of all dominating edges covering the circle of U as *the set of dominating edge vertices.*

Lemma 15. *Let L be a hole and V_L be the set of hole vertices of a proper U-graph G.*

1. *If G is not chordal and not a proper Helly U-graph, one can identify the set of non-Helly clique vertices of G in polynomial time given V_L.*
2. *One can in polynomial time identify the set of dominating edge vertices given V_L.*
3. *If v is a vertex of G adjacent to at least 3 vertices of L, v must be placed (not necessarily completely) on the circle of U in every (proper) U-representation.*

We give the following procedure to obtain the set $V(L)$ of hole vertices, and the possibly empty set $V(R)$ of *revealed vertices*. Informally, the revealed vertices do not appear in any hole but also must be represented (not necessarily completely) on the circle of U due to their neighborhoods in $V(L)$ if G is a proper U-graph.

Procedure 16. Given a connected graph G which is not chordal, we identify the sets $V(L)$ and $V(R) \subseteq V(G) \setminus V(L)$ of hole and revealed vertices as follows:

1. Let $V(L)$ and $V(R)$ be the initially empty sets of hole and revealed vertices of G, respectively.
2. For each $u \in V(G)$ and each pair of its neighbors $v \ne w \in N_G(v)$ such that $\{v, w\} \notin E(G)$:
   - Identify the shortest path P between v and w in $G - (N_G[u] \setminus \{v, w\})$ using Dijkstra's algorithm [13]. If $P \ne \emptyset$, $V(L) \leftarrow V(L) \cup V(P) \cup \{u, v, w\}$.
3. For each vertex $u \in V(G) \setminus V(L)$:
   - If u has at least three neighbors in $V(L)$ forming an induced path in G, then $V(R) \leftarrow V(R) \cup u$. This can be checked by identifying whether there exist three distinct vertices $v, w, x \in (N_G[u] \setminus u) \cap V(L)$ forming an induced path of length 2 in G.
4. Return $V(L)$ and $V(R)$ as the sets of hole and revealed vertices, respectively.

Lemma 17. *Procedure 16 correctly identifies the set $V(L)$ of hole vertices and the set $V(R)$ of revealed vertices in polynomial time.*

Proof. The correctness and complexity of identifying the set $V(L)$ of hole vertices depends on the Dijkstra's algorithm [13], and greedily checking the neighborhood of each vertex of G. We now prove that $V(R)$ is indeed a revealed set. It is clear that $V(R)$ can be computed in polynomial time. Since only the consecutive vertices of holes are connected by an edge, each vertex u having at least three neighbors $v, w, x \in V(L)$ appearing as an induced path in G has to be placed on the circle by Lemma 15. Also, if u has more that three such neighbors, some triple among them still form an induced path of length 2, and therefore, it is sufficient to check all three neighbors of u in polynomial time. $\qquad\square$

Note that since U is unicyclic, every vertex found using Procedure 16 lies on the circle of U in every proper U-representation if the input is a proper U-graph.

Lemma 18. *For a connected proper U-graph G, consider the following sets:*

- *V_L and V_R: The sets of hole and revealed vertices found using Procedure 16.*
- *V_C: The union of sets of non-Helly clique vertices and dominating edge vertices found using Lemma 15 given $V_L \cup V_R$.*

Then, $G[V_L \cup V_R \cup V_C]$ is a circular-arc graph with fixed cyclic ordering.

Proof. As G is a proper U-graph, the vertices in $V_L \cup V_R$ must be placed (not necessarily completely) on the circle of U by Lemma 17, and V_C must be placed (not necessarily completely) on the circle of U since every T-graph satisfies the Helly property for a tree T, both $G[V_L \cup V_R]$ and $G[V_L \cup V_R \cup V_C]$ form circular-arc graphs. In addition, the cyclic order of $G[V_L \cup V_R]$ is fixed (up to reversal) since V_L only contains the vertices of holes and V_R only contains the vertices which neighbors in V_L form induced paths. Since, holes and induced paths have fixed circular orderings, $G[V_L \cup V_R]$ has a fixed circular ordering. Also, each vertex of V_C has neighbors in $V_L \cup V_R$ which have a fixed ordering in $G[V_L \cup V_R]$. This also means that $G[V_L \cup V_R \cup V_C]$ has a fixed circular ordering. $\qquad\square$

We introduce the following notion that helps in recognizing proper U-graphs.

Definition 19. *Let V_L, V_R and V_C be as in the setting of Lemma 18 for a connected graph G. Let $\mathcal{X}$ be the set of components in $G - (V_L \cup V_R \cup V_C)$ such that $|\mathcal{X}| \leq |U|$ and each component of $\mathcal{X}$ has at most two upper attachments in $G[V_L \cup V_R \cup V_C]$. Let $\mathcal{F}$ be the family of all those upper attachments of the components in $\mathcal{X}$. Let $\mathcal{B}$ be the branching nodes on the circle of U. We call a partition of $\mathcal{X}$ into subsets $\mathcal{X}_1, \ldots, \mathcal{X}_{|\mathcal{B}|}$ a Helly partition if the following hold:*

a) *For every pair $1 \leq i < j \leq |\mathcal{B}|$, it holds that $\mathcal{X}_i \cap \mathcal{X}_j = \emptyset$.*
b) *For each $1 \leq i \leq |\mathcal{B}|$, $\mathcal{X}_i$ contains at most $deg(b_i) - 2$ components with exactly 1 upper attachment where $deg(b_i)$ denotes the degree of $b_i \in \mathcal{B}$.*
c) *Let $\mathcal{F}_1, \ldots, \mathcal{F}_{|\mathcal{B}|}$ be the subfamilies of $\mathcal{F}$ such that for every $1 \leq i \leq |\mathcal{B}|$, $\mathcal{F}_i$ consists of exactly 1 upper attachment of each component $X_j \in \mathcal{X}_i$ with respect to the fixed cyclic order of $G[V_L \cup V_R]$.*
d) *For each $b_i \in \mathcal{B}$, it holds that $|\mathcal{X}_i| = |\mathcal{F}_i| \leq deg(b_i)$.*

314   D. Ağaoğlu Çağırıcı and P. Zeman

e) *The union $C_{\mathcal{F}_i}$ of all vertices of G contained in each $\mathcal{F}_i$ forms a clique in G.*
f) *The circular-arc graph $G[V_L \cup V_R \cup V_C]$ has a circular-ones property with $C_{\mathcal{F}_1}, \ldots, C_{\mathcal{F}_{|\mathcal{B}|}}$ appearing in this order.*
g) *For every pair $1 \le i \ne j \le |\mathcal{B}|$, there exists no $X_k \in \mathcal{X}_i$ such that any of the (at most 2) upper attachments of X_k has an intersection with $C_{\mathcal{F}_j}$ which is a strict superset of the intersection of the upper attachment of some $X_l \in \mathcal{X}_j$.*

Lemma 20. *There are at most $(|U|+4)^{|U|}$ Helly partitions of $\mathcal{X}$ into distinct choices of subfamilies $\mathcal{F}_1, \ldots, \mathcal{F}_{|\mathcal{B}|}$ and they can be computed in FPT-time (Fig. 2).*

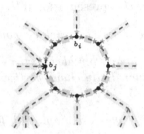

Fig. 2. After the sets described in Lemma 18 are found, teal and orange components are obtained without their deterministic placements on U. Teal components have two upper attachments, marked with blue stars, and must be placed on the circle of U. Conversely, orange components have one upper attachment, marked with red stars, and can not be placed on the circle of U. (Color figure online)

We now give Procedure 21 to recognize general proper U-graphs.

Procedure 21. Given a connected graph G on n vertices and a fixed unicyclic graph U, we decide whether G is a proper U-graph as follows:

1. If G is chordal, then run Procedure 5 and return its output.
2. Check whether G is a proper Helly U-graph using Procedure 5 with Theorem 13. If it returns that G is a proper Helly U-graph, return its output.
3. Using Procedure 16, find the sets V_L and V_R of hole and revealed vertices, resp.
4. Using Lemma 15, given the set $V_L \cup V_R$, identify the union V_C of sets of non-Helly clique vertices and dominating edge vertices. If $G[V_L \cup V_R \cup V_C]$ is not a circular-arc graph, return that G is not a proper U-graph by Lemma 18.
5. Find the set $\mathcal{X}$ of components of $G - (V_L \cup V_R \cup V_C)$. If $\mathcal{X}$ has more than $|U|$ components, return that G is not a proper U-graph.
6. Identify the upper attachments of each component of $\mathcal{X}$ in $G[V_L \cup V_L \cup V_C]$ with respect to the fixed cyclic order of $G[V_L \cup V_R]$. If there is a component with at least 3 upper attachments, return that G is not a proper U-graph.
7. Every component in $\mathcal{X}$ has at most 2 upper attachments. Let $\mathcal{B}$ denote the branching nodes on the circle of U, $deg(\mathcal{B})$ be the sum of degrees of nodes in $\mathcal{B}$ and $\mathcal{F}$ be the family of all upper attachments. If there are more than $deg(\mathcal{B})$ such attachments in $\mathcal{F}$, return that G is not a proper U-graph.

8. Compute all Helly partitions of $\mathcal{X}$ into $\mathcal{X}_1, \ldots, \mathcal{X}_{|\mathcal{B}|}$, and the (possibly maximal) cliques $\{C_{\mathcal{F}_1}, \ldots, C_{\mathcal{F}_{|\mathcal{B}|}}\}$ for distinct choices of subfamilies $\{\mathcal{F}_1, \ldots, \mathcal{F}_{|\mathcal{B}|}\}$.
9. For each partition $\mathcal{X}_1, \ldots, \mathcal{X}_{|\mathcal{B}|}$ and choice of subfamilies of $\mathcal{F}$, use Procedure 5 from Step (a) with $\mathcal{C}_f = \{C_{\mathcal{F}_1}, \ldots, C_{\mathcal{F}_{|\mathcal{B}|}}\}$ and the components $\mathcal{X}$ of $G - \mathcal{C}_f$.
   (a) If Procedure 5 returns that G is a proper U-graph, return its output.
10. Return that G is not a proper U-graph.

Lemma 22. *One can in* FPT*-time identify a set of (maximal) Helly cliques in Step 8 of Procedure 21 if the input graph has (such) a proper U-representation.*

We shortly mention that the proof of Lemma 22 follows from the bound on the number of components. Thus, we get the following.

Theorem 23. *Proper U-graph recognition is in* FPT *parameterized by $|U|$.*

5   Isomorphism Testing for Proper U-Graphs

In this section, we show how to test proper U-graph isomorphism in *FPT*-time parameterized by $|U|$. We assume connectedness since there are polynomially many pairs of connected components if we are given disconnected graphs, and utilize Procedure 5 and Procedure 21 considering the same assumptions given on the corresponding sections. We first modify Procedure 5 as follows:

1. On the input, we are given two connected proper U-graphs G_1 and G_2 on n vertices which are both chordal or proper Helly U-graphs, along with U.
2. Let $\mathcal{C}_i$ be the isomorphism invariant set of rich cliques of G_i for $i \in \{1, 2\}$ by Lemma 4. In Step 2, we check whether $|\mathcal{C}_1| = |\mathcal{C}_2|$ holds.
3. Using Procedure 5, we fix a proper U-representation of G_1 where $f : \mathcal{C}_1 \to \mathcal{B}$ denotes the assignment of branching cliques on the circle. Let $\mathcal{C}_1^f \subseteq \mathcal{C}_1$ be those cliques and C_1^i be the maximal clique of $\mathcal{C}_1^f$ placed on the branching node $b_i \in \mathcal{B}$ in the current assignment f.
4. We denote each assignment $\mathcal{C}_2 \to \mathcal{B}$ by g, and consider those g that match to f as being in Step iii, iv and v of Procedure 5. Thus, P is the same for both graphs in every step. Let $\mathcal{C}_2^g \subseteq \mathcal{C}_2$ be the branching cliques on the circle w.r.t. g, and C_2^i be the maximal clique of $\mathcal{C}_2^g$ placed on the branching node $b_i \in \mathcal{B}$ in the current assignment g. We require that $|C_1^i| = |C_2^i|$ for every $b_i \in \mathcal{B}$.
5. For each branching node b_i of U, we compute the subtree $Y_1 \subsetneq U$ and the graph $\mathcal{Y}_1^*$ obtained by G_1 as in Procedure 5. Let $Y_2 \subsetneq U$ be the corresponding subtree for G_2 and $\mathcal{Y}_2^*$ be the graph obtained by G_2 w.r.t. the assignment g.
6. We first check whether $Y_1 \simeq Y_2$, and instead of checking for proper Y-graph recognition, we test whether $G_1[\mathcal{Y}_1^*] \simeq G_2[\mathcal{Y}_2^*]$ using the proper Y-graph isomorphism testing for the tree $Y_1 \simeq Y_2$ given in [5].
7. In Step iv, we additionally check if $|C_j^1 \cap C_{j+1}^1| = |C_j^2 \cap C_{j+1}^2|$.
8. In Step v, we check whether the defined interval graphs are isomorphic proper interval graphs using the algorithm given in [7].

9. We return either $G_1 \simeq G_2$ or $G_1 \not\simeq G_2$ analogous to Procedure 5.

Theorem 24. *The above modifications to Procedure 5 result in a correct FPT-time isomorphism testing for chordal proper and proper Helly U-graphs.*

With analogous modifications to Procedure 21, we get the following.

Theorem 25. *Proper U-graph isomorphism can be tested in* FPT-*time.*

6 *GI*-Completeness for *H*-Graphs and Proper *H*-Graphs

In this section, we prove that if H is not unicyclic nor a tree, then proper H-graph isomorphism is *GI-complete*. Since the class of proper H-graphs is a subclass of H-graphs for every H, this also implies the *GI*-completeness of H-graph isomorphism in general. The key is the following theorem where B denotes the 5-vertex graph consisting of two triangles intersecting at exactly one vertex.

Theorem 26. *Proper B-graph isomorphism is* GI-*complete.*

Note that if a graph H is a minor of a graph H', then every H-graph is an H'-graph. Clearly, every B-graph can be represented as a D-graph where D denotes the diamond graph, i.e. K_4 minus one edge. This immediately gives the following.

Corollary 27. *If H is not unicyclic nor a tree, then the isomorphism problem for the class of all proper H-graphs and the class of all H-graphs is* GI-*complete.*

To prove Theorem 26, we give the following lemmas.

Lemma 28. *If G is a disjoint union of stars, then its complement $\overline{G}$ is a proper circular-arc graph.*

From a connected graph $G = (V, E)$, we construct a new graph G^* as follows:

1. Let $G' = (V \cup V_i, E')$ and $G'' = (V \cup V_i \cup V_j, E'')$ be the graphs resulting from subdividing each edge of G and G', respectively.
2. Let G''' be the graph obtained from G'' by adding all edges between V and V_i.
3. Finally, we set the graph G^* to be the complement of G'''.

Lemma 29. *Let G be a connected graph. Then, G^* is a proper B-graph.*

Lemma 30. *Let G_1 and G_2 be connected graphs with minimum degree at least three. Then, G_1 and G_2 are isomorphic if and only if G_1^* and G_2^* are isomorphic.*

Proof of Theorem 26. $G_1{}^*$ can be constructed from G_1 in polynomial time, and by Lemma 29, $G_1{}^*$ is a proper B-graph. Clearly, the isomorphism problem for the class of all graphs with minimum degree three is *GI*-complete, and by Lemma 30, G_1 and G_2 with minimum degree three are isomorphic if and only if $G_1{}^* \simeq G_2{}^*$. It follows that proper B-graph isomorphism is *GI*-complete.   □

7   Conclusions

We have shown that proper $\mathcal{U}$-graph recognition is *NP*-hard where $\mathcal{U}$ is the class of all unicyclic graphs, and proper H-graph isomorphism is *GI*-complete when H is not unicyclic nor a tree. For a fixed U, we gave *FPT*-procedures to recognize and test the isomorphism of proper U-graphs parameterized by $|U|$. What remains to be shown is how to solve the same problems in *FPT*-time for general U-graphs.

Acknowledgments. We would like to thank Steven Chaplick and Petr Hliněný for their helpful ideas concerning the hardness results.

References

1. Ağaoğlu Çağırıcı, D., Hliněný, P.: Isomorphism testing for T-graphs in FPT. In: Mutzel, P., Rahman, M.S., Slamin (eds.) WALCOM 2022. LNCS, vol. 13174, pp. 239–250. Springer, Cham (2022). https://doi.org/10.1007/978-3-030-96731-4_20
2. Arvind, V., Nedela, R., Ponomarenko, I., Zeman, P.: Testing isomorphism of chordal graphs of bounded leafage is fixed-parameter tractable. In: Bekos, M.A., Kaufmann, M. (eds.) WG 2022. LNCS, vol. 13453, pp. 29–42. Springer, Cham (2022). https://doi.org/10.1007/978-3-031-15914-5_3
3. Ağaoğlu, D., Hliněný, P.: Isomorphism problem for S_d-graphs. In: MFCS, pp. 4:1–4:14 (2020)
4. Ağaoğlu Çağırıcı, D., et al.: Recognizing H-graphs - beyond circular-arc graphs. In: MFCS, pp. 8:1–8:14 (2023)
5. Ağaoğlu Çağırıcı, D., Hliněný, P.: Efficient isomorphism for S_d-graphs and T-graphs. Algorithmica **85**, 352–383 (2023). https://doi.org/10.1007/s00453-022-01033-8
6. Biró, M., Hujter, M., Tuza, Z.: Precoloring extension. I. Interval graphs. Discrete Math. **100**, 267–279 (1992)
7. Booth, K.S., Lueker, G.S.: Testing for the consecutive ones property, interval graphs, and graph planarity using PQ-tree algorithms. J. Comput. Syst. Sci. **13**, 335–379 (1976)
8. Chaplick, S., Golovach, P.A., Hartmann, T.A., Knop, D.: Recognizing proper tree-graphs. In: IPEC, pp. 8:1–8:15 (2020)
9. Chaplick, S., Töpfer, M., Voborník, J., Zeman, P.: On H-topological intersection graphs. In: Bodlaender, H., Woeginger, G. (eds.) WG 2017. LNCS, vol. 10520, pp. 167–179. Springer, Cham (2017). https://doi.org/10.1007/978-3-319-68705-6_13
10. Chaplick, S., Zeman, P.: Combinatorial problems on H-graphs. Electron. Notes Discrete Math. **61**, 223–229 (2017)
11. Curtis, A.R., et al.: Isomorphism of graph classes related to the circular-ones property. Discrete Math. Theor. Comput. Sci. **15**, 157–182 (2013)
12. Deng, X., Hell, P., Huang, J.: Linear-time representation algorithms for proper circular-arc graphs and proper interval graphs. SIAM J. Comput. **25**, 390–403 (1996)
13. Dijkstra, E.W.: A note on two problems in connexion with graphs. Numer. Math. **1**, 269–271 (1959)
14. Dinur, I., Regev, O., Smyth, C.D.: The hardness of 3 - uniform hypergraph coloring. In: FOCS, p. 33 (2002)

15. Fomin, F.V., Golovach, P.A., Raymond, J.: On the tractability of optimization problems on H-graphs. In: ESA, vol. 112, pp. 30:1–30:14 (2018)
16. Gavril, F.: The intersection graphs of subtrees in trees are exactly the chordal graphs. J. Comb. Theory Ser. B **16**, 47–56 (1974)
17. Keil, J.M.: Finding Hamiltonian circuits in interval graphs. Inf. Process. Lett. **20**, 201–206 (1985)
18. Klavík, P., Kratochvíl, J., Otachi, Y., Saitoh, T.: Extending partial representations of subclasses of chordal graphs. Theor. Comput. Sci. **576**, 85–101 (2015)
19. Krawczyk, T.: Testing isomorphism of circular-arc graphs - Hsu's approach revisited. CoRR abs/1904.04501 (2019)
20. Lin, M.C., Szwarcfiter, J.L.: Characterizations and linear time recognition of Helly circular-arc graphs. In: Chen, D.Z., Lee, D.T. (eds.) COCOON 2006. LNCS, vol. 4112, pp. 73–82. Springer, Heidelberg (2006). https://doi.org/10.1007/11809678_10
21. Lin, M.C., Szwarcfiter, J.L.: Characterizations and recognition of circular-arc graphs and subclasses: a survey. Discrete Math. **309**, 5618–5635 (2009)
22. McConnell, R.M.: Linear-time recognition of circular-arc graphs. In: FOCS, pp. 386–394 (2001)
23. Rose, D.J., Tarjan, R.E., Lueker, G.S.: Algorithmic aspects of vertex elimination on graphs. SIAM J. Comput. **5**, 266–283 (1976)
24. Tsukiyama, S., Ide, M., Ariyoshi, H., Shirakawa, I.: A new algorithm for generating all the maximal independent sets. SIAM J. Comput. **6**, 505–517 (1977)
25. Zemlyachenko, V.N., Korneenko, N.M., Tyshkevich, R.I.: Graph isomorphism problem. J. Sov. Math. **29**, 1426–1481 (1985). https://doi.org/10.1007/BF02104746

Canonization of a Random Circulant Graph by Counting Walks

Oleg Verbitsky[1][(✉)] and Maksim Zhukovskii[2]

[1] Institut für Informatik, Humboldt-Universität zu Berlin, Berlin, Germany
verbitsk@informatik.hu-berlin.de
[2] Department of Computer Science, University of Sheffield, Sheffield, UK

Abstract. It is well known that almost all graphs are canonizable by a simple combinatorial routine known as color refinement. With high probability, this method assigns a unique label to each vertex of a random input graph and, hence, it is applicable only to asymmetric graphs. The strength of combinatorial refinement techniques becomes a subtle issue if the input graphs are highly symmetric. We prove that the combination of color refinement with vertex individualization produces a canonical labeling for almost all circulant digraphs (Cayley digraphs of a cyclic group). To our best knowledge, this is the first application of combinatorial refinement in the realm of vertex-transitive graphs. Remarkably, we do not even need the full power of the color refinement algorithm. We show that the canonical label of a vertex v can be obtained just by counting walks of each length from v to an individualized vertex.

1 Introduction

As it is well known, the graph isomorphism problem is very efficiently solvable in the average case by a simple combinatorial method known as *color refinement* (also *degree refinement* or *naive vertex classification*). When a random graph $\mathbf{G}_n$ on n vertices is taken as an input, this algorithm produces a canonical labeling of all vertices in $\mathbf{G}_n$ by coloring them initially by their degrees and then by refining the color classes as follows: Two equally colored vertices u and v get new distinct colors if one of the initial colors occurs in the neighborhoods of u and v with different multiplicity. In this way, every vertex gets a unique color with probability $1 - O(n^{-1/7})$ (Babai, Erdős, and Selkow [4]). Thus, the method produces a canonical labeling for almost all graphs on a fixed set of n vertices.

This approach is not applicable to regular graphs, even with many refinement rounds, because if all vertices have the same degree, then the refinement step makes obviously no further vertex separation. Weisfeiler and Leman [33] came up with a more powerful refinement algorithm which colors pairs of vertices instead of single vertices. The idea can be lifted to k-tuples of vertices, for each integer parameter k, and the general approach is referred to as the k-dimensional

O. Verbitsky was supported by DFG grant KO 1053/8–2. He is on leave from the IAPMM, Lviv, Ukraine.

Weisfeiler-Leman algorithm, abbreviated as k-WL. Thus, 2-WL is the original algorithm in [33], and 1-WL corresponds to color refinement. Remarkably, 2-WL is powerful enough to produce a canonical labeling for almost all regular graphs of a given vertex degree (Bollobás [8]; see also [23]).

Further restriction of regular input graphs to vertex-transitive graphs is challenging for combinatorial refinement because no vertex classification is at all possible in this case. Indeed, 2-WL assigns the same color to any two vertices u and v because u is mapped to v by an automorphism of the graph. The same holds for any dimension k. In fact, the Cai-Fürer-Immerman construction [10] of non-isomorphic graphs X_k and Y_k indistinguishable by k-WL can be modified so that these graphs become vertex-transitive [17]. A natural way to enhance combinatorial refinement is to combine it with vertex individualization [25]. This algorithmic approach proves to be advantageous in many contexts (see, e.g., [3]) but still cannot overcome the obstacle posed by the CFI graphs. Indeed, $(k + k')$-WL supersedes any combination of k-WL with prior individualization of k' vertices. As a consequence, isomorphism of vertex-transitive graphs cannot be solved by a constant-dimensional WL algorithm even under individualization of any constant number of vertices.

Motivated by the question of whether or not these basic obstacles persist in the average case setting, we focus in this paper on Cayley graphs and, more specifically, on circulant graphs, that is, Cayley graphs of a cyclic group. While the canonization problem for this class of graphs is known to be solvable in polynomial time [16] by advanced algebraic methods, it is an open question whether this can be done by using k-WL for some dimension k; see [30]. This poses an ongoing challenge for the combinatorial refinement method, especially because the research on isomorphism of circulant graphs has a long history with many deep results (see [2,16,27,28] and the references therein) and because k-WL with small dimension k is known to be applicable to many other natural graph classes (like, e.g., planar graphs [21]). The recent paper [22] investigates the round complexity of 2-WL on circulant graphs, exploiting the close connections of the subject with intricate mathematical concepts. Circulant graphs are also interesting on their own right as they naturally appear and are intensively investigated in many other theoretical and applied areas; see, e.g., the books [9,12,14]. After all, our primary motivation for the study of circulant graphs is that this is the most natural first choice of a graph class for benchmarking of combinatorial refinement in the realm of vertex-transitive graphs.

Our goal is to show that the individualization-refinement approach can be used to canonize almost all circulants at minimal computational costs. Our treatment covers also circulant directed graphs, which is advantageous for expository purposes as the case of digraphs is technically somewhat simpler. We show that the individualization of a single vertex suffices for random circulant digraphs, and that two individualized vertices are enough in the undirected case (in fact, we just perform color refinement twice, each time with a single individualized vertex). In both the directed and the undirected cases we keep the overall running time within $O(n^2 \log n)$, which is the standard running time of color refinement

[11]. This is possible due to the fact that our input graphs are vertex-transitive and, hence, it does not actually matter which one of the vertices is individualized. Here, n denotes the number of vertices. Thus, our time bound is actually linearithmic, that is, it is $O(N \log N)$ for the input length N where the input (di)graph is presented by the adjacency matrix and the cyclic structure is not explicitly given (see below the discussion of different representation concepts). Note also that, as one can expect, our average-case bound $O(n^2 \log n)$ is substantially better than the worst-case bound resulting from [16]. We summarize our main result in a somewhat condensed form as follows.

Theorem 1. *A uniformly distributed random circulant (di)graph is with probability at least $1 - n^{-1/2+o(1)}$ canonizable by color refinement combined with vertex individualization in running time $O(n^2 \log n)$.*

Theorem 1 includes two statements, one for undirected graphs (where all undirected circulant graphs are equiprobable) and the similar statement for directed graphs. Note that the concept of a uniform distribution of n-vertex circulants is somewhat ambiguous as the notion of an *n-vertex circulant* alone can be defined in three different natural ways:

- as a Cayley (di)graph of the cyclic group $\mathbb{Z}_n$,
- as an isomorphism class of Cayley (di)graphs of $\mathbb{Z}_n$ (which we call an *unlabeled circulant*),
- as a (di)graph on the vertex set $\{0, 1, \ldots, n-1\}$ isomorphic to a Cayley (di)graph of $\mathbb{Z}_n$ (which we call a *labeled circulant*).

We prove Theorem 1 first for random Cayley (di)graphs and then extend it to the other two concepts. The transition from one distribution to another is quite general and is based on known results in algebraic graph theory [7, 13, 27].

Remarkably, we show that canonization of a random circulant does not even need the full power of color refinement and can actually be accomplished by a weaker algorithmic tool. Let G be an arbitrary (di)graph on the vertex set $V = \{0, 1, \ldots, n-1\}$, and let $T \subseteq V$. The *walk matrix* $W_T = (w_{ij})_{i,j \in V}$ is defined by setting w_{ij} to be the number of walks of length j from the vertex i to a vertex in T. If A is the adjacency matrix of G and χ_T denotes the characteristic vector of the subset T, then W_T is formed by the columns $\chi_T, A\chi_T, A^2\chi_T, \ldots, A^{n-1}\chi_T$. The theory of walk matrices, including their applicability to isomorphism testing, has been developed by Godsil [18] and by Liu and Siemons [24]. Let G_T be obtained from G by coloring all vertices in T by the same color. We call G_T *walk-discrete* if the rows of W_T are pairwise different. For any walk-discrete G_T, the walk matrix W_T yields a canonical labeling of the vertices of G_T. This purely algebraic canonization method can be superseded by the purely combinatorial method of color refinement because if $w_{uj} \neq w_{vj}$ for some j, then color refinement assigns distinct colors to the vertices u and v in G_T (see Sect. 2.1 for details).

Let $W = W_V$ be the *standard* walk matrix of $G = G_V$. Obviously, G is walk-discrete whenever W is non-singular. Noteworthy, the rank of W for an undirected graph G is equal to the number of different *main* eigenvalues of the adjacency matrix A; see [19]. As shown by O'Rourke and Touri [29], a random undirected graph $\mathbf{G}_n$ has non-singular walk matrix with high probability. As a consequence, $\mathbf{G}_n$ is with high probability canonizable by computing its standard walk matrix.

The above theory essentially exploits the fact that the adjacency matrices of undirected graphs are symmetric and, by this reason, does not apply to directed graphs. Nevertheless, we obtain the following spectral criterion for circulant digraphs.

Lemma 2. *Let X be a Cayley digraph of a cyclic group and $X_0 = X_{\{0\}}$ be its version with one individualized vertex. Let W_0 be the walk matrix of X_0. Then W_0 is non-singular (implying that X_0 is walk-discrete) if and only if X has simple spectrum, that is, all eigenvalues of X are pairwise distinct.*

Suppose now that X is an undirected Cayley graph of $\mathbb{Z}_n$. In this case, the map $x \mapsto (-x) \bmod n$ is an automorphism of X_0, which implies that the walk matrix of X_0 has at most $\lceil (n+1)/2 \rceil$ different rows. If this bound is achieved, we call X_0 *walk-saturated*. As we will show, this property is sufficient for efficient canonization of X. On the other hand, the spectrum of X has at most $\lceil (n+1)/2 \rceil$ different eigenvalues. If there are exactly so many eigenvalues, we say that X has *saturated spectrum*. We have the following analog of Lemma 2 for the undirected case.

Lemma 3. *Let X be a Cayley graph of the cyclic group $\mathbb{Z}_n$. Then W_0 has the maximum possible rank $\lceil (n+1)/2 \rceil$ (implying that X_0 is walk-saturated) if and only if X has saturated spectrum.*

Lemmas 2 and 3 imply that the walk matrix can be used for canonization of a circulant whenever it has simple (for digraphs) or saturated (for graphs) spectrum. The following theorem, therefore, estimates the probability that this canonization method is successful on a random circulant.

Theorem 4.

1. *A uniformly distributed random Cayley digraph of $\mathbb{Z}_n$ has simple spectrum with probability at least $1 - n^{-1/2+o(1)}$.*
2. *A uniformly distributed random Cayley graph of $\mathbb{Z}_n$ has saturated spectrum with probability at least $1 - n^{-1/2+o(1)}$.*

Theorem 1 follows from Theorem 4 on the account of Lemmas 2 and 3. Theorem 4 is our main technical contribution, which can be of independent interest in the context of the research on random circulant matrices [9,26].

The paper is organized as follows. A detailed description of our canonical labeling algorithm is given in Sect. 2. Lemmas 2 and 3 are proved in Sect. 3, where they are restated as Lemma 7. The proofs of Theorems 4 and 1 are presented in Sects. 4 and 5 respectively.

2   The Walk Matrix and Color Refinement

2.1   Definitions and a Relationship

Speaking of a directed graph or, for short, *digraph* $G = (V, E)$, we always assume that G is loopless, that is, the adjacency relation $E \subset V^2$ is irreflexive. Without loss of generality we suppose that G is defined on the vertex set $V = \{0, 1, \ldots, n-1\}$. If E is symmetric, G is referred to as an (undirected) *graph*. The definitions given below for digraphs apply, as a special case, also to graphs.

For $t \in V$, we write G_t to denote the digraph G with distinguished vertex t. The vertex t is called *terminal* or *individualized*. We consider G_t to be a vertex-colored digraph where all vertices are equally colored with the exception of t which has a special color. An isomorphism from G_t to another vertex-individualized digraph H_u is defined as a digraph isomorphism from G to H taking t to u.

A *walk* in G is a sequence of vertices $x_0 x_1 \ldots x_k$ such that $(x_i, x_{i+1}) \in E$ for every $0 \leq i < k$. We say that $x_0 x_1 \ldots x_k$ is a walk of length k from x_0 to x_k. Note that a one-element sequence x_0 is a walk of length 0. Given a digraph G_t with terminal vertex t, we define its *walk matrix* $W_t = (w_{x,k})_{0 \leq x, k < n}$ by setting $w_{x,k}$ to be the number of walks of length k from x to t. Let

$$W_t(x) = (w_{x,0}, w_{x,1}, \ldots w_{x,n-1})$$

be the row of W_t corresponding to the vertex x. If ϕ is an isomorphism from G_t to H_u, then clearly $W_u(\phi(x)) = W_t(x)$. This means that $W_t(x)$ can be used as a canonical label for a vertex x in G_t. We call G_t *walk-discrete* if $W_t(x) \neq W_t(x')$ for all $x \neq x'$. Thus, the walk matrix yields a canonical labeling for the class of walk-discrete digraphs with an individualized vertex.

As it was mentioned in Sect. 1, the walk matrix is efficiently computable by linear algebraic operations. For walk-discrete digraphs, the corresponding canonical labeling can also be obtained combinatorially by the *color refinement* algorithm (CR). Let C_0 be the initial coloring of G_t, that is, $C_0(x) = C_0(x')$ and $C_0(x) \neq C_0(t)$ for all $x \neq t$ and $x' \neq t$. CR iteratively computes new colorings

$$C_{i+1}(x) = \left(C_i(x), \{\!\!\{ C_i(y) \}\!\!\}_{y \in N(x)} \right),$$

where $\{\!\!\{\}\!\!\}$ denotes a multiset and $N(x) = \{y : (x, y) \in E\}$ is the out-neighborhood of x. The color classes of C_{i+1} refine the color classes of C_i. As soon as the color partition stabilizes, CR terminates and outputs the current coloring at that point. To prevent the exponential increase of the color encoding, the colors are renamed after each refinement step (we never need more than n color names). A relationship between CR and the walk counts was observed in [31]. We use the following adaptation of this result for our purposes.

Lemma 5. *If $w_{x,k} \neq w_{x',k}$, then $C_k(x) \neq C_k(x')$.*

Proof. Using the induction on k, we prove that $w_{x,k} = w_{x',k}$ whenever $C_k(x) = C_k(x')$. In the base case of $k = 0$, the equalities $w_{x,0} = w_{x',0}$ and $C_0(x) = C_0(x')$ are equivalent by definition. Assume that $C_k(y) = C_k(y')$ implies $w_{y,k} = w_{y',k}$ for all y and y'. Let $C_{k+1}(x) = C_{k+1}(x')$. By the definition of the refinement step, we have $\{\!\{C_k(y)\}\!\}_{y \in N(x)} = \{\!\{C_k(y)\}\!\}_{y \in N(x')}$. Using the induction assumption, from here we derive the equality $\{\!\{w_{y,k}\}\!\}_{y \in N(x)} = \{\!\{w_{y,k}\}\!\}_{y \in N(x')}$. The equality $w_{x,k+1} = w_{x',k+1}$ now follows by noting that $w_{x,k+1} = \sum_{y \in N(x)} w_{y,k}$. □

Lemma 5 shows that a canonical labeling of a walk-discrete digraph G_t can be obtained by running CR on G_t rather than by directly computing the walk matrix. Recall that CR can be implemented in time $O(n^2 \log n)$ [6,11,20].

2.2   Cayley Graphs of a Cyclic Group

Let $\mathbb{Z}_n$ denote a cyclic group of order n. More specifically, we let $\mathbb{Z}_n = \{0, 1, \ldots, n-1\}$ and consider the addition modulo n on this set. The *Cayley digraph* $X = \mathrm{Cay}(\mathbb{Z}_n, S)$ is defined by a *connection set* $S \subseteq \mathbb{Z}_n \setminus \{0\}$ as follows: $V(X) = \mathbb{Z}_n$ and $(x, y) \in E(X)$ if and only if $y - x \in S$. Note that $S = N(0)$, the out-neighborhood of 0. If S is *inverse-closed*, i.e., $S = -S$, then $E(X)$ is symmetric and we speak of a *Cayley graph*. Cayley (di)graphs of $\mathbb{Z}_n$ are also called *circulant (di)graphs* or *circulants*.

For $u \in \mathbb{Z}_n$, let X_u be the vertex-individualized version of X. Since X is vertex-transitive, all X_u are isomorphic to each other, and we can speak about X_0 without loss of generality. Clearly, in order to canonize X, it is sufficient to canonize X_0. Therefore, the canonization method in the preceding subsection applies to any Cayley digraph $X = \mathrm{Cay}(\mathbb{Z}_n, S)$ provided that X_0 is walk-discrete. We just have to individualize an arbitrary vertex of X and then run CR.

This method does not work for circulant *graphs*. Indeed, define $\rho : \mathbb{Z}_n \to \mathbb{Z}_n$ by $\rho(x) = -x$. If $S = -S$, then ρ is an automorphism of $X = \mathrm{Cay}(\mathbb{Z}_n, S)$ and, hence, $W_0(\rho(x)) = W_0(x)$. This implies that the walk matrix W_0 has at most $\lceil (n+1)/2 \rceil$ different rows, and X_0 cannot be walk-discrete. If this maximum is attained, we call X_0 *walk-saturated*.

Lemma 6. *Let $X = \mathrm{Cay}(\mathbb{Z}_n, S)$, where $S = -S$, and suppose that X_0 is walk-saturated. Fix $u \neq 0$ such that $u \neq n/2$ if n is even. Then*

$$(W_0(x), W_u(x)) \neq (W_0(y), W_u(y))$$

for any two different vertices x and y of X.

Proof. Since X_0 is walk-saturated, the equality $W_0(x) = W_0(y)$ for $x \neq y$ is possible only if $y = \rho(x)$, i.e., $y = -x$ in $\mathbb{Z}_n$. Note that $W_u(x) = W_0(x - u)$. Therefore, the equality $W_u(x) = W_u(y)$ implies that $y = 2u - x$. The equalities $y = -x$ and $y = 2u - x$ can be fulfilled simultaneously only if $2u = 0$, which is excluded. □

Lemma 6 justifies the correctness of the following algorithm for the class of walk-saturated circulant graphs.

CANONICAL LABELING ALGORITHM

INPUT: a circulant graph X.

1. Individualize an arbitrary vertex of X. By vertex-transitivity, we can without loss of generality assume that the individualized vertex is 0.
2. Run CR on X_0. Let C be the obtained coloring of the vertex set.
3. Let c be the lexicographically least color such that there are exactly two vertices u_1 and u_2 with $C(u_1) = C(u_2) = c$. If such c does not exist, then give up. Let u be any of u_1 and u_2. Individualize u in X.
4. Run CR on X_u. Let C' be the obtained coloring.
5. To each vertex x, assign the label $L(x) = (C(x), C'(x))$.
6. Check that all labels $L(x)$ are pairwise distinct. If not, then give up.

For each circulant input graph, our canonization algorithm either produces a vertex labeling L or explicitly gives up (doing always the same for isomorphic inputs). The labeling L is canonical because it does not depend on which vertex is individualized in Step 1 (by vertex-transitivity) nor in Step 3 (because u_1 and u_2 are interchangeable by an automorphism of X_0). Lemma 6 ensures that the algorithm succeeds whenever X_0 is walk-saturated, and this will allow us to estimate the success probability based on Lemma 3 (cf. Lemma 7) and Theorem 4.

Finally, we remark that if the algorithm is run on a non-circulant input graph and outputs a vertex labeling, then this labeling does not need to be canonical. To make it canonical in all cases, Steps 1 and 3 have to be performed for all possible individualized vertices, which can yield $2n$ different labelings $L_1, \ldots, L_{2n}$. Of all these candidate labelings, the algorithm chooses that which yields the isomorphic copy of X with lexicographically least adjacency matrix. The running time of this algorithm variant is $O(n^3 \log n)$. The similar modification is as well possible in the case of digraphs.

3   The Walk Matrix and the Spectrum of Circulants

3.1   The Spectrum of a Circulant

Let $A = (a_{ij})$ be the adjacency matrix of a circulant digraph X, that is, A is the 0–1 matrix whose rows and columns are indexed by the elements $0, 1, \ldots, n-1$ of $\mathbb{Z}_n$ such that $a_{ij} = 1$ exactly when $(i, j) \in E(X)$, that is, $j - i \in S$. Note that A is a *circulant matrix*, which means that the $(i+1)$-th row of A is obtained from its i-th row by the cyclic shift in one element to the right.

Let ω be an n-th root of unity, i.e., $\omega \in \mathbb{C}$ and $\omega^n = 1$. For the vector $V_\omega = (1, \omega, \omega^2, \ldots, \omega^{n-1})^\top$, the definition of a circulant matrix easily implies the equality

$$A V_\omega = \left(a_0 + a_1\omega + a_2\omega^2 + \cdots + a_{n-1}\omega^{n-1}\right) V_\omega = \left(\sum_{j \in S} \omega^j\right) V_\omega,$$

where $(a_0, a_1, a_2, \ldots, a_{n-1})$ is the first row of A, that is, the characteristic vector of $S \subset \mathbb{Z}_n$. We conclude that V_ω is an eigenvector of A corresponding to the eigenvalue $\lambda_{\omega, S} = \sum_{j \in S} \omega^j$.

Now, let $\omega = \zeta_n$ be a primitive n-th root of unity. To be specific, we fix $\zeta_n = e^{-2\pi i/n}$. The n vector-columns V_ω for $\omega = \zeta_n^0, \zeta_n^1, \zeta_n^2, \ldots, \zeta_n^{n-1}$ form a Vandermonde matrix with non-zero determinant. It follows that these n vectors are linearly independent and, therefore, $\lambda_{\zeta_n^0, S}, \lambda_{\zeta_n^1, S}, \ldots, \lambda_{\zeta_n^{n-1}, S}$ is the full spectrum of A. The i-th eigenvalue in this sequence will be denoted by

$$\lambda_{i,S} = \sum_{j \in S} \zeta_n^{ij}. \tag{1}$$

3.2   Discrete Fourier Transform

Let $\mathbb{C}^{\mathbb{Z}_n}$ denote the vector space of all functions from $\mathbb{Z}_n$ to the field of complex numbers $\mathbb{C}$ with pointwise addition and pointwise scalar multiplication. The pointwise multiplication on $\mathbb{C}^{\mathbb{Z}_n}$ will be denoted by $\circ$. Another way to introduce a product on $\mathbb{C}^{\mathbb{Z}_n}$ is to consider the convolution $\alpha * \beta$ of two functions α, β : $\mathbb{Z}_n \to \mathbb{C}$, which is defined by $(\alpha * \beta)(x) = \sum_{y \in \mathbb{Z}_n} \alpha(x - y)\beta(y)$ for each $x \in \mathbb{Z}_n$. Both $\circ$ and $*$ are bilinear and, therefore, both $(\mathbb{C}^{\mathbb{Z}_n}, \circ)$ and $(\mathbb{C}^{\mathbb{Z}_n}, *)$ are n-dimensional algebras over $\mathbb{C}$. The algebra $(\mathbb{C}^{\mathbb{Z}_n}, *)$ can alternatively be seen as the *group algebra* of $\mathbb{Z}_n$ over $\mathbb{C}$ and, as such, it is semisimple by Maschke's theorem; see, e.g., [15, Section 7.1]. Like any two n-dimensional commutative semisimple $\mathbb{C}$-algebras, the algebras $(\mathbb{C}^{\mathbb{Z}_n}, *)$ and $(\mathbb{C}^{\mathbb{Z}_n}, \circ)$ are isomorphic (see [15, Corollary 2.4.2]). We now describe an explicit algebra isomorphism from $(\mathbb{C}^{\mathbb{Z}_n}, *)$ to $(\mathbb{C}^{\mathbb{Z}_n}, \circ)$.

For $T \subseteq \mathbb{Z}_n$, let $\chi_T \in \mathbb{C}^{\mathbb{Z}_n}$ be the characteristic function of T. In particular, $\chi_{\mathbb{Z}_n}$ is the identically one function. For $x \in \mathbb{Z}_n$, we set $\delta_x = \chi_{\{x\}}$.

For $\alpha : \mathbb{Z}_n \to \mathbb{C}$, define $\widehat{\alpha} : \mathbb{Z}_n \to \mathbb{C}$ by

$$\widehat{\alpha}(i) = \sum_{j=0}^{n-1} \zeta_n^{ij} \alpha(j). \tag{2}$$

The *discrete Fourier transform (DFT)* is the linear operator $\mathcal{F} : \mathbb{C}^{\mathbb{Z}_n} \to \mathbb{C}^{\mathbb{Z}_n}$ defined by $\mathcal{F}(\alpha) = \widehat{\alpha}$. In the standard basis $\delta_0, \delta_1, \ldots, \delta_{n-1}$, the DFT is represented by the matrix $F = (\zeta_n^{ij})_{i,j \in \mathbb{Z}_n}$. Since F is the familiar Vandermonde matrix with non-zero determinant, the map F is a linear isomorphism from $\mathbb{C}^{\mathbb{Z}_n}$ onto itself. It is well known and easy to derive from the definitions that

$$\mathcal{F}(\alpha * \beta) = \mathcal{F}(\alpha) \circ \mathcal{F}(\beta). \tag{3}$$

3.3   The Rank of the Walk Matrix

An obvious sufficient condition for a digraph X_0 to be walk-discrete is the non-singularity of its walk matrix W_0. In the case of Cayley graphs $X = \mathrm{Cay}(\mathbb{Z}_n, S)$, this observation has the following analog: The vertex-individualized graph X_0 is walk-saturated whenever its walk matrix W_0 has maximum possible rank $\lceil (n+1)/2 \rceil$. In both cases, the condition that W_0 has maximum possible rank admits a spectral criterion.

Recall that an $n \times n$ matrix has *simple spectrum* if all n of its eigenvalues are pairwise distinct. In the case of an undirected graph $X = \text{Cay}(\mathbb{Z}_n, S)$, where $S = -S$, Equality (1) implies that $\lambda_{a,S} = \lambda_{b,S}$ for $a \neq b$ whenever $a + b = n$. If the eigenvalues of A are unequal otherwise, that is, A has $\lceil (n+1)/2 \rceil$ distinct eigenvalues, then we say that X has *saturated spectrum* (i.e., λ_i has multiplicity 1 if $i = 0$ or if n is even and $i = n/2$, and multiplicity 2 otherwise).

Lemma 7. *Let $X = \text{Cay}(\mathbb{Z}_n, S)$ and W_0 be the walk matrix of X_0.*

1. $\text{rk}\, W_0 = n$ *if and only if X has simple spectrum.*
2. *Let $S = -S$. Then* $\text{rk}\, W_0 = \lceil (n+1)/2 \rceil$ *if and only if X has saturated spectrum.*

Proof. 1. A vector-column $(a_0, a_1, \ldots, a_{n-1})^\top$ will be identified in a natural way with the function $\alpha \in \mathbb{C}^{\mathbb{Z}_n}$ such that $\alpha(x) = a_x$ for $x \in \mathbb{Z}_n$. In this way, the columns of the walk matrix W_0 correspond to the functions $\eta_0, \eta_1, \ldots, \eta_{n-1}$ where $\eta_k(x) = w_{x,k}$. Thus, the rank of W_0 is equal to the dimension of the linear subspace U of $\mathbb{C}^{\mathbb{Z}_n}$ spanned by these functions.

Note that

$$\eta_{k+1}(x) = \sum_{y \in N(x)} \eta_k(y) = \sum_{y \in \mathbb{Z}_n} \chi_S(y - x)\eta_k(y)$$

$$= \sum_{y \in \mathbb{Z}_n} \chi_{-S}(x - y)\eta_k(y) = (\chi_{-S} * \eta_k)(x).$$

It follows that $\eta_0 = \delta_0$, $\eta_1 = \chi_{-S}$, $\eta_2 = \chi_{-S} * \chi_{-S}$ and, generally, $\eta_k = \chi_{-S}^{*(k)}$ is the $(k-1)$-fold convolution of k copies of the characteristic function χ_{-S} of the set $-S$.

Let us apply the discrete Fourier transform $\mathcal{F}$. Note that $\mathcal{F}(\delta_0)$ is the all-ones vector. As easily seen from (1) and (2), $\mathcal{F}(\chi_{-S})$ is the vector whose entries are the eigenvalues $\lambda_{0,-S}, \lambda_{1,-S}, \ldots, \lambda_{n-1,-S}$ of $X' = \text{Cay}(\mathbb{Z}_n, -S)$, the transpose of the digraph X where all arcs are reversed. If X' has at least two equal eigenvalues, then Equality (3) readily implies that, seen as vectors, $\mathcal{F}(\eta_0), \mathcal{F}(\eta_1), \ldots, \mathcal{F}(\eta_{n-1})$ have a common pair of equal coordinates. Therefore, $\dim U = \dim \mathcal{F}(U) \leq n - 1$.

Conversely, suppose that all eigenvalues of X' are pairwise distinct. Recall that $\mathcal{F}(U)$ contains the vectors $\mathcal{F}(\delta_0) = (1, 1, \ldots, 1)^\top$ and $\mathcal{F}(\chi_S) = (\lambda_{0,-S}, \lambda_{1,-S}, \ldots, \lambda_{n-1,-S})^\top$. By (3), $\mathcal{F}(U)$ contains also the vectors $\mathcal{F}(\chi_S^{*(k)}) = \left((\lambda_{0,-S})^k, (\lambda_{1,-S})^k, \ldots, (\lambda_{n-1,-S})^k \right)^\top$ for all $k < n$. These n vectors form a Vandermonde matrix, which is non-degenerate because all $\lambda_{i,-S}$ are pairwise distinct. It follows that $\dim U = \dim \mathcal{F}(U) = n$.

We have shown that $\text{rk}\, W_0 = n$ exactly when the transpose digraph X' has simple spectrum. It remains to note that the spectra of X and X' are simultaneously simple or not. This follows from (1) by applying the automorphism of the cyclotomic field $\mathbb{Q}(\zeta_n)$ from the Galois group $\mathrm{Aut}(\mathbb{Q}(\zeta_n)/\mathbb{Q})$ mapping ζ_n to ζ_n^{-1} (and fixing $\mathbb{Q}$).[1]

2. This part is proved by virtually the same argument (which is even somewhat simpler because we do not need to consider the transpose of X).   □

4   Proof of Theorem 4

We set $\zeta_n = e^{-2\pi i/n}$. As discussed in Subsect. 3.1, a circulant $X = \mathrm{Cay}(\mathbb{Z}_n, S)$ has eigenvalues $\lambda_0, \lambda_1, \ldots, \lambda_{n-1}$ where

$$\lambda_a = \sum_{j \in S} \zeta_n^{aj} = \sum_{j=0}^{n-1} \chi_S(j) \zeta_n^{aj}.$$

Let $\sigma_j = \chi_S(j)$. If X is a random digraph, i.e., the connection set S is chosen uniformly at random among all subsets of $\mathbb{Z}_n \setminus \{0\}$, then $\sigma_1, \sigma_2, \ldots, \sigma_{n-1}$ is a Bernoulli process, that is, these $n - 1$ random variables are independent and identically distributed with σ_j taking each of the values 0 and 1 with probability $1/2$. If X is a random graph, i.e., the connection set $S = -S$ is chosen randomly among all inverse-closed subsets, the values $\sigma_1, \sigma_2, \ldots, \sigma_{\lfloor n/2 \rfloor}$ form a Bernoulli process, and the remaining values are determined by the equality $\sigma_j = \sigma_{n-j}$. For each $a = 0, 1, \ldots, n - 1$, the eigenvalue

$$\lambda_a = \sum_{j=1}^{n-1} \sigma_j \zeta_n^{aj} \tag{4}$$

becomes a random variable taking its values in the cyclotomic field $\mathbb{Q}(\zeta_n)$.

We will use the following observation. As usually, $\phi(n)$ stands for Euler's totient function.

Lemma 8. *No two different subsets of $\{\zeta_n^j : 1 \le j \le n/\ln n\}$ have equal sums of their elements.*

Proof. The known lower bounds for $\phi(n)$ (see, e.g., [5]) imply that $\phi(n) > n/\ln n$ for $n \ge 3$. The existence of such subsets would therefore yield a non-trivial linear combinations with rational coefficient of $1, \zeta_n, \zeta_n^2, \ldots, \zeta_n^{\phi(n)-1}$, contradicting the fact that these numbers form a basis of $\mathbb{Q}(\zeta_n)$ considered as a vector space over $\mathbb{Q}$.
□

Our overall strategy for proving Theorem 4 will be to use the union bound

$$\mathsf{P}[\lambda_a = \lambda_b \text{ for some } 0 \le a, b \le n - 1] \le \sum_{0 \le a, b \le n-1} \mathsf{P}[\lambda_a = \lambda_b]$$

---

[1] Recall that $\mathbb{Q}(\zeta_n)$ is obtained by adjoining ζ_n to the field of rationals $\mathbb{Q}$. In other words, this is the smallest subfield of $\mathbb{C}$ containing $\mathbb{Q}$ and ζ_n.

and to show that the right hand side is bounded by $n^{-1/2+o(1)}$. The summation goes through the pairs of unequal a and b, which in the undirected case also have to satisfy the condition $a + b \neq n$. We split the set of all such pairs a, b in three classes P_1, P_2, P_3 and prove that $\sum_{(a,b) \in P_\ell} P[\lambda_a - \lambda_b = 0]$ is sufficiently small for each $\ell = 1, 2, 3$. To define the sets P_1, P_2, P_3, let us introduce some notation.

Given an integer a, let

$$g(a) = n/\gcd(a, n) \text{ and } a' = a/\gcd(a, n).$$

Note that

$$\zeta_n^{aj} = \zeta_{g(a)}^{a'j}. \tag{5}$$

The integers a' and $g(a)$ are coprime, and we set $r = (a')^{-1} \bmod g(a)$.

Furthermore, let $g = g(a)$ and $\xi = \zeta_g$. Since ξ is a g-th primitive root of unity, the $g - 1$ numbers

$$\xi = \zeta_n^{ar}, \ \xi^2 = \zeta_n^{2ar}, \ \ldots, \xi^{g-1} = \zeta_n^{(g-1)ar} \tag{6}$$

(where the equalities are due to (5)) are pairwise distinct. It is useful to note that they appear in the right hand side of (4) for the $g-1$ indices $j = r, 2r, \ldots, (g-1)r$, which are understood modulo n.

Given another integer b, set $\eta = \zeta_n^{br}$ and note that the $g - 1$ numbers (where $g = g(a)$ as defined above)

$$\eta = \zeta_n^{br}, \ \eta^2 = \zeta_n^{2br}, \ \ldots, \eta^{g-1} = \zeta_n^{(g-1)br} \tag{7}$$

appear in the same places in the corresponding expression for λ_b. Let $h = g(b)$.

Considering a pair of different a and b, we will always assume without loss of generality that $g \geq h$. Let $g' = g'(a, b)$ be the minimum positive integer such that $\eta^{g'} = 1$. Note that $g' \leq h \leq g$. Thus, the number of different numbers in the sequence (7) is equal to g' if $g' < g$ and to $g' - 1$ if $g' = g$.

Note that $\eta \neq \xi$. Otherwise we would have $\zeta_n^{ar} = \zeta_n^{br}$ and $h = g$. By (5), this would imply $\zeta_g^{a'r} = \zeta_g^{b'r}$ and, as a consequence, $a' = b'$ and $a = b$.

Let $\varepsilon > 0$ be an arbitrarily small constant. Once this parameter is fixed, n will be supposed to be sufficiently large. We divide the set of all pairs a, b into three parts P_1, P_2, P_3 and, correspondingly, split our analysis into three cases. We give a detailed argument for digraphs (Part 1 of the theorem), which with minor changes works also for graphs (Part 2). We comment on these changes in the end of the proof.

Case 1: $P_1 = \{(a, b) : g \geq n^{10\varepsilon} \text{ and } g' \geq n^\varepsilon\}$. Note that $\eta \neq 1$ for else $g' = 1$. For the argument $\arg(z)$ of a complex number z, we suppose that $\arg(z) \in [0, 2\pi)$. Note that $\arg(\eta) > \arg(\xi)$ because $\eta \neq \xi$ and $g(b) \leq g$. We claim that there exists s such that $1 \leq s \leq \ln^{10} n$ and

$$\text{either } \arg(\eta^s) < 2\pi/\ln^{10} n \text{ or } \arg(\eta^{-s}) < 2\pi/\ln^{10} n.$$

Indeed, there exist s_1 and s_2 such that $1 \leq s_1 < s_2 \leq \ln^{10} n \leq g'$ and the distance between η^{s_2} and η^{s_1} in the circle $\{z \in \mathbb{C} : |z| = 1\}$ is at most $2\pi/\ln^{10} n$ and non-zero. We can take $s = s_2 - s_1$.

We now consider three subcases. If $\arg(\eta^{-s}) < 2\pi/\ln^{10} n$, then we set $K = \{s, 2s, \ldots, \lfloor \ln^2 n \rfloor s\}$. Note that

$$\left\{\arg(\xi^k)\right\}_{k \in K} \subset (0, 2\pi/\ln^8 n) \text{ and } \left\{\arg(\eta^k)\right\}_{k \in K} \subset (2\pi - 2\pi/\ln^8 n, 2\pi), \quad (8)$$

which also implies

$$\left\{\xi^k\right\}_{k \in K} \cap \left\{\eta^k\right\}_{k \in K} = \emptyset. \quad (9)$$

The second case we consider is that $\arg(\eta^s) < 2\pi/\ln^{10} n$ and $\eta^s \neq \xi^s$. As easily seen, there exists a subset $K \subset \{s, 2s, \ldots, \lfloor \ln^2 n \rfloor s\}$ of size at least $\lfloor \ln^2 n \rfloor/2$ such that, while

$$\left\{\arg(\xi^k)\right\}_{k \in K} \subset (0, 2\pi/\ln^8 n) \text{ and } \left\{\arg(\eta^k)\right\}_{k \in K} \subset (0, 2\pi/\ln^8 n), \quad (10)$$

Equality (9) holds anyway.

We now show that in the above two cases the equality $\lambda_a = \lambda_b$, i.e.,

$$\sum_{j=1}^{n-1} \sigma_j \zeta_n^{aj} = \sum_{j=1}^{n-1} \sigma_j \zeta_n^{bj} \quad (11)$$

has small probability. Let us expose the values of all random variables σ_j excepting σ_{kr} with $k \in K$ (recall that the indices are considered modulo n). Equality (11) can now be written as

$$c_1 + \sum_{k \in K} \sigma_{kr} \xi^k = c_2 + \sum_{k \in K} \sigma_{kr} \eta^k$$

for some constants c_1 and c_2. In other words, we estimate the probability of the event that for a random subset $K' \subseteq K$ the random variable $\sum_{k \in K'}(\xi^k - \eta^k)$ is equal to a fixed number. Consider two different subsets $K_1, K_2 \subseteq K$. The equality

$$\sum_{k \in K_1}(\xi^k - \eta^k) = \sum_{k \in K_2}(\xi^k - \eta^k)$$

is equivalent to

$$\sum_{k \in K_1} \xi^k + \sum_{k \in K_2} \eta^k = \sum_{k \in K_2} \xi^k + \sum_{k \in K_1} \eta^k, \quad (12)$$

which is impossible due to (9) and Lemma 8. Lemma 8 is applicable due to (8) and (10) (in the former case, both sides of (12) have to be multiplied by ζ_n^t for a small t). We conclude that $\lambda_a = \lambda_b$ with probability at most $2^{-|K|} = n^{-\Theta(\ln n)}$.

There remains the case that $\arg(\eta^s) < 2\pi/\ln^{10} n$ and $\eta^s = \xi^s$. Consider the complex numbers $\eta^{ks+1} - \xi^{ks+1}$ for $0 \leq k \leq \ln^2 n$. Their absolute values are equal as $|\eta^{ks+1} - \xi^{ks+1}| = |\xi^{ks}(\eta - \xi)| = |\eta - \xi|$. Moreover, these numbers are pairwise different and close to each other in the circle $\{z \in \mathbb{C} : |z| = |\eta - \xi|\}$ because

$$\left|\arg(\eta^{ks+1} - \xi^{ks+1}) - \arg(\eta^{(k-1)s+1} - \xi^{(k-1)s+1})\right|$$

$$= \left|\arg(\xi^{ks}(\eta - \xi)) - \arg(\xi^{(k-1)s}(\eta - \xi))\right| = s\arg(\xi) \leq \frac{2\pi}{\ln^{10} n}.$$

In order to estimate the probability of (11), let us expose all random variables σ_j excepting $\sigma_{(ks+1)r}$ with $0 \leq k \leq \ln^2 n$. Lemma 8 implies that there is at most one assignment of the unexposed variables satisfying (11). This yields the probability bound $n^{-\Theta(\ln n)}$. We conclude that

$$\sum_{(a,b)\in P_1} \mathsf{P}[\lambda_a = \lambda_b] = |P_1|n^{-\Theta(\ln n)} = n^{-\Theta(\ln n)}.$$

Case 2: $P_2 = \{(a,b) : g < n^{10\varepsilon}\}$. Let $J = \{j < n : \zeta_n^{aj} = \xi\}$. Note that $|J| = \frac{n}{g} > n^{1-10\varepsilon}$. Moreover, J contains a subset $J' \subseteq J$ of size $|J'| \geq |J|/h > n^{1-20\varepsilon}$ such that the numbers ζ_n^{bj} for all $j \in J'$ are equal to the same h-th root of unity η'. If $h < g$, then clearly $\xi \neq \eta'$. If $h = g$, then we have $\eta' = \eta$ and, hence, $\xi \neq \eta'$ as well.

Let us expose all random variables σ_j except those for $j \in J'$. Equality (11) then implies that $\sum_{j \in J'} \sigma_j = c$ for a constant c. The probability of the last event is bounded by $\binom{|J'|}{\lfloor|J'|/2\rfloor}2^{-|J'|} = O(|J'|^{-1/2}) = O(n^{-1/2+10\varepsilon})$.

We now estimate the number of pairs (a, b) in P_2. Recall that $a = \gcd(a, n)a'$, where $a' \leq n/\gcd(a, n) = g < n^{10\varepsilon}$. The factor $\gcd(a, n)$ of a can be chosen in at most $d(n)$ ways, where $d(n)$ denotes the total number of divisors of n. It is known [1, Theorem 13.12] that $d(n) = n^{O(1/\ln\ln n)}$. Since there are less than $n^{10\varepsilon}$ possibilities to choose the factor a', the element a can be chosen in at most $n^{11\varepsilon}$ ways, and the same holds true as well for b. It follows that $|P_2| \leq n^{22\varepsilon}$, and we conclude that

$$\sum_{(a,b)\in P_2} \mathsf{P}[\lambda_a = \lambda_b] = O(|P_2|n^{-1/2+10\varepsilon}) = O(n^{-1/2+32\varepsilon}).$$

Case 3: $P_3 = \{(a,b) : g \geq n^{10\varepsilon}$ while $g' < n^{\varepsilon}\}$. Set m to be the largest integer such for the corresponding element of the sequence (6) we have $\arg(\xi^m) < 2\pi/\ln^2 n$. Let $J = \{r, 2r, \ldots, mr\}$. Note that $|J| = m \geq n^{8\varepsilon}$. Moreover, J contains a subset $J' \subseteq J$ of size $|J'| \geq |J|/g' > n^{7\varepsilon}$ such that the numbers ζ_n^{bj} for all $j \in J'$ are equal to the same h-th root of unity η'.

Let us expose all σ_j except for $j \in J'$. By Chernoff's bound, we have $\frac{1}{2}|J'| - |J'|^{2/3} < \sum_{j\in J'} \sigma_j < \frac{1}{2}|J'| + |J'|^{2/3}$ with probability no less than $1 - 2\exp(-|J'|^{1/3}) > 1 - \exp(-n^{\varepsilon})$. Consider the event (11) conditioned on $\sum_{j\in J'} \sigma_j = m$ for a fixed integer m such that $\frac{1}{2}|J'|-|J'|^{2/3} < m < \frac{1}{2}|J'|+|J'|^{2/3}$. Lemma 8 implies that Equality (11) can be satisfied by at most one assignment to the variables σ_j for $j \in J'$. Therefore, (11) is fulfilled with probability at most $1/\binom{|J'|}{m}$, which does not exceed $2^{-n^{6\varepsilon}}$. This readily implies that $\sum_{(a,b)\in P_3} \mathsf{P}[\lambda_a = \lambda_b] \leq \exp(-(1 + o(1))n^{\varepsilon})$.

Summing up, we see that $\sum_{0\leq a,b\leq n-1} \mathsf{P}[\lambda_a = \lambda_b] = O(n^{-1/2+\varepsilon})$ for each $\varepsilon > 0$ and sufficiently large n. This proves Part 1 of Theorem 4. The proof of Part 2 is virtually the same. It should only be noted that for an unexposed random variable σ_j, in place of the term $\sigma_j\zeta_n^{aj}$ we now have to deal with $\sigma_j\zeta_n^{aj} +$

$\sigma_{n-j}\zeta_n^{a(n-j)} = \sigma_j(\zeta_n^{aj} + \zeta_n^{-aj})$. Lemma 8 is still applicable after multiplication of the whole sum by ζ_n^t for a small t. In the first subcase of Case 1, there is a possibility that $\eta^{-s} = \xi^s$, which is treated similarly to the third subcase of Case 1.

Remark 9. The probability bound in Theorem 4 is nearly optimal. As can be shown, a random digraph $\mathrm{Cay}(\mathbb{Z}_n, S)$ for $n = 3p$ with p prime has repeated eigenvalues with probability $\Omega(n^{-1/2})$. Also, the spectrum of a random graph $\mathrm{Cay}(\mathbb{Z}_n, S)$ for $n = 5p$ is not saturated with the same probability bound.

5   Proof of Theorem 1

We are now ready to prove our main result. Theorem 1 holds true for each of the three concepts of a circulant:

- A Cayley (di)graph $X = \mathrm{Cay}(\mathbb{Z}_n, S)$. The uniform probability distribution of X means that the connection set S is equiprobably chosen among all subsets of $\mathbb{Z}_n \setminus \{0\}$ in the case of digraphs and among all inverse-closed subsets in the case of graphs.
- An unlabeled circulant, i.e., an isomorphism class of Cayley (di)graphs $X = \mathrm{Cay}(\mathbb{Z}_n, S)$. The uniform distribution means that each isomorphism class on $\mathbb{Z}_n$ is chosen equiprobably. In the algorithmic setting, an isomorphism class is presented by its representative (a (di)graph from the class). Alternatively, we can think of the probability distribution on all Cayley (di)graphs $X = \mathrm{Cay}(\mathbb{Z}_n, S)$ in which each X appears with probability $1/(\ell_n\, s(X))$, where ℓ_n is the total number of n-vertex unlabeled circulants and $s(X)$ is the number of connection sets S such that $\mathrm{Cay}(\mathbb{Z}_n, S) \cong X$.
- A labeled circulant, i.e., a (di)graph on the vertex set $\{0, 1, \ldots, n-1\}$ isomorphic to some $X = \mathrm{Cay}(\mathbb{Z}_n, S)$. The uniform distribution is considered on all n-vertex (di)graphs in this class.

In each of the three cases, we use the same canonization algorithm presented in Subsect. 2.2. For digraphs, the algorithm is extremely simple: We just individualize one vertex in an input digraph X and run CR on the obtained vertex-colored graph X_0. In this way we either get an individual label for each vertex of X or the algorithm gives up. The labeling is canonical for all circulants X, and it is successfully produced whenever X_0 is walk-discrete. For graphs, the algorithm is a little bit more complicated and is discussed in detail in Subsect. 2.2. It succeeds whenever X_0 is walk-saturated.

In Subsect. 3.3 we noted two sufficient spectral conditions: X_0 is walk-discrete whenever X has simple spectrum, and X_0 is walk-saturated whenever X has saturated spectrum. This reduces our task to estimating the probability that the random digraph X has simple spectrum and, respectively, that the random graph X has saturated spectrum. In the case of Cayley (di)graphs, the proof is completed by applying Theorem 4.

It remains to show that the estimate of Theorem 4 stays as well true for the uniformly distributed labeled and unlabeled circulants. A complete treatment of these two cases can be found in the full version of the paper [32] where we present a general way to convert an estimate for one distribution into an estimate for another distribution with a small overhead cost.

References

1. Apostol, T.M.: Introduction to Analytic Number Theory. Springer, New York (1998)
2. Babai, L.: Isomorphism problem for a class of point-symmetric structures. Acta Math. Acad. Sci. Hung. **29**(3–4), 329–336 (1977). https://doi.org/10.1007/BF01895854
3. Babai, L., Chen, X., Sun, X., Teng, S., Wilmes, J.: Faster canonical forms for strongly regular graphs. In: 54th Annual IEEE Symposium on Foundations of Computer Science (FOCS 2013), pp. 157–166. IEEE Computer Society (2013). https://doi.org/10.1109/FOCS.2013.25
4. Babai, L., Erdős, P., Selkow, S.M.: Random graph isomorphism. SIAM J. Comput. **9**(3), 628–635 (1980)
5. Bach, E., Shallit, J.: Algorithmic Number Theory: Efficient Algorithms. The MIT Press, Cambridge (1996)
6. Berkholz, C., Bonsma, P.S., Grohe, M.: Tight lower and upper bounds for the complexity of canonical colour refinement. Theory Comput. Syst. **60**(4), 581–614 (2017). https://doi.org/10.1007/s00224-016-9686-0
7. Bhoumik, S., Dobson, T., Morris, J.: On the automorphism groups of almost all circulant graphs and digraphs. Ars Math. Contemp. **7**(2), 499–518 (2014). https://doi.org/10.26493/1855-3974.315.868
8. Bollobás, B.: Distinguishing vertices of random graphs. Ann. Discrete Math. **13**, 33–49 (1982). https://doi.org/10.1016/S0304-0208(08)73545-X
9. Bose, A., Saha, K.: Random Circulant Matrices. CRC Press, Boca Raton (2019). https://doi.org/10.1201/9780429435508
10. Cai, J., Fürer, M., Immerman, N.: An optimal lower bound on the number of variables for graph identifications. Combinatorica **12**(4), 389–410 (1992). https://doi.org/10.1007/BF01305232
11. Cardon, A., Crochemore, M.: Partitioning a graph in $O(|A| \log_2 |V|)$. Theor. Comput. Sci. **19**, 85–98 (1982)
12. Davis, P.J.: Circulant Matrices, 2nd edn. AMS Chelsea Publishing, New York (1994)
13. Dobson, E., Spiga, P., Verret, G.: Cayley graphs on abelian groups. Combinatorica **36**(4), 371–393 (2016). https://doi.org/10.1007/s00493-015-3136-5
14. Dobson, T., Malnič, A., Marušič, D.: Symmetry in Graphs. Cambridge Studies in Advanced Mathematics, vol. 198. Cambridge University Press, Cambridge (2022). https://doi.org/10.1017/9781108553995
15. Drozd, Y.A., Kirichenko, V.V.: Finite Dimensional Algebras. Springer, Berlin (1994). https://doi.org/10.1007/978-3-642-76244-4
16. Evdokimov, S., Ponomarenko, I.: Circulant graphs: recognizing and isomorphism testing in polynomial time. St. Petersburg Math. J. **15**(6), 813–835 (2004)
17. Fuhlbrück, F., Köbler, J., Ponomarenko, I., Verbitsky, O.: The Weisfeiler-Leman algorithm and recognition of graph properties. Theor. Comput. Sci. **895**, 96–114 (2021). https://doi.org/10.1016/j.tcs.2021.09.033

18. Godsil, C.: Controllable subsets in graphs. Ann. Comb. **16**(4), 733–744 (2012). https://doi.org/10.1007/s00026-012-0156-3
19. Hagos, E.M.: Some results on graph spectra. Linear Algebra Appl. **356**(1–3), 103–111 (2002). https://doi.org/10.1016/S0024-3795(02)00324-5
20. Immerman, N., Lander, E.: Describing graphs: a first-order approach to graph canonization. In: Selman, A.L. (ed.) Complexity Theory Retrospective, pp. 59–81. Springer, New York (1990). https://doi.org/10.1007/978-1-4612-4478-3_5
21. Kiefer, S., Ponomarenko, I., Schweitzer, P.: The Weisfeiler-Leman dimension of planar graphs is at most 3. J. ACM **66**(6), 44:1–44:31 (2019). https://doi.org/10.1145/3333003
22. Kluge, L.: Combinatorial refinement on circulant graphs. Technical report (2022). arxiv.org/abs/2204.01054
23. Kucera, L.: Canonical labeling of regular graphs in linear average time. In: 28th Annual Symposium on Foundations of Computer Science (FOCS 1987), pp. 271–279 (1987). https://doi.org/10.1109/SFCS.1987.11
24. Liu, F., Siemons, J.: Unlocking the walk matrix of a graph. J. Algebraic Comb. **55**(3), 663–690 (2022). https://doi.org/10.1007/s10801-021-01065-3
25. McKay, B.D., Piperno, A.: Practical graph isomorphism, II. J. Symb. Comput. **60**, 94–112 (2014). https://doi.org/10.1016/j.jsc.2013.09.003
26. Meckes, M.W.: Some results on random circulant matrices. In: High Dimensional Probability. V: The Luminy Volume, pp. 213–223. IMS, Institute of Mathematical Statistics, Beachwood (2009). https://doi.org/10.1214/09-IMSCOLL514
27. Muzychuk, M.: A solution of the isomorphism problem for circulant graphs. Proc. Lond. Math. Soc. **88**(1), 1–41 (2004). https://doi.org/10.1112/S0024611503014412
28. Muzychuk, M.E., Klin, M.H., Pöschel, R.: The isomorphism problem for circulant graphs via Schur ring theory. In: Codes and Association Schemes. DIMACS Series in Discrete Mathematics and Theoretical Computer Science, vol. 56, pp. 241–264. DIMACS/AMS (1999). https://doi.org/10.1090/dimacs/056/19
29. O'Rourke, S., Touri, B.: On a conjecture of Godsil concerning controllable random graphs. SIAM J. Control. Optim. **54**(6), 3347–3378 (2016). https://doi.org/10.1137/15M1049622
30. Ponomarenko, I.: On the WL-dimension of circulant graphs of prime power order. Technical report (2022). http://arxiv.org/abs/2206.15028
31. Powers, D.L., Sulaiman, M.M.: The walk partition and colorations of a graph. Linear Algebra Appl. **48**, 145–159 (1982). https://doi.org/10.1016/0024-3795(82)90104-5
32. Verbitsky, O., Zhukovskii, M.: Canonization of a random circulant graph by counting walks. Technical report (2023). arxiv.org/abs/2310.05788
33. Weisfeiler, B., Leman, A.: The reduction of a graph to canonical form and the algebra which appears therein. NTI, Ser. **2**(9), 12–16 (1968). English translation is available at https://www.iti.zcu.cz/wl2018/pdf/wl_paper_translation.pdf

Counting Vanishing Matrix-Vector Products

Cornelius Brand[1], Viktoriia Korchemna[2], Kirill Simonov[3],
and Michael Skotnica[4($\boxtimes$)]

[1] Algorithms and Complexity Theory Group, Regensburg University, Regensburg,
Germany
cornelius.brand@informatik.uni-regensburg.de
[2] Algorithms and Complexity Group, TU Wien, Vienna, Austria
vkorchemna@ac.tuwien.ac.at
[3] Hasso Plattner Institute, University of Potsdam, Potsdam, Germany
Kirill.Simonov@hpi.de
[4] Department of Applied Mathematics, Charles University, Prague, Czech Republic
skotnica@kam.mff.cuni.cz

Abstract. Consider the following parameterized counting variation of
the classic subset sum problem, which arises notably in the context of
higher homotopy groups of topological spaces: Let $\mathbf{v} \in \mathbb{Q}^d$ be a rational vector, $(T_1, T_2 \ldots T_m)$ a list of $d \times d$ rational matrices, $S \in \mathbb{Q}^{h \times d}$ a
rational matrix not necessarily square and k a parameter. The goal is to
compute the number of ways one can choose k matrices $T_{i_1}, T_{i_2}, \ldots, T_{i_k}$
from the list such that $S T_{i_k} \cdots T_{i_1} \mathbf{v} = \mathbf{0} \in \mathbb{Q}^h$.

In this paper, we show that this problem is #W[2]-hard for parameter k. As a consequence, computing the k-th homotopy group of a d-
dimensional 1-connected topological space for $d > 3$ is #W[2]-hard for
parameter k. We also discuss a decision version of the problem and its
several modifications for which we show W[1]/W[2]-hardness. This is in
contrast to the parameterized k-sum problem, which is only W[1]-hard
(Abboud-Lewi-Williams, ESA'14). In addition, we show that the decision
version of the problem without parameter is an undecidable problem, and
we give a fixed-parameter tractable algorithm for matrices of bounded
size over finite fields, parameterized by the matrix dimensions and the
order of the field.

Keywords: parameterized complexity · W[2]-hardness ·
undecidability · k-sum · homotopy group

1 Introduction

Topology is one of the most important and active areas of mathematics, emerging
from vast generalizations of geometry (see, e.g., [13] for a gentle introduction
along this path). In full generality, it studies fundamental properties of *topological
spaces*, which generalize a broad array of geometric objects (including manifolds,

© The Author(s), under exclusive license to Springer Nature Singapore Pte Ltd. 2024
R. Uehara et al. (Eds.): WALCOM 2024, LNCS 14549, pp. 335–349, 2024.
https://doi.org/10.1007/978-981-97-0566-5_24

Hilbert spaces, algebraic varieties and even embeddings of graphs). The concept of a topological space allows to speak in a very general manner about the "shape" of a space, and a prime goal of topology consists in classifying spaces according to their shapes. For instance, it is intuitively obvious that a mug with a handle and a football should belong to distinct classes of shapes, for instance because one has a hole in it and the other, preferably, does not. Whether or not, then, a mug with sharp edges and a doughnut belong to the same class is a different question, and good reasons exist for choosing either way of answering it.

Thus, clearly, any such classification depends on the precise way in which the classes are defined and the structures provided on top of purely topological information (such as differential information, i.e., about "sharp edges"); one particularly important way of doing so is to make a single class out of all those shapes that can be deformed into each other according to specific rules retaining. The usual notion of equivalence under deformation of shapes corresponding to general topological spaces is furnished by *homotopy*, which, very roughly speaking, identifies any two shapes that can be obtained from one another through arbitrary deformations without "tearing" or "cutting" (and hence identifying the mug with the doughnut, while differentiating both from the football).

Associated to this notion are the so-called *homotopy groups* of a topological space, denoted π_k, for $k \geq 1$. The most intuitive of them is the group π_1, which is often called the *fundamental group* of the space. It captures certain data about the different ways that loops (that is, closed curves in the space) can pass through the space. The higher homotopy groups ($k > 1$) correspond to ways of routing higher-dimensional "loops" in the space, and Whitehead's Theorem provides a crucial equivalence between the structure of homotopy groups and the homotopy class of a broad category of topological spaces called CW-complexes [20, 21]. The present paper deals with an intermediate problem related to the computation of homotopy groups, which allows to show lower bounds for the complexity of computing the higher homotopy groups of a topological space.

Before speaking about computational tasks associated with topological spaces, one needs to define how a topological space is even represented. While the generality of the concept may make it seem hard to come up with such a representation in general, the usual path taken in computational topology is as follows: Many topological spaces can be described by finite structures, e.g., by abstract simplicial complexes, which are simply collections of point sets closed under taking subsets, and it hence suffices to provide the maximal subsets of a simplicial complex to specify it in full. Such structure can then be used as an input for a computer and therefore, it is natural to ask how hard it is to compute these homotopy groups of a given topological space, represented by an abstract simplicial complex.

Novikov in 1955 [17] and independently Boone in 1959 [5] showed undecidability of the word problem for groups. Their result also implies undecidability of computing the fundamental group. In fact, even determining whether the fundamental group of a given topological space is trivial is undecidable.

On the other hand, for 1-connected spaces (for those, whose π_1 is trivial) it is known that their π_k for $k > 1$ are finitely generated abelian groups which are always isomorphic to groups of the form $\mathbb{Z}^n \oplus \mathbb{Z}_{p_1} \oplus \mathbb{Z}_{p_2} \oplus \cdots \oplus \mathbb{Z}_{p_m}$, where $p_1, \ldots, p_m$ are powers of prime numbers.[1] An algorithm for computing π_k of a 1-connected space, where $k > 1$, was first introduced by Brown in 1957 [7].

In 1989, Anick [4] proved that computing the rank of π_k, that is, the number of direct summands isomorphic to $\mathbb{Z}$ (represented by n in the expression above) is #P-hard for 4-dimensional 1-connected spaces.[2] Another computational problem called VEST, which we define below, was used in Anick's proof as an intermediate step. Briefly said, #P-hardness of VEST implies #P-hardness of computing the rank of π_k, which is the motivation for studying the problem in the present article.

Vector Evaluated After a Sequence of Transformations (VEST). The input of this problem defined by Anick [4] is a vector $\mathbf{v} \in \mathbb{Q}^d$, a list $(T_1, \ldots, T_m)$ of rational $d \times d$ matrices and a rational matrix $S \in \mathbb{Q}^{h \times d}$ where $d, m, h \in \mathbb{N}$.

For an instance of VEST let an *M-sequence* be a sequence of integers $M_1, M_2, M_3, \ldots$, where

$$M_k := |\{(i_1, \ldots, i_k) \in \{1, \ldots, m\}^k; ST_{i_k} \cdots T_{i_1} \mathbf{v} = \mathbf{0}\}|.$$

Given an instance of VEST and $k \in \mathbb{N}$, the goal is to compute M_k.

From an instance of VEST, it is possible to construct a corresponding algebraic structure called *123H-algebra* in polynomial time whose *Tor-sequence* is equal to the M-sequence of the original instance of a VEST. This is stated in [4, Theorem 3.4] and it follows from [2, Theorem 1.3] and [3, Theorem 7.6].

Given a presentation of a 123H-algebra, one can construct a corresponding 4-dimensional simplicial complex in polynomial time whose sequence of ranks (rk π_2, rk $\pi_3, \ldots$) is related to the Tor-sequence of the 123H-algebra. In particular, it is possible to compute that Tor-sequence from the sequence of ranks using an FPT algorithm.[3] This follows from [19] and [8]. To sum up, hardness of computing M_k of VEST implies hardness of computing π_k.

Parameterized Complexity and the W-Hierarchy. Parameterized complexity classifies decision or counting computational problems with respect to a given parameter(s). For instance, one can ask if there exists an independent set of size k in a given graph or how many independent sets of size k (for counting version) are in a given graph, respectively, where k is the parameter. From this viewpoint, we can divide decision problems into several groups which form the W-*hierarchy*:

$$\mathsf{FPT} \subseteq \mathsf{W}[1] \subseteq \mathsf{W}[2] \subseteq \cdots \subseteq \mathsf{XP}.$$

The class FPT consists of decision problems solvable in time $f(k)n^c$, where $f(k)$ is a computable function of the parameter k, n is the size of input and c is a

---

[1] Note that $\mathbb{Z}^n$ is a direct sum of n copies of $\mathbb{Z}$ while $\mathbb{Z}_{p_i}$ is a finite cyclic group of order p_i.

[2] When k is a part of the input and represented in unary.

[3] The notion of FPT algorithm is defined in the next paragraph.

constant, while the class XP consists of decision problems solvable in time $cn^{f(k)}$. We also refer to algorithms with running time $f(k)n^c$ ($cn^{f(k)}$ correspondingly) adis FPT (XP) algorithms. The class W[1] consists of all problems which admit a parameterized reduction to the satisfiability problem of a boolean circuit of constant depth with AND, OR and NOT gates such that there is at most 1 gate of higher input size than 2 on each path from the input gate to the final output gate (this number of larger gates is called *weft*), where the parameter is the number of input gates set to TRUE. Here, a parameterized reduction from a parameterized problem A to a parameterized problem B is an algorithm that, given an instance (x, k) of A, in time $f(k)n^c$ produces an equivalent instance (x, k') of B such that $k' \le g(k)$, for some computable functions $f(\cdot)$, $g(\cdot)$, and a constant c.

The class W[i] then consists of problems that admit a parameterized reduction to the satisfiablity problem of a boolean circuit of a constant depth and weft at most i, parameterized by the number of input gates set to TRUE.

It is only known that FPT $\subsetneq$ XP, while the other inclusions in the W-hierarchy are not known to be strict. However, it is strongly believed that FPT $\subsetneq$ W[1]. *Therefore, one cannot expect existence of an algorithm solving a W[1]-hard problem in time $f(k)n^c$ where $f(k)$ is a computable function of k and c is a constant.* For the detailed presentation of W-hierarchy and parameterized complexity in general we refer the reader to [15].

Analogously, one can define classes FPT and XP for counting problems. That is, a class of counting problems solvable in time $f(k)n^c$ or $cn^{f(k)}$, respectively. Problems for which there is a parameterized counting reduction to a problem of counting solutions for a boolean circuit of constant depth and weft at most i then form class #W[i]. Note that there are decision problems from FPT whose counting versions are #W[1]-hard, e.g., counting paths or cycles of length k parameterized by k [14]. Similarly to the decision case, *if a counting problem is shown to be #W[i]-hard for some i one should not expect existence of an algorithm solving this problem in time $f(k)n^c$.* For more details on parameterized counting we refer the reader to [14].

In our case, the number k of the homotopy group π_k plays the role of the parameter. In 2014 Čadek et al. [9] proved that computing π_k (and thus, also computing the rank of π_k) is in XP parameterized by k.

A lower bound for the complexity from the parameterized viewpoint was obtained by Matoušek in 2013 [16]. He proved that computing M_k of a VEST instance is #W[1]-hard. This also implies #W[1]-hardness for the original problem of computing the rank of higher homotopy groups π_k (for 4-dimensional 1-connected spaces) for parameter k. Matoušek's proof also works as a proof for #P-hardness and it is shorter and considerably easier than the original proof of Anick in [4].

In this paper, we strengthen the result of Matoušek and show that computing M_k of a VEST instance is #W[2]-hard. Our proof is even simpler than the previous proof of #W[1]-hardness.

Theorem 1. *Given a* VEST *instance, computing M_k is #W[2]-hard when parameterized by k.*

Theorem 1 together with the result of Anick [4] implies the following.

Corollary 1. *Computing the rank of the k-th homotopy group of a d-dimensional 1-connected space for $d > 3$ is #W[2]-hard for parameter k.*

Remark 1. Note that computing M_k of a VEST instance is an interesting natural self-contained problem even without the topological motivation. We point out that our reduction showing #W[2]-hardness of this problem uses only 0, 1 values in the matrices and the initial vector $\mathbf{v}$. Moreover, each matrix will have at most one 1 in each row. Therefore, such construction also shows #W[2]-hardness of computing M_k of a VEST instance in the $\mathbb{Z}_2$ setting. That is, for the case when $T_1, T_2, \ldots T_m \in \mathbb{Z}_2^{d \times d}, S \in \mathbb{Z}_2^{h \times d}$ and $\mathbf{v} \in \mathbb{Z}_2^d$.

The Decision Version of VEST. We also provide a comprehensive overview of the parameterized complexity of VEST as a decision problem, where given an instance of VEST one needs to determine whether $M_k > 0$. In addition to the standard variant of the problem, we consider several modifications of VEST: when the matrices have constant size, when the matrix S is the identity matrix, when we omit the initial vector and the target is identity/zero matrix etc.

Unfortunately, even considering the simplifications above, we show that nearly all versions in our consideration are W[1]- or W[2]-hard. The following table is an overview of our results.

Size of matrices	a) $\mathbf{v}$ and S	b) only $\mathbf{v}$		c) only S	d) no $\mathbf{v}$, no S
1. 1×1	P	P	0	P	P
			I	W[1]-hard	W[1]-hard
2. 2×2	W[1]-hard	W[1]-hard	0	W[1]-hard	W[1]-hard
			I		
3. input size	W[2]-hard	W[2]-hard	0	W[2]-hard	W[2]-hard
			I	W[1]-hard	W[1]-hard

The first column stands for the standard VEST while the second stands for the VEST without the special matrix S or alternatively, for the case when S is the identity matrix. Therefore, the hardness results for the first column follow from the second.

The third and the fourth columns are without the initial vector $\mathbf{v}$. In this case, we want to choose k matrices $T_{i_1}, T_{i_2}, \ldots, T_{i_k}$ such that $ST_{i_k} \cdots T_{i_1}$ (or $T_{i_k} \cdots T_{i_1}$, respectively) is equal to a given target matrix. It is natural to assume the following two target matrices: the zero matrix (the rows labeled by $\mathbf{0}$) and the identity matrix (the rows labeled by I). Again, the hardness results for the third column follow from the fourth.

Regarding the 1×1 case, the only nontrivial case is when the target is $I = 1$. The W[1]-hardness results for the 1×1 case also implies W[1]-hardness for the 2×2 case and the input size case when the target is the identity matrix.

Therefore, in Sect. 3 we prove hardness for

- "1 d) I" (Theorem 3),
- "2 b)" (Theorem 5),
- "2 d) $\mathbf{0}$" (Theorem 4).

The W[2]-hardness for "3 c)" follows from the proof of Theorem 1 (see Remark 3) and we show that "3 b)" and "3 d) $\mathbf{0}$" are equivalent to "3 a)" under parameterized reduction (Theorems 6, 7).

Remark 2. We should also point out that we actually do not know what is the perameterized complexity class containing VEST. However, we conjecture that it is W[2] when there is no restriction for the size of matrices, and W[1] when the size of matrices is bounded.

Fixed-Parameter Tractability over Finite Fields. Our results, summarized in the table above, show that VEST remains hard even on highly restricted instances, such as binary matrices with all the ones located along the main diagonal, or matrices of a constant size. However, it turns out that combination of these two restrictions – on the field size and the matrix sizes – makes even the counting version of VEST tractable.

We proceed by lifting tractability to the matrices of unbounded size but with all non-zero entries occurring in at most the p first rows.

Theorem 2. *Given an instance of VEST and $k \in \mathbb{N}$, computing M_k is FPT with respect to $|\mathbb{F}|$ and p, if all non-zero entries of matrices belong to p first rows.*

The problem remains FPT with respect to $|\mathbb{F}|$ and p even if the task is to find the minimal k for which $M_k > 0$, or to report that there is no such k.

Undecidability of VEST Without Parameter. In contrast, we show in the last section (Sect. 5) that for $\mathbb{F} = \mathbb{Q}$ the problem of determining whether there exists k such that $M_k > 0$ for an instance of VEST is an undecidable problem.[4]

2   The Proof of #W[2]-Hardness of VEST

In this section, we prove that computing M_k of a VEST instance is #W[2]-hard (Theorem 1). Our reduction is from the problem of counting dominating sets of size k which is known to be #W[2]-complete (see [14]) and which we recall in the paragraph below.

For a graph $G(V, E)$ and its vertex $v \in V$ let $N[v]$ denote the closed neighborhood of a vertex v. That is, $N[v] := \{u \in V; \{u, v\} \in E\} \cup \{v\}$. A *dominating*

---

[4] One of the reviewers thankfully pointed out related results about the matrix mortality problem (see, e.g., [10]) which give a shorter proof of this fact and which also give a proof of undecidability of VEST without the initial vector and the parameter.

set of a graph $G(V, E)$ is a set $U \subseteq V$ such that for each $v \in V$ there is $u \in U$ such that $v \in N[u]$.

NUMBER OF DOMINATING SETS OF SIZE k

Input: A graph $G(V, E)$ and a parameter k.
Question: How many dominating sets of size k are in G?

$$\begin{pmatrix} 0 & 0 & 0 & 0 \\ 0 & 0 & 1 & 0 \\ 0 & 0 & 0 & 1 \\ 0 & 0 & 0 & 1 \end{pmatrix}$$

Fig. 1. The submatrix of T_u consisting of rows and columns $u_1, \ldots, u_4$. The rest of the non-diagonal entries of T_u are zeros. The diagonal entries $T_u^{w_1, w_1}$ for $w \in N[u]$ are zeros, the rest of the diagonal entries are ones.

Proof (of Theorem 1). As we said, we show an FPT counting reduction from the problem of counting dominating sets of size k to VEST.

Let $G(V, E)$ be the input graph and let $n = |V|$. The corresponding instance of VEST will consist of n matrices $\{T_u : u \in V\}$ of size $4n \times 4n$, one for each vertex, and a matrix S of the same size. Whence, the initial vector $\mathbf{v}$ must be of size $4n$. For each vertex $u \in V$, we introduce four new coordinates $u_1, \ldots, u_4$ and set $\mathbf{v}_{u_1} = 1, \mathbf{v}_{u_2} = \mathbf{v}_{u_3} = 0$ and $\mathbf{v}_{u_4} = 1$.

We define the matrices $\{T_u : u \in V\}$ and S by describing their behavior. Let $\mathbf{x}$ be a vector which is going to be multiplied with a matrix T_u (that is, some intermediate vector obtained from $\mathbf{v}$ after potential multiplications). The matrix T_u sets $\mathbf{x}_{w_1}$ to zero for each $w \in N[u]$, which corresponds to domination of vertices in $N[u]$ by the vertex u, and also sets $\mathbf{x}_{u_2}$ to $\mathbf{x}_{u_3}$ and $\mathbf{x}_{u_3}$ to $\mathbf{x}_{u_4}$. The rest of the entries of $\mathbf{x}$ including $\mathbf{x}_{u_4}$ are kept, see Fig. 1. The matrix S then nullifies coordinates u_3, u_4 and keeps the coordinates u_1 and u_2 for each $u \in V$. In other words, S is diagonal such that $S^{u_1, u_1} = S^{u_2, u_2} = 1$ and $S^{u_3, u_3} = S^{u_4, u_4} = 0$.

The parameter remains equal to k.

For correctness, let $u^1, \ldots, u^k$ be any vertices from V, and let $\mathbf{r}$ be the vector obtained from $\mathbf{v}$ after multiplying by the matrices $T_{u^1}, \ldots, T_{u^k}$ (observe that the order of multiplication does not matter since all T_u, $u \in V$, pairwise commute). By construction, for every vertex $u \in V$, the entry $\mathbf{r}_{u_1} = 0$ if and only if u is dominated by some u^i, $i \in [k]$, and $\mathbf{r}_{u_2} = 0$ if and only if T_u appears among $T_{u^1}, \ldots, T_{u^k}$ at most once. Indeed, if T_u is selected once then $\mathbf{r}_{u_2} = \mathbf{v}_{u_3} = 0$ while if it is selected more than once then $\mathbf{r}_{u_2} = \mathbf{v}_{u_4} = 1$. If T_u is not among $T_{u^1}, \ldots, T_{u^k}$ then $\mathbf{r}_{u_2} = \mathbf{v}_{u_2} = 0$.

Therefore, $\mathbf{r} = S T_{u^1} \cdots T_{u^k} \mathbf{v}$ is a zero vector if and only if $u^1, \ldots, u^k$ are pairwise distinct and form the dominating set in G. This provides a one-to-one correspondence between subsets of matrices yielding the solution of VEST and dominating sets of size k in G. It remains to note that every such subset of matrices gives rise to $k!$ sequences that have to be counted in M_k. Hence, $M_k = k! D_k$ where D_k is the number of dominating sets of size k in G. The reduction is clearly FPT since the construction does not use parameter k and is polynomial in size of the input. □

Remark 3. Note that the decision version of the problem of DOMINATING SETS OF SIZE k is W[2]-hard. For showing W[2]-hardness of the decision version of VEST we need not deal with the repetition of matrices. In particular, we do not need the special coordinates u_2, u_3, u_4 and therefore, the corresponding instance of VEST can consist only of diagonal 0, 1 matrices of size $n \times n$.

3   Modifications of VEST

In this section, we prove hardness for the variants of the decision version of VEST we have discussed in the introduction. First of all, we recall a well-known W[1]-hard k-SUM problem [1].

---

k-SUM

Input: A set A of integers and a parameter k.
Question: Is it possible to choose k distinct integers from A such that their sum is equal to zero?

We note that in the versions of k-SUM studied in the literature the goal is to pick *distinct* elements of the input set in order to achieve 0 or eventually another number. However, the motivation for VEST does not suggest that the matrices applied to the vector have to be distinct. Thus, in order to model VEST by k-SUM, it is more natural to also allow repetition of numbers. For our particular proofs, we will use the following version with target number 1.

AT-MOST-k-SUM WITH REPETITIONS AND TARGET 1

Input: A set A of integers and parameter k.
Question: Is it possible to choose *at most* k integers from A (possibly with repetition) such that their sum is equal to 1?

We are not aware of any previous studies on parameterized complexity of AT-MOST-k-SUM WITH REPETITIONS AND TARGET 1, nor does it seem that there exists a simple parameterized reduction from the original variant of the problem to the one with repetitions. The proof of W[1]-hardness of this problem is deferred to the full version [6]. When we assume multiplication instead of addition, the following problem arises.

---

k-PRODUCT WITH REPETITIONS

Input: A set A of rational numbers and a parameter k.
Question: Is it possible to choose k numbers from A (possibly with repetitions) such that their product is equal to 1?

W[1]-hardness for this problem might be a folklore result but we present a complete proof using a reduction from k-EXACT COVER, which is known to be W[1]-hard (see [12]).

---

k-EXACT COVER

Input: A universe U, a collection $\mathcal{C}$ of subsets of U and a parameter k.
Question: Can U be partitioned into k sets from $\mathcal{C}$?

Theorem 3. k-PRODUCT WITH REPETITIONS *is* W[1]-*hard parameterized by* k.

Proof. We show a parameterized reduction from k-EXACT COVER. For each element $u \in U$ we associate one prime p_u, then for each $C \in \mathcal{C}$ we set $i_C := p \prod_{c \in C} p_c$ where p is a prime which is not used for any element from U and $s := \frac{1}{p^k \prod_{u \in U} p_u}$.

The integers i_C for each $C \in \mathcal{C}$ and s then form the input for $(k+1)$-PRODUCT WITH REPETITIONS

If $C_1, C_2, \ldots, C_k \in \mathcal{C}$ is a solution of k-EXACT COVER then $s \prod_{i=1}^{k} i_{C_i} = 1$.

Conversely, let $q_1, q_2, \ldots, q_{k+1}$ be a solution of the constructed $(k+1)$-PRODUCT WITH REPETITIONS. First of all, note that s must be chosen precisely once. Indeed, all numbers except for s are greater than 1 and thus, s must be chosen at least once. If it were chosen more than once it would not be possible to cancel a power of p^k in the denominator since the numerator would contain at most p^{k-1}. Therefore, the product of $q_1, q_2, \ldots, q_{k+1}$ is of the form $s i_{C_{j_k}} i_{C_{j_{k-1}}} \cdots i_{C_{j_1}} = 1$ which means that each prime representing an element of U in the denominator is canceled. In other words, each element of U is covered. Note also that since s is chosen precisely once there cannot be any repetition within $i_{C_{j_k}} i_{C_{j_{k-1}}} \cdots i_{C_{j_1}}$.

The reduction is parameterized since we only need the parameter k for k multiplications of $\frac{1}{p}$ and first $n+1$ primes, where $n = |U|$, can be generated in time $\mathcal{O}(n^3)$ using, e.g., the Sieve of Eratosthenes for $(n+1)^2$. This follows from the fact, that the first n primes lie among $1, \ldots, n^2$. For more details we refer the reader to the full version [6].   □

Let us now call the variant of VEST without S and $\mathbf{v}$ MATRIX k-PRODUCT WITH REPETITIONS. As we have mentioned in the introduction we consider two cases regarding the target matrix, namely, the Identity matrix and the Zero matrix.

MATRIX k-PRODUCT WITH REPETITIONS RESULTING TO ZERO MATRIX

Input: A list of $d \times d$ rational matrices and a parameter k.
Question: Is it possible to choose k matrices from the list (possibly with repetitions) such that their product is the $d \times d$ zero matrix?

> MATRIX k-PRODUCT WITH REPETITIONS RESULTING TO IDENTITY MATRIX
>
> Input: A list of $d \times d$ rational matrices and a parameter k.
>
> Question: Is it possible to choose k matrices from the list (possibly with repetitions) such that their product is the $d \times d$ identity matrix?

Note that MATRIX k-PRODUCT WITH REPETITIONS RESULTING TO IDENTITY MATRIX for 1×1 matrices is exactly k-PRODUCT WITH REPETITIONS. Therefore W[1]-hardness for MATRIX k-PRODUCT WITH REPETITIONS RESULTING TO IDENTITY MATRIX for all matrix sizes follows from Theorem 3.

Regarding MATRIX k-PRODUCT WITH REPETITIONS RESULTING TO ZERO MATRIX, we can easily see that it is solvable in linear time for 1×1 matrices. Indeed, it is sufficient to check whether $T_i = 0$ for some i. However, already for 2×2 matrices the problem becomes hard.

Theorem 4. MATRIX k-PRODUCT WITH REPETITIONS RESULTING TO ZERO MATRIX *is* W[1]-*hard for parameter k even for 2×2 integer matrices.*

Proof. We reduce from AT-MOST-k-SUM WITH REPETITIONS AND TARGET 1.

For every integer x let us define the matrix $U_x := \begin{pmatrix} 1 & x \\ 0 & 1 \end{pmatrix}$. It is easy to see that $U_x U_y = U_{x+y}$. Let $\mathcal{I}$ be an instance of AT-MOST-k-SUM WITH REPETITIONS AND TARGET 1 with the set of integers A and parameter k. We create an equivalent instance $\mathcal{I}'$ of MATRIX $(k+2)$-PRODUCT WITH REPETITIONS RESULTING TO ZERO MATRIX with the set of matrices $\{U_a : a \in A\} \cup \{X\}$, where $X = \begin{pmatrix} 0 & 0 \\ -1 & 1 \end{pmatrix}$.

For correctness, assume that $\mathcal{I}$ is a YES-instance and $a_1, \ldots, a_\ell \in A$ are such that $\ell \le k$ and $\sum_{i=1}^{\ell} a_i = 1$. Consider the following product of $\ell + 2$ matrices:

$$X \cdot \prod_{i=1}^{l} U_{a_i} \cdot X = X \cdot U_{\sum_{i=1}^{\ell} a_i} \cdot X = XU_1X = \begin{pmatrix} 0 & 0 \\ -1 & 1 \end{pmatrix}\begin{pmatrix} 1 & 1 \\ 0 & 1 \end{pmatrix}\begin{pmatrix} 0 & 0 \\ -1 & 1 \end{pmatrix} = \mathbf{0}.$$

For the other direction, assume that $\mathcal{I}'$ is a YES-instance. Let ℓ, $1 \le \ell \le k + 2$, be the minimal integer such that there are matrices $T_1, \ldots, T_\ell$ from $\{U_a : a \in A\} \cup \{X\}$ with $T_\ell T_{\ell-1} \cdots T_1 = \mathbf{0} \in \mathbb{Q}^{2 \times 2}$. Since the matrix X is idempotent (i.e. $X^2 = X$), it does not appear two times in a row, otherwise we could reduce the length of the product. Notice that X should appear at least once, since the determinants of all U_a are non-zero. Assume that there is precisely one occurrence of X, then the product has form:

$$U_r X U_s = \begin{pmatrix} 1 & r \\ 0 & 1 \end{pmatrix}\begin{pmatrix} 0 & 0 \\ -1 & 1 \end{pmatrix}\begin{pmatrix} 1 & s \\ 0 & 1 \end{pmatrix} = \begin{pmatrix} -r & -rs + r \\ -1 & 1 - s \end{pmatrix} \neq \mathbf{0}.$$

Hence, X appears at least twice. Let us fix any two consequent occurrences and consider the partial product between them:

$$XU_rX = \begin{pmatrix} 0 & 0 \\ -1 & 1 \end{pmatrix}\begin{pmatrix} 1 & r \\ 0 & 1 \end{pmatrix}\begin{pmatrix} 0 & 0 \\ -1 & 1 \end{pmatrix} = \begin{pmatrix} 0 & 0 \\ r-1 & 1-r \end{pmatrix} = (1 - r) \cdot X.$$

If $r \neq 1$, we would get a shorter product resulting in zero, which contradicts to minimality of ℓ. Hence $r = 1$, so the product of U_a that appear between two occurrences of X is equal to U_1. Since there are at most k of such U_a and the sum of corresponding indices a is equal to 1, we obtain a solution to $\mathcal{I}$.   □

We can use a similar approach to establish hardness of the VEST problem without S (or alternatively when S is the identity matrix). Recall that here the task is to obtain not necessarily a zero matrix but any matrix which contains a given vector $\mathbf{v}$ in a kernel.

Theorem 5. VEST *is* W[1]*-hard for parameter k even for 2×2 integer matrices and when S is the identity matrix.*

The proof of this theorem is very similar to the proof of Theorem 4 and it can be found in the full version [6].

At the end of this section, we show that VEST is equivalent to VEST without S (in other words, when $S = I_d$) and to MATRIX k-PRODUCT WITH REPETITIONS RESULTING TO ZERO MATRIX.

Theorem 6. *There is a parameterized reduction from VEST to the special case of VEST where S is the identity matrix, and the other way around.*

Proof. One direction is trivial since the case when $S = I$ is just a special case of VEST.

Regarding the other, let $\left(S \in \mathbb{Q}^{h \times d}, T_1, T_2, \ldots, T_m \in \mathbb{Q}^{d \times d}, \mathbf{v} \in \mathbb{Q}^d, k\right)$ be an instance of VEST. First, we observe that without loss of generality we can suppose that S is a square matrix (in other words, $h = d$). Indeed, if $h < d$ then we just add $d - h$ zero rows to S. If $h > d$ we add $h - d$ zero columns to S, $h - d$ zero entries to $\mathbf{v}$ and $h - d$ zero rows as well as $h - d$ zero columns to each T_i, obtaining a new instance $\left(S' \in \mathbb{Q}^{h \times h}, T_1', T_2', \ldots, T_m' \in \mathbb{Q}^{h \times h}, \mathbf{v}' \in \mathbb{Q}^h, k\right)$ of VEST, where $S' = \left(S\ 0\right)$, $T_i' = \begin{pmatrix} T_i\ 0 \\ 0\ 0 \end{pmatrix}$, $i = 1, \ldots, m$, and $\mathbf{v}' = \begin{pmatrix} \mathbf{v} \\ 0 \end{pmatrix}$.

Now, assuming $d = h$, we add 2 dimensions. To the vector $\mathbf{v}$ we add k on the $(d+1)$-st position and 1 on the $(d+2)$-nd position. To each matrix T_i we add a 2×2 submatrix which subtracts the $(d+2)$-nd component of a vector from the $(d+1)$-st. To the matrix S we add a submatrix which nullifies the $(d+2)$-nd component and multiplies the $(d+1)$-th component by 10. Thus, we obtain:

$$\mathbf{v}' = \begin{pmatrix} \mathbf{v} \\ k \\ 1 \end{pmatrix}, \quad S' = \begin{pmatrix} S\ 0\ 0 \\ 0\ 10\ 0 \\ 0\ 0\ 0 \end{pmatrix}, \quad T_i' = \begin{pmatrix} T_i\ 0\ 0 \\ 0\ 1\ -1 \\ 0\ 0\ 1 \end{pmatrix}, \quad i = 1, \ldots, m.$$

The new parameter is set to $k + 1$. If there is a solution of the original problem, that is, there are k matrices $T_{i_1}, \ldots, T_{i_k}$ such that $S T_{i_k} T_{i_{k-1}} \cdots T_{i_1} \mathbf{v} = \mathbf{0}$, then $S' T_{i_k}' T_{i_{k-1}}' \cdots T_{i_1}' \mathbf{v}' = \mathbf{0}$, since 1 is k times subtracted from the $(d+1)$-st component of $\mathbf{v}'$ and the $(d+2)$-nd component is then nullified by S'.

Conversely, if there are $k + 1$ matrices $Y_1, Y_2, \ldots, Y_{k+1}$, where each Y_i is either S' or T_j' for some j, such that $\mathbf{r} = Y_{k+1} Y_k \cdots Y_1 \mathbf{v}' = \mathbf{0}$ then Y_{k+1} must be equal to S' and the rest of the matrices are of type T'_j, otherwise $\mathbf{r}_{d+1} \neq 0$ or $\mathbf{r}_{d+2} \neq 0$. Indeed, at first k matrices of type T'_j must be selected to nullify the $(d+1)$-st component: if $Y_i = S'$ for some $i \leq k$, this would increase the non-zero $(d+1)$-st component, so there would be no way to nullify it by remaining matrices $Y_{i+1}, \ldots, Y_{k+1}$. At the same time, S' should be necessarily selected once to nullify the $(d+2)$-nd component, so $Y_{k+1} = S'$. Therefore, by restricting the matrices $Y_1, \ldots, Y_k$ to the first d coordinates we obtain a solution to VEST with matrix S.   □

Theorem 7. VEST *and* MATRIX k-PRODUCT WITH REPETITIONS RESULTING TO ZERO MATRIX *are equivalent under parameterized reduction.*

Note that one implication is relatively straightforward. Regarding the other, the idea is to again "simulate" the special matrix S and the vector $\mathbf{v}$ by an ordinary matrix and force them to be selected as the leftmost and the rightmost, respectively. For the complete proof, please see the full version [6].

4   Fixed-Parameter Tractability over Finite Fields

While most of the hardness results for VEST and its variations in the previous section use constant-sized matrices, the entries of these matrices can be arbitrarily large. Here, we study the variation of the problem when all the matrices have entries from some finite field. Notice that restricting the field size by itself does not make the problem tractable: recall the reduction from dominating set from Sect. 1 which also works over $\mathbb{Z}_2$. However, along with a bound on the matrix sizes this makes the problem tractable.

Lemma 1. *Computing M_k for a given instance of* VEST *over a finite field $\mathbb{F}$ is* FPT *when parameterized by the size of $\mathbb{F}$ and the size of matrices.*

Proof. Let $\mathcal{M}_{\mathbb{F}}^d$ be the set of all $d \times d$ matrices with entries from $\mathbb{F}$, then $|\mathcal{M}_{\mathbb{F}}^d| = |\mathbb{F}|^{d^2}$. For every $X \in \mathcal{M}_{\mathbb{F}}^d$ and every integer $i \in [k]$ we will compute the number a_X^i of sequences of i matrices from the input such that their product is equal to X. In particular, this allows to obtain $M_k = \sum_{X \in \mathcal{M}_{\mathbb{F}}^d : SX\mathbf{v}=\mathbf{0}} a_X^k$.

For $i = 1$ the computation can be done simply by traversing the input matrices. Assume that a_X^i have been computed for all the matrices X and all $i \in [j]$. We initiate by setting $a_X^{j+1} = 0$ for every $X \in \mathcal{M}_{\mathbb{F}}^d$. Then, for every pair (X, q), where $X \in \mathcal{M}_{\mathbb{F}}^d$ and $q \in [m]$, we increment $a_{XT_q}^{j+1}$ by a_X^j. In the end we will then have a correctly computed value $a_Y^{j+1} = \sum_{q=1}^m \sum_{X : XT_q = Y} a_X^j$.   □

Our next step is to consider the matrices of unbounded size, but with at most p first rows containing non-zero entries. In particular, if $\mathbb{F} = \mathbb{Z}_2$, we can associate to every such matrix T a graph with the vertex set $[d]$ such that there exists an edge between the vertices i and j, $i \leq j$, if and only if $T^{i,j} = 1$. Conversely, a graph with the vertex set $[d]$ can be represented by such a matrix if and only if the vertices in $[p]$ form its vertex cover.

Every matrix with all but the first p rows equal to zero has form $\begin{pmatrix} A & B \\ 0 & 0 \end{pmatrix}$, where A is $p \times p$ matrix and B is $p \times (d-p)$ matrix. Further, we will denote matrices of this form by $(A|B)$. Observe that the product of two such matrices has the same form: $(A_1|B_1)(A_2|B_2) = (A_1 A_2 | A_1 B_2)$.

Corollary 2. $\prod_{i=1}^{k} (A_i|B_i) = \left(\left(\prod_{i=1}^{k} A_i \right) \mid \left(\prod_{i=1}^{k-1} A_i \cdot B_k \right) \right) = (X A_k | X B_k)$, where $X = \prod_{i=1}^{k-1} A_i$. In particular, the product does not depend on B_i for $i < k$.

Theorem 2. *Given an instance of* VEST *and* $k \in \mathbb{N}$, *computing* M_k *is* FPT *with respect to* $|\mathbb{F}|$ *and* p, *if all non-zero entries of matrices belong to* p *first rows.*

Proof. We slightly modify the definition of a_X^i from the proof of Lemma 1. Now, for every $i \in [k]$ and every matrix $X \in \mathcal{M}_{\mathbb{F}}^p$, let a_X^i be the number of sequences of i matrices $T_j = (A_j|B_j)$ from the input such that corresponding product of A_j is equal to X.

The values of a_X^i for every $i \in [k-1]$ can be computed in the same way as in the proof of Lemma 1. Given this information, we can count the sequences of length k that nullify $\mathbf{v}$. Indeed, by Corollary 2, the number of such sequences with the rightmost matrix $T_j = (A_j|B_j)$ is precisely $b_j = \sum_{X \in \mathcal{M}_{\mathbb{F}}^p : \, S \cdot (X A_j | X B_j) \cdot \mathbf{v} = 0} a_X^{k-1}$, and M_k is then equal to $\sum_{j=1}^{m} b_j$. $\square$

We remark that the algorithm for computing a_X^i from the last proof can be exploited to determine minimal k such that $M_k > 0$, or to report that there is no such k. For this, we run the algorithm with $k = 1$, then with $k = 2$ and so on. If after some iteration $k = j+1$ we obtain that $M_i = 0$ for all $i \in [j]$ and there is no $X \in \mathcal{M}_{\mathbb{F}}^p$ such that $a_X^1 = \cdots = a_X^j = 0$ and $a_X^{j+1} \neq 0$, we may conclude that $M_k = 0$ for all $k \in \mathbb{N}$, since every product of length more than j can be obtained as a product of length at most j, and none of the latter nullify $\mathbf{v}$. Otherwise, there exists at least one $X \in \mathcal{M}_{\mathbb{F}}^p$ such that $a_X^1 = \cdots = a_X^j = 0$ and $a_X^{j+1} \neq 0$. Note that every $X \in \mathcal{M}_{\mathbb{F}}^p$ can play this role only for one value of k. Therefore, it always suffices to make $|\mathcal{M}_{\mathbb{F}}^p|$ iterations of the algorithm.

5   Undecidability of VEST

In this section, we show that determining whether there exists $k \in \mathbb{N}$ such that $M_k > 0$ for an instance of VEST is an undecidable problem. We reduce from POST'S CORRESPONDENCE PROBLEM which is known to be undecidable [18].

(BINARY) POST'S CORRESPONDENCE PROBLEM

Input: m pairs $(v_1, w_2), (v_2, w_2), \ldots, (v_m, w_m)$ of words over alphabet $\{0, 1\}$.

Question: Is it possible to choose k pairs $(v_{i_1}, w_{i_k}), (v_{i_2}, w_{i_2}) \ldots, (v_{i_k}, w_{i_k})$, for some $k \in \mathbb{N}$, such that $v_{i_1} v_{i_2} \cdots v_{i_k} = w_{i_1} w_{i_2} \cdots w_{i_k}$?

For a word $v \in \{0, 1\}^*$ let $|v|$ be its length and let $(v)_2$ be the integer value of v interpreting it as a binary number. Moreover, we associate with v the matrix

$$T_v = \begin{pmatrix} 2^{|v|} - (v)_2 & (v)_2 \\ 2^{|v|} - (v)_2 - 1 & (v)_2 + 1 \end{pmatrix},$$

which allows to model concatenations as the following lemma suggests.

Lemma 2. *Let v, w be binary words. Then, $T_v T_w = T_{wv}$ where wv is the concatenation of w and v.*

The construction of T_v is a based on [11][Satz 28, p. 157] which we are aware of thanks to Günter Rote. For the complete proof of Lemma 2, see the full version [6].

Reduction. Given an instance of POST'S CORRESPONDENCE PROBLEM we describe what an instance of VEST may look like. For each pair (v, w) we define $T_{(v,w)} = \begin{pmatrix} T_v & 0 \\ 0 & T_w \end{pmatrix}$, we set the initial vector $\mathbf{v} := (0, 1, 0, 1)^T$ and $S := (1, 0, -1, 0)$. The undecidability of VEST then follows from the following lemma.

Lemma 3. *Let $(v_{i_1}, w_{i_1}), (v_{i_2}, w_{i_2}) \ldots, (v_{i_k}, w_{i_k})$ be k pairs of binary words. Then $ST_{(v_{i_k}, w_{i_k})} \cdots T_{(v_{i_1}, w_{i_1})}\mathbf{v} = \mathbf{0}$ if and only if $v_{i_1} \cdots v_{i_k} = w_{i_1} \cdots w_{i_k}$.*

Proof. By Lemma 2, $T_{(v_{i_k}, w_{i_k})} \cdots T_{(v_{i_2}, w_{i_2})} T_{(v_{i_1}, w_{i_1})} = T_{(v_{i_1} v_{i_2} \cdots v_{i_k}, w_{i_1} w_{i_2} \cdots w_{i_k})}$. The vector $\mathbf{v}$ selects the second column of the submatrix $T_{v_{i_1} v_{i_2} \cdots v_{i_k}}$ and the second column of the submatrix $T_{w_{i_1} w_{i_2} \cdots w_{i_k}}$. Therefore, the result is equal to

$$\left((v_{i_1} v_{i_2} \cdots v_{i_k})_2, (v_{i_1} v_{i_2} \cdots v_{i_k})_2 + 1, (w_{i_1} w_{i_2} \cdots w_{i_k})_2, (w_{i_1} w_{i_2} \cdots w_{i_k})_2 + 1 \right)^T.$$

The final result after multiplying S with the vector above is the following 1-dimensional vector $(v_{i_1} v_{i_2} \cdots v_{i_k})_2 - (w_{i_1} w_{i_2} \cdots w_{i_k})_2$. $\square$

Acknowledgments C.B. and V.K. were supported by Austrian Science Fund (FWF, project Y1329). C.B. is also funded by the European Union (ERC, CountHom, 101077083). Views and opinions expressed are however those of the author(s) only and do not necessarily reflect

European Research Council
Established by the European Commission

those of the European Union or the European Research Council Executive Agency. Neither the European Union nor the granting authority can be held responsible for them.

K.S. was supported by DFG Research Group ADYN via grant DFG 411362735, M.S. acknowledges support by the project "Grant Schemes at CU" (reg. no. CZ.02.2.69/0.0/0.0/19_073/0016935) and GAČR grant 22-19073S.

We also thank the anonymous reviewers for their useful comments.

Disclosure of Interests. The authors have no competing interests to declare that are relevant to the content of this article.

References

1. Abboud, A., Lewi, K., Williams, R.: Losing weight by gaining edges. In: Schulz, A.S., Wagner, D. (eds.) ESA 2014. LNCS, vol. 8737, pp. 1–12. Springer, Cham (2014). https://doi.org/10.1007/978-3-662-44777-2_1
2. Anick, D.J.: Diophantine equations, Hilbert series, and undecidable spaces. Ann. Math. **122**, 87–112 (1985)
3. Anick, D.J.: Generic algebras and CW complexes. In: Algebraic Topology and Algebraic K-Theory, pp. 247–321 (1987)
4. Anick, D.J.: The computation of rational homotopy groups is #℘-hard. Computers in geometry and topology. In: Proceedings Conference Chicago/Ill (1986). Lect. Notes Pure Appl. Math. **114**, 1–56 (1989)
5. Boone, W.W.: The word problem. Ann. Math. **70**, 207–265 (1959)
6. Brand, C., Korchemna, V., Simonov, K., Skotnica, M.: Counting vanishing matrix-vector products (2023). arXiv: 2309.13698
7. Brown, E.H.: Finite computability of Postnikov complexes. Ann. Math. **65**, 1 (1957)
8. Čadek, M., Krčál, M., Matoušek, J., Vokřínek, L., Wagner, U.: Extendability of continuous maps is undecidable. Discrete Comput. Geom. **51**(1), 24–66 (2014). https://doi.org/10.1007/s00454-013-9551-8
9. Čadek, M., Krčál, M., Matoušek, J., Vokřínek, L., Wagner, U.: Polynomial-time computation of homotopy groups and Postnikov systems in fixed dimension. SIAM J. Comput. **43**(5), 1728–1780 (2014)
10. Cassaigne, J., Halava, V., Harju, T., Nicolas, F.: Tighter undecidability bounds for matrix mortality, zero-in-the-corner problems, and more (2014). arXiv: 1404.0644
11. Claus, V.: Stochastische Automaten. Vieweg+Teubner Verlag (1971). (in German)
12. Downey, R.G., Fellows, M.R.: Fixed-parameter tractability and completeness II: on completeness for W[1]. Theor. Comput. Sci. **141**(1), 109–131 (1995)
13. Flegg, G.: From Geometry to Topology. Courier Corporation (2001)
14. Flum, J., Grohe, M.: The parameterized complexity of counting problems. SIAM J. Comput. **33**(4), 892–922 (2004)
15. Flum, J., Grohe, M.: Parameterized Complexity Theory. Springer, Heidelberg (2004). https://doi.org/10.1007/3-540-29953-X
16. Matoušek, J.: Computing higher homotopy groups is W[1]-hard (2013). arXiv:1304.7705
17. Novikov, P.S.: On the algorithmic unsolvability of the word problem in group theory. Trudy Mat. Inst. Steklova **44**, 1–143 (1955). (in Russian)
18. Post, E.L.: A variant of a recursively unsolvable problem. J. Symb. Log. **12**(2), 55–56 (1947)
19. Roos, J.E.: Relations between the poincaré-betti series of loop spaces and of local rings. In: Malliavin, M.P. (ed.) Séminaire d'Algèbre Paul Dubreil. LNM, vol. 740, pp. 285–322. Springer, Heidelberg (1979). https://doi.org/10.1007/BFb0071068
20. Whitehead, J.H.C.: Combinatorial homotopy. I. Bull. Am. Math. Soc. **55**(3), 213–245 (1949)
21. Whitehead, J.H.C.: Combinatorial homotopy. II. Bull. Am. Math. Soc. **55**(5), 453–496 (1949)

Efficient Enumeration of Drawings and Combinatorial Structures for Maximal Planar Graphs

Giordano Da Lozzo, Giuseppe Di Battista, Fabrizio Frati$^{(\boxtimes)}$, Fabrizio Grosso, and Maurizio Patrignani

Roma Tre University, Rome, Italy
{giordano.dalozzo,giuseppe.dibattista,fabrizio.frati,fabrizio.grosso,
maurizio.patrignani}@uniroma3.it

Abstract. We propose efficient algorithms for enumerating the celebrated combinatorial structures of maximal planar graphs, called canonical orderings and Schnyder woods, and the related classical graph drawings by de Fraysseix, Pach, and Pollack [Combinatorica, 1990] and by Schnyder [SODA, 1990], called canonical drawings and Schnyder drawings, respectively. To this aim (i) we devise an algorithm for enumerating special e-bipolar orientations of maximal planar graphs, called *canonical orientations*; (ii) we establish bijections between canonical orientations and canonical drawings, and between canonical orientations and Schnyder drawings; and (iii) we exploit the known correspondence between canonical orientations and canonical orderings, and the known bijection between canonical orientations and Schnyder woods. All our enumeration algorithms have $\mathcal{O}(n)$ setup time, space usage, and delay between any two consecutively listed outputs, for an n-vertex maximal planar graph.

Keywords: Enumeration algorithms · Planar graphs · Canonical orderings · Schnyder woods · Worst-case delay

1 Introduction and Overview

An *enumeration algorithm* lists all the solutions of a problem, without duplicates, and then stops. Its efficiency is measured in terms of setup time, space usage, and *delay* between the outputs of two consecutive solutions; see, e.g., [3,42,50,58]. In this paper, we present efficient algorithms to enumerate: (i) straight-line grid drawings produced with the algorithms by de Fraysseix, Pach, and Pollack [25,26] (*FPP-algorithm*) and by Schnyder [52] (*S-algorithm*), and (ii) the corresponding combinatorial structures. To the best of our knowledge, these are the first enumeration algorithms for drawings of graphs.

Notable applications of graph drawing enumeration algorithms include:

Research partially supported by PRIN projects no. 2022ME9Z78 "NextGRAAL" and no. 2022TS4Y3N "EXPAND".

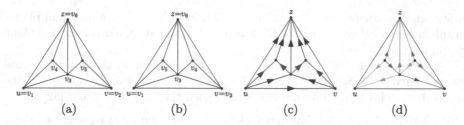

Fig. 1. (a), (b) The two canonical orderings with first vertex u of a maximal planar graph G. (c) The unique canonical orientation with first vertex u of G. (d) The unique Schnyder wood of G.

(1) Users can benefit from having multiple alternative drawings that highlight different features of the graph, enabling them to choose the most suitable for their specific needs; this strategy has been used already in [8].
(2) Machine-learning-based graph drawing tools profit from multiple drawings of a graph; small-delay enumeration algorithms may fuel the training process of these tools.
(3) Computer-aided systems that aim to verify geometric and topological statements can leverage enumeration algorithms to explore the solution space of graph drawing problems.

Preliminary Definitions. We consider graphs and digraphs with multiple edges. A *maximal planar graph* is a planar graph without parallel edges to which no edge can be added without losing planarity or simplicity. A *plane graph* is a planar graph with a prescribed embedding.

Let G be a maximal plane graph and let (u, v, z) be the cycle delimiting its outer face, where u, v, and z appear in this counter-clockwise order along the cycle. A *canonical ordering of G with first vertex u* is a labeling of the vertices $v_1 = u, v_2 = v, v_3, \ldots, v_{n-1}, v_n = z$ such that, for every $3 \le k \le n-1$ (see Figs. 1(a), 1(b), and [26]):

(CO-1) The plane subgraph $G_k \subseteq G$ induced by $v_1, v_2, \ldots, v_k$ is 2-connected; let C_k be the cycle bounding its outer face;
(CO-2) v_{k+1} is in the outer face of G_k, and its neighbors in G_k form an (at least 2-element) subinterval of the path $C_k - (u, v)$.

A *canonical ordering* of G is a canonical ordering of G with first vertex x, where $x \in \{u, v, z\}$. If G' is a maximal planar graph, a *canonical ordering* of G' is a canonical ordering of a maximal plane graph isomorphic to G'. Let $\pi = (v_1, \ldots, v_n)$ be a canonical ordering of G with first vertex v_1. Orient every edge (v_i, v_j) of G from v_i to v_j if and only if $i < j$. The resulting orientation is the *canonical orientation of G with respect to π*. An orientation $\mathcal{D}$ of G is a *canonical orientation with first vertex u* if there exists a canonical ordering π of G with first vertex u such that $\mathcal{D}$ is the canonical orientation of G with respect to π; see Fig. 1(c). A *canonical orientation* of G is a canonical orientation with first vertex x, where x is a vertex in $\{u, v, z\}$. Finally, if G' is a maximal planar graph, a *canonical orientation* of G' is a canonical orientation of a maximal plane graph isomorphic to G'.

A vertex or edge of G is *internal* if it is not incident to the outer face and *outer* otherwise. A *Schnyder wood* of G is an assignment of directions and of the colors 1, 2 and 3 to the internal edges of G such that (see Fig. 2 and [52]):

(S-1) For $i = 1, 2, 3$, each internal vertex x has one outgoing edge e_i of color i. The outgoing edges e_1, e_2, and e_3 appear in this counter-clockwise order at x. Further, for $i = 1, 2, 3$, all the incoming edges at x of color i appear in the clockwise sector between e_{i+1} and e_{i-1}, where $i + 1 = 1$ if $i = 3$ and $i - 1 = 3$ if $i = 1$.

(S-2) At the outer vertices u, v, and z, all the internal edges are incoming and of color 1, 2, and 3, respectively. If G' is a maximal planar graph, a *Schnyder wood* of G' is a Schnyder wood of a maximal plane graph isomorphic to G'.

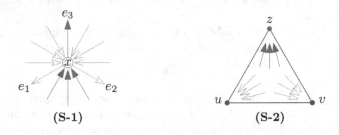

(S-1)   (S-2)

Fig. 2. Illustration for the properties of a Schnyder wood.

Our Contributions. First, we present an algorithm that enumerates all canonical orientations of an n-vertex maximal plane graph G by applying edge contraction or removals. This results in smaller graphs, whose canonical orientations are recursively enumerated and modified to obtain canonical orientations of G by orienting the contracted or removed edges. To achieve polynomial delay, contractions and removals should be applied only if the corresponding branch of computation produces at least one canonical orientation of G. We determine necessary and sufficient conditions for a subgraph of G to allow for an orientation that can be extended to a canonical orientation of G. We establish topological properties that determine whether applying a contraction or a removal results in a graph satisfying these conditions. Additionally, we create efficient data structures for testing and applying the operations based on these properties.

Second, we prove that canonical orderings are topological sortings of canonical orientations. This allows our algorithm for enumerating the former to be used for enumerating the latter. Moreover, since canonical orientations are in correspondence with Schnyder woods [22, Theorem 3.3], our algorithm for enumerating canonical orientations can also be used to enumerate all Schnyder woods of G.

Third, we show that applying the FPP-algorithm with different canonical orderings corresponding to the same canonical orientation yields the same drawing of G. This establishes a correspondence between the canonical orientations of G and the drawings produced by the FPP-algorithm. Together with our algorithm for enumerating canonical orientations, this allows us to enumerate such drawings.

Finally, we show that the drawings generated by the S-algorithm are in bijection with the Schnyder woods. This, the bijection between canonical orientations and Schnyder woods, and our algorithm for enumerating canonical orientations enable us to enumerate the drawings of G produced by the S-algorithm.

All our enumeration algorithms have $\mathcal{O}(n)$ setup time, space usage, and worst-case delay.

Related Results. The planar straight-line drawings of maximal planar graphs generated by the FPP-algorithm [25, 26] and by the S-algorithm [52] are fundamental in graph drawing [28, 45, 57] and find applications in other fields, e.g., knot theory [15, 36, 37] and computational complexity [7, 34, 51]. Further, the combinatorial structures conceived for these algorithms, i.e., canonical orderings and Schynder woods, are used for a plethora of problems in graph drawing [1, 2, 4, 20, 23, 27, 30, 31, 33, 41, 46] and beyond [5, 11–13, 18, 38, 39]. Canonical orderings and Schnyder woods appear to be distant concepts. However, Schnyder [52] has shown how to get a Schnyder wood from a canonical ordering. Also, their relationship is explained by the concept of canonical orientations, which in [22] are proved to be in bijection with the Schnyder woods; see also [44].

While enumerating graph drawings is a novel subject, the enumeration of graph orientations has a rich literature. The enumeration of acyclic orientations and k-arc-connected orientations are studied in [6, 19, 55] and [10], respectively. An *st-orientation* of a graph G is an acyclic orientation of G such that s and t are its unique source and sink, respectively. In [24] a polynomial-delay algorithm is provided for enumerating st-orientations. In [53], the algorithm in [24] is refined to obtain linear delay, if the input is biconnected and planar. Our paper is connected to these algorithms through a result by de Fraysseix and Ossona de Mendez [22]: There is a bijection between the canonical orientations and the bipolar orientations *such that every internal vertex has at least two incoming edges*. Our enumeration algorithm for canonical orientations follows the strategy of [24, 53] for enumerating bipolar orientations of biconnected planar graphs. However, requiring that every internal vertex has at least two incoming edges dramatically increases the complexity of the problem and reveals new and, in our opinion, interesting topological properties of the desired orientations.

The canonical orientations of a maximal plane graph form a distributive lattice $\mathcal{L}$ [32]. By the fundamental theorem of finite distributive lattices [9], there is a finite poset P whose order ideals correspond to the elements of $\mathcal{L}$ and it is known that $|P|$ is polynomial in n [32, page 10]. Enumerating the order ideals of P is a studied problem. In [35] an algorithm is presented that lists all order ideals of P in $\mathcal{O}(\Delta(P))$ delay, where $\Delta(P)$ is the maximum indegree of the covering graph of P. However, the algorithm has three drawbacks that make it unsuitable for solving our problems. First, the guaranteed delay of the algorithm is amortized, and not worst-case. Second, the algorithm uses $\mathcal{O}(w(P) \cdot |P|) = \mathcal{O}(n^3)$ space, where $w(P) = \mathcal{O}(n)$ is the width of P, and $\mathcal{O}(|P|^2) = \mathcal{O}(n^4)$ preprocessing time. Third and most importantly, each order ideal is produced twice by the algorithm, rather than just once as required by an enumeration algorithm. Similarly, the algorithms in [48,54,56] are affected by all or by part of the above drawbacks.

Open Problems. Our research sparkles new questions. In general, for a graph G and a drawing style $\mathcal{D}$, we may ask for algorithms to enumerate the drawings of G respecting $\mathcal{D}$. Examples include: (i) enumerating the planar straight-line drawings of a planar graph on a given grid; (ii) enumerating the orthogonal representations of a plane graph with at most b bends; and (iii) enumerating the upward planar embeddings of a single-source or triconnected DAG.

Full details of omitted or sketched proofs can be found in the full version of the paper [21].

2   Canonical Orientations

In [22, Lemma 3.6–3.7, Theorem 3.3], the following characterization has been shown, for which we provide an alternative proof in the full version of the paper.

Theorem 1. ([22]). *Let G be an n-vertex maximal plane graph and let (u, v, z) be the cycle delimiting its outer face, where u, v, and z appear in this counterclockwise order along the cycle. An orientation $\mathcal{D}$ of G is a canonical orientation with first vertex u if and only if $\mathcal{D}$ is a uz-orientation in which every internal vertex has at least two incoming edges.*

Our proof of Theorem 1 also implies the following.

Lemma 1. *Any topological sorting of a canonical orientation with first vertex u of a maximal plane graph G is a canonical ordering of G with first vertex u.*

Let h_1 and h_2 be parallel edges in a plane graph. We denote by $\ell(h_1, h_2)$ the open region bounded by h_1 and h_2, and say that $\ell(h_1, h_2)$ is a *multilens* if it contains no vertices in its interior. Note that $\ell(h_1, h_2)$ might contain edges parallel to h_1 and h_2 in its interior, or it might coincide with an internal face of the graph; see Fig. 3. We now provide some crucial definitions.

Definition 1. *A biconnected plane graph G with two distinguished vertices s and t is called* well-formed *if it satisfies the following conditions (refer to Fig. 3):*

(1) s and t (which are called the poles *of G) are incident to the outer face of G and s immediately precedes t in clockwise order along the cycle C_o bounding such a face;*

(2) all the internal faces of G have either two or three incident vertices;

(3) multiple edges, if any, are all incident to s; and

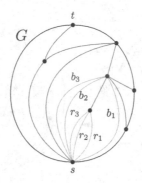

Fig. 3. Any two of the edges r_1, r_2, and r_3 form a multilens, the edges b_2 and b_3 form a multilens, whereas neither b_2 nor b_3 forms a multilens with b_1; the multilenses $\ell(r_1,r_2)$, $\ell(r_2,r_3)$, and $\ell(b_2,b_3)$ are also faces.

(4) if two parallel edges h_1 and h_2 with end-vertices s and x exist such that $\ell(h_1,h_2)$ is not a multilens, then two parallel edges h'_1 and h'_2 between s and a vertex $y \neq x$ exist such that $\ell(h'_1,h'_2)$ is a multilens and such that $\ell(h'_1,h'_2) \subset \ell(h_1,h_2)$.

Definition 2. *An st-orientation $\mathcal{D}$ of a well-formed biconnected plane graph G with poles s and t is* inner-canonical *if every internal vertex has at least two incoming edges in $\mathcal{D}$.*

Definition 3. *Let G be a plane graph. The* contraction *of an edge $e = (u,v)$ removes e from G and "merges" u and v into a new vertex w. Let $e, e_1^u, \ldots, e_h^u$ and $e, e_1^v, \ldots, e_k^v$ be the clockwise order of the edges incident to u and to v, respectively. Then the clockwise order of the edges incident to w is $e_1^u, \ldots, e_h^u, e_1^v, \ldots, e_k^v$.*

Note that the contraction of an edge may introduce parallel edges or self-loops.

Let G be a well-formed biconnected plane graph with poles s and t. Let $e_1, e_2, \ldots, e_m$ be the counter-clockwise order of the edges incident to s, where e_1 and e_m are the rightmost and leftmost edge incident to s, respectively, and let $v_1, \ldots, v_m$ be the end-vertices of $e_1, \ldots, e_m$ different from s, respectively. Moreover, let G^* be the plane multigraph resulting from the contraction of e_1 in G; see Fig. 4(a). Also, if G contains parallel edges, let $j \in \{1, \ldots, m-1\}$ be the smallest index such that e_j and e_{j+1} define a multilens of G; denote by G^- the plane graph resulting from the removal of $e_1, \ldots, e_j$ from G; see Fig. 4(b). The next lemmas prove that, under certain conditions, G^* and G^- are well-formed multigraphs and can be used to obtain inner-canonical orientations of G.

Lemma 2. *Suppose that G does not contain parallel edges between s and w_1, where w_1 is the vertex that follows s in counter-clockwise direction along the outer face of G. Then G^* is a well-formed biconnected plane graph with poles s and t. Also, let $\mathcal{D}^*$ be an inner-canonical orientation of G^*. The orientation $\mathcal{D}$ of G*

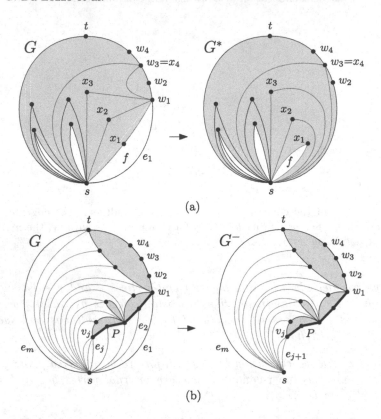

Fig. 4. Illustration for the contraction of e_1 (a) and the removal of $e_1, \dots, e_j$ (b).

obtained from $\mathcal{D}^*$ by orienting the edge (s, w_1) away from s and by keeping the orientation of all other edges unchanged is inner-canonical.

Lemma 3. *Suppose that G contains parallel edges and let $j \in \{1, \dots, m-1\}$ be the smallest index such that e_j and e_{j+1} define a multilens of G. Suppose also that either $j = 1$, or $j > 1$ and $v_2, \dots, v_j$ are not incident to the outer face of G. Then the graph G^- is a well-formed biconnected plane graph with poles s and t. Also, let $\mathcal{D}^-$ be an inner-canonical orientation of G^-. The orientation $\mathcal{D}$ of G obtained from $\mathcal{D}^-$ by orienting the edges $e_1, e_2, \dots, e_j$ away from s and by keeping the orientation of all other edges unchanged is inner-canonical.*

By induction on $|E(G)|$ and using Lemmas 2 and 3 we can prove the following.

Lemma 4. *Every well-formed biconnected plane graph G with poles s and t has at least one inner-canonical orientation.*

Section 2.1 is devoted to the proof of the following main result.

Theorem 2. *Let G be a well-formed biconnected plane graph with φ edges. There is an algorithm with $\mathcal{O}(\varphi)$ setup time and space usage listing all the inner-canonical orientations of G with $\mathcal{O}(\varphi)$ delay.*

Provided that Theorem 2 holds, we can prove the following.

Lemma 5. *Let G be an n-vertex maximal plane graph and (u, v, z) be the cycle delimiting its outer face f_o, where u, v, and z appear in this counter-clockwise order along f_o. There is an algorithm with $\mathcal{O}(n)$ setup time and space usage listing all the canonical orientations of G with first vertex u with $\mathcal{O}(n)$ delay.*

Proof (sketch): We have that G is a well-formed biconnected plane graph with poles u and z. By Theorem 1, any canonical orientation of G with first vertex u is a uz-orientation such that every internal vertex has at least two incoming edges, i.e., an inner-canonical orientation. Also, any inner-canonical orientation of G is canonical. This, combined with G having $\mathcal{O}(n)$ edges, implies that the algorithm in Theorem 2 enumerates all canonical orientations of G within the stated bounds. □

Theorem 3. *There is an algorithm $\mathcal{A}_1$ (resp. $\mathcal{A}_2$) with $\mathcal{O}(n)$ setup time and space usage listing all canonical orientations of an n-vertex maximal plane (planar) graph with $\mathcal{O}(n)$ delay.*

Proof (sketch): Algorithm $\mathcal{A}_1$ uses the algorithm for the proof of Lemma 5 three times, i.e., once for each choice of the first vertex among the three vertices incident to the outer face of the input graph G. Algorithm $\mathcal{A}_2$ applies $4n - 8$ times algorithm $\mathcal{A}_1$, since there are $4n - 8$ maximal plane graphs that are isomorphic to G. □

2.1 The Inner-Canonical Enumerator Algorithm

We now describe the INNER-CANONICAL ENUMERATOR (ICE) algorithm that enumerates all the inner-canonical orientations of a well-formed biconnected plane graph G with poles s and t (see Theorem 2). In the full version of the paper [21], we provide implementation details, data structures, and pseudocode for achieving the claimed worst-case bounds.

The ICE algorithm works recursively as follows. In the base case, G is the single edge $e_m = (s, t)$, and its unique inner-canonical orientation is the one with the edge e_m directed from s to t. Otherwise, the algorithm considers four cases.

In Cases 1 and 2, G contains parallel edges and e_1 is the unique edge between s and w_1. Let $j \in \{2, \ldots, m - 1\}$ be the smallest index such that e_j and e_{j+1} define a multilens of G. In Case 1, there exists an index $i \in \{2, \ldots, j\}$ such that v_i is incident to the outer face of G, while in Case 2 such an index does not exist. In Case 3, G does not contain parallel edges. Finally, in Case 4, G contains parallel edges between s and w_1. Note that exactly one of Cases 1–4 applies to G.

In Cases 1 and 3, **we contract** the edge (s, w_1). Let G^* be the resulting plane graph which, by Lemma 2, is biconnected and well-formed. Thus, the algorithm can be applied recursively to enumerate all inner-canonical orientations of G^*. The algorithm obtains all inner-canonical orientations of G from the ones of G^* as in Lemma 2.

In Case 4, **we remove** the edges $e_1, e_2, \ldots, e_j$. Let G' be the resulting plane graph which, by Lemma 3, is biconnected and well-formed. Thus, the algorithm can be applied recursively in order to enumerate all inner-canonical orientations of G'. The algorithm obtains all inner-canonical orientations of G from the ones of G' as in Lemma 3.

In Case 2, the algorithm branches and applies **both the contraction and the removal operations**. Precisely, first we contract the edge (s, w_1), obtaining a well-formed biconnected plane graph G^*. After all inner-canonical orientations of G^* have been used to produce inner-canonical orientations of G as in Lemma 2, we remove the edges $e_1, e_2, \ldots, e_j$ from G, obtaining a well-formed biconnected plane graph G', from which the remaining inner-canonical orientations of G are produced as in Lemma 3.

Note that the ICE algorithm outputs an inner-canonical orientation each time the base case applies. The next lemma summarizes its correctness.

Lemma 6. *The* ICE *algorithm outputs all and only the inner-canonical orientations of G without repetitions.*

3   Enumeration of Canonical Orderings and Drawings

We show how to efficiently enumerate the canonical orderings and drawings of a maximal plane or planar graph G. By Theorem 3, the canonical orientations of G can be generated efficiently. By Lemma 1, for every canonical orientation $\mathcal{D}$ of G, the canonical orderings π of G such that $\mathcal{D}$ is the canonical orientation of G with respect to π are the topological sortings of $\mathcal{D}$. Since there exist $\mathcal{O}(1)$-delay algorithms [47,49] for listing all such topological sortings, we get the following.

Theorem 4. *There is an algorithm with $\mathcal{O}(n)$ setup time and space usage listing all canonical orderings of an n-vertex maximal plane/planar graph with $\mathcal{O}(n)$ delay.*

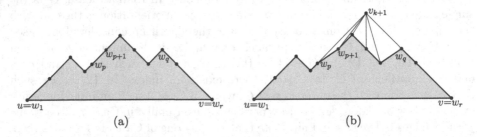

Fig. 5. Illustrations for the FPP-algorithm. (a) Γ_k. (b) Γ_{k+1}.

We show how to enumerate the planar straight-line drawings produced by the *FPP-algorithm* [26]. Its input is an n-vertex maximal plane graph G, whose outer face is delimited by a cycle (u, v, z) and a canonical ordering $\pi = (v_1 = u, v_2 = v, v_3, \ldots, v_n = z)$ of G. The FPP-algorithm works in steps.

The first step constructs a planar straight-line drawing Γ_3 of G_3 with v_1, v_2, and v_3 at $(0,0)$, $(2,0)$, and $(1,1)$, respectively, and defines sets $M_3(v_1) = \{v_1, v_2, v_3\}$, $M_3(v_3) = \{v_2, v_3\}$, and $M_3(v_2) = \{v_2\}$.

For $k = 3, \ldots, n-1$, step $k-1$ constructs a planar straight-line drawing Γ_{k+1} of G_{k+1} by modifying Γ_k as follows; see Fig. 5. Let $w_1 = u, w_2, \ldots, w_r = v$ be the clockwise order of the vertices along the outer face of G_k. Assume that step $k-2$ has defined, for $i = 1, \ldots, r$, a subset $M_k(w_i)$ of the vertices of G_k, where $M_k(w_1) \supset \cdots \supset M_k(w_r)$. Let $w_p, w_{p+1}, \ldots, w_q$ be the neighbors of v_{k+1} in G_k, where $p < q$. Then Γ_{k+1} is obtained from Γ_k by increasing the x-coordinate of each vertex in $M_k(w_{p+1})$ by one unit, the x-coordinate of each vertex in $M_k(w_q)$ by one additional unit, and placing v_{k+1} at the intersection of the line through w_p with slope $+1$ and the line through w_q with slope -1. Then step $k-1$ proceeds to define the sets:

(i) $M_{k+1}(w_i) = M_k(w_i) \cup \{v_{k+1}\}$, for $i = 1, \ldots, p$;
(ii) $M_{k+1}(v_{k+1}) = M_k(w_{p+1}) \cup \{v_{k+1}\}$; and
(iii) $M_{k+1}(w_i) = M_k(w_i)$, for $i = q, \ldots, r$.

We call *canonical drawing with base edge* (u,v) the drawing Γ_n of G constructed by the FPP-algorithm; we often say that Γ_n *corresponds* to π. The following is the main tool for our enumeration algorithm.

Theorem 5. *Let G be an n-vertex maximal plane graph and (u, v, z) be the cycle delimiting the outer face of G, with u, v, and z in this counter-clockwise order along the cycle. There is a bijective function f from the canonical orientations of G with first vertex u to the canonical drawings of G with base edge (u, v). Also, given a canonical orientation with first vertex u, the corresponding canonical drawing with base edge (u, v) can be constructed in $\mathcal{O}(n)$ time.*

Proof (sketch): The function f is as follows. Consider any canonical orientation $\mathcal{D}$ of G with first vertex u and let π be any canonical ordering with first vertex u that *extends* $\mathcal{D}$ (that is, the canonical orientation of G with respect π is $\mathcal{D}$). Then $f(\mathcal{D})$ is the canonical drawing with base edge (u, v) that corresponds to π. Since π can be computed as any topological sorting of $\mathcal{D}$ in $\mathcal{O}(n)$ time [40] and the FPP-algorithm can be implemented in $\mathcal{O}(n)$ time [17], the second part of the statement follows. Clearly, f is injective. Indeed, any distinct canonical orientations $\mathcal{D}_1$ and $\mathcal{D}_2$ of G differ on the orientation of some edge (a, b). Thus, for any two canonical orderings π_1 and π_2 that extend $\mathcal{D}_1$ and $\mathcal{D}_2$, respectively, we have that b follows a in π_1 and precedes a in π_2 (or vice versa). Hence, the y-coordinate of b is larger than the one of a in $f(\mathcal{D}_1)$ and smaller than the one of a in $f(\mathcal{D}_2)$ (or vice versa), thus $f(\mathcal{D}_1)$ and $f(\mathcal{D}_2)$ are not the same drawing. The core of the proof that f is surjective is in the proof of the following statement.

Claim 1. *Any two canonical orderings π and τ that extend $\mathcal{D}$ are such that the canonical drawings of G corresponding to π and τ are the same drawing.*

The proof of Claim 1 is by induction on $|V(G)|$ and it relies on a natural extension of the concepts of canonical ordering, orientation, and drawing to biconnected internally-triangulated plane graphs; hence, in the following, the outer face of G might have more than three incident vertices. The proof of Claim 1 exploits the following claim, which is also proved by induction on the size of G.

Claim 2. *Let $z_1, \ldots, z_r$ be the clockwise order of the vertices along the outer face of G, let $M^\pi(z_1), \ldots, M^\pi(z_r)$ (let $M^\tau(z_1), \ldots, M^\tau(z_r)$) be the sets associated to $z_1, \ldots, z_r$, respectively, by the FPP-algorithm, when applied with canonical ordering π (resp. τ). For $i = 1, \ldots, r$, the sets $M^\pi(z_i)$ and $M^\tau(z_i)$ coincide.*

The inductive proof of Claim 1 distinguishes two cases.

In Case 1, π and τ have the same last vertex. This can be removed from both, resulting in canonical orderings λ and ξ, respectively, of a smaller graph G'. By induction, the canonical drawings of G' corresponding to λ and ξ coincide. This and the fact that the sets associated to the vertices along the boundary of G' by the FPP-algorithm when applied with canonical orderings λ and ξ coincide imply that the canonical drawings of G corresponding to π and τ also coincide.

In Case 2, π and τ do not have the same last vertex. Then we define a sequence of canonical orderings of G such that: (i) the first canonical ordering in the sequence is τ; (ii) any two canonical orderings consecutive in the sequence coincide, except for two vertices, whose positions are adjacent and swapped in the two canonical orderings; and (iii) the last canonical ordering in the sequence has the same last vertex as π. Note that the last canonical ordering in the sequence and π are such that the corresponding canonical drawings of G are the same drawing, by Case 1. The proof that the canonical drawings of G corresponding to two consecutive canonical orderings in the sequence are the same drawing relies on the similarity of such canonical orderings. By transitivity, we get that the canonical drawings of G corresponding to π and τ are the same drawing. □

Theorem 3 and Theorem 5 imply the following.

Theorem 6. *There is an algorithm with $\mathcal{O}(n)$ setup time and space usage listing all canonical drawings of an n-vertex maximal plane/planar graph with $\mathcal{O}(n)$ delay.*

4   Enumeration of Schnyder Woods and Drawings

We now show how to efficiently enumerate the Schnyder woods and drawings of an n-vertex maximal plane graph G. The Schnyder woods and the canonical orientations of G are in bijection [22]. Further, given a canonical orientation of G, the corresponding Schnyder wood of G can be constructed in $\mathcal{O}(n)$ time [14, 16, 26, 27, 29, 43, 52]. This, together with Theorem 3, implies the following.

Theorem 7. *There is an algorithm with $\mathcal{O}(n)$ setup time and space usage listing all Schnyder woods of an n-vertex maximal plane/planar graph with $\mathcal{O}(n)$ delay.*

We now deal with the enumeration of the planar straight-line drawings produced by the algorithm by Schnyder [52], known as *Schnyder drawings*. The S-algorithm takes as input (see Fig. 6(a)) an n-vertex maximal plane graph G, whose outer face is delimited by a 3-cycle (u_1, u_2, u_3), where u_1, u_2, and u_3 appear in this counter-clockwise order along the cycle, and a Schnyder wood $\mathcal{W} = (T_1, T_2, T_3)$ of G, where T_i contains u_i, for $i = 1, 2, 3$.

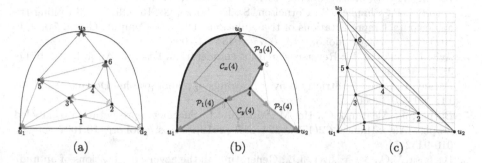

Fig. 6. (a) A Schnyder wood $\mathcal{W}$ of a maximal plane graph G. (b) Paths $\mathcal{P}_1(4)$, $\mathcal{P}_2(4)$, and $\mathcal{P}_3(4)$, and cycles $\mathcal{C}_x(4)$ and $\mathcal{C}_y(4)$. (c) The Schnyder drawing $s(\mathcal{W})$.

The S-algorithm assigns coordinates $(0, 0)$, $(2n - 5, 0)$, and $(0, 2n - 5)$ to u_1, u_2, and u_3, respectively. For a cycle $\mathcal{C}$, let $\#_f(\mathcal{C})$ be the number of internal faces of G inside $\mathcal{C}$. For $i = 1, 2, 3$, properties (S-1) and (S-2) of $\mathcal{W}$ imply that T_i contains a directed path $\mathcal{P}_i(w)$ from any internal vertex w to u_i; see Fig. 6(b). Also, $\mathcal{P}_1(w)$, $\mathcal{P}_2(w)$, and $\mathcal{P}_3(w)$ only share w [52]. Let $\mathcal{C}_x(w)$ and $\mathcal{C}_y(w)$ be the cycles $\mathcal{P}_1(w) \cup \mathcal{P}_3(w) \cup (u_1, u_3)$ and $\mathcal{P}_1(w) \cup \mathcal{P}_2(w) \cup (u_1, u_2)$, respectively. Then the algorithm assigns coordinates $(\#_f(\mathcal{C}_x(w)), \#_f(\mathcal{C}_y(w)))$ to w; see Fig. 6(c).

The following is the main tool for our enumeration algorithm.

Theorem 8. *Let G be an n-vertex maximal plane graph. There is a bijective function s from the Schnyder woods of G to the Schnyder drawings of G. Also, given a Schnyder wood of G, the corresponding Schnyder drawing of G can be constructed in $\mathcal{O}(n)$ time.*

Proof (sketch): The function s is the S-algorithm, which can be implemented in $\mathcal{O}(n)$ time [52], from which the second part of the statement follows. The core of the proof that s is bijective consists of proving that a Schnyder drawing Γ uniquely determines the Schnyder wood $\mathcal{W} = (T_1, T_2, T_3)$ such that $s(\mathcal{W}) = \Gamma$. This follows from the fact that in Γ, for each vertex v of G, the edges of T_1, T_2, and T_3 incoming into v have slopes in the intervals $(0°, 90°)$, $(135°, 180°)$, and $(270°, 315°)$, respectively, while the edges of T_1, T_2, and T_3 outgoing from v have slopes in the intervals $(180°, 270°)$, $(315°, 360°)$, and $(90°, 135°)$, respectively; see [27]. Thus, whether each edge (u, v) of G belongs to T_1, T_2, or T_3 and whether it is directed from u to v or vice versa is uniquely determined by its slope in Γ. □

Theorem 7 and Theorem 8 imply the following.

Theorem 9. *There is an algorithm with $\mathcal{O}(n)$ setup time and space usage listing all Schnyder drawings of an n-vertex maximal plane graph with $\mathcal{O}(n)$ delay.*

References

1. Alam, M.J., Biedl, T., Felsner, S., Kaufmann, M., Kobourov, S., Ueckerdt, T.: Computing cartograms with optimal complexity. Discrete Comput. Geom. **50**(3), 784–810 (2013). https://doi.org/10.1007/s00454-013-9521-1
2. Angelini, P., Chaplick, S., Cornelsen, S., Da Lozzo, G., Roselli, V.: Morphing triangle contact representations of triangulations. Discrete Comput. Geom. **70**, 991–1024 (2023). https://doi.org/10.1007/s00454-022-00475-9
3. Avis, D., Fukuda, K.: Reverse search for enumeration. Discrete Appl. Math. **65**(1–3), 21–46 (1996)
4. Bárány, I., Rote, G.: Strictly convex drawings of planar graphs. Documenta Math. **11**, 369–391 (2006)
5. Barbay, J., Aleardi, L.C., He, M., Munro, J.I.: Succinct representation of labeled graphs. Algorithmica **62**(1–2), 224–257 (2012). https://doi.org/10.1007/s00453-010-9452-7
6. Barbosa, V.C., Szwarcfiter, J.L.: Generating all the acyclic orientations of an undirected graph. Inf. Process. Lett. **72**(1–2), 71–74 (1999)
7. Been, K., Daiches, E., Yap, C.: Dynamic map labeling. IEEE Trans. Vis. Comput. Graph. **12**(5), 773–780 (2006)
8. Biedl, T., Marks, J., Ryall, K., Whitesides, S.: Graph multidrawing: finding nice drawings without defining nice. In: Whitesides, S.H. (ed.) GD 1998. LNCS, vol. 1547, pp. 347–355. Springer, Heidelberg (1998). https://doi.org/10.1007/3-540-37623-2_26
9. Birkhoff, G.: Rings of sets. Duke Math. J. **3**(3), 443–454 (1937)
10. Blind, S., Knauer, K., Valicov, P.: Enumerating k-arc-connected orientations. Algorithmica **82**(12), 3588–3603 (2020). https://doi.org/10.1007/s00453-020-00738-y
11. Bonichon, N., Gavoille, C., Hanusse, N., Poulalhon, D., Schaeffer, G.: Planar graphs, via well-orderly maps and trees. Graphs Comb. **22**(2), 185–202 (2006). https://doi.org/10.1007/s00373-006-0647-2
12. Bose, P., Dujmovic, V., Hurtado, F., Langerman, S., Morin, P., Wood, D.R.: A polynomial bound for untangling geometric planar graphs. Discrete Comput. Geom. **42**(4), 570–585 (2009). https://doi.org/10.1007/s00454-008-9125-3
13. Bose, P., Gudmundsson, J., Smid, M.H.M.: Constructing plane spanners of bounded degree and low weight. Algorithmica **42**(3–4), 249–264 (2005). https://doi.org/10.1007/s00453-005-1168-8
14. Brehm, E.: 3-orientations and Schnyder 3-tree-decompositions. Master's thesis, Freie Universität Berlin (2000)
15. Cantarella, J., Kusner, R.B., Sullivan, J.M.: On the minimum ropelength of knots and links. Invent. Math. **150**, 257–286 (2002). https://doi.org/10.1007/s00222-002-0234-y
16. Castelli Aleardi, L.: Algorithms for graphs on surfaces: from graph drawing to graph encoding. Habilitation thesis, Université de Paris (2021)
17. Chrobak, M., Payne, T.H.: A linear-time algorithm for drawing a planar graph on a grid. Inf. Process. Lett. **54**(4), 241–246 (1995)
18. Chuang, R.C., Garg, A., He, X., Kao, M., Lu, H.: Compact encodings of planar graphs via canonical orderings and multiple parentheses. In: Larsen, K.G., Skyum, S., Winskel, G. (eds.) ICALP 1998. LNCS, vol. 1443, pp. 118–129. Springer, Heidelberg (1998). https://doi.org/10.1007/BFb0055046
19. Conte, A., Grossi, R., Marino, A., Rizzi, R.: Efficient enumeration of graph orientations with sources. Discrete Appl. Math. **246**, 22–37 (2018)

20. Da Lozzo, G., D'Angelo, A., Frati, F.: On the area requirements of planar greedy drawings of triconnected planar graphs. In: Kim, D., Uma, R., Cai, Z., Lee, D. (eds.) COCOON 2020. LNCS, vol. 12273, pp. 435–447. Springer, Cham (2020). https://doi.org/10.1007/978-3-030-58150-3_35
21. Da Lozzo, G., Di Battista, G., Frati, F., Grosso, F., Patrignani, M.: Efficient enumeration of drawings and combinatorial structures for maximal planar graphs. CoRR, abs/2310.02247 (2023)
22. de Fraysseix, H., de Mendez, P.O.: On topological aspects of orientations. Discrete Math. 229(1–3), 57–72 (2001)
23. de Fraysseix, H., de Mendez, P.O., Rosenstiehl, P.: On triangle contact graphs. Comb. Probab. Comput. 3, 233–246 (1994)
24. de Fraysseix, H., de Mendez, P.O., Rosenstiehl, P.: Bipolar orientations revisited. Discrete Appl. Math. 56(2–3), 157–179 (1995)
25. de Fraysseix, H., Pach, J., Pollack, R.: Small sets supporting Fáry embeddings of planar graphs. In: Simon, J. (ed.) STOC 1998, pp. 426–433 (1988)
26. de Fraysseix, H., Pach, J., Pollack, R.: How to draw a planar graph on a grid. Combinatorica 10(1), 41–51 (1990). https://doi.org/10.1007/BF02122694
27. Dhandapani, R.: Greedy drawings of triangulations. Discrete Comput. Geom. 43(2), 375–392 (2010). https://doi.org/10.1007/s00454-009-9235-6
28. Di Battista, G., Eades, P., Tamassia, R., Tollis, I.G.: Graph drawing: algorithms for the visualization of graphs (1999)
29. Di Battista, G., Tamassia, R., Vismara, L.: Output-sensitive reporting of disjoint paths. Algorithmica 23(4), 302–340 (1999). https://doi.org/10.1007/PL00009264
30. Dujmovic, V., Eppstein, D., Suderman, M., Wood, D.R.: Drawings of planar graphs with few slopes and segments. Comput. Geom. 38(3), 194–212 (2007)
31. Felsner, S.: Convex drawings of planar graphs and the order dimension of 3-polytopes. Order 18(1), 19–37 (2001)
32. Felsner, S.: Lattice structures from planar graphs. Electron. J. Comb. 11(1), R15 (2004)
33. Felsner, S., Zickfeld, F.: Schnyder woods and orthogonal surfaces. Discrete Comput. Geom. 40(1), 103–126 (2008). https://doi.org/10.1007/s00454-007-9027-9
34. Ganian, R., et al.: Are there any good digraph width measures? J. Comb. Theory Ser. B 116, 250–286 (2016)
35. Habib, M., Medina, R., Nourine, L., Steiner, G.: Efficient algorithms on distributive lattices. Discrete Appl. Math. 110(2–3), 169–187 (2001)
36. Hass, J., Lagarias, J.C.: The number of Reidemeister moves needed for unknotting. J. Am. Math. Soc. 14, 399–428 (2001)
37. Hass, J., Lagarias, J.C., Pippenger, N.: The computational complexity of knot and link problems. J. ACM 46(2), 185–211 (1999)
38. Hayes, T.P.: A simple condition implying rapid mixing of single-site dynamics on spin systems. In: FOCS 2006, pp. 39–46 (2006)
39. He, X., Kao, M., Lu, H.: Linear-time succinct encodings of planar graphs via canonical orderings. SIAM J. Discrete Math. 12(3), 317–325 (1999)
40. Kahn, A.: Topological sorting of large networks. Commun. ACM 5(11), 558–562 (1962)
41. Kant, G.: Drawing planar graphs using the canonical ordering. Algorithmica 16(1), 4–32 (1996). https://doi.org/10.1007/BF02086606
42. Knuth, D.E.: The Art of Computer Programming, Volume 4A: Combinatorial Algorithms, Part 1 (2011)
43. Kobourov, S.G.: Canonical orders and Schnyder realizers. In: Encyclopedia of Algorithms, pp. 277–283 (2016)

44. Miura, K., Azuma, M., Nishizeki, T.: Canonical decomposition, realizer, Schnyder labeling and orderly spanning trees of plane graphs. Int. J. Found. Comput. Sci. **16**(1), 117–141 (2005)
45. Nishizeki, T., Rahman, M.S.: Planar Graph Drawing, Volume 12 of Lecture Notes Series on Computing (2004)
46. Nöllenburg, M., Prutkin, R., Rutter, I.: On self-approaching and increasing-chord drawings of 3-connected planar graphs. J. Comput. Geom. **7**(1), 47–69 (2016)
47. Ono, A., Nakano, S.: Constant time generation of linear extensions. In: Liskiewicz, M., Reischuk, R. (eds.) FCT 2005. LNCS, vol. 3623, pp. 445–453. Springer, Heidelberg (2005). https://doi.org/10.1007/11537311_39
48. Pruesse, G., Ruskey, F.: Gray codes from antimatroids. Order **10**(3), 239–252 (1993). https://doi.org/10.1007/BF01110545
49. Pruesse, G., Ruskey, F.: Generating linear extensions fast. SIAM J. Comput. **23**(2), 373–386 (1994)
50. Ruskey, F.: Combinatorial generation. University of Victoria, Victoria BC, Canada (2003). Preliminary working draft
51. Schaefer, M., Stefankovic, D.: Decidability of string graphs. J. Comput. Syst. Sci. **68**(2), 319–334 (2004)
52. Schnyder, W.: Embedding planar graphs on the grid. In: Johnson, D.S. (ed.) SODA 1990, pp. 138–148 (1990)
53. Setiawan, A., Nakano, S.-I.: Listing all *st*-orientations. IEICE Trans. Fund. Electr. Comm. Comp. Sci. **94**(10), 1965–1970 (2011)
54. Squire, M.B.: Gray codes and efficient generation of combinatorial structures. Ph.D. thesis, North Carolina State University (1995)
55. Squire, M.B.: Generating the acyclic orientations of a graph. J. Algorithms **26**(2), 275–290 (1998)
56. Steiner, G.: An algorithm to generate the ideals of a partial order. Oper. Res. Lett. **5**(6), 317–320 (1986)
57. Tamassia, R. (ed.): Handbook on Graph Drawing and Visualization (2013)
58. Wasa, K.: Enumeration of enumeration algorithms. CoRR, abs/1605.05102 (2016)

(t, s)-Completely Independent Spanning Trees

Shin-ichi Nakano[✉️] [ID]

Gunma University, Kiryu 376-8515, Japan
nakano@gunma-u.ac.jp

Abstract. In this paper we first define (t, s)-completely independent spanning trees, which is a generalization of completely independent spanning trees. A set of t spanning trees of a graph is (t, s)-*completely independent* if, for any pair of vertices u and v, among the set of t paths from u to v in the t spanning trees, at least $s \leq t$ paths are internally disjoint. By (t, s)-completely independent spanning trees, one can ensure any pair of vertices can communicate each other even if at most $s - 1$ vertices break down. We prove that every maximal planar graph has a set of $(3, 2)$-completely independent spanning trees, every tri-connected planar graph has a set of $(3, 2)$-completely independent spanning trees, and every 3D grid graph has a set of $(3, 2)$-completely independent spanning trees. Also one can compute them in linear time.

Keywords: Independent Spanning Trees · Spanning Tree

1 Introduction

Two paths from vertex u to v are *internally disjoint* if they have no common internal vertex.

A set of t spanning trees of a graph is *completely independent* if, for any pair of vertices u and v, the set of t paths from u to v in the t spanning trees are internally disjoint (and edge disjoint) [5]. A necessary and sufficient condition for the existence of a set of t completely independent spanning trees is known [4,5].

In this paper we generalize the concept of completely independent spanning trees as follows. A set of t spanning trees of a graph is (t, s)-*completely independent* if, for any pair of vertices u and v, among the set of t paths from u to v in the t spanning trees, at least $s \leq t$ paths are internally disjoint. By (t, s)-completely independent spanning trees, one can ensure any pair of vertices can communicate each other even if at most $s - 1$ vertices break down. The original completely independent spanning trees are (t, t)-complete spanning trees.

Intuitively, when we have t interconnection (spanning tree) networks, we want to ensure $s \leq t$ of separate (independent) routes for each pair of vertices. The original completely spanning tree concept may be too strong for some applications and may fail to construct them, however (t, s)-completely independent tree concept may be a flexible choice for some applications and may increase the chance to construct them.

R. Uehara et al. (Eds.): WALCOM 2024, LNCS 14549, pp. 365–376, 2024.
https://doi.org/10.1007/978-981-97-0566-5_26

In this paper, we first design an algorithm to construct a set of $(3,2)$-completely independent spanning trees in a given maximal planar graph based on the realizer [18], then design an algorithm to construct a set of $(3,2)$-completely independent spanning trees in a given tri-connected planar graphs based on the canonical decomposition [11], then design an algorithm to construct a set of $(3,2)$-completely independent spanning trees in a given 3D grid graph. Those algorithms are simple and run in $O(n)$ time, where n is the number of vertices of the given graph.

The remainder of this paper is organized as follows. Section 2 gives some definitions and two basic lemmas. In Sect. 3 we design our first algorithm which constructs a set of $(3,2)$-completely independent spanning trees in a given maximal planar graph. In Sect. 4 we design our second algorithm which constructs a set of $(3,2)$-completely independent spanning trees in a given tri-connected planar graph. In Sect. 5 we design our third algorithm which constructs a set of $(3,2)$-completely independent spanning trees in a given 3D grid graph. Finally Sect. 6 is a conclusion.

2   Preliminaries

A *tree* is a connected graph with no cycle. A *rooted tree* is a tree with a designated vertex as *the root*. Given a graph G, a *spanning tree* of G is a subgraph of G which is a tree and contains all vertices of G.

A graph is *planar* if it can be embedded on the plane so that no two edges intersect geometrically except at a vertex to which they are both incident. A *plane* graph is a planar graph with a fixed plane embedding.

A graph G with more than k vertices is *k-connected* if removal of any $k-1$ vertices results in a connected graph.

A 3D *grid graph* with size $L_x \times L_y \times L_z$ is the graph consisting of vertex set $\{(x,y,z)|0 \le x \le L_x, 0 \le y \le L_y, 0 \le z \le L_z$, and x,y,z are integers $\}$ and edge set $\{\{(x_1,y_1,z_1),(x_2,y_2,z_2)\} \mid |x_1 - x_2| + |y_1 - y_2| + |z_1 - z_2| = 1\}$.

Independent Spanning Trees

Let n be the number of vertices of a given graph G. A set of t rooted spanning trees with a common root r of a graph G is *independent* if, for any vertex v, the set of t paths from r to v in the t spanning trees are internally disjoint. It is conjectured that, for any $k \ge 1$, every k-connected graph G has a set of k independent spanning trees rooted at any vertex [12,19]. If G is bi-connected then one can find two independent spanning trees in linear time by the st-numbering [1,10]. If G is tri-connected then one can find three independent spanning trees in $O(n^2)$ time by the ear-decomposition [1,2]. If G is four-connected then one can find four independent spanning trees in $O(n^3)$ time by the chain-decomposition [3]. If G is a tri-connected planar graph then one can find three independent spanning trees in linear time by the canonical decomposition [1]. If G is a four-connected planar graph then one can find four independent spanning trees in

$O(n^3)$ time [8] then in linear time [13,14]. If G is a five-connected planar graph then one can find five independent spanning trees in polynomial time [9]. If G is a five-connected maximal planar graph then one can find five independent spanning trees in linear time by the 5-canonical decomposition [15,16].

Completely Independent Spanning Trees
A set of spanning trees is *completely independent* if, for any pair of vertices u and v, the set of paths from u to v in the spanning trees are internally disjoint (and edge disjoint) [5]. A necessary and sufficient condition for the existence of k completely independent spanning trees is known [4,5].

(t, s)-Completely Independent Spanning Trees
A set of t spanning trees is (t, s)-*completely independent* if, for any pair of vertices u and v, among the set of t paths from u to v in the t spanning trees, at least s paths are internally disjoint.

Realizer
Every maximal planar graph with $n \geq 4$ vertices is tri-connected, and has a unique embedding on a sphere only up to mirror copy [6]. In the embedding each face has exactly three vertices on the boundary. Given a maximal planar graph G with n vertices, we can compute a maximal plane graph G' corresponding to G in linear time [7]. Let r_r, r_b, r_y be the three vertices on the outer face of G', and assume that they appear on the outer face clockwise in this order. A partition $\{E_r, E_b, E_y\}$ of inner edges of G' is called a *realizer* of G' if the following conditions (re1)–(re3) are satisfied [18]. See an example in Fig. 1(b). Let T_r be the tree induced by all edges in E_r. Similarly, let T_b and T_y be the trees induced by all edges in E_b and E_y, respectively.

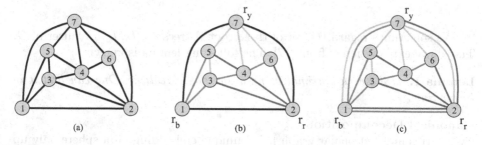

Fig. 1. (a) A maximal plane graph G (b) a realizer of G (c) three spanning thees of G.

(**re1**) T_r is a tree spanning all inner vertices of G and r_r. Similarly, T_b is a tree spanning all inner vertices of G and r_b, and T_y is a tree spanning all inner vertices of G and r_y.

(**re2**) Every inner edge incident to r_r is in T_r. Similarly, every inner edge incident to r_b is in T_b, and every inner edge incident to r_y is in T_y.

(**re3**) Define the orientation of each inner edge as follows. In tree T_r, we regard r_r as the root of T_r, and orient each edge in T_r from a child to its parent. Similarly, we regard r_b and r_y as the roots of T_b and T_y, respectively, and define the orientation of each inner edge in T_b and T_y from a child to its parent.

Then, for each inner vertex v, all edges incident to v appear around v clockwise in the following order. (See Fig. 2)

Exactly one outgoing edge in T_r.
Zero or more incoming edges in T_y.
Exactly one outgoing edge in T_b.
Zero or more incoming edges in T_r.
Exactly one outgoing edge in T_y.
Zero or more incoming edges in T_b.

Fig. 2. Illustration for the condition of a realizer.

We sometimes regard the set of three rooted trees T_r, T_b, T_y a realizer of G. The above explanation is from [17]. The following lemma is known.

Lemma 1. *[18] Every maximal plane graph has a realizer. One can find it in linear time.*

Canonical Decomposition

Every tri-connected planar graph has a unique embedding on a sphere only up to mirror copy [6]. Given a tri-connected planar graph G with n vertices, we can compute a plane graph G' corresponding to G in linear time [7]. Let v_1, v_2, v_n be the three consecutive vertices on the outer face of G', and they appear on the outer face counterclockwise in order (v_1, v_2, v_n). A partition $V_1, V_2, \cdots, V_h$ of vertices of G' is called a *canonical decomposition* of G' if the following conditions (cd1)–(cd4) are satisfied [11]. See an example in Fig. 6(a). Let G_i be the subgraph of G' induced by $V_1 \cup V_2 \cup \cdots \cup V_i$, and Let $\overline{G_i}$ be the subgraph of G' induced by $V_{i+1} \cup V_{i+2} \cup \cdots \cup V_h$.

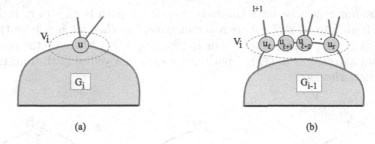

Fig. 3. An illustration for (cd3).

(**cd1**) $V_1 = \{v_1, v_2\}$.

(**cd2**) For each $i = 2, 3, \cdots, h$, G_i is bi-connected.

(**cd3**) For each $i = 2, 3, \cdots, h - 1$, V_i is either (1) a vertex u on the outer face of G_i having at least one neighbor in $\overline{G_i}$ (See Fig. 3(a)), or (2) consecutive vertices $\{u_\ell, u_{\ell+1}, \cdots, u_r\}$ on the outer face of G_i such that each vertex has degree two in G_i and has at least one neighbor in $\overline{G_i}$ (See Fig. 3(b)).

(**cd4**) $V_h = \{v_n\}$.

One can regard the canonical decomposition of a maximal plane graph is a realizer.

The following lemma is known.

Lemma 2. *[11] Every tri-connected plane graph has a canonical decomposition. One can find it in linear time.*

3   Algorithm I

In this section we design a linear time algorithm to construct a set of $(3, 2)$-completely independent spanning trees in a given maximal planar graph with n vertices. The algorithm is based on the realizer [18].

Let T_r, T_b, T_y be a realizer of a maximal planar graph. We have the following lemma.

Lemma 3. *[18][Theorem 4.6] For any inner vertex v, let S be the set of three paths consisting of (1) the path from v to r_r in T_r, (2) the path from v to r_b in T_b and (3) the path from v to r_y in T_y. Then any two paths in S share only v.*

Proof. Assume otherwise for a contradiction. If the path from v to r_r in T_r and the path from v to r_y in T_y share a vertex except v, then let $u \neq v$ be the first such vertex in the path from v to r_r in T_r (See Fig. 4 (b)), then, by the planarity, (re3) is not satisfied at u. (A red path never crosses a yellow path from right to left.) A contradiction.

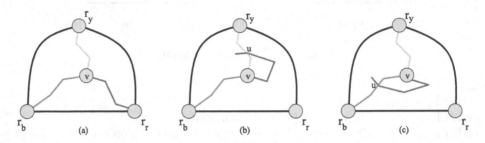

Fig. 4. Illustration for Lemma 3.

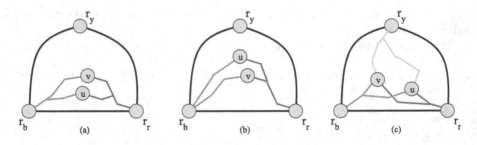

Fig. 5. Illustration for Theorem 1.

Similar for the other cases. See Fig. 4 (c).   □

Given a realizer of a plane graph G' corresponding to a maximal planer graph G, let T_r' be the spanning tree of G' rooted at r_r consisting of T_r and two edges (r_y, r_r) and (r_b, r_r). Similarly, let T_b' be the spanning tree of G' rooted at r_b consisting of T_b and two edges (r_r, r_b) and (r_y, r_b), and T_y' be the spanning tree of G' rooted at r_y consisting of T_y and two edges (r_b, r_y) and (r_r, r_y). See an example in Fig. 1(c).

We have the following theorem.

Theorem 1. T_r', T_b', T_y' *are* $(3, 2)$-*completely independent spanning trees.*

Proof. For an inner vertex $\dot{v}$, let $Y(v)$ be the region surrounded by the path from v to r_r in T_r', the path from v to r_b in T_b' and edge (r_r, r_b). Similarly, let $R(v)$ be the region surrounded by the path from v to r_b in T_b', the path from v to r_y

in T_y' and edge (r_b, r_y) and $B(v)$ be the region surrounded by the path from v to r_y in T_y', the path from v to r_r in T_r' and edge (r_y, r_r).

Given two vertices u and v in G', let S be the set of three paths consisting of the path from u to v in T_r', the path from u to v in T_b' and the path from u to v in T_y'. Then we show that some pair of paths in S are internally disjoint.

If $\{u, v\} \subset \{r_r, r_b, r_y\}$ then the claim holds. Assume otherwise.

We have the following three cases to consider.

Case 1: $Y(v)$ contains u. See Fig. 5(a).

The path from u to v in T_r' and the path from u to v in T_b' are internally disjoint. (If the path from u to v in T_r' and the path from u to v in T_b' are not internally disjoint, then, similar to the proof of Lemma 3, we can show that there is a vertex where (re3) does not satisfied. A contradiction.)

Case 2: $Y(u)$ contains v. See Fig. 5(b).

The path from u to v in T_r' and the path from u to v in T_b' are internally disjoint.

Similar to Case 1.

Case 3: Otherwise.

Then either $B(v)$ contains u (See Fig. 5(c)) or $R(v)$ contains u.

The path from u to v in T_b' and the path from u to v in T_y' are internally disjoint. (Also the path from u to v in T_r' and the path from u to v in T_y' are internally disjoint.) Similar to the proof of Lemma 3.

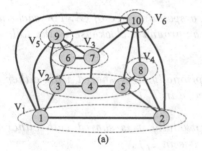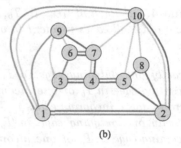

(a)   (b)

Fig. 6. (a) A canonical decomposition of a tri-connected plane graph. (b) Three spanning trees.

4   Algorithm II

In this section we design a linear time algorithm to construct a set of $(3, 2)$-completely independent spanning trees in a given tri-connected planar graph with n vertices. The algorithm is based on the canonical decomposition [11].

Given a tri-connected planar graph G, let G' be the corresponding plane graph, and $V_1, V_2, \cdots V_h$ its canonical decomposition. We define, for each vertex

v of G, three outgoing edges $ll(v)$, $rl(v)$ and $h(v)$ from v, as follows. We call those edges left leg, right leg and head of v, and intuitively each left leg points lower left, each right leg points lower right and each head points upward.

For vertex in $V_1 = \{v_1, v_2\}$ we define as follows. $rl(v_1) = (v_1, v_2)$ and $ll(v_2) = (v_2, v_1)$. And v_1 has no left leg and v_2 has no right leg.

For each vertex $v \in V_2 \cup V_3 \cup \cdots \cup V_h$, we have the following two cases.

If $|V_i| = 1$ then let $V_i = \{v\}$ and $u_\ell, u_{\ell+1}, \cdots, u_r$ be the neighbor of v on the outer face of G_{i-1} and assume that they appear in this order clockwise. We define $ll(v) = (v, u_\ell)$ and $rl(v) = (v, u_r)$.

If $|V_i| > 1$ then let $V_i = \{v_1, v_2, \cdots, v_k\}$, and they appear in this order clockwise on the outer face of G_i and v_0 and v_{k+1} be the neighbor of v_1 and the neighbor of v_k on the outer face of G_{i-1}, respectively. Then, for each $j = 1, 2, \cdots k$, we define $ll(v_i) = (v_i, v_{i-1})$ and $rl(v_i) = (v_i, v_{i+1})$.

For $v_n \in V_h$ we define $ll(v_n) = v_1$ and $rl(v_n) = v_2$.

Also, for each vertex $v \in V_1 \cup V_2 \cup \cdots \cup V_{h-1}$, let $u \in V_{h'}$ be the neighbor of v with the maximum h'. (For tie we choose the rightmost vertex.) We define $h(v) = (v, u)$. For $v_n \in V_h$, v_n has no head.

We regard v_1, v_2, v_n as the three roots r_b, r_r, r_y, respectively.

Let T_r be the tree rooted at r_r consisting of all right legs. Similarly, let T_b be the tree rooted at r_b consisting of all left legs and let T_y be the tree rooted at r_y consisting of all heads. The set of those three trees is called a realizer of a triconnected plane graph [1].

We have the following lemma.

Lemma 4. *[1] Each of trees T_r, T_b, T_y is a spanning tree of G'. For each inner vertex v of G' all edges incident to v appear around v clockwise in the following order.*

- *Exactly one outgoing edge in T_r. (Optionally it is shared with either one incoming edge in T_b or one incoming edge in T_y)*
- *Zero or more incoming edges in T_y.*
- *Exactly one outgoing edge in T_b. (Optionally it is shared with either one incoming edge in T_y or one incoming edge in T_r)*
- *Zero or more incoming edges in T_r.*
- *Exactly one outgoing edge in T_y. (Optionally it is shared with either one incoming edge in T_r or one incoming edge in T_b).*
- *Zero or more incoming edges in T_b.*

Proof. We denote the claim by (cd).

We can prove (cd) by induction on V_i, that is, for each i, the following (1)–(5) holds. (1) (cd) holds on each vertex of G_i having no neighbor in $\overline{G_i}$, (2) a relax version of (cd) holds on each vertex of G_i having a neighbor in $\overline{G_i}$, (3) the right legs induce a spanning tree of G_i rooted at r_r, (4) the left legs induce a spanning tree of G_i rooted at r_b, (5) the heads induce a spanning forest of G_i with each root on the outer face of G_i and each root has a neighbor in $\overline{G_i}$.

Lemma 5. *[1][Lemma 6] For any vertex v, let S be the set of three paths consisting of (1) the path from v to r_r in T_r, (2) the path from v to r_b in T_b and (3) the path from v to r_y in T_y. Then any two paths in S share only v.*

Proof. Assume otherwise for a contradiction. If the path from v to r_r in T_r and the path from v to r_y in T_y share vertex $u \neq v$, then, by the planarity, the condition (cd) of lemma 4 is not satisfied at u. A contradiction.

Similar for other cases.

Theorem 2. T_r, T_b, T_y *are* $(3, 2)$*-completely independent spanning trees.*

Proof. Similar to Theorem 1 we can prove the following.

Given two vertices u and v in G, let S be the set of three paths consisting of the path from u to v in T_r, the path from u to v in T_b and the path from u to v in T_y. Then some pair of paths in S are internally disjoint.

□

5 Algorithm III

In this section we design a linear time algorithm to construct a set of $(3, 2)$-completely independent spanning trees in a given 3D grid graph with size $L_x \times L_y \times L_z$. We assume $L_x \geq 1, L_y \geq 1, L_z \geq 1$.

Let G be a grid graph with size $L_x \times L_y \times L_z$. For a vertex (x, y, z) with $x < L_x$ we define its parent vertex as $(x + 1, y, z)$, and for a vertex (x, y, z) with $x = L_z$ and $y > 0$ we define its parent vertex as $(L_z, y - 1, z)$, and for a vertex (x, y, z) with $x = L_x, y = 0$ and $z > 0$ we define its parent vertex as $(L_x, 0, z - 1)$. The root r_x is the vertex at $(L_x, 0, 0)$ and it has no parent. Then for each vertex of G except the root r_x we append the edge connecting v and its parent. Those edges induces the spanning tree of G and we denote it as T_{xyz}. The path from a vertex (x, y, z) to r_x in T_{xyz} consists of three line segments, those are (1) the line segment from (x, y, z) to (L_x, y, z), (2) the line segment from (L_x, y, z) to $(L_x, 0, z)$, and (3) the line segment from $(L_x, 0, z)$ to $(L_x, 0, 0)$.

Similarly, we define the spanning tree T_{yzx} with the root r_y at $(0, L_y, 0)$ and the spanning tree T_{zxy} with the root r_z at $(0, 0, L_z)$. The path from a vertex (x, y, z) to the root r_y in T_{yzx} consists of three line segments, those are (1) the line segment from (x, y, z) to (x, L_y, z), (2) the line segment from (x, L_y, z) to $(x, L_y, 0)$, and (3) the line segment from $(x, L_y, 0)$ to $(0, L_y, 0)$. Similarly the path from a vertex (x, y, z) to the root r_z in $Tzxy$ consists of three line segments, those are (1) the line segment from (x, y, z) to (x, y, L_z), (2) the line segment from (x, y, L_z) to $(0, y, L_z)$, and (3) the line segment from $(0, y, L_z)$ to $(0, 0, L_z)$.

Note that the second part and the third part of $P_{xyz}(u, r_x)$, which is the path from u to r_x in T_{xyz}, locate on the plane with $x = L_x$, and the second part and the third part of $P_{yzx}(u, r_y)$ locate on the plane with $y = L_y$.

We have the following theorem.

Theorem 3. $T_{xyz}, T_{yzx}, T_{zxy}$ *are* $(3, 2)$*-completely independent spanning trees.*

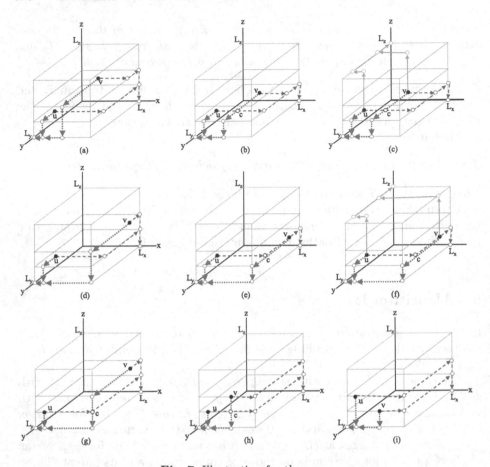

Fig. 7. Illustration for the cases.

Proof. For any pair of two vertices u and v in G we show if the path $P_{xyz}(u,v)$ connecting u and v in T_{xyz} and the path $P_{yzx}(u,v)$ connecting u and v in T_{yzx} are not internally disjoint, then either (a) $P_{xyz}(u,v)$ and the path $P_{zxy}(u,v)$ connecting u and v in T_{zxy} are internally disjoint, or (b) $P_{yzx}(u,v)$ and the path $P_{zxy}(u,v)$ connecting u and v in T_{zxy} are internally disjoint.

Assume that $P_{xyz}(u,v)$ and $P_{yzx}(u,v)$ are not internally disjoint.

We have the following four cases.

Case 1: $x(u) \le L_x, x(v) < L_x, y(u) < L_y$ and $y(v) \le L_y$ hold.

If $z(u) \ne z(v)$ then $P_{xyz}(u,v)$ and $P_{yzx}(u,v)$ are internally disjoint. A contradiction. See Fig. 7 (a). If $z(u) = z(v)$ and either $x(u) = x(v)$ or $y(u) = y(v)$ then $P_{xyz}(u,v)$ and $P_{yzx}(u,v)$ are internally disjoint. A contradiction.

So assume otherwise. Now $z(u) = z(v)$, $x(u) \ne x(v)$ and $y(u) \ne y(v)$ hold. See Fig. 7 (b). If $P_{xyz}(u,r)$ and $P_{yzx}(v,r)$ cross at a vertex c on the plane $z = z(u)$. Then $P_{xyz}(u,v)$ and $P_{zxy}(u,v)$ are internally disjoint. See Fig. 7 (c).

Case 2: $x(u) \le L_x, x(v) = L_x, y(u) < L_y$ and $y(v) \le L_y$ hold.

If $z(u) \neq z(v)$ then $P_{xyz}(u, v)$ and $P_{yzx}(u, v)$ are internally disjoint. See Fig. 7 (d). So assume otherwise.

If $z(u) = z(v)$ and $P_{xyz}(u, v)$ and $P_{yzx}(u, v)$ share some vertex c on the plane $z = z(u)$. See Fig. 7 (e), then $P_{xyz}(u, v)$ and $P_{zxy}(u, v)$ are internally disjoint. See Fig. 7 (f).

Case 3: $x(u) \leq L_x, x(v) = L_x, y(u) = L_y$ and $y(v) \leq L_y$ hold.

If $x(u) = L_x$ then $P_{xyz}(u, v)$ and $P_{yzx}(u, v)$ are internally disjoint. So assume otherwise.

If $P_{xyz}(u, v)$ and $P_{yzx}(u, v)$ cross at a vertex c on the line with $x = L_x$ and $y = L_y$. See Fig. 7 (g). Then $P_{xyz}(u, v)$ and $P_{zxy}(u, v)$ are internally disjoint.

Case 4: $x(u) \leq L_x, x(v) \leq L_x, y(u) = L_y$ and $y(v) = L_y$ hold.

If $P_{xyz}(u, v)$ and $P_{yzx}(u, v)$ share some vertex c on the plane $y = L_y$ then $P_{xyz}(u, v)$ and $P_{zxy}(u, v)$ are internally disjoint. See Fig. 7 (h).

Otherwise $P_{xyz}(u, v)$ and $P_{yzx}(u, v)$ are internally disjoint. See Fig. 7 (i).

Each of other cases is symmetric to one of above cases.

$\square$

6 Conclusion

In this paper we have defined (t, s)-completely independent spanning trees which is a generalization of completely independent spanning trees. Then we have designed an algorithm to construct a set of $(3, 2)$-completely independent spanning trees in a given maximal planar graph, an algorithm to construct a set of $(3, 2)$-completely independent spanning trees in a given tri-connected planar graph, and an algorithm to construct a set of $(3, 2)$-completely independent spanning trees in a given 3D grid graph. Those algorithms are simple and run in $O(n)$ time, where n is the number of vertices of the given graph.

Can we design an algorithm to construct a set of (t, s)-completely independent spanning trees for other classes of graphs and some other choices of t and s?

References

1. Di Battista, G., Tamassia, R., Vismara, L.: Output-sensitive reporting of disjoint paths. Algorithmica **23**(4), 302–340 (1999)
2. Cheriyan, J., Maheshwari, S.N.: Finding nonseparating induced cycles and independent spanning trees in 3-connected graphs. J. Algorithms **9**(4), 507–537 (1988)
3. Curran, S., Lee, O., Xingxing, Yu.: Finding four independent trees. SIAM J. Comput. **35**(5), 1023–1058 (2006)
4. Hasunuma, T.: Completely independent spanning trees in the underlying graph of a line digraph. Discrete Math. **234**, 149–157 (2001)
5. Hasunuma, T.: Completely independent spanning trees in maximal planar graphs. In: Goos, G., Hartmanis, J., van Leeuwen, J., Kučera, L. (eds.) WG 2002. LNCS, vol. 2573, pp. 235–245. Springer, Heidelberg (2002). https://doi.org/10.1007/3-540-36379-3_21

6. Hopcroft, J.E., Wong, K.: Linear time algorithm for isomorphism of planar graphs. In: Proceeding of the 6th Annual ACM Symposium on Theory of Computing, pp. 172–184 (1974)
7. John, E.: Hopcroft and Robert Endre Tarjan, efficient planarity testing. J. ACM **21**(4), 549–568 (1974)
8. Huck, A.: Independent trees in graphs. Graphs Comb. **10**(1), 29–45 (1994)
9. Huck, A.: Independent trees in planar graphs independent trees. Graphs Comb. **15**(1), 29–77 (1999)
10. Itai, A., Rodeh, M.: The multi-tree approach to reliability in distributed networks. Inf. Comput. **79**(1), 43–59 (1988)
11. Kant, G.: Drawing planar graphs using the canonical ordering. Algorithmica **16**(1), 4–32 (1996)
12. Khuller, S., Schieber, B.: On independent spanning trees. Inf. Process. Letters **42**(6), 321–323 (1992)
13. Miura, K., Takahashi, D., Nakano, S., Nishizeki, T.: A linear-time algorithm to find four independent spanning trees in four-connected planar graphs. In: Hromkovič, J., Sýkora, O. (eds.) WG 1998. LNCS, vol. 1517, pp. 310–323. Springer, Heidelberg (1998). https://doi.org/10.1007/10692760_25
14. Miura, K., Takahashi, D., Nakano, S.-I., Nishizeki, T.: A linear-time algorithm to find four independent spanning trees in four connected planar graphs. Int. J. Found. Comput. Sci. **10**(2), 195–210 (1999)
15. Nagai, S., Nakano, S.: A linear-time algorithm to find independent spanning trees in maximal planar graphs. In: Brandes, U., Wagner, D. (eds.) WG 2000. LNCS, vol. 1928, pp. 290–301. Springer, Heidelberg (2000). https://doi.org/10.1007/3-540-40064-8_27
16. Nagai, S., Nakano, S.: A linear-time algorithm for five-partitioning five-connected internally triangulated plane graphs. IEICE Trans. Fund. **E84–A**(9), 2330–2337 (2001)
17. Nakano, S.: Planar drawings of plane graphs. IEICE Trans. Inf. Syst. **E83–D**(3), 384–391 (2000)
18. Schnyder, W.: Embedding planar graphs on the grid. In: Proceedings of SODA, pp. 138–148 (1990)
19. Zehavi, A., Itai, A.: Three tree-paths. J. Graph Theory **13**(2), 175–188 (1989)

Orientable Burning Number of Graphs

Julien Courtiel[1] , Paul Dorbec[1] , Tatsuya Gima[2,3] , Romain Lecoq[1],
and Yota Otachi[2(✉)]

[1] Normandie Univ, UNICAEN, ENSICAEN, CNRS, GREYC, 14000 Caen, France
{julien.courtiel,paul.dorbec,romain.lecoq}@unicaen.fr
[2] Nagoya University, Nagoya, Japan
{gima,otachi}@nagoya-u.jp
[3] JSPS Research Fellow, Tokyo, Japan

Abstract. In this paper, we introduce the problem of finding an orientation of a given undirected graph that maximizes the burning number of the resulting directed graph. We show that the problem is polynomial-time solvable on Kőnig–Egerváry graphs (and thus on bipartite graphs) and that an almost optimal solution can be computed in polynomial time for perfect graphs. On the other hand, we show that the problem is NP-hard in general and W[1]-hard parameterized by the target burning number. The hardness results are complemented by several fixed-parameter tractable results parameterized by structural parameters. Our main result in this direction shows that the problem is fixed-parameter tractable parameterized by cluster vertex deletion number plus clique number (and thus also by vertex cover number).

Keywords: Burning number · Graph orientation · Fixed-parameter algorithm

1 Introduction

The *burning number* of a directed or undirected graph G, denoted $b(G)$, is the minimum number of steps for burning all vertices of G in the following way: in each step, we pick one vertex and burn it; and then between any two consecutive steps, the fire spreads to the neighbors (to the out-neighbors, in the directed setting) of the already burnt vertices. In other words, $b(G)$ is the minimum integer b such that there exists a sequence $\langle w_0, \ldots, w_{b-1} \rangle$ of vertices such that for each vertex v of G, there exists i ($0 \le i \le b-1$) such that the distance from w_i to v is at most i. Note that each w_i corresponds to the vertex that we picked in the $(b-i)$th step.

The concept of burning number is introduced by Bonato, Janssen, and Roshanbin [6,7] as a model of information spreading, while the same concept

Partially supported by JSPS KAKENHI Grant Numbers JP18H04091, JP20H05793, JP21K11752, JP22H00513, JP23KJ1066. The full version of this paper is available at https://arxiv.org/abs/2311.13132.

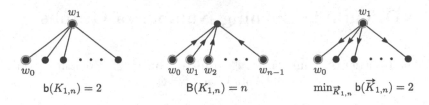

Fig. 1. The star graph $K_{1,n}$ with $n \geq 2$.

was studied already in 1992 by Alon [1]. The central question studied so far on this topic is the so-called *burning number conjecture*, which is about the worst case for a burning process and states that $\mathsf{b}(G) \leq \lceil \sqrt{n} \rceil$ for every connected undirected graph with n vertices. The conjecture has been studied intensively but it is still open (see [5] and the references therein). Recently, it has been announced that the conjecture holds asymptotically, that is, $\mathsf{b}(G) \leq (1 + o(1))\sqrt{n}$ [41]. For the directed case, the worst cases are completely understood in both weakly and strongly connected settings [29]. Since the problem of computing the burning number is hard [4,36,40], several approximation algorithms [34,35,38] and parameterized algorithms [3,30,32] are studied.

In this paper, we investigate the worst case for a directed graph in the setting where we only know the underlying undirected graph. That is, given an undirected graph, which is assumed to be the underlying graph of a directed graph, we want to know how bad the original directed graph can be in terms of burning number. This concept is represented by the following new graph parameter: the *orientable burning number* of an undirected graph G, denoted $\mathsf{B}(G)$, is the maximum burning number over all orientations of G; that is,

$$\mathsf{B}(G) = \max_{\text{orientation } \vec{G} \text{ of } G} \mathsf{b}(\vec{G}).$$

See Fig. 1. Recall that an orientation $\vec{G}$ of an undirected graph G is a directed graph that gives exactly one direction to each edge of G. Now the main problem studied in this paper is formalized as follows.

Problem: ORIENTABLE BURNING NUMBER (OBN)
Input: An undirected graph $G = (V, E)$ and an integer b.
Question: Is $\mathsf{B}(G) \geq b$?

In the setting of information spreading applications, this new problem can be seen as the one determining directions of links in a given underlying network structure to make the spread of something bad as slow as possible. Note that the dual problem of minimizing the burning number by an orientation is equivalent to the original graph burning problem on undirected graphs since $\mathsf{b}(G) = \min_{\text{orientation } \vec{G} \text{ of } G} \mathsf{b}(\vec{G})$. To see this equality, observe that each edge is used at most once and only in one direction to spread the fire (see Fig. 1).

1.1 Our Results

We first present, in Sect. 3, several lower and upper bounds connecting the orientable burning number of a graph with other parameters such as the independence number. In particular, for perfect graphs, we present almost tight lower and upper bounds that differ only by 2 and can be computed in polynomial time. We also consider Kőnig–Egerváry graphs, which generalize bipartite graphs. Although our bounds for them are not exact, we show that the orientable burning number of a Kőnig–Egerváry graph can be computed in polynomial time.

Next we consider the computational intractability of OBN in Sect. 4. We show that OBN is W[1]-hard parameterized by the target burning number b. Although the proof of this result implies the NP-hardness of OBN for general graphs as well, we present another NP-hardness proof that can be applied to restricted graph classes that satisfy a couple of conditions. For example, this shows that OBN is NP-hard on planar graphs of maximum degree 3.

To cope with the hardness of OBN, we study structural parameterizations in Sect. 5. We first observe that some sparseness parameters (e.g., treewidth) combined with b make the problem fixed-parameter tractable. The main question there is the tractability of structural parameterizations *not* combined with b. We show that OBN parameterized by cluster vertex deletion number plus clique number is fixed-parameter tractable. As a corollary to this result, we can see that OBN parameterized by vertex cover number is fixed-parameter tractable.

Due to the space limitation, some parts are shortened or omitted. The proofs of the statements marked with ★ are given in the full version.

1.2 Related Problems

Although the problem studied in this paper is new, the concept of orientable number has long history in the settings of some classical graph problems.

The most relevant is the orientable domination number. The *orientable domination number* of an undirected graph G, denoted $\mathsf{DOM}(G)$, is the maximum domination number over all orientations $\vec{G}$ of G. That is,

$$\mathsf{DOM}(G) = \max_{\text{orientation } \vec{G} \text{ of } G} \gamma(\vec{G}),$$

where $\gamma(\vec{G})$ is the size of a minimum dominating set of the directed graph $\vec{G}$.[1] Erdős [19] initiated (under a different formulation) the study of orientable domination number by showing that $\mathsf{DOM}(K_n) \simeq \log_2 n$, where K_n is the complete graph on n vertices. Later, the concept of orientable domination number is explicitly introduced by Chartrand et al. [12]. We can show that orientable domination number (plus 1) is an upper bound of orientable burning number (see Observation 3.1).

There are two other well-studied problems. One is to find an orientation that minimizes the length of a longest path, which is equivalent to the graph coloring

---

[1] In a directed graph, a vertex dominates itself and its out-neighbors.

problem by the Gallai–Hasse–Roy–Vitaver theorem [20,25,42,43]. The other one is to find a strong orientation that minimizes or maximizes the diameter. It is NP-complete to decide if a graph admits a strong orientation with diameter 2 [13] and the maximum diameter of a strong orientation is equal to the length of a longest path in the underlying 2-edge connected graph [23].

2   Preliminaries

We assume that the readers are familiar with the theory of parameterized algorithms. (For standard concepts, see [16].)

Terms in Graph Burning. Let $D = (V, A)$ be a directed graph. By $N_{\ell,D}^+[v]$, we denote the set of vertices with distance at most ℓ from v in D. We often omit D in the subscript and write $N_\ell^+[v]$ instead when it is clear from the context. A *burning sequence* of D with length b is a sequence $\langle w_0, w_1, \ldots, w_{b-1} \rangle \in V^b$ such that $\bigcup_{0 \le i \le b-1} N_i^+[w_i] = V$. Note that the burning number of D is the minimum integer b such that D has a burning sequence of length b. We call the ith vertex w_i in a burning sequence the *fire* of *radius* i and say that w_i *burns* $N_i^+[w_i]$.

Some Basic Graph Terms. Let $G = (V, E)$ be a graph. The *complement* of G is denoted by $\overline{G}$. For $S \subseteq V$, let $G[S]$ denote the subgraph of G induced by S. For $S \subseteq V$, let $G - S = G[V \setminus S]$. For a graph G, let $\alpha(G)$ denote the *independence number*, $\chi(G)$ the *chromatic number*, $\omega(G)$ the *clique number*, $\theta(G)$ the *clique cover number*, $\mu(G)$ the *matching number*. (See the full version for their definitions.) Note that $\alpha(G) = \omega(\overline{G})$ and $\chi(G) = \theta(\overline{G})$.

A graph $G = (V, E)$ is a *perfect graph* if $\omega(G[S]) = \chi(G[S])$ holds for all $S \subseteq V$. Equivalently, G is a perfect graph if $\alpha(G[S]) = \theta(G[S])$ holds for all $S \subseteq V$ since the class of perfect graphs is closed under taking complements [37]. The class of perfect graphs contains several well-studied classes of graphs such as bipartite graphs and chordal graphs (see, e.g., [8]). A graph $G = (V, E)$ is a *Kőnig–Egerváry graph* if $\alpha(G) = |V| - \mu(G)$. It is known that every bipartite graph is a Kőnig–Egerváry graph [17,31].

Structural Parameters of Graphs. Let $G = (V, E)$ be a graph. A *vertex cover* of G is a set $S \subseteq V$ such that $G - S$ has no edge. The *vertex cover number* of G, denoted $\mathsf{vc}(G)$, is the minimum size of a vertex cover of G. A *cluster vertex deletion set* of G is a set $S \subseteq V$ such that each connected component of $G - S$ is a complete graph. The *cluster vertex deletion number* of G, denoted $\mathsf{cvd}(G)$, is the minimum size of a cluster vertex deletion set of G.

Observation 2.1 (★). $\chi(G) \le \mathsf{cvd}(G) + \omega(G) \le 2\mathsf{vc}(G) + 1$ *for any graph* G.

We can see that $\mathsf{cvd}(G) + \omega(G)$ is an upper bound of vertex integrity, and thus of treedepth, pathwidth, treewidth, and clique-width. We are not going to define these parameters as we do not explicitly use them in this paper. For their definitions, see [21,26].

3   General Lower and Upper Bounds

In this section, we present lower and upper bounds of orientable burning number, which are useful in presenting algorithmic and computational results in the rest of the paper. We start with a simple observation that orientable burning number is bounded from above by orientable domination number plus 1.

Observation 3.1 (★). $\mathsf{B}(G) \leq \mathsf{DOM}(G) + 1$ for every graph G.

Since $\mathsf{DOM}(G) \in O(\alpha \cdot \log |V(G)|)$ [10,24], Observation 3.1 implies that $\mathsf{B}(G) \in O(\alpha \cdot \log |V(G)|)$.

For orientable domination number, it is known that $\alpha(G) \leq \mathsf{DOM}(G) \leq n - \mu(G)$ for every n-vertex graph G [11,12]. The following counterpart for orientable burning number can be shown in almost the same way.

Lemma 3.2 (★). For every n-vertex graph G, $\alpha(G) \leq \mathsf{B}(G) \leq n - \mu(G) + 1$.

The equality $\alpha(G) = n - \mu(G)$ for Kőnig–Egerváry graphs implies that $\mathsf{DOM}(G) = \alpha(G)$ for them [11]. On the other hand, because of the additive factor $+1$ in Lemma 3.2, the bounds only give that $\mathsf{B}(G) \in \{\alpha(G), \alpha(G) + 1\}$ for Kőnig–Egerváry graphs. By taking a closer look at their structure, we can present a characterization that determines which is the case.

Theorem 3.3 (★). Let G be a Kőnig–Egerváry graph with more than four vertices, then
$$
\mathsf{B}(G) = \begin{cases} \alpha(G) + 1 & \text{if } G = mP_2, \\ \alpha(G) & \text{otherwise}, \end{cases}
$$
where $m = |E|$ and mP_2 is the disjoint union of m edges.

This immediately gives the complexity of ORIENTABLE BURNING NUMBER for Kőnig–Egerváry graphs.

Corollary 3.4. ORIENTABLE BURNING NUMBER on Kőnig–Egerváry graphs is solvable in polynomial time.

A *tournament* is an orientation of a complete graph. A *king* of a tournament $T = (V, A)$ is a vertex $v \in V$ such that $N_2^+[v] = V$ [39]. The following fact due to Landau [33] is well known.

Proposition 3.5 ([33]). In a tournament, every vertex with the maximum out-degree is a king.

By using Proposition 3.5, we can show the following upper bound of orientable burning number in terms of clique cover number.

Lemma 3.6 (★). For every graph G, $\mathsf{B}(G) \leq \theta(G) + 2$.

Recall that $\theta(G) = \alpha(G)$ holds for every perfect graph G. Hence, Lemmas 3.2 and 3.6 imply the following almost tight bounds for perfect graphs.

Corollary 3.7. For every perfect graph G, $\alpha(G) \leq \mathsf{B}(G) \leq \alpha(G) + 2$.

Since $\alpha(G)$ of a perfect graph G can be computed in polynomial time [22], one can compute in polynomial time a value b such that $b \leq \mathsf{B}(G) \leq b + 2$. We left the complexity of OBN on perfect graphs unsettled.

4   Hardness of the Problem

Now we demonstrate that ORIENTABLE BURNING NUMBER is intractable in general. We show that OBN is NP-hard, and OBN is W[1]-hard parameterized by the target burning number b.

We can see that our reduction showing the W[1]-hardness also shows NP-hardness in general. However, we present the separate reduction for NP-hardness as it has a wider range of applications. Basically, our reduction for W[1]-hardness works only for dense graphs, while the one for NP-hardness works also for sparse graphs like planar graphs.

In this short version, we only give the proof of the W[1]-hardness and the proofs of the following results are given in the full version.

Theorem 4.1 (★). *Let $\mathcal{G}$ be a graph class such that* INDEPENDENT SET *is NP-complete on $\mathcal{G}$. If $\mathcal{G}$ is closed under additions of isolated vertices, then* ORIENTABLE BURNING NUMBER *on $\mathcal{G}$ is NP-hard.*

Corollary 4.2 (★). ORIENTABLE BURNING NUMBER *is NP-hard on planar graphs of maximum degree* 3.

4.1   W[1]-Hardness Parameterized by b

Given an undirected graph $G = (V, E)$ and a partition $(V_1, \ldots, V_k)$ of V into cliques, MULTICOLORED INDEPENDENT SET (MCIS) asks whether G contains an independent set of size k. It is known that MCIS parameterized by k is W[1]-complete [16]. We reduce MCIS parameterized by k to OBN parameterized by b.

Theorem 4.3. ORIENTABLE BURNING NUMBER *on connected graphs is* W[1]-*hard parameterized by the target burning number b.*

Proof. Let $(G, V_1, \ldots, V_k)$ be an instance of MCIS. Let H be the connected graph obtained from G by first adding a set I of four isolated vertices and then adding a universal vertex u. We prove that $(H, k + 4)$ is a yes-instance of OBN if and only if (G, k) is a yes-instance of MCIS.

To show the if direction, assume that G has an independent set S of size k. Since $S \cup I$ is an independent set of H, there is an orientation $\overrightarrow{H}$ of H such that each vertex in $S \cup I$ has in-degree 0, and thus $\mathsf{b}(\overrightarrow{H}) \geq k + 4$.

In the following, we show the only-if direction. Assume that $(H, k + 4)$ is a yes-instance of OBN and an orientation $\overrightarrow{H}$ of H satisfies $\mathsf{b}(\overrightarrow{H}) \geq k + 4$.

We construct a sequence $\sigma = \langle w_0, \ldots, w_{k+3} \rangle$ as follows. If all vertices in I are of in-degree 0, then we set w_0, w_1, w_2, w_3 to the vertices in I. Otherwise, we set w_3 to u and set w_0, w_1, w_2 to three vertices of I including the ones of in-degree 0 (if any exist). For $1 \leq i \leq k$, we set w_{i+3} to a king of the tournament $\overrightarrow{H}[V_i]$. Recall that a king of a tournament can reach the other vertices in the tournament in at most two steps. Recall also that every tournament has a king, which can be found in polynomial time (see Proposition 3.5). We can see that σ is a burning sequence of $\overrightarrow{H}$ (with length $k + 4$) as follows. Each vertex of in-degree 0 in $\{u\} \cup I$, if any exists, is burned by itself, and the other vertices in $\{u\} \cup I$ are burned by w_3. For $1 \leq i \leq k$, w_{i+3} burns V_i as $i + 3 > 2$ and w_{i+3} is a king of $\overrightarrow{H}[V_i]$. Since $\mathsf{b}(\overrightarrow{H}) \geq k + 4$, σ is a shortest burning sequence of $\overrightarrow{H}$.

Now we show that $\{w_4, \dots, w_{k+3}\}$ is an independent set of G, which implies that $(G, V_1, \dots, V_k)$ is a yes-instance of MCIS. Suppose to the contrary that G has an edge between vertices $w_p, w_q \in \{w_4, \dots, w_{k+3}\}$. By symmetry, we may assume that $(w_p, w_q) \in A(\overrightarrow{H})$. Let $\sigma' = \langle w_0', \dots, w_{k+2}' \rangle$ be the sequence obtained from σ by skipping w_q, that is, a sequence σ' defined as

$$w_i' = \begin{cases} w_i & 0 \leq i \leq q - 1, \\ w_{i+1} & q \leq i \leq k + 2. \end{cases}$$

We show that σ' is a burning sequence of $\overrightarrow{H}$, which contradicts that σ is shortest.

As $\{w_0', w_1', w_2', w_3'\} = \{w_0, w_1, w_2, w_3\}$, we have $\bigcup_{0 \leq i \leq 3} N_i^+[w_i'] \supseteq \{u\} \cup I$. For $4 \leq i \leq q - 1$, we have $w_i' = w_i$, and thus $N_i^+[w_i'] = N_i^+[w_i] \supseteq V_{i-3}$. For $q \leq i \leq k + 2$, $w_i' = w_{i+1}$ is a king of V_{i-2}, and thus $N_i^+[w_i'] \supseteq V_{i-2}$. The discussion so far implies that $V(G) \setminus V_{q-3} \subseteq \bigcup_{0 \leq i \leq k+2} N_i^+[w_i']$.

Now it suffices to show that V_{q-3} is also burned by σ'. Since w_q is a king of V_{q-3}, we have $V_{q-3} \subseteq N_2^+[w_q]$. As $(w_p, w_q) \in A(\overrightarrow{H})$, it holds that $N_2^+[w_q] \subseteq N_3^+[w_p]$. Since $p \geq 4$, $N_3^+[w_p] \subseteq N_{p-1}^+[w_p]$ holds. The chain of inclusions implies that $V_{q-3} \subseteq N_{p-1}^+[w_p]$. Since $w_p \in \{w_{p-1}', w_p'\}$, we have $V_{q-3} \subseteq N_{p-1}^+[w_{p-1}']$ or $V_{q-3} \subseteq N_{p-1}^+[w_p'] \subseteq N_p^+[w_p']$. This implies that V_{q-3} is burnt by w_{p-1}' or w_p'. $\square$

5 Structural Parameterizations

In this section, we consider some structural parameterizations of ORIENTABLE BURNING NUMBER. Given Theorem 4.3, which shows that OBN is intractable when parameterized by the target burning number b, it is natural to consider the problem parameterized by some structural parameters of the input graph.

The first observation is that some sparseness parameters combined with b make the problem tractable. In particular, we can show the following.

Corollary 5.1 ($\bigstar$). ORIENTABLE BURNING NUMBER *is fixed-parameter tractable parameterized by b plus average degree.*

Corollary 5.1 implies that OBN is fixed-parameter tractable parameterized by b + treewidth, and OBN on planar graphs is fixed-parameter tractable parameterized by b. Recall that OBN is NP-hard on planar graphs even if the maximum degree is 3 (Corollary 4.2). On the other hand, the parameterized complexity of OBN parameterized solely by treewidth remains unsettled.

Next we consider structural parameterizations *not* combined with b. As the first step in this direction, we consider parameters less general than treewidth such as vertex cover number. In some sense, vertex cover number is one of the most restricted parameters that is an upper bound of treewidth (see [21]).

We show that OBN parameterized solely by vertex cover number is fixed-parameter tractable. Our proof is actually for a slightly more general case, where the parameter is cluster vertex deletion number plus clique number. In the rest of this section, we prove the following theorem.

Theorem 5.2. ORIENTABLE BURNING NUMBER *is fixed-parameter tractable parameterized by cluster vertex deletion number plus clique number.*

Theorem 5.2 and Observation 2.1 imply the fixed-parameter tractability parameterized by vertex cover number.

Corollary 5.3. ORIENTABLE BURNING NUMBER *is fixed-parameter tractable parameterized by vertex cover number.*

Proof of Theorem 5.2. In the proof, we use the theory of monadic second-order logic on graphs (MSO$_2$), which will be introduced right before we use it. If we allow an MSO$_2$ formula to have length depending on b, it is not difficult to express OBN. However, this only implies the fixed-parameter tractability of OBN parameterized by a parameter combined with b. To avoid the dependency on b, we have to bound the length of an MSO$_2$ formula with a function not depending on b. To this end, we make a series of observations to bound the number of *parts* not used in a *good* burning sequence, then represent the problem by expressing the unused parts instead of the used parts.

Useful Observations. In the following, let (G, b) be an instance of OBN. Let ω be the clique number of G; that is, $\omega = \omega(G)$. Let S be a cluster vertex deletion set of G with size $s = \mathsf{cvd}(G)$. Our parameter is $k := \omega + s$. Note that finding S is fixed-parameter tractable parameterized by s [27], and thus by k as well. Let $C_1, \ldots, C_p$ be the connected components of $G - S$, which are complete graphs. When we are dealing with an orientation $\vec{G}$ of G, we sometimes mean by C_i the tournament $\vec{G}[V(C_i)]$. For example, we may say "a king of C_i."

Claim 5.4. If $b \leq p$, then (G, b) is a yes-instance.

Proof. By picking arbitrary one vertex from each C_i, we can construct an independent set of size p. By Lemma 3.2, $\mathsf{B}(G) \geq \alpha(G) \geq p \geq b$. ☐

Claim 5.5. If $b > p + s + 2$, then (G, b) is a no-instance.

Proof. Let $\vec{G}$ be an orientation of G. It suffices to show that $\mathsf{b}(\vec{G}) \leq p + s + 2$. For each C_i, we place a fire of radius at least 2 at a king of C_i. For each vertex in S, we place a fire of arbitrary radius. If we have not used the fires of radii 0 and 1, then we place them at arbitrary vertices. ☐

By Claims 5.4 and 5.5, we may assume that $p < b \leq p + s + 2$.

Let ℓ be the length of a longest path in G. We assume that $\ell \geq 1$ since otherwise G cannot have any edge and the problem becomes trivial. Note that in every orientation $\vec{G}$ or G, the length of a longest directed path is at most ℓ.

Claim 5.6. $\ell \leq s\omega + s + \omega - 1.$

Proof. Let P be a longest path in G. Since P can visit at most $|S| + 1$ connected components of $G - S$, we have $|V(P)| \leq |S| + (|S| + 1)\omega$ as each C_i is a complete graph. The claim follows as $|S| = s$ and $|E(P)| = |V(P)| - 1$.   □

In a burning sequence of an orientation of G, we call a fire of radius at least ℓ a *large fire*. Note that a large fire w burns all vertices that can be reached from w in the orientation as no directed path in the orientation is longer than ℓ.

In the following, we focus on burning sequences of length $b - 1$ since we are going to express the *non-existence* of such sequences. Let $L = \max\{0, b - 1 - \ell\}$; that is L is the number of large fires in a sequence of length $b - 1$. Observe that $L \leq p + s$ as $b - 1 - \ell \leq b - 2 \leq p + s$.

A burning sequence of an orientation of G is *good* if the following conditions are satisfied:

1. two large fires do not have the same position;
2. each C_j contains at most one large fire;
3. if a large fire is placed in some C_h, then it is placed at a king of C_h.

Claim 5.7. Let $\vec{G}$ be an orientation of G. If $\vec{G}$ admits a burning sequence with length $b - 1$, then there is a good burning sequence of $\vec{G}$ with the same length.

Proof. From a burning sequence σ of $\vec{G}$ with length $b - 1$, we first construct a sequence σ_1 that satisfies the first condition of the goodness. We repeatedly find two large fires placed at the same vertex and then replace arbitrary one of them with another vertex not occupied by any large fire. The replacement is possible as L is not larger than the number of vertices. Since two large fires placed at the same vertex burn the same set of vertices, the obtained sequence is still a burning sequence of $\vec{G}$. When there is no pair of large fires occupying the same vertex, we stop this phase and call the resultant sequence σ_1.

Next we modify σ_1 to obtain a sequence σ_2 that satisfies the first and second conditions. Assume that two large fires w_i and w_j are placed in the same connected component C_h of $G - S$ and that $(w_i, w_j) \in A(\vec{G})$. (Recall that C_h is a complete graph.) Since w_i is a large fire, it burns every vertex that can be reached from w_i. In particular, w_i burns all vertices reachable from w_j. Hence, w_j is useless for burning the graph. We replace w_j with another vertex v such that v is not occupied by any large fire and if v belongs to some $C_{h'}$, then there is no large fire belonging to $C_{h'}$ prior to the replacement. This is always possible as $L \leq p + s$. We exhaustively apply this replacement procedure and get σ_2, which satisfies the first and second conditions of the goodness.

Finally, we obtain a sequence σ_3 from σ_2 by replacing each large fire that is placed in some C_h with a king of C_h. We can see that σ_3 is a burning sequence of $\vec{G}$ since the new large fire placed at a king of C_h burns all vertices reachable form the king and the king can reach all vertices in C_h. Since σ_3 satisfies all conditions of goodness and has the same length as σ, the claim holds.   □

If σ is a good burning sequence of an orientation $\vec{G}$ of G with length $b-1$, then the sum of the following two numbers is $p+s-L$: (1) the number of vertices in S not occupied by the large fires of σ, and (2) the number of connected components of $G-S$ not including large fires of σ. Since $L = \max\{0, b-1-\ell\}$ and $p < b$, it holds that $p + s - L < s + \ell + 1$. Since ℓ can be bounded from above by a function of $k = s + \omega$, so is $p + s - L$. Thus our MSO_2 formula can have length depending on $p + s - L$, the number of *unused parts*.

MSO_2 Expressions. We now express the problem in the monadic second-order logic on graphs. A formula in the monadic second-order logic on graphs, denoted MSO_2, can have variables representing vertices, vertex sets, edges, and edge sets. As atomic formulas, we can use the equality $x = y$, the inclusion $x \in X$, the adjacency relation $\mathtt{adj}(x,y)$ meaning that vertices x and y are adjacent, and the incidence relation $\mathtt{inc}(e,x)$ meaning that a vertex x is an endpoint of an edge e. Atomic formulas can be recursively combined by the usual Boolean connectives $\neg$, $\wedge$, $\vee$, $\Rightarrow$, and $\Leftarrow$ to form an MSO_2 formula. Furthermore, variables in an MSO_2 formula can be quantified by $\exists$ and $\forall$. If an MSO_2 formula $\phi(X)$ with one free (vertex-set or edge-set) variable X is evaluated to true on a graph G and a subset S of $V(G)$ or $E(G)$, we write $G \models \phi(S)$. It is known that, given an MSO_2 formula $\phi(X)$, a graph G, and a subset S of $V(G)$ or $E(G)$, the problem of deciding whether $G \models \phi(S)$ is fixed-parameter tractable parameterized by the length of ϕ plus the treewidth of G [2,9,14].

In the following, we construct an MSO_2 formula $\phi(X)$ such that $G \models \phi(S)$ if and only if (G, b) is a yes-instance of OBN, S is a minimum cluster vertex deletion set, $p < b \leq p + s + 2$, and the sum of the numbers of unused vertices in S and unused connected components of $G - S$ is $p + s - L$. Recall that $s = \mathtt{cvd}(G) = |S|$, $\omega = \omega(G)$, p is the number of connected components in $G-S$, and L is the number of large fires in a sequence of length $b - 1$. Also, we show that the length of $\phi(X)$ is bounded from above by a function of $k = s + \omega$. Since the treewidth of G is at most $s + \omega$ (see [21]), this implies Theorem 5.2.

The formula $\phi(X)$ asks whether there exists an orientation $\vec{G}$ of G such that no sequence $\langle w_0, \ldots, w_{b-2} \rangle$ of length $b - 1$ is a good burning sequence of $\vec{G}$. In MSO_2, we can handle orientations of k-colorable graphs by first assuming a *default* orientation using a k-coloring and then represent the *reversed* edges by a set of edges [15,28].[2] More precisely, such a formula first expresses a k-coloring and a set of reversed edge, and then it considers each edge as oriented from the vertex with a smaller label to the one with a larger label if and only if the edge is not a reversed one. Note that Observation 2.1 implies that G is k-colorable. We use this technique and thus $\phi(X)$ has the following form:

$$\phi(X) = \exists V_1, \ldots, V_k \subseteq V, \exists F \subseteq E: \mathtt{proper\text{-}coloring}(V_1, \ldots, V_k) \wedge \phi_1,$$

where $\mathtt{proper\text{-}coloring}(V_1, \ldots, V_k)$ expresses that $V_1, \ldots, V_k$ is a proper k-coloring of G. We define $\mathtt{proper\text{-}coloring}(V_1, \ldots, V_k)$ in the full version ($\bigstar$).

---

[2] There is another way for handling orientation by using a variant of MSO_2 defined for directed graphs, where we can fix an arbitrary orientation first (without using a k-coloring) and then represent reversed edges by an edge set. See e.g., [18].

The subformula ϕ_1 can use the formula $\texttt{arc}(u,v)$ expressing that there is an arc from u to v in the orientation defined by $V_1, \ldots, V_k$ and F, where

$$\texttt{arc}(u,v) = \texttt{adj}(u,v) \wedge \big(((u < v) \wedge \neg\texttt{rev}(u,v)) \vee (\neg(u < v) \wedge \texttt{rev}(u,v))\big),$$

with $(u < v) = \bigvee_{1 \leq i < j \leq k}(u \in V_i \wedge v \in V_j)$ and $\texttt{rev}(u,v) = \exists e \in F\colon \texttt{inc}(e,u) \wedge \texttt{inc}(e,v)$.

Given an orientation defined above, the subformula ϕ_1 expresses that there is no good burning sequence of length $b-1$ for this orientation. Indeed, we set $\phi_1 = \neg\phi_2$ and give a definition of ϕ_2 that expresses there is a good burning sequence of length $b-1$. We assume that $b-1 > \ell$ since the other case can be easily obtained from the expression of this case. The subformula ϕ_2 has the following form

$$\phi_2 = \exists w_0, \ldots, w_{\ell-1} \in V, \exists u_1, \ldots, u_{p+s-L} \in V\colon \bigwedge_{1 \leq i < j \leq p+s-L}(u_i \neq u_j)$$
$$\wedge \bigwedge_{1 \leq i < j \leq p+s-L}((u_i \notin X) \wedge (u_j \notin X) \Rightarrow \neg\texttt{adj}(u_i, u_j)) \wedge \phi_3,$$

where $w_0, \ldots, w_{\ell-1}$ simply correspond to the first ℓ fires in a (good) burning sequence and $u_1, \ldots, u_{p+s-L}$ correspond to the representatives of unused parts. More precisely, if $u_i \in X$, then it means that u_i is not used by any large fire; if $u_i \notin X$ and thus u_i belongs to some connected component C of $G - X$, then it means that no vertex in C is used by large fires. Note that the second line of the formula forces that $u_1, \ldots, u_{p+s-L}$ are distinct and not chosen multiple times from one connected component of $G - X$. (Recall that X is promised to be a cluster vertex deletion set.) Now ϕ_3 expresses that every vertex is burned. Hence, it can be expressed as $\phi_3 = \forall v \in V\colon \texttt{burned}(v)$, where the definition of $\texttt{burned}(v)$ is given below.

To define $\texttt{burned}(v)$, observe that v is burned if and only if one of the following conditions is satisfied:

1. some w_i ($0 \leq i \leq \ell - 1$) has a directed path of length at most i to v;
2. some large fire has a directed path to v.

We express the first case as $\texttt{burned-small}(v)$ and the second as $\texttt{burned-large}(v)$, and thus $\texttt{burned}(v) = \texttt{burned-small}(v) \vee \texttt{burned-large}(v)$. The first case is easy to state as

$$\texttt{burned-small}(v) = \bigvee_{0 \leq i \leq \ell-1} \texttt{reachable}_i(w_i, v),$$

where $\texttt{reachable}_d(x, y)$ means that there is a directed path of length at most d from x to y, which can be defined as

$$\texttt{reachable}_d(x, y) = \exists z_0, \ldots, z_d \in V\colon (z_0 = x) \wedge (z_d = y) \wedge$$
$$\bigwedge_{0 \leq j \leq d-1}((z_j = z_{j+1}) \vee \texttt{arc}(z_j, z_{j+1})).$$

On the other hand, the second case is a bit tricky as the large fires are not explicitly handled. Recall that the vertices $u_1, \ldots, u_{p+s-L}$ tell us which vertices in X are not large fires and which connected components of $G - X$ include no large fires. From this information, we can determine whether a vertex x is used as a large fire by setting $\texttt{large-fire}(x) = \neg\texttt{unused}(x)$, where

$$\texttt{unused}(x) = \bigvee_{1 \leq i \leq p+s-L}(x = u_i)$$
$$\vee \left((x \notin X) \wedge \bigvee_{1 \leq i \leq p+s-L}((u_i \notin X) \wedge \texttt{adj}(x, u_i))\right).$$

Note that the correctness of the second line depends on the assumption that each connected component of $G - X$ is a complete graph. Now $\texttt{burned-large}(v)$ can be expressed as follows.

$$\texttt{burned-large}(v) = \exists x \in V : \texttt{large-fire}(x) \wedge \texttt{reachable}_\ell(x, v).$$

The length of the entire formula $\phi(X)$ depends only on k, ℓ, and $p + s - L$, where ℓ and $p + s - L$ can be bounded from above by function depending only on k. This completes the proof of Theorem 5.2.

6   Concluding Remarks

In this paper, we initiated the study of ORIENTABLE BURNING NUMBER (OBN), which is the problem of finding an orientation of a graph that maximizes the burning number. We first observed some graph-theoretic bounds and then showed algorithmic and complexity results.

We showed that OBN is NP-hard even on some classes of sparse graphs (Theorem 4.1). However, we do not know whether OBN belongs to NP. We can see that it belongs to Σ_2^P since it is an $\exists\forall$-problem that asks for the existence of an orientation of a given graph such that all short sequences of fires are not burning sequences of the oriented graph (see [44] for a friendly introduction to Σ_2^P). It would be natural to suspect that the problem is indeed Σ_2^P-hard.

Question 6.1. Does OBN belong to NP, or is it Σ_2^P-complete?

In contrast to the NP-hardness of the general case, we showed that the problem is solvable in polynomial time on bipartite graphs or more generally on Kőnig–Egerváry graphs (Corollary 3.4). We also showed that for perfect graphs, which form a large superclass of bipartite graphs, we can compute the orientable burning number with an additive error of 2 (Corollary 3.7). Given these facts, we would like to ask whether the problem can be solved in polynomial time on perfect graphs or on some of its subclasses such as chordal graphs.

Question 6.2. Is OBN polynomial-time solvable on perfect graphs, or on some of its (non-bipartite) subclasses such as chordal graphs?

In the parameterized setting, we showed that OBN parameterized by the target burning number b is W[1]-hard in general (Theorem 4.3), while it is fixed-parameter tractable on some sparse graphs such as planar graphs (Corollary 5.1). We then studied the setting where b is not part of the parameter. In this case, we showed that OBN parameterized solely by vertex cover number (or more generally by cluster vertex deletion number plus clique number) is fixed-parameter tractable (Theorem 5.2). It would be interesting to study the complexity of parameterizations by more general parameters, e.g., vertex integrity [21].

Question 6.3. Is OBN fixed-parameter tractable when parameterized solely by treewidth, pathwidth, treedepth, vertex integrity, or other related parameters?

Finally, we ask a graph-theoretic question. Most of the algorithmic and complexity results in this paper directly or indirectly used the relations between the orientable burning number and the independence number shown in Sect. 3. As shown there, we have $\alpha(G) \leq \mathsf{B}(G)$ and $\mathsf{B}(G) \in O(\alpha(G) \cdot \log n)$. Now the question would be the maximum difference between $\alpha(G)$ and $\mathsf{B}(G)$. At this point, we only know that the maximum gap is at least 2 as $\mathsf{B}(K_n) = 3 = \alpha(K_n) + 2$ for $n \geq 5$ (★).

Question 6.4. Is there a graph G with $\mathsf{B}(G) > \alpha(G) + 2$? Is there a function f such that $\mathsf{B}(G) \leq f(\alpha(G))$ for every graph G?

References

1. Alon, N.: Transmitting in the n-dimensional cube. Discret. Appl. Math. **37**(38), 9–11 (1992). https://doi.org/10.1016/0166-218X(92)90121-P
2. Arnborg, S., Lagergren, J., Seese, D.: Easy problems for tree-decomposable graphs. J. Algorithms **12**(2), 308–340 (1991). https://doi.org/10.1016/0196-6774(91)90006-K
3. Ashok, P., Das, S., Kanesh, L., Saurabh, S., Tomar, A., Verma, S.: Burn and win. In: Hsieh, S.Y., Hung, L.J., Lee, C.W. (eds.) IWOCA 2023. LNCS, vol. 13889, pp. 36–48. Springer, Cham (2023). https://doi.org/10.1007/978-3-031-34347-6_4
4. Bessy, S., Bonato, A., Janssen, J.C.M., Rautenbach, D., Roshanbin, E.: Burning a graph is hard. Discret. Appl. Math. **232**, 73–87 (2017). https://doi.org/10.1016/j.dam.2017.07.016
5. Bonato, A.: A survey of graph burning. Contributions Discret. Math. **16**(1), 185–197 (2021). https://doi.org/10.11575/cdm.v16i1.71194
6. Bonato, A., Janssen, J., Roshanbin, E.: Burning a graph as a model of social contagion. In: Bonato, A., Graham, F.C., Prałat, P. (eds.) WAW 2014. LNCS, vol. 8882, pp. 13–22. Springer, Cham (2014). https://doi.org/10.1007/978-3-319-13123-8_2
7. Bonato, A., Janssen, J.C.M., Roshanbin, E.: How to burn a graph. Internet Math. **12**(1–2), 85–100 (2016). https://doi.org/10.1080/15427951.2015.1103339
8. Bonomo-Braberman, F., Durán, G., Safe, M.D., Wagler, A.K.: On some graph classes related to perfect graphs: a survey. Discret. Appl. Math. **281**, 42–60 (2020). https://doi.org/10.1016/j.dam.2019.05.019
9. Borie, R.B., Parker, R.G., Tovey, C.A.: Automatic generation of linear-time algorithms from predicate calculus descriptions of problems on recursively constructed graph families. Algorithmica **7**(5&6), 555–581 (1992). https://doi.org/10.1007/BF01758777
10. Caro, Y., Henning, M.A.: A greedy partition lemma for directed domination. Discret. Optim. **8**(3), 452–458 (2011). https://doi.org/10.1016/j.disopt.2011.03.003

11. Caro, Y., Henning, M.A.: Directed domination in oriented graphs. Discret. Appl. Math. **160**(7–8), 1053–1063 (2012). https://doi.org/10.1016/j.dam.2011.12.027
12. Chartrand, G., VanderJagt, D.W., Yue, B.Q.: Orientable domination in graphs. Congr. Numer. **119**, 51–63 (1996)
13. Chvátal, V., Thomassen, C.: Distances in orientations of graphs. J. Comb. Theory Ser. B **24**(1), 61–75 (1978). https://doi.org/10.1016/0095-8956(78)90078-3
14. Courcelle, B.: The monadic second-order logic of graphs. I. Recognizable sets of finite graphs. Inf. Comput. **85**(1), 12–75 (1990). https://doi.org/10.1016/0890-5401(90)90043-H
15. Courcelle, B.: The monadic second-order logic of graphs VIII: orientations. Ann. Pure Appl. Log. **72**(2), 103–143 (1995). https://doi.org/10.1016/0168-0072(95)94698-V
16. Cygan, M., et al.: Parameterized Algorithms. Springer, Cham (2015). https://doi.org/10.1007/978-3-319-21275-3
17. Egerváry, J.: Matrixok kombinatorius tulajdonságairól. Matematikai és Fizikai Lapok **38**, 16–28 (1931)
18. Eggemann, N., Noble, S.D.: The complexity of two graph orientation problems. Discret. Appl. Math. **160**(4–5), 513–517 (2012). https://doi.org/10.1016/j.dam.2011.10.036
19. Erdős, P.: On a problem in graph theory. Math. Gaz. **47**(361), 220–223 (1963). https://doi.org/10.2307/3613396
20. Gallai, T.: On directed graphs and circuits. In: Theory of Graphs (Proceedings of the Colloquium held at Tihany), pp. 115–118 (1968)
21. Gima, T., Hanaka, T., Kiyomi, M., Kobayashi, Y., Otachi, Y.: Exploring the gap between treedepth and vertex cover through vertex integrity. Theor. Comput. Sci. **918**, 60–76 (2022). https://doi.org/10.1016/j.tcs.2022.03.021
22. Grötschel, M., Lovász, L., Schrijver, A.: Geometric Algorithms and Combinatorial Optimization. Springer, Heidelberg (1988). https://doi.org/10.1007/978-3-642-97881-4
23. Gutin, G.Z.: Minimizing and maximizing the diameter in orientations of graphs. Graphs Comb. **10**(2–4), 225–230 (1994). https://doi.org/10.1007/BF02986669
24. Harutyunyan, A., Le, T., Newman, A., Thomassé, S.: Domination and fractional domination in digraphs. Electron. J. Comb. **25**(3), 3 (2018). https://doi.org/10.37236/7211
25. Hasse, M.: Zur algebraischen begründung der graphentheorie. i. Mathematische Nachrichten **28**(5–6), 275–290 (1965). https://doi.org/10.1002/mana.19650280503
26. Hliněný, P., Oum, S., Seese, D., Gottlob, G.: Width parameters beyond tree-width and their applications. Comput. J. **51**(3), 326–362 (2008). https://doi.org/10.1093/comjnl/bxm052
27. Hüffner, F., Komusiewicz, C., Moser, H., Niedermeier, R.: Fixed-parameter algorithms for cluster vertex deletion. Theory Comput. Syst. **47**(1), 196–217 (2010). https://doi.org/10.1007/s00224-008-9150-x
28. Jaffke, L., Bodlaender, H.L., Heggernes, P., Telle, J.A.: Definability equals recognizability for k-outerplanar graphs and l-chordal partial k-trees. Eur. J. Comb. **66**, 191–234 (2017). https://doi.org/10.1016/j.ejc.2017.06.025
29. Janssen, R.: The burning number of directed graphs: bounds and computational complexity. Theory Appl. Graphs **7**(1), Article 8 (2020). https://doi.org/10.20429/tag.2020.070108
30. Kare, A.S., Vinod Reddy, I.: Parameterized algorithms for graph burning problem. In: Colbourn, C.J., Grossi, R., Pisanti, N. (eds.) IWOCA 2019. LNCS, vol.

11638, pp. 304–314. Springer, Cham (2019). https://doi.org/10.1007/978-3-030-25005-8_25

31. Kőnig, D.: Gráfok és mátrixok. Matematikai és Fizikai Lapok **38**, 116–119 (1931)
32. Kobayashi, Y., Otachi, Y.: Parameterized complexity of graph burning. Algorithmica **84**(8), 2379–2393 (2022). https://doi.org/10.1007/s00453-022-00962-8
33. Landau, H.G.: On dominance relations and the structure of animal societies: III the condition for a score structure. Bull. Math. Biophys. **15**, 143–148 (1953). https://doi.org/10.1007/BF02476378
34. Lieskovský, M., Sgall, J.: Graph burning and non-uniform k-centers for small treewidth. In: Chalermsook, P., Laekhanukit, B. (eds.) WAOA 2022. LNCS, vol. 13538, pp. 20–35. Springer, Cham (2022). https://doi.org/10.1007/978-3-031-18367-6_2
35. Lieskovský, M., Sgall, J., Feldmann, A.E.: Approximation algorithms and lower bounds for graph burning. In: APPROX/RANDOM 2023. LIPIcs, vol. 275, pp. 9:1–9:17 (2023). https://doi.org/10.4230/LIPIcs.APPROX/RANDOM.2023.9
36. Liu, H., Hu, X., Hu, X.: Burning number of caterpillars. Discret. Appl. Math. **284**, 332–340 (2020). https://doi.org/10.1016/j.dam.2020.03.062
37. Lovász, L.: Normal hypergraphs and the perfect graph conjecture. Discret. Math. **2**(3), 253–267 (1972). https://doi.org/10.1016/0012-365X(72)90006-4
38. Martinsson, A.: On the approximability of the burning number. CoRR abs/2308.04390 (2023). https://doi.org/10.48550/arXiv.2308.04390
39. Maure, S.B.: The king chicken theorems. Math. Mag. **53**(2), 67–80 (1980). https://doi.org/10.2307/2689952
40. Mondal, D., Rajasingh, A.J., Parthiban, N., Rajasingh, I.: APX-hardness and approximation for the k-burning number problem. Theor. Comput. Sci. **932**, 21–30 (2022). https://doi.org/10.1016/J.TCS.2022.08.001
41. Norin, S., Turcotte, J.: The burning number conjecture holds asymptotically. CoRR abs/2207.04035 (2022). https://doi.org/10.48550/arXiv.2207.04035
42. Roy, B.: Nombre chromatique et plus longs chemins d'un graphe. Revue française d'informatique et de recherche opérationnelle **1**(5), 129–132 (1967). https://doi.org/10.1051/m2an/1967010501291
43. Vitaver, L.M.: Determination of minimal coloring of vertices of a graph by means of boolean powers of the incidence matrix. In: Doklady Akademii Nauk. vol. 147, pp. 758–759. Russian Academy of Sciences (1962), https://www.mathnet.ru/eng/dan27289
44. Woeginger, G.J.: The trouble with the second quantifier. 4OR **19**(2), 157–181 (2021). https://doi.org/10.1007/s10288-021-00477-y

Dichotomies for Tree Minor Containment with Structural Parameters

Tatsuya Gima[1,3], Soh Kumabe[2,3], Kazuhiro Kurita[1], Yuto Okada[1(✉)],
and Yota Otachi[1]

[1] Nagoya University, Nagoya, Japan
{gima,otachi}@nagoya-u.jp, kurita@i.nagoya-u.ac.jp,
okada.yuto.b3@s.mail.nagoya-u.ac.jp
[2] The University of Tokyo, Tokyo, Japan
soh_kumabe@mist.i.u-tokyo.ac.jp
[3] JSPS Research Fellow, Tokyo, Japan

Abstract. The problem of determining whether a graph G contains another graph H as a minor, referred to as the *minor containment problem*, is a fundamental problem in the field of graph algorithms. While it is NP-complete when G and H are general graphs, it is sometimes tractable on more restricted graph classes. This study focuses on the case where both G and H are trees, known as the *tree minor containment problem*. Even in this case, the problem is known to be NP-complete. In contrast, polynomial-time algorithms are known for the case when both trees are caterpillars or when the maximum degree of H is a constant. Our research aims to clarify the boundary of tractability and intractability for the tree minor containment problem. Specifically, we provide dichotomies for the computational complexities of the problem based on three structural parameters: the diameter, pathwidth, and path eccentricity.

Keywords: Minor containment · Tree · Diameter · Path eccentricity · Pathwidth

1 Introduction

In the field of graph algorithms, given two graphs G and H, the problem of determining whether G contains H is a fundamental problem. This type of problem, such as (induced) subgraph isomorphism [4], minor containment [13], and topological embedding [12], is often NP-complete when G and H are general graphs. Therefore, extensive research has been conducted on whether these problems can be efficiently solved on more restricted classes of graphs [3,7,8,13]. The class of

Partially supported by JSPS KAKENHI Grant Numbers JP23KJ1066, JP21J20547, JP21K17812, JP22H03549, JP21K11752, and JP22H00513, and JST ACT-X Grant Number JPMJAX2105. The full version of this paper is available at https://arxiv.org/abs/2311.03225.

R. Uehara et al. (Eds.): WALCOM 2024, LNCS 14549, pp. 392–405, 2024.
https://doi.org/10.1007/978-981-97-0566-5_28

Table 1. In these tables, diam, pe, and pw denote the diameter, path eccentricity, and pathwidth, respectively. The first row represents the values that a tree T has, and the first column represents the values that a tree P has. There is no need to consider problems in these areas marked "meaningless".

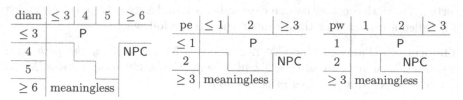

trees is the most fundamental one among such graph classes. For all the problems listed above, except the minor containment problem, there are polynomial time algorithms [1,14,18], even for generalized versions [6].

We focus on the *minor containment problem*, which is the problem of determining whether graph G contains graph H as a minor. Even when both G and H are trees, in which case we call the problem TREE MINOR CONTAINMENT, it remains NP-complete [13]. Furthermore, it remains NP-complete even if the diameters of both trees are constant [13]. However, polynomial-time algorithms are known for cases where the maximum degrees of H is constant [2,11,16] or when both trees are caterpillars [7,15]. Therefore, what condition makes TREE MINOR CONTAINMENT tractable is a natural question. In what follows, we denote G and H as T and P, respectively, since both graphs are trees.

1.1 Our Contributions

In this paper, we show dichotomies for three different structural parameters, diameter, pathwidth, and path eccentricity. We summarize dichotomies with respect to each parameter in Table 1.

Even when the diameters of T and P are constant, it is known that TREE MINOR CONTAINMENT is NP-complete [13]. Although they did not clarify the exact value of the constant, it can easily be observed that the constant is 8, which is not tight. Our first contribution is to provide the tight diameter requirement for TREE MINOR CONTAINMENT to be NP-complete.

Theorem 1. TREE MINOR CONTAINMENT *is* NP-*complete if the diameters of T and P are at least 6 and 4, respectively. Otherwise,* TREE MINOR CONTAINMENT *can be solved in polynomial time.*

When the pathwidths of both trees are 1 (or equivalently, both trees are *caterpillars*), TREE MINOR CONTAINMENT can be solved in polynomial time [7, 15]. Our second contribution is extending the positive result to the case where the pathwidth of T is arbitrary, and proving tight NP-completeness.

Theorem 2. TREE MINOR CONTAINMENT *is* NP-*complete if the pathwidths of both trees are at least* 2. *Otherwise,* TREE MINOR CONTAINMENT *can be solved in polynomial time.*

As evident from the theorem above, a caterpillar is an important class to consider when studying the tractability of TREE MINOR CONTAINMENT. The *path eccentricity* is known as a more direct parameter to express "caterpillar-likeness," [5,9] which is defined as the distance from a specific path to the farthest vertex. The path eccentricity of a caterpillar is 1, and a tree of a path eccentricity 2 is called a *lobster*. Our third contribution is the following.

Theorem 3. TREE MINOR CONTAINMENT *is* NP-*complete if the path eccentricities of T and P are at least* 3 *and* 2, *respectively. Otherwise,* TREE MINOR CONTAINMENT *can be solved in polynomial time.*

By definition, for a tree, the path eccentricity is at most the pathwidth. Therefore, the positive result for the case where both T and P have path eccentricity of 2 can be seen as encompassing cases that were not covered by considering the dichotomy for pathwidth.

1.2   Related Work

The most significant result concerning the minor containment problem is probably the Graph Minor Theory developed by Robertson and Seymour [17]. They proved that the minor containment problem can be solved in $f(H) \cdot O(|V(G)|^3)$-time, where f is some computable function. Using this algorithm, they proved the existence of an algorithm that determines whether a graph G satisfies any minor-closed property in $O(|V(G)|^3)$ time. Kawarabayashi, Kobayashi, and Reed improved this time complexity to $O(|V(G)|^2)$ [10].

Matoušek and Thomas proved that this problem remains NP-complete even on trees with bounded diameters [13]. Furthermore, they addressed the minor containment problem on graphs with treewidth k, and provided a polynomial-time algorithm for cases where H is connected and its degree is bounded and Gupta et al. provided a polynomial-time algorithm for the case where both G and H are k-connected and have pathwidth at most k [7]. Their results can also be applied to the subgraph isomorphism problem and the topological embedding problem.

A generalization of TREE MINOR CONTAINMENT, called the *tree inclusion problem*, has also been investigated. In this problem, we are given two rooted trees, T and P, with labeled vertices, and the objective is to determine whether it is possible to repeatedly contract vertices of T towards their parent until T matches P, including the labels. The special case where all vertices have the same label corresponds to TREE MINOR CONTAINMENT. Kilpeläinen and Mannila showed that there is an FPT-time algorithm parameterized by the maximum degree of P. It runs in $O(4^{\deg(P)} \cdot \text{poly}(n))$ time [11], and Akutsu et al. improved this result to $O(2^{\deg(P)} \cdot \text{poly}(n))$ time, where $\deg(P)$ is the maximum degree of P [2]. Miyazaki, Hagihara, and Hirata have provided a polynomial-time algorithm for the case where both T and P are caterpillars [15]. Additionally, Kilpeläinen and Mannila have proved that the problem remains NP-complete even when T has depth 3 [11]. However, it should be noted that their proof relies on the existence of labels, so it does not directly imply our NP-completeness result for the TREE MINOR CONTAINMENT for trees with bounded diameters.

As another generalization of TREE MINOR CONTAINMENT, the problem of finding the smallest tree containing two trees as minors is also investigated. For this problem, Nishimura, Ragde, and Thilikos gave an FPT-time algorithm parameterized by the maximum degree [16].

2   Preliminaries

Let T be a tree and n be the number of vertices or nodes in T. We denote the set of vertices and edges of T as $V(T)$ and $E(T)$, respectively. For a vertex v, the set of vertices adjacent to v is the *neighbors of v* and denoted by $N_T(v)$. The cardinality of the neighbor of v is the *degree of v* and is denoted by $\deg_G(v)$. Moreover, the *degree of G* is defined by $\max_{v \in V} \deg_G(v)$ and denoted by $\deg(G)$. For two vertices $u, v \in V$, the *distance* between u and v is the length of a shortest u-v path. We denote the distance between u and v as $\text{dist}(u, v)$. The *diameter* of a tree T, denoted by $\text{diam}(T)$, is the maximum distance between two vertices in T. For a set of edges F, we denote an edge-induced subgraph $T[E \setminus F]$ as $T - F$. Similarly, we denote an induced subgraph $T[V \setminus U]$ as $T - U$. For a tree T and a set of vertices U, *vertex contraction T/U* is the graph obtained by considering all vertices in U identical. More precisely, $V(T/U) := (V \setminus U) \cup \{w\}$ and $E(T/U) := \{\{u, v\} \mid \{u, v\} \in E(T) \ \wedge u, v \in V(T/U)\} \cup \{\{w, v\} \mid v \in V(T/U) \wedge \exists u \in U, \{u, v\} \in E(T)\}$. For two disjoint trees $T = (V, E)$ and $P = (U, F)$, we denote the forest $(V \cup U, E \cup F)$ as $T \cup P$.

A tree T is *caterpillar* if T becomes a path by removing all leaves in T. Moreover, T is *lobster* if T becomes a caterpillar by removing all leaves in T. As a generalization of lobsters, a tree T is *k-caterpillar* if T becomes a path by removing all leaves k times. We call the minimum value of k *path eccentricity* of T. Therefore, T is a path if and only if $k = 0$, T is a caterpillar if and only if $k \leq 1$, and T is a lobster if and only if $k \leq 2$. A path P is a *backbone* of a k-caterpillar T if for any $v \in T$, P has a vertex u such that $\text{dist}(u, v) \leq k$.

We next define the *pathwidth* of $T = (V, E)$. The pathwidth of T is defined by a *path decomposition* of T. A path decomposition of T is a pair $(\mathcal{X}, P)$, where $P = (V_P, E_P)$ is a path and $\mathcal{X} = \{\mathcal{X}_i \mid i \in V_P\}$ is a family of subsets of V, called *bags* that satisfies the following conditions. (I) $\bigcup_{i \in V_P} \mathcal{X}_i = V$, (II) for each edge $e \in E$, there is a bag $\mathcal{X}_i$ such that $T[\mathcal{X}_i]$ contains e, and (III) for all $v \in V$, we define the set of vertices $U := \{i \in V_P \mid v \in \mathcal{X}_i\}$ and $P[U]$ is connected. For a path decomposition $(\mathcal{X}, P)$, the width of this decomposition is defined by $\max_{\mathcal{X}_i \in \mathcal{X}} |\mathcal{X}_i| - 1$. Moreover, the *pathwidth* of T is the minimum width of any path decomposition. We denote it as $\text{pw}(T)$.

A tree P is a *minor* of a tree T if there exists a surjective map called *minor embedding* (or simply embedding) $f : T \to P$ such that

- for all $v \in V(P)$, the subgraph of T induced by $f^{-1}(v)$ is connected, and
- for all $e = (u, v) \in E(P)$, there exists an edge $e' = (u', v')$ of T such that $f(u') = u$ and $f(v') = v$.

If P is a minor of T, we say that T *contains* P as a minor.

Finally, we give the definition of the problem addressed in this paper.

TREE MINOR CONTAINMENT

Input: Two trees T and P.

Question: Is P a minor of T?

Theorems and lemmas marked with (∗) are shown in the appendix due to space limitation.

3   NP-completeness of Tree Minor Containment

We show that TREE MINOR CONTAINMENT is NP-complete even if diameters of T and P are at least 6 and 4, respectively, or pathwidths of T and P are at least 2. In Sect. 3.1, we show that TREE MINOR CONTAINMENT is NP-complete if diameters of T and P are at least 6 and 4, respectively. Moreover, in Sect. 3.2, we show that TREE MINOR CONTAINMENT is NP-complete if pathwidths of T and P are at least 2.

3.1   Bounded Diameter and Bounded Path Eccentricity

In this subsection, we improve the previous bound in [13]. To this end, we show the NP-completeness of INCLUSIVE SET COVER, a variant of SET COVER. To define INCLUSIVE SET COVER, we introduce some notations. The *disjoint union* of two sets A and B is, denoted by $A \sqcup B$, $\{(a,0) : a \in A\} \cup \{(b,1) : b \in B\}$. The *disjoint union* of a family of sets $\mathcal{A} = (A_i)_{i \in \lambda}$ is $\bigcup_{i \in \lambda}\{(a,i) : a \in A_i\}$, and denoted by $\bigsqcup_{i \in \lambda} A_i$ or simply $\bigsqcup \mathcal{A}$. We often consider an element $(x,i) \in A \sqcup B$ (or $(x,i) \in \bigsqcup_{i \in \lambda} A_i$) simply as an element $x \in A \cup B$ (or $x \in \bigcup_{i \in \lambda} A_i$ respectively). We are ready to define INCLUSIVE SET COVER.

INCLUSIVE SET COVER

Input: A set $U = \{1, 2, \ldots, n\}$, a collection of m sets $\mathcal{S} \subseteq 2^U$, and an integer $k \in \mathbb{N}$.

Question: Does there exist $\mathcal{R} \subseteq \mathcal{S}$ such that $|\mathcal{R}| \leq k$ and there is a surjection $f : \bigsqcup \mathcal{R} \to U$ such that $v \geq f((v,i))$ for each $(v,i) \in \bigsqcup \mathcal{R}$?

Lemma 4. (∗)INCLUSIVE SET COVER *is* NP-*complete.*

Theorem 5. TREE MINOR CONTAINMENT *is* NP-*complete even if the diameters of T and P are at least 6 and 4, respectively.*

Proof. It is clear that this problem is in NP. We show the NP-completeness of TREE MINOR CONTAINMENT by providing a reduction from INCLUSIVE SET COVER. From an instance $\langle U, \mathcal{S}, k \rangle$, we construct trees T and P as follows (see also Fig. 1). We first explain how to construct P. We consider stars $R_1, \ldots, R_n$, $X_1, \ldots, X_{m-k}$, and $Y_1 \ldots, Y_k$. Each star R_i, X_i, and Y_i have i, n^3, and n^2 leaves, respectively. Moreover, we add one vertex p that connects all the centers in

$R_1, \ldots, R_n$, $X_1, \ldots, X_{m-k}$, and $Y_1 \ldots, Y_k$. Finally, we add $3n^4$ leaves to p and obtain a tree P with the diameter 4.

We next explain how to construct T. We construct m rooted trees $T_1, \ldots, T_m$ as follows. Let t_i be the root of T_i and S_i be a set of integers $\{s_1^i, \ldots, s_\ell^i\}$ in $\mathcal{S}$. Each T_i has n^3 leaves as children of t_i. For each s_j^i, we add the star with s_j^i leaves as a child of t_i. Moreover, we add one star with n^2 leaves as a child of t_i and one vertex t that connects all the roots in $T_1, \ldots, T_m$. Finally, we add $3n^4$ leaves to t and obtain a tree T with the diameter 6. In what follows, for each T_i, R_i, X_i, and Y_i, we denote the root of T_i, R_i, X_i, and Y_i as t_i, r_i, x_i, y_i, respectively. Moreover, we denote the set of subtrees $\{T_1, \ldots, T_m\}$, $\{R_1, \ldots, R_n\}$, $\{X_1, \ldots, X_{m-k}\}$, and $\{Y_1, \ldots, Y_k\}$ as $\mathcal{T}$, $\mathcal{R}$, $\mathcal{X}$, and $\mathcal{Y}$, respectively.

Completeness. Let $\{S_{a_1}, \ldots, S_{a_k}\}$ be a subset of $\mathcal{S}$ that has a surjection f from $\bigsqcup \mathcal{S}$ to U satisfying $v \geq f((v, i))$ for each $(v, i) \in \bigsqcup \mathcal{S}$. In what follows, we assume that T and P are rooted at t and p, respectively. We give an embedding g from T to P that satisfies $g(t) = p$. We pick a subtree T_{a_i} for each a_i and define $g(t_{a_i}) = p$. For each integer in $S_{a_i} = \{s_1^{a_i}, \ldots, s_\ell^{a_i}\}$, we obtain the set of integers $\bigcup_{s \in S_{a_i}} \{f((s, i))\}$. From the construction of T_{a_i}, T_{a_i} has ℓ stars as subtrees. Moreover, j-th star has $s_j^{a_i}$ leaves. Therefore, we can embed a subtree in T_{a_i} with $s_j^{a_i}$ leaves into a subtree in P with $f((s_j^{a_i}, i))$ leaves since $s_j^{a_i} \geq f((s_j^{a_i}, i))$. Moreover, for each T_{a_i}, we can embed one subtree in $\mathcal{Y}$ since $g(t_{a_i}) = p$. Therefore we can embed all subtrees in P without each $\mathcal{X}$. For each $j \in \{1, \ldots, n\} \setminus \{a_1, \ldots, a_k\}$, T_j has a subtree with n^3 leaves. Therefore, each X can be embedded in each T_j. Finally, since both t and p have $3n^4$ neighbors with the degree 1, T has a P as a minor.

Soundness. We first show that any embedding $g : T \to P$ satisfies $g(t) = p$. Suppose that $g(t) \neq p$. Since $g^{-1}(p)$ does not contain t, $g^{-1}(p)$ is contained in a connected component in $T - \{t\}$. However, each connected component has at most $n^3 + 3n^2/2$ leaves despite p having $3n^4$ leaves. Therefore, each connected component does not contain a star with $3n^4$ leaves as a minor, and $g(t) = p$. In what follows, we regard T and P as rooted trees rooted at t and p, respectively.

We next show that T has $m - k$ trees T_i that satisfies $g(t_i) = x$ for some $X \in \mathcal{X}$, where x is the root of X. Since $g(t) = p$, $g^{-1}(x)$ is contained in some T_i. If $g(t_i) \neq x$, $g^{-1}(X)$ is contained in a connected component in $T_i - x$. However, each connected component in $T_i - x$ has at most n^2 leaves despite X having n^3 leaves. Therefore $g(t_i) = x$ holds. Moreover, since $g(t) = p$ and $g(t_i) = x$, T_i has no vertices v such that $g(v) \notin V(X)$. Since $\mathcal{X}$ has $m - k$ subtrees, T has $m - k$ subtrees as above.

Let $\{T_{a_1}, \ldots, T_{a_k}\}$ be the subtrees in $\mathcal{T}$ that satisfies $g(t_i) \neq x$ for any $X \in \mathcal{X}$. We show that for any T_{a_i}, either $g(t_{a_i}) = p$ or $g(t_{a_i}) = y$ for some $Y \in \mathcal{Y}$, where y is the root of Y. If $g(t_{a_i}) \neq p$ and $g(t_{a_i}) \neq y$ for any $Y \in \mathcal{Y}$, $V(T_{a_i}) \setminus \{t_{a_i}\}$ has no vertices v such that $g(v) = y$ since any v does not adjacent to t even if y adjacent to p. Moreover, T_{a_i} contains at most one subtree in $\mathcal{Y}$ even if $g(t_{a_i}) = p$ or $g(t_{a_i}) = y$. Since $\mathcal{Y}$ has k subtrees, any embedding satisfies either $g(t_{a_i}) = p$ or $g(t_{a_i}) = y$.

398   T. Gima et al.

From the above discussion, for each $X \in \mathcal{X}$, $g^{-1}(x)$ contains a child of t and for each $Y \in \mathcal{Y}$, $g^{-1}(y)$ contains a child of t. Moreover, when $g(t_i) = x$, T_i has no vertex v such that $g(v) \notin V(X)$. Similarly, when $g(t_i) = y$, T_i has no vertex v such that $g(v) \notin V(Y)$. Therefore, for any $R \in \mathcal{R}$, $g^{-1}(R)$ consists of vertices in T_i satisfying $g(t_i) = p$. From the definition of T_i, $T_i - \{t_i\}$ has $|S_i| + 1$ stars. Since $g(t_i) = p$, $g^{-1}(R)$ is contained in a star in $T_i - \{t_i\}$. Therefore, the number of leaves of this star is greater than or equal to the number of leaves of R. Since $\mathcal{T}$ has at most k subtrees such that $g(t_i) \neq x$ for any $X \in \mathcal{X}$, if we select S_i if and only if $g(t_i) \neq x$ for any $X \in \mathcal{X}$, the number of sets is at most k. Moreover, since g is an embedding from T to P, these selections from $\mathcal{S}$ are a solution of $\langle U, \mathcal{S}, k \rangle$. Therefore, $\langle U, \mathcal{S}, k \rangle$ is a yes-instance if T contains P as a minor.   □

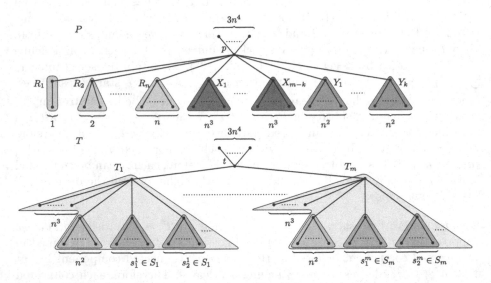

Fig. 1. An example of the construction of T and P in the proof of Theorem 5

Since $pe(T) \le k$ if $diam(T) = 2k$, we obtain the following corollary.

Corollary 1. TREE MINOR CONTAINMENT *is* NP-*complete even if the path eccentricities of T and P are 3 and 2, respectively.*

3.2 Bounded Pathwidth

In this subsection, we show that TREE MINOR CONTAINMENT is NP-complete even if the pathwidths of T and P are 2. To prove this, we first consider the following problem, which we call INCLUSIVE POSET PAIR COVER.

INCLUSIVE POSET PAIR COVER

Input: A partial ordered set $\langle U, \leq_U \rangle$, a subset X of U^2, and a pair (Y, Z) where Y is a subset of U^2 and Z is a subset of U.

Question: Does there exists two injections $f\colon Y \to X$ and $g\colon Z \to X \times \{1, 2\}$ such that

- $f(Y) \cap \{x \in X : (x, i) \in g(Z)\} = \emptyset$,
- if $f((y_1, y_2)) = (x_1, x_2)$ then $(y_1 \leq_U x_1) \wedge (y_2 \leq_U x_2)$ or $(y_2 \leq_U x_1) \wedge (y_1 \leq_U x_2)$, and
- if $g(z) = ((x_1, x_2), i)$ then $z \leq_U x_i$.

Lemma 6. (∗)INCLUSIVE POSET PAIR COVER *is* NP-*complete.*

From here, we provide a proof of the following theorem.

Theorem 7. TREE MINOR CONTAINMENT *is* NP-*complete even if the pathwidths of T and P are 2.*

We show the NP-completeness by presenting a reduction from INCLUSIVE POSET PAIR COVER. Let $\langle \langle U, \leq_U \rangle, X, (Y, Z) \rangle$ be an instance of INCLUSIVE POSET PAIR COVER. Let $U = \{u_0, u_1, \ldots, u_{n-1}\}$. Without loss of generality, we can assume that X, Y, and Z are not empty, and U contains exactly all of the elements that appear in X, Y, and Z. We can also assume $2|X| = 2|Y| + |Z|$ without loss of generality, because creating a new element u of U which is smaller than any element of U and adding u to Z does not change the solution as long as $2|X| > 2|Y| + |Z|$. First of all, we define the following notation, to describe an element of the partial order $\langle U, \leq_U \rangle$ into a caterpillar.

Definition 1. *The* order caterpillar *of $a \in U$ is a graph* $\mathrm{OCat}(a)$ *such that*

- *the vertex set is the union of* $V^a = \{v_0^a, v_1^a, \ldots, v_{n-1}^a, v_n^a, v_{n+1}^a\}$, $L^a = \{l_i^a : u_i \leq_U a\}$, *and*
- *the edge set is* $\bigcup_{0 \leq i \leq n} \{v_i^a, v_{i+1}^a\} \cup \bigcup_{l_i^a \in L^a} \{v_i^a, l_i^a\}$.

An example of order caterpillar is shown in Fig. 2. Note that v_n^a and v_{n+1}^a do not correspond to any elements in U, and guarantees that the maximum path length from v_0 in any order caterpillar is exactly n. Since the degree of every vertex l_i^a is 1, and since an edge set $\bigcup_{0 \leq i \leq n-2} \{v_i^a, v_{i+1}^a\}$ forms a path graph, every order caterpillar is a caterpillar. Note that the number of vertices is at most $2n + 2$ for every order caterpillar.

Observation 8. *Every order caterpillar is a caterpillar, and hence its pathwidth is 1.*

Definition 2. *Let $a, b \in U$. When $l_i^b \in L^b$ if $l_i^a \in L^a$ for all $i \in \{0, \cdots, n-1\}$, we can define the mapping $f\colon \mathrm{OCat}(b) \to \mathrm{OCat}(a)$ such that $f(v_i^b) = v_i^a$ for all $i \in \{0, \ldots, n+1\}$ and $f(l_i^b) = l_i^a$ if $u_i \leq_U a$, $f(l_i^b) = v_i^a$ if $u_i \not\leq_U a$, and we call this mapping f the* natural embedding *from $\mathrm{OCat}(b)$ to $\mathrm{OCat}(a)$. If there exists i such that $l_i^b \notin L^b$ and $l_i^a \in L^a$, we say that the natural embedding from $\mathrm{OCat}(b)$ to $\mathrm{OCat}(a)$ does not exists.*

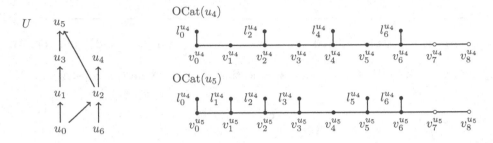

Fig. 2. Examples of a partial order set $U = \{u_0, \ldots, u_6\}$ and the order caterpillars of u_4 and u_5. Partial order $\leq_U$ denoted by the Hasse diagram of $\langle U, \leq_U \rangle$, i.e., an arrow from a to b indicates $a \leq_U b$ and there is no c such that $a <_U c <_U b$. In pictures of order caterpillars, a white node denotes a node such that there is no corresponding vertex in U.

Clearly, for $a, b \in U$, the natural embedding from OCat(b) to OCat(a) is an embedding from OCat(b) to OCat(a) if it exists. By the transitivity and reflexivity of $\leq_U$ relation, we have the following.

Observation 9. (∗) *Let $a, b \in U$. There exists the natural embedding from* OCat(b) *to* OCat(a) *if and only if $a \leq_U b$.*

Construction of TREE MINOR CONTAINMENT *instance* $\langle T, P \rangle$. See Fig. 3 for the whole image of TREE MINOR CONTAINMENT instance $\langle T, P \rangle$. For a pair $x = (a, b)$, we write x_1 for the first element a and x_2 for the second element b.

We first define a family of trees $(T_x)_{x \in X}$ to describe T. Let $x \in X$. Each tree T_x consists of three part, two subtrees T_x^L and T_x^R, and a root vertex r_x. A subtree T_x^L is just OCat(x_1), and T_x^R is just OCat(x_2). Connect r_x to $v_0^{x_1}$ in T_x^L and $v_0^{x_2}$ in T_x^R. Then we obtain a family of trees $(T_x)_{x \in X}$. Note that each T_x is a caterpillar such that its backbone has $2n + 5$ vertices and the number of vertices is at most $4n + 5$. Add a new vertex r_T and connect r_T to all r_x in T_x, then we obtain a tree T. Note that the number of vertices of T is at most $(4n + 5) \cdot |X| + 1$. Since each connected component of $T - \{r_T\}$ is a caterpillar, its pathwidth is 1, and the pathwidth of T is at most 2.

We next explain how to construct P. First, we construct a family of trees $(Q_y)_{y \in Y}$ by an analogous way to $(T_x)_{x \in X}$. That is, for $y \in Y$, tree Q_y is a tree has a root vertex r_y and two subtrees Q_y^L and Q_y^R such that Q_y^L is just OCat(y_1), Q_y^R just OCat(y_2), and r_y is connected to $v_0^{y_1}$ in Q_y^L and $v_0^{y_2}$ in Q_y^R. Next, we define a family of trees $(R_z)_{z \in Z}$, where each tree R_z is just OCat(z). Finally, we add a new vertex r_P and connect r_P to each r_y in Q_y and each v_0^z in R_z, and then we obtain P. Note that the number of vertices of P is at most $(4n + 5) \cdot |Y| + (2n + 2) \cdot |Z| + 1$. Since each connected component of $P - \{r_P\}$ is a caterpillar, its pathwidth is 1, and the pathwidth of P is at most 2.

Lemma 10. (∗) *If $\langle \langle U, \leq_U \rangle, X, (Y, Z) \rangle$ is a yes-instance then P is a minor of T.*

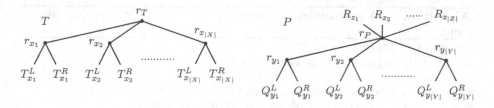

Fig. 3. An example of the reduction in the proof of Theorem 7.

Lemma 11. (*) *If P is a minor of T then* $\langle\langle U, \leq_U\rangle, X, (Y, Z)\rangle$ *is a yes-instance.*

This completes the proof of Theorem 7.

4 Polynomial-Time Algorithms with Small Path Eccentricity and Its Application for the Other Positive Results

We give two polynomial-time algorithms for TREE MINOR CONTAINMENT with a small path eccentricity. The former algorithm determines whether a tree T contains a caterpillar P. The latter algorithm determines whether a lobster T contains a lobster P. In Sect. 4.3, we give polynomial-time algorithms for all cases in Table 1 using the above two algorithms. Throughout this section, we assume $|V(P)| \geq 2$; otherwise, the problem is trivial.

4.1 Tree-Caterpillar Containment

We begin by considering the case where P is a caterpillar. The algorithm is given in Algorithm 1. It is easy to verify that Algorithm 1 works in polynomial time. Briefly, Algorithm 1 first guesses a backbone C of T (u-v path) that corresponds to the backbone of P, and then finds out how to contract C to form the backbone of P by a greedy method. Moreover, the algorithm is based on the fact that contracting an internal vertices in $V(T) \setminus C$ to a vertex in C does not affect whether there is a minor embedding $f\colon T \to P$ such that $f(C)$ maps to the backbone of P since P is a caterpillar. Thus, it can be computed whether the backbone of P can be embedded into C by focusing only on the number of leaves.

Theorem 12. *If P is a caterpillar, Algorithm 1 returns yes if and only if P is a minor of T.*

Proof. Assume that P is embedded into T by a mapping f. Let e_i be the edge connecting $f^{-1}(b_i)$ and $f^{-1}(b_{i+1})$ for $i = 1, \ldots, s - 1$. Then, there exists a path in T in which $e_1, \ldots, e_{s-1}$ appear in this order. Take a minimal such path $C = (c_1, \ldots, c_t)$. Consider the case we have $u = c_1$ and $v = c_t$ in the loop starting from line 6.

Algorithm 1: A polynomial-time algorithm for tree-caterpillar containment.

```
1  Procedure CatInTree(T, P)
2      Let B = (b_1, ..., b_s) be a backbone of P;
3      for i = 1, ..., s do
4          Let P_i be the connected component of P − E[B] containing b_i;
5          Let p_i be the number of leaves in P_i other than b_i;
6      for u, v ∈ V(T) do
7          Let C = (u = c_1, ..., c_t = v) be the u-v path in T;
8          for i = 1, ..., t do
9              Let T_i be the connected component of T − E[C] containing c_i;
10             Let l_i be the number of leaves in T_i other than c_i;
11         x ← 0, flag ← true;
12         for i = 1, ..., s do
13             Let j be the smallest index such that p_i ≤ Σ_{k=x+1}^{j} l_k;
14             if There is no such j then
15                 flag ← false;
16                 break;
17             x ← j;
18         if flag = true then return yes ;
19     return no;
```

Using the integers $1 = z_1, \ldots, z_{s+1} = t + 1$, for each $i = 1, \ldots, s$, we define $f^{-1}(b_i) \cap C = \{c_{z_i}, \ldots, c_{z_{i+1}-1}\}$. Let $T_{j',j} = \left(\bigcup_{k=j'}^{j} T_k\right) / \{c_{j'}, \ldots, c_j\}$ for $1 \leq j' \leq j \leq s$. Then, the mapping naturally induced by f embeds P_i into $T_{c_{z_i}, c_{z_{i+1}-1}}$. In particular, the number of leaves in $T_{c_{z_i}, c_{z_{i+1}-1}}$, denoted as $\sum_{k=c_{z_i}}^{c_{z_{i+1}-1}} l_k$, is greater than or equal to p_i. Therefore, for $j \leq c_{z_i}$ and $c_{z_{i+1}-1} \leq j'$, it holds that $p_i \leq \sum_{k=j}^{j'} l_k$. In particular, considering the i-th iteration of the loop starting from line 12 and denoting the value of x at the end of that iteration as x_i, it is clear by induction that $x_i \leq c_{z_{i+1}-1}$ always holds. Hence, the algorithm returns "yes".

Conversely, assuming that the algorithm returns "yes", we consider the corresponding $C = (c_1, \ldots, c_t)$. We define x_i as above for each $i = 1, \ldots, s$. Then, it holds that $p_i \leq \sum_{k=c_{z_i}}^{c_{z_{i+1}-1}} l_k$. For each $i = 1, \ldots, s$, let $v_{i,1}, \ldots, v_{i,p_i}$ be the vertices in P_i other than b_i, and let $v'_{i,1}, \ldots, v'_{i,p_i}$ be p_i selected leaves (not on C) in $T_{c_{z_i}}, \ldots, T_{c_{z_{i+1}}}$, respectively. Define $f(w)$ as follows:

$$f(w) = \begin{cases} v_{i,j} & (w = v'_{i,j}) \\ b_i & \left(w \in \bigcup_{k=c_{z_i}}^{c_{z_{i+1}-1}} V(T_k) \setminus \{v'_{i,1}, \ldots, v'_{i,p_i}\}\right) \end{cases}.$$

Then, f is a mapping that embeds P into T. □

4.2   Lobster-Lobster Containment

In this section, we provide a polynomial-time algorithm for TREE MINOR CONTAINMENT when both T and P are lobsters, i.e., have path eccentricity 2. The overall strategy of Algorithm 2 is the same as Algorithm 1, first guess a backbone of T, and decide where to contract it to form the backbone of P. However, deciding whether the (partial) minor relation holds after contracting the vertices in the guessed backbone is not as simple as when P is a caterpillar. This means we need to solve the following subproblem.

DEPTH 2 TREE MINOR EMBEDDING FROM LOBSTER (D2M)
Input: Lobster T, vertex r_T of T, and tree P such that the distance of each vertex of P is at most 2 from r_P.
Question: Is there an embedding from P into T such that $f(r_T) = r_P$?

Lemma 13. $(*)$D2M *can be solved in polynomial time.*

Proof (sketch). The essential case is when r_T is in a backbone $C = (c_1, \ldots, c_k = r_T, \ldots, c_t)$ of T. Since $P - r_P$ is a disjoint union of stars (here, a graph with a single vertex is also called a star), if we fix an interval of the backbone $\{c_l, \ldots, c_r\} \ni r_T$ that is contracted to r_T, then D2M can be reduced to the problem to determine whether a disjoint union of stars contains a disjoint union of stars as a minor, and this problem can be solved in polynomial time. Thus, D2M can be solved by trying all intervals $\{c_l, \ldots, c_r\}$. Otherwise, except for some special cases, we can show that r_T must be contracted with a vertex that is closer to the backbone of T, and thus D2M is finally reduced to the case that r_T is on a backbone of T. $\square$

We denote a polynomial time algorithm that computes the solution of D2M by EMBEDFULL(T, r_T, P, r_P).

We present an algorithm that computes the solution of TREE MINOR CONTAINMENT when both trees are lobsters by using EMBEDFULL$(\cdot)$ as a subroutine.

Theorem 14. $(*)$ *Algorithm 2 returns yes if and only if P is a minor of T.*

4.3   Applications of Algorithms 1 and 2

As shown in Table 1, we give polynomial-time algorithms for TREE MINOR CONTAINMENT with small diameter, path eccentricity, and pathwidth. Since we already show the case $pe(P) \leq 1$ in Theorem 12 and $pe(T) \leq 2$ in Theorem 14, we show the cases with small diameters and pathwidths. These results can be easily shown by using the results in previous subsections.

Theorem 15. TREE MINOR CONTAINMENT *can be solved in polynomial time when* $\mathrm{diam}(P) \leq 3$ *or* $\mathrm{diam}(T) \leq 5$.

Algorithm 2: An algorithm for the case that both of trees are lobsters

---

1 **Procedure** LOBINLOB(T, P)
2   Let $B = (b_1, \ldots, b_s)$ be a backbone of P;
3   **for** $i = 1, \ldots, s$ **do**
4   Let P_i be the connected component of $P - E[B]$ containing b_i;

5   **for** $u, v \in V(T)$ **do**
6   Let $C = (u = c_1, \ldots, c_t = v)$ be the $u - v$ path in T;
7   **for** $i = 1, \ldots, t$ **do**
8   Let T_i be the connected component of $T - E[C]$ containing c_i;

9        $x \leftarrow 0$, flag $\leftarrow$ true;
10   **for** $i = 1, \ldots, s$ **do**
11   Let j be the smallest index such that
            EMBEDFULL$(T_{x+1,j}, c_{x+1}, P_i, b_i)$ returns yes, where
            $T_{x+1,j} = \left(\bigcup_{k=x+1}^{j} T_k \right) / \{c_{x+1}, \ldots, c_j\}$;
12   **if** *There is no such j* **then**
13   flag $\leftarrow$ false;
14   **break**;
15          $x \leftarrow j$;

16   **if** flag $=$ true **then return** yes ;

17   **return** no;

Proof. Since a tree with a diameter at most 3 is a caterpillar, we can solve TREE MINOR CONTAINMENT when diam$(P) \leq 3$ from Theorem 12. Moreover, when diam$(P) >$ diam(T), T does not contains P as a minor obviously. Therefore, we can assume that diam$(P) \leq$ diam(T). Since a tree with a diameter at most 5 is a lobster, TREE MINOR CONTAINMENT can be solved in polynomial time when diam$(P) \leq$ diam$(T) \leq 5$ from Theorem 14. $\square$

Theorem 16. TREE MINOR CONTAINMENT *can be solved in polynomial time when* pw$(P) \leq 1$.

Proof. Since a tree with pathwidth 1 is a caterpillar, we obtain a polynomial-time algorithm from Theorem 12. $\square$

References

1. Abboud, A., Backurs, A., Hansen, T.D., Vassilevska Williams, V., Zamir, O.: Subtree isomorphism revisited. ACM Trans. Algorithms **14**(3) (2018). https://doi.org/10.1145/3093239
2. Akutsu, T., Jansson, J., Li, R., Takasu, A., Tamura, T.: New and improved algorithms for unordered tree inclusion. Theoret. Comput. Sci. **883**, 83–98 (2021). https://doi.org/10.1016/j.tcs.2021.06.013
3. Bodlaender, H.L., et al.: Subgraph isomorphism on graph classes that exclude a substructure. Algorithmica **82**(12), 3566–3587 (2020). https://doi.org/10.1007/s00453-020-00737-z

4. Garey, M.R., Johnson, D.S.: Computers and Intractability; A Guide to the Theory of NP-Completeness. W. H. Freeman and Company, New York (1990)
5. Gómez, R., Gutiérrez, J.: Path eccentricity of graphs. Discret. Appl. Math. **337**, 1–13 (2023). https://doi.org/10.1016/j.dam.2023.04.012
6. Gupta, A., Nishimura, N.: Finding largest subtrees and smallest supertrees. Algorithmica **21**(2), 183–210 (1998). https://doi.org/10.1007/PL00009212
7. Gupta, A., Nishimura, N., Proskurowski, A., Ragde, P.: Embeddings of k-connected graphs of pathwidth k. Discret. Appl. Math. **145**(2), 242–265 (2005). https://doi.org/10.1016/j.dam.2002.12.005
8. Hajiaghayi, M., Nishimura, N.: Subgraph isomorphism, log-bounded fragmentation, and graphs of (locally) bounded treewidth. J. Comput. Syst. Sci. **73**(5), 755–768 (2007). https://doi.org/10.1016/j.jcss.2007.01.003
9. Hedetniemi, S.M., Cockayne, E.J., Hedetniemi, S.T.: Linear algorithms for finding the Jordan center and path center of a tree. Transp. Sci. **15**(2), 98–114 (1981). https://doi.org/10.1287/trsc.15.2.98
10. Kawarabayashi, K.I., Kobayashi, Y., Reed, B.: The disjoint paths problem in quadratic time. J. Comb. Theory Ser. B **102**(2), 424–435 (2012). https://doi.org/10.1016/j.jctb.2011.07.004
11. Kilpeläinen, P., Mannila, H.: Ordered and unordered tree inclusion. SIAM J. Comput. **24**(2), 340–356 (1995). https://doi.org/10.1137/S0097539791218202
12. LaPaugh, A.S., Rivest, R.L.: The subgraph homeomorphism problem. In: Proceedings of the 10th Annual ACM symposium on Theory of Computing (STOC 1978), pp. 40–50 (1978). https://doi.org/10.1145/800133.804330
13. Matoušek, J., Thomas, R.: On the complexity of finding ISO- and other morphisms for partial k-trees. Discret. Math. **108**(1), 343–364 (1992). https://doi.org/10.1016/0012-365X(92)90687-B
14. Matula, D.W.: Subtree Isomorphism in $O(n^{5/2})$. Algorithmic Aspects Comb. Ann. Discrete Math. **2**, 91–106 (1978). https://doi.org/10.1016/S0167-5060(08)70324-8
15. Miyazaki, T., Hagihara, M., Hirata, K.: Caterpillar inclusion: inclusion problem for rooted labeled caterpillars. In: Proceedings of the 11th International Conference on Pattern Recognition Applications and Methods (ICPRAM 2022), pp. 280–287 (2022). https://doi.org/10.5220/0010826300003122
16. Nishimura, N., Ragde, P., Thilikos, D.M.: Finding smallest supertrees under minor containment. Int. J. Found. Comput. Sci. **11**(03), 445–465 (2000). https://doi.org/10.1142/S0129054100000259
17. Robertson, N., Seymour, P.D.: Graph minors. XIII. The disjoint paths problem. J. Comb. Theory Ser. B **63**(1), 65–110 (1995). https://doi.org/10.1006/jctb.1995.1006
18. Shamir, R., Tsur, D.: Faster subtree isomorphism. J. Algorithms **33**(2), 267–280 (1999). https://doi.org/10.1006/jagm.1999.1044

Structural Parameterizations of Vertex Integrity [Best Paper]

Tatsuya Gima[1,2], Tesshu Hanaka[3], Yasuaki Kobayashi[4], Ryota Murai[1],
Hirotaka Ono[1], and Yota Otachi[1(✉)]

[1] Nagoya University, Nagoya, Japan
{gima,ono,otachi}@nagoya-u.jp, murai.ryota.r6@s.mail.nagoya-u.ac.jp
[2] JSPS Research Fellow, Tokyo, Japan
[3] Kyushu University, Fukuoka, Japan
hanaka@inf.kyushu-u.ac.jp
[4] Hokkaido University, Sapporo, Japan
koba@ist.hokudai.ac.jp

Abstract. The graph parameter *vertex integrity* measures how vulnerable a graph is to a removal of a small number of vertices. More precisely, a graph with small vertex integrity admits a small number of vertex removals to make the remaining connected components small. In this paper, we initiate a systematic study of structural parameterizations of the problem of computing the unweighted/weighted vertex integrity. As structural graph parameters, we consider well-known parameters such as clique-width, treewidth, pathwidth, treedepth, modular-width, neighborhood diversity, twin cover number, and cluster vertex deletion number. We show several positive and negative results and present sharp complexity contrasts.

Keywords: Vertex integrity · Vulnerability of graphs · Structural graph parameter · Parameterized complexity

1 Introduction

Barefoot et al. [3] introduced the concept of vertex integrity as a vulnerability measure of communication networks. Intuitively, having small vertex integrity implies that one can remove a small set of vertices to make the remaining components small [2]. The *vertex integrity* (or just the *integrity*) of a graph $G = (V, E)$, denoted $\mathsf{vi}(G)$, is defined as

$$\mathsf{vi}(G) = \min_{S \subseteq V} \left\{ |S| + \max_{C \in \mathsf{cc}(G-S)} |V(C)| \right\},$$

Partially supported by JSPS KAKENHI Grant Numbers JP18H04091, JP20H00595, JP20H05793, JP20H05967, JP21H05852, JP21K11752, JP21K17707, JP21K19765, JP22H00513, JP23H03344, JP23KJ1066. The full version of this paper is available at http://arxiv.org/abs/2311.05892.

where $G - S$ denotes the graph obtained from G by deleting all vertices in S, and $\mathsf{cc}(G - S)$ is the set of connected components of $G - S$. For a vertex-weighted graph $G = (V, E, \mathsf{w})$ with weight function $\mathsf{w} \colon V \to \mathbb{Z}^+$, we can also define the *weighted vertex integrity* of G, denoted $\mathsf{wvi}(G)$, by replacing $|S|$ and $|V(C)|$ with $\mathsf{w}(S)$ and $\mathsf{w}(V(C))$ in the definition, respectively, where $\mathsf{w}(X) = \sum_{v \in X} \mathsf{w}(v)$ for $X \subseteq V$.[1] Note that the unweighted vertex integrity $\mathsf{vi}(G)$ can be still defined for a vertex-weighted graph $G = (V, E, \mathsf{w})$ by ignoring w (or using a unit-weight function).

In this paper, we study the problems of computing the unweighted/weighted vertex integrity of a graph, which can be formalized as follows:

Problem. UNWEIGHTED/WEIGHTED VERTEX INTEGRITY
Input. A graph G (with $\mathsf{w} \colon V \to \mathbb{Z}^+$ in the weighted version) and an integer k.
Question. Is $\mathsf{vi}(G) \leq k$?/Is $\mathsf{wvi}(G) \leq k$?

Note that the complexity of WEIGHTED VERTEX INTEGRITY may depend on the representation of the vertex-weight function w. We denote by BINARY/UNARY WEIGHTED VERTEX INTEGRITY the two cases where w is encoded in binary and unary, respectively.

UNWEIGHTED VERTEX INTEGRITY has been studied on several graph classes. It is NP-complete on planar graphs [6], and on co-bipartite graphs and chordal graphs [9]. On the other hand, the problem becomes tractable when the input is restricted to some classes of graphs: trees and cactus graphs [2, without proofs], graphs with linear structures (such as interval graphs, circular-arc graphs, permutation graphs, trapezoid graphs, and co-comparability graphs of bounded dimension) [20], and on split graphs [23]. The parameterized complexity of UNWEIGHTED VERTEX INTEGRITY with the natural parameter k was first addressed by Fellows and Stueckle [12], who showed that it can be solved in $\mathcal{O}(k^{3k}n)$ time. Drange et al. [9] later generalized and improved the result by presenting an algorithm for WEIGHTED VERTEX INTEGRITY with the running time $\mathcal{O}(k^{k+1}n)$. The existence of an $\mathcal{O}(c^k n^{\mathcal{O}(1)})$-time algorithm for a constant c is open even in the unweighted case. Drange et al. [9] also presented an $\mathcal{O}(k^3)$-vertex kernel for WEIGHTED VERTEX INTEGRITY. Using an approximation algorithm by Lee [22, Theorem 1] for a related problem as a subroutine, the vertex integrity can be approximated within a factor of $\mathcal{O}(\log \mathsf{opt})$, where opt is the vertex integrity.

The $\mathcal{O}(k^{k+1}n)$-time algorithm for WEIGHTED VERTEX INTEGRITY by Drange et al. [9] implies that WEIGHTED VERTEX INTEGRITY is fixed-parameter tractable parameterized by weighted vertex integrity and UNWEIGHTED VERTEX INTEGRITY is fixed-parameter tractable parameterized by unweighted vertex integrity. To the best of our knowledge, there has been no other result deal-

---

[1] We consider positive weights only since a vertex of non-positive weight is safely removed from the graph.

ing with structural parameterizations of the UNWEIGHTED/WEIGHTED VERTEX INTEGRITY.

Recently, vertex integrity has been studied as a structural graph parameter since it is an upper bound of treedepth and a lower bound of vertex cover number plus 1. This line of research has been studied extensively [5,16,17,21].

1.1   Our Results

In this paper, we study UNWEIGHTED/WEIGHTED VERTEX INTEGRITY parameterized by well-studied structural parameters and show sharp complexity contrasts. Our results can be summarized as follows (see Fig. 1):

(1) UNWEIGHTED VERTEX INTEGRITY is
    – FPT parameterized by cluster vertex deletion number (Theorem 3.5),
    – W[2]-hard parameterized by pathwidth (Corollary 4.5),
(2) UNARY WEIGHTED VERTEX INTEGRITY is
    – FPT parameterized by modular-width (Theorem 3.3),
    – W[1]-hard parameterized by cluster vertex deletion number and unweighted vertex integrity or by feedback vertex set number and unweighted vertex integrity (Theorem 4.1),
    – XP parameterized by clique-width (Theorem 3.2) (and thus polynomial-time solvable on distance hereditary graphs);
(3) BINARY WEIGHTED VERTEX INTEGRITY is
    – FPT parameterized by neighborhood diversity and by twin cover number (Corollary 3.1),
    – NP-complete on subdivided stars (and thus paraNP-complete parameterized simultaneously by cluster vertex deletion number, unweighted vertex integrity, and feedback vertex set number) (Theorem 4.2).

Since we focus on the classification of the parameterized complexity with respect to different parameters, we do not optimize or specify the running time of our algorithms.

We also show that UNARY WEIGHTED VERTEX INTEGRITY is NP-complete on planar bipartite graphs of maximum degree 4 and line graphs. We include these results in the full version.

1.2   Related Graph Parameters

Since the concept of vertex integrity is natural, there are a few other parameters defined in similar ways. The *fracture number* [10] of a graph is the minimum k such that one can remove at most k vertices to make the maximum size of a remaining component at most k. The *starwidth* [11] is the minimum width of a tree decomposition restricted to be a star. The *safe number* [13] of G is the minimum size of a non-empty vertex set S such that each connected component in $G[S]$ is not smaller than any adjacent connected component in $G - S$. As structural graph parameters, these parameters including vertex integrity are

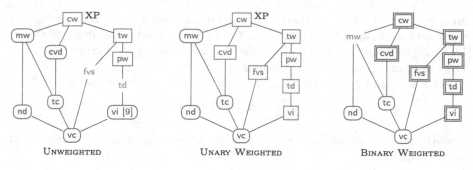

Fig. 1. The complexity of UNWEIGHTED/WEIGHTED VERTEX INTEGRITY with structural parameters. (See Sect. 2 for the definitions of the acronyms.) All but the one with a reference are shown in this paper or implied by other results. The double, single, and rounded rectangles indicate paraNP-complete, W[∗]-hard, and fixed-parameter tractable cases, respectively. A connection between two parameters implies that the one above generalizes the one below; that is, one below is lower-bounded by a function of the one above.

equivalent in the sense that one is small if and only if the others are small. More precisely, the starwidth, the fracture number, and the vertex integrity of a graph differ only by a constant factor, while the safe number may be as small as the square root of the vertex integrity. Note however that computing them exactly would be quite different tasks. For example, computing the safe number is NP-hard on split graphs [1], while computing the unweighted vertex integrity is polynomial-time solvable on them [23]. Furthermore, in the binary-weighted setting, we can see that computing the fracture number or the safe number is NP-hard on complete graphs as it generalizes the classical NP-complete problem PARTITION [15], while the computation of the vertex integrity becomes trivial as it is equal to the sum of the vertex weights in this case.

The ℓ-*component order connectivity* is the minimum size p of a vertex set such that the removal of the vertex set makes the maximum size of a remaining connected component at most ℓ [9]. Although we cannot directly compare this parameter to vertex integrity, some of the techniques used in this paper would be useful for studying ℓ-component order connectivity as well.

2   Preliminaries

We assume that the readers are familiar with the theory of fixed-parameter tractability. See standard textbooks (e.g., [7]) for the definitions omitted in this paper. The proofs of the statements marked with ★ are omitted here and given in the full version.

For a graph G, we denote its clique-width by $cw(G)$, treewidth by $tw(G)$, pathwidth by $pw(G)$, treedepth by $td(G)$, vertex cover number by $vc(G)$, feedback vertex set number by $fvs(G)$, modular-width by $mw(G)$, neighborhood diversity by $nd(G)$, cluster vertex deletion set number by $cvd(G)$, and twin cover number by tc(G). We defer their definitions until needed and if a full definition is not necessary, we omit it and present only the properties needed. (See, e.g., [14,18,25] for their definitions.) Note that for a vertex-weighted graph, we define these structural parameters on the underlying unweighted graph.

For a weighted graph $G = (V, E, \mathsf{w})$, a set $S \subseteq V$ is a wvi(k)-*set* if

$$\mathsf{w}(S) + \max_{C \in \mathsf{cc}(G-S)} \mathsf{w}(V(C)) \le k.$$

A wvi(k)-set is a wvi-*set* if $k = $ wvi(G). In the analogous ways, vi(k)-set and vi-set are defined.

Irredundant Set. Let G be a graph and $S \subseteq V(G)$. A vertex $v \in S$ is *redundant* if at most one connected component of $G - S$ contains neighbors of v. The set S is *irredundant* if it contains no redundant vertices. A vertex is *simplicial* if its neighborhood is a clique. Since no vertex set can separate a clique into two or more connected components, simplicial vertices cannot belong to irredundant sets.

Observation 2.1. *An irredundant set contains no simplicial vertex.*

The following is shown in [9, Lemma 3.1] for simplicial vertices, but their proof works for this generalization to redundant vertices.

Lemma 2.2 (★, [9, Lemma 3.1]). *Let $G = (V, E, \mathsf{w})$ be a vertex-weighted graph. If $S \subseteq V$ contains a redundant vertex v, then*

$$\mathsf{w}(S \setminus \{v\}) + \max_{C \in \mathsf{cc}(G-(S \setminus \{v\}))} \mathsf{w}(V(C)) \le \mathsf{w}(S) + \max_{C \in \mathsf{cc}(G-S)} \mathsf{w}(V(C)).$$

Corollary 2.3. *A graph with a wvi(k)-set has an irredundant wvi(k)-set.*

We denote by $N(v)$ and $N[v]$ the (*open*) *neighborhood* and the *closed neighborhood* of v, respectively. For a set S of vertices, we define $N(S) = \bigcup_{v \in S} N(v) \setminus S$. Two vertices $u, v \in V(G)$ are *twins* if they have the same neighborhood except for themselves; that is, if $N(u) \setminus \{v\} = N(v) \setminus \{u\}$ holds. A *twin class* T of G is a maximal set of twin vertices in G. We can show that an irredundant set contains either none or all of a twin class.

Lemma 2.4 (★). *Let G be a graph. If $S \subseteq V(G)$ is an irredundant set, then for each twin class T of G it holds that $T \cap S = \emptyset$ or $T \subseteq S$.*

3 Positive Results

Here we present our algorithmic results. In the descriptions of algorithms, we sometimes use a phrase like "to guess something." For example, when an integer ℓ belongs to $\{1, \ldots, c\}$, we may say that "we guess ℓ from $\{1, \ldots, c\}$." This means that we try all possibilities $\ell = 1, \ldots, c$, and output the optimal one, where the optimality depends on (and should be clear from) the context. We thus need a multiplicative factor of c in the running time. After "guessing" some object, we assume that the object is fixed to one of the candidates.

The proofs of the following results are given in the full version.

Corollary 3.1 (★). BINARY WEIGHTED VERTEX INTEGRITY *is fixed-parameter tractable parameterized by neighborhood diversity and by twin cover number.*

Theorem 3.2 (★). UNARY WEIGHTED VERTEX INTEGRITY *belongs to* XP *parameterized by clique-width.*

3.1   The Unary-Weighted Problem Parameterized by mw

Now we consider modular-width [14], which is a generalization of both neighborhood diversity and twin cover number, and show that UNARY WEIGHTED VERTEX INTEGRITY is fixed-parameter tractable with this parameter.

Let H be a graph with two or more vertices $v_1, \ldots, v_c$, and let $H_1, \ldots, H_c$ be c disjoint graphs. The *substitution* $H(H_1, \ldots, H_c)$ of the vertices of H by $H_1, \ldots, H_c$ is the graph with

$$V(H(H_1, \ldots, H_c)) = \bigcup_{1 \le i \le c} V(H_i),$$

$$E(H(H_1, \ldots, H_c)) = \bigcup_{1 \le i \le c} E(H_i)$$
$$\cup \{\{u, w\} \mid u \in V(H_i), w \in V(H_j), \{v_i, v_j\} \in E(H)\}.$$

That is, $H(H_1, \ldots, H_c)$ is obtained from the disjoint union of $H_1, \ldots, H_c$ by adding all possible edges between $V(H_i)$ and $V(H_j)$ for each edge $\{v_i, v_j\}$ of H. Each $V(H_i)$ is a *module* of $H(H_1, \ldots, H_c)$.

A *modular decomposition* is a rooted ordered tree T such that each non-leaf node with c children is labeled with a graph of c vertices and each node represents a graph as follows:

- a leaf node represents the one-vertex graph;
- a non-leaf node labeled with a graph H with c vertices $v_1, \ldots, v_c$ (and thus with exactly c children) represents the graph $H(H_1, \ldots, H_c)$, where H_i is the graph represented by the ith child.

A *modular decomposition of a graph* G is a modular decomposition whose root represents a graph isomorphic to G. The *width* of a modular decomposition is the maximum number of children of an inner node. The *modular-width* of a graph G, denoted $\mathsf{mw}(G)$, is the minimum width of a modular decomposition of G. It is known that a modular decomposition of the minimum width can be computed in linear time (and thus has a liner number of nodes) [24].

Theorem 3.3. UNARY WEIGHTED VERTEX INTEGRITY *is fixed-parameter tractable parameterized by modular-width.*

Proof. Let $G = (V, E, \mathsf{w})$ be a vertex-weighted graph. We guess an integer $\ell \in \{1, \ldots, \mathsf{w}(V)\}$ such that G has an irredundant wvi-set S such that

$$\max_{C \in \mathsf{cc}(G-S)} \mathsf{w}(V(C)) = \ell.$$

As there are only $\mathsf{w}(V)$ candidates of ℓ, which is polynomial in the input size, we can assume that ℓ is correctly guessed.[2]

We compute a modular decomposition of G with width $\mathsf{mw}(G)$ in linear time [24] and then proceed in a bottom-up manner. For each graph G' represented by a node in the modular decomposition, we compute the minimum weight of a set S irredundant in G' such that $\max_{C \in \mathsf{cc}(G'-S)} \mathsf{w}(V(C)) \leq \ell$. Let $\mu(G')$ denote the minimum weight of such S.

We first consider the case where G' has only one vertex v. If $\mathsf{w}(v) \leq \ell$, then we set $\mu(G') = 0$ as the empty set is irredundant. Otherwise we set $\mu(G') = \infty$ since the entire vertex set $\{v\}$ is redundant.

In the following, we consider the case of $G' = H(H_1, \ldots, H_c)$. For simplicity, we identify the vertices $v_1, \ldots, v_c$ of H with the integers $1, \ldots, c$. For a hypothetical irredundant set S that we are looking for, we guess a partition $(I_\mathsf{f}, I_\mathsf{p}, I_\emptyset)$ of $\{1, \ldots, c\}$ such that

- if $i \in I_\mathsf{f}$, then $V(H_i) \subseteq S$;
- if $i \in I_\mathsf{p}$, then $V(H_i) \cap S \neq \emptyset$ and $V(H_i) \not\subseteq S$;
- if $i \in I_\emptyset$, then $V(H_i) \cap S = \emptyset$.

As there are only $3^c \leq 3^{\mathsf{mw}(G)}$ candidates for the partition $(I_\mathsf{f}, I_\mathsf{p}, I_\emptyset)$, we assume that we have correctly guessed it. For each connected component C of $H - I_\mathsf{f}$, we compute the minimum weight $\mu(C)$ of vertices that we need to remove from the modules of G' corresponding to C.

Claim 3.4. Each $i \in I_\mathsf{p}$ has degree 0 in $H - I_\mathsf{f}$.

Proof (Claim 3.4). Let C_i be the connected component of $H - I_\mathsf{f}$ that contains i. Suppose to the contrary that $|V(C_i)| \geq 2$. Observe that for each edge $\{j, h\} \in E(C_i)$, the subgraph of G' induced by $(V(H_j) \cup V(H_h)) \setminus S$ is connected since both $V(H_j) \setminus S$ and $V(H_h) \setminus S$ are nonempty as $j, h \notin I_\mathsf{f}$ and there are all possible edges between them as $\{j, h\} \in E(C_i) \subseteq E(H)$. This fact and the connectivity of C_i imply that $(\bigcup_{j \in V(C_i)} V(H_j)) \setminus S$ induces a connected component D of $G' - S$.

Since $i \in I_\mathsf{p}$, there are vertices $u \in V(H_i) \cap S$ and $v \in V(H_i) \setminus S$ ($\subseteq V(D)$). Let w be a neighbor of u such that $w \notin S$. If $w \in V(H_i)$, then w belongs to D. If $w \notin V(H_i)$, then v is also adjacent to w since H_i is a module. This implies that w belongs to D in this case as well. Hence, no connected component of $G' - S$ other than D may contain a neighbor of u. This implies that u is redundant, a contradiction. ◇

Claim 3.4 implies that a component C of $H - I_\mathsf{f}$ is either a singleton formed by some $i \in I_\mathsf{p}$, or one formed by a subset of $I_\emptyset$. In the second case, $\mu(C) = 0$ if the total weight of the vertices in the corresponding modules is at most ℓ; and $\mu(C) = \infty$ otherwise. For the first case, let $i \in I_\mathsf{p}$ be the vertex forming C. By Claim 3.4, every neighbor j of i in H satisfies $j \in I_\mathsf{f}$ (if any exists). Thus,

[2] Note that this is the only part that requires the unary representation of weights. Note also that we cannot binary-search ℓ as the irredundancy makes the problem non-monotone.

$S \cap V(H_i)$ has to be irredundant for H_i. This implies that $\mu(C) = \mu(H_i)$ as we can consider H_i independently from the rest of the graph. Thus we can compute $\mu(G')$ as

$$\mu(G') = \sum_{i \in I_f} \mathsf{w}(V(H_i)) + \sum_{i \in I_p} \mu(H_i).$$

Since $\mu(H_i)$ for every $i \in I_p$ is already computed in the lower layers of the bottom-up computation, the computation of $\mu(G')$ can be done in polynomial time. This completes the proof.   □

3.2   The Unweighted Problem Parameterized by cvd

The *cluster vertex deletion number* [8] of a graph G, denoted $\mathsf{cvd}(G)$, is the minimum size of a *cluster vertex deletion set*; that is, a set of vertices whose removal makes the remaining graph a disjoint union of complete graphs. Finding a minimum cluster vertex deletion set is fixed-parameter tractable parameterized by $\mathsf{cvd}(G)$ [19], and thus we assume that such a set is given when $\mathsf{cvd}(G)$ is part of the parameter. Note that the definition directly implies that $\mathsf{cvd}(G) \leq \mathsf{tc}(G)$. By considering subdivided stars and complete bipartite graphs, one can see that cluster vertex deletion number is incomparable with neighborhood diversity and modular-width.

We show that UNWEIGHTED VERTEX INTEGRITY is fixed-parameter tractable parameterized by cluster vertex deletion number. This result provides an interesting contrast with the W[1]-hardness (Theorem 4.1) and NP-completeness (Theorem 4.2) of UNARY/BINARY WEIGHTED VERTEX INTEGRITY, respectively, parameterized by the same parameter.

Theorem 3.5. UNWEIGHTED VERTEX INTEGRITY *is fixed-parameter tractable parameterized by cluster vertex deletion number.*

We split the proof of Theorem 3.5 into two parts. In the first part, we show that the problem can be solved by solving $2^{\mathcal{O}(2^k)}$ instances of a subproblem (SUBVICVD) defined below, where $k = \mathsf{cvd}(G)$. In the second part, we present an algorithm that solves the subproblem in time $g(k) \cdot \mathrm{poly}(|V(G)|)$ with some g.

A vertex set S is (D,k)-*feasible* if $S \cap D = \emptyset$ and $|S| \leq k$. An (D,k)-*feasible vi-set* of a graph G is a set $S \subseteq V(G)$ that minimizes $|S| + \max_{C \in \mathsf{cc}(G-S)} |V(C)|$ under the condition that S is (D,k)-feasible. Since making a set smaller never loses (D,k)-feasibility, we can directly use Lemma 2.2 to obtain the (D,k)-feasible counterparts of Corollary 2.3, Observation 2.1, and Lemma 2.4. Thus we can use them also for the (D,k)-feasible setting.

Now the subproblem is defined as follows.

Problem. SUBVICVD
Input. A graph G, an integer k, and a cluster vertex deletion set D of G with $|\mathsf{cc}(G[D])| \leq k$.
Output. A (D,k)-feasible vi-set of G.
Parameter. k.

Note that in an instance (G, k, D) of SubViCvd, the size of D is not necessarily bounded by a function of k.

Reduction to Nice Instances of SubViCvd. Let G be a graph and D be a cluster vertex deletion set of G with $\mathsf{cvd}(G) = k$. We find a vi-set S of G such that each twin class of G is either completely contained in S or has no intersection with S. By Lemma 2.4, such a vi-set exists.

Let C be a connected component of $G - D$ that has the maximum number of vertices. Since C is a complete graph, the vertices of $C - S$ are included in the same connected component of $G - S$. Thus, $\mathsf{vi}(G) \geq |S| + |V(C) \setminus S| = |S \setminus V(C)| + |V(C)|$. On the other hand, as D is a $\mathsf{vi}(|D| + |V(C)|)$-set, we have $\mathsf{vi}(G) \leq |D| + |V(C)| = k + |V(C)|$. Hence, we have

$$|S \setminus V(C)| \leq k. \tag{1}$$

For each twin class $T \subseteq V(C)$ of G with $|T| > k$, we guess whether $T \subseteq S$ or not. Since C is a complete graph and $N(V(C)) \subseteq D$, there are at most 2^k different twin classes in C, and thus there are at most 2^{2^k} possible ways for this guess. Let S_C be the union of the twin classes $T \subseteq V(C)$ guessed as $T \subseteq S$. Now, by Eq. (1), it suffices to find an $(\emptyset, k)$-feasible vi-set of $G - S_C$.

We now guess which vertices of D are included in S. We call the set of guessed vertices S_D. There are at most 2^k possible options for this guess. Note that $D \setminus S_D$ is a cluster vertex deletion set of $G - (S_C \cup S_D)$.

Let $G' = G - (S_C \cup S_D)$ and $D' = D \setminus S_D$. Assuming that S_C and S_D are correctly guessed, the remaining problem is to find a (D', k)-feasible vi-set of the graph G'. That is, our task is to solve the instance (G', k, D') of SubViCvd.

Before solving the instance of SubViCvd, we enlarge the cluster vertex deletion set. Let D'_+ be the union of D' and all twin classes T in $G' - D'$ satisfying both $|T| > k$ and $N(T) \cap D' \neq \emptyset$. Adding such large twin classes is safe as we are finding a set of size at most k. To see that (G', k, D'_+) is an instance of SubViCvd, observe that $|\mathsf{cc}(G'[D'_+])| \leq |\mathsf{cc}(G'[D'])| \leq k$ since each vertex in $D'_+ \setminus D'$ has neighbors in D'.

The discussion so far shows that it suffices to solve $2^{2^k + k}$ instances (G', k, D'_+) of SubViCvd such that each twin class T of $G - D'_+$ satisfies $|T| \leq k$ or $N(T) \cap D'_+ = \emptyset$. We call such an instance *nice*.

Solving a Nice Instance of SubViCvd. Let (G, k, D) be a nice instance of SubViCvd. (Notice that we renamed the objects.)

For a vertex $v \in V(G) \setminus D$, let $\mathcal{N}_D(v)$ be the set of connected components of $G[D]$ that include a neighbor of v; that is, $\mathcal{N}_D(v) = \{C \in \mathsf{cc}(G[D]) \mid N(v) \cap V(C) \neq \emptyset\}$. Since D is a cluster vertex deletion set of G, two vertices v, v' with $\mathcal{N}_D(v) = \mathcal{N}_D(v')$ in the same connected component of $G - D$, which is a complete graph, play the same role in SubViCvd. Namely, if a (D, k)-feasible vi-set S contains v but not v', then $(S \setminus \{v\}) \cup \{v'\}$ is also a (D, k)-feasible vi-set.

We say that $K, K' \in \mathrm{cc}(G - D)$ are of the *same type* if (1) for every non-empty $\mathcal{C} \subseteq \mathrm{cc}(G[D])$, $|\{v \in K \mid \mathcal{N}_D(v) = \mathcal{C}\}| = |\{v \in K' \mid \mathcal{N}_D(v) = \mathcal{C}\}|$, and (2) $\{v \in K \mid \mathcal{N}_D(v) = \emptyset\}$ and $\{v \in K' \mid \mathcal{N}_D(v) = \emptyset\}$ are both empty or both non-empty. The relation of having the same type is an equivalence relation among the connected components of $G - D$. Observe that there are at most $2(k+1)^{2^k - 1}$ equivalence classes (or *types*) as $|\mathrm{cc}(G[D])| \le k$ and every twin class of $G - D$ with neighbors in D has size at most k.

Claim 3.6. There is a (D, k)-feasible vi-set S of G satisfying the conditions that (1) $N(S) \subseteq D$ and (2) if P and Q are connected components of $G - D$ that have the same type and $|V(P)| < |V(Q)|$, then S intersects P only if S intersects Q as well.

Proof (Claim 3.6). Let S be a (D, k)-feasible vi-set of G that contains no simplicial vertex. (Recall that Observation 2.1 can be used also in the (D, k)-feasible setting.) We can see that $N(S) \subseteq D$ since each vertex in $V(G) \setminus D$ is either a vertex with a neighbor in D or a simplicial vertex. (Recall that D is a cluster vertex deletion set.) We assume that, under this condition, S maximizes $\beta(S) := \sum_{C \in \mathrm{cc}(G - D),\, V(C) \cap S \ne \emptyset} |V(C)|$. Let P and Q be connected components of $G - D$ that have the same type and $|V(P)| < |V(Q)|$. Observe that P and Q both contain vertices with no neighbors in D since they have the same type but different sizes. Suppose to the contrary that S intersects P but not Q.

Since S contains no simplicial vertex, each vertex in $S \cap V(P)$ is adjacent to some vertices in D, and thus $V(P) \setminus S \ne \emptyset$. Since P and Q have the same type, for every non-empty subset $\mathcal{C} \subseteq \mathrm{cc}(G[D])$, the numbers of vertices in P and Q having neighbors exactly in $\mathcal{C}$ are the same. Thus there is an injection $f \colon S \cap V(P) \to V(Q)$ such that for each $v \in S \cap V(P)$, the neighbors of v and $f(v)$ belong to the same set of connected components of $G[D]$. Let S' be the set obtained from S by swapping $S \cap V(P)$ with $f(S \cap V(P))$; that is, $S' = (S \setminus V(P)) \cup f(S \cap V(P))$. Note that $|S| = |S'|$ and $|V(P) \setminus S| < |V(Q) \setminus S'|$.

Let C_P and C_Q be the (possibly identical) connected components of $G - S$ that contain $V(P) \setminus S\ (\ne \emptyset)$ and $V(Q) \setminus S\ (= V(Q))$, respectively. Similarly, let C'_P, C'_Q be the connected components of $G - S'$ that contain $V(P) \setminus S'\ (= V(P))$ and $V(Q) \setminus S'\ (\ne \emptyset)$, respectively. Note that such components are well defined as P and Q are complete graphs and $V(P) \setminus S$ and $V(Q) \setminus S'$ are non-empty. We show that

$$\max\{|V(C_P)|, |V(C_Q)|\} \ge \max\{|V(C'_P)|, |V(C'_Q)|\}. \tag{2}$$

Observe that $C_P = C_Q$ if and only if $C'_P = C'_Q$, and in such a case, they have the same number of vertices. In the following, we assume that $C_P \ne C_Q$ (and thus $C'_P \ne C'_Q$). This implies that C_Q and C'_Q have no intersection with P. Let p and q be the number of vertices in P and Q, respectively, that have no neighbors in D. Note that $p < q$. Since $V(P) \setminus S$ and $V(Q) \setminus S'$ have the same adjacency to $\mathrm{cc}(G[D])$, we have $|V(C_P)| - p = |V(C'_Q)| - q$ and $|V(C'_P)| - p = |V(C_Q)| - q$, and thus $|V(C_P)| < |V(C'_Q)|$ and $|V(C'_P)| < |V(C_Q)|$. We can also see that

$V(C'_Q) \subseteq V(C_Q)$ since $S \setminus V(P) = S' \setminus V(Q)$ and C_Q and C'_Q have no intersection with P. Thus Eq. (2) holds.

Observe that $\mathsf{cc}(G - S) \setminus \{C_P, C_Q\} = \mathsf{cc}(G - S') \setminus \{C'_P, C'_Q\}$. This implies that S' is also a (D, k)-feasible vi-set. This contradicts the assumption that $\beta(S)$ is maximum as $\beta(S') = \beta(S) - |V(P)| + |V(Q)| > \beta(S)$.   ◇

Let S be a (D, k)-feasible vi-set satisfying the conditions in Claim 3.6. Observe that the second condition guarantees that if S intersects k' ($\leq k$) connected components $K_1, \ldots, K_{k'}$ of $G - D$ with the same type, then $K_1, \ldots, K_{k'}$ are the largest ones in that type. In other words, for each type of connected components of $G - D$, only the k largest ones (where ties are broken arbitrarily) can intersect S. This observation leads to the following algorithm for solving SUBVICVD on the nice instance (G, k, D).

1. Select at most k connected components of $G - D$.
    - When selecting multiple components of the same type, pick them in non-decreasing order of their sizes (by breaking ties arbitrarily).
    - There are at most $(2(k + 1)^{2^k - 1})^k$ options for this phase as there are at most $2(k + 1)^{2^k - 1}$ types.
2. From each of the selected connected components, select at most k vertices (while keeping the total number of selected vertices to be at most k).
    - The vertices of a connected component of $G - D$ with neighbors in D can be partitioned into at most $2^k - 1$ equivalence classes by $\mathcal{N}_D$. Thus there are at most $k^{2^k - 1}$ options for one selected connected component.
    - In total, there are at most $(k^{2^k - 1})^k$ possible options for this phase.

The number of candidates for (D, k)-feasible vi-set S enumerated in the algorithm above is bounded by a function $g(k)$ that depends only on k. Since they can be enumerated in time polynomial per candidate, the algorithm runs in $g(k) \cdot \mathrm{poly}(|V(G)|)$ time.

4   Negative Results

The proofs of the following results are given in the full version.

Theorem 4.1 (★). UNARY WEIGHTED VERTEX INTEGRITY *is* W[1]-*hard parameterized by cluster vertex deletion number plus unweighted vertex integrity or by feedback vertex number plus unweighted vertex integrity.*

Theorem 4.2 (★). BINARY WEIGHTED VERTEX INTEGRITY *is* NP-*complete on subdivided stars.*

4.1   The Unweighted Problem Parameterized by pw

Given a graph G, and integers ℓ and p, COMPONENT ORDER CONNECTIVITY asks whether the ℓ-component order connectivity of G is at most p. The special case of COMPONENT ORDER CONNECTIVITY with $\ell = p$ is called FRACTURE NUMBER. (See Sect. 1.2 for the definitions of ℓ-component order connectivity and fracture number.)

To show the W[2]-hardness of UNWEIGHTED VERTEX INTEGRITY parameterized by pathwidth, we first show that FRACTURE NUMBER (and thus COMPONENT ORDER CONNECTIVITY as well) is W[2]-hard parameterized by pathwidth, and then we present a pathwidth-preserving reduction from COMPONENT ORDER CONNECTIVITY to UNWEIGHTED VERTEX INTEGRITY.

Let $G = (V, E)$ be a connected graph. A set $S \subseteq V$ is a *connected safe set* if $G[S]$ is connected and $|S| \geq |V(C)|$ for each $C \in cc(G - S)$. Given a graph G and an integer k, CONNECTED SAFE SET asks whether G contains a connected safe set of size at most k. Belmonte et al. [4] showed that CONNECTED SAFE SET is W[2]-hard parameterized by pathwidth even if the input graph G contains a universal vertex. This almost directly implies that FRACTURE NUMBER is W[2]-hard parameterized by pathwidth as we show below. Here we omit the definition of pathwidth as it is not necessary.

Proposition 4.3. FRACTURE NUMBER *is* W[2]-*hard parameterized by pathwidth.*

Proof. Let $G = (V, E)$ a graph that contains a universal vertex u. We show that (G, k) is a yes-instance of CONNECTED SAFE SET if and only if (G, k) is a yes-instance of FRACTURE NUMBER.

To show the only-if direction, assume that (G, k) is a yes-instance of CONNECTED SAFE SET. A minimum connected safe set $S \subseteq V$ satisfies that $|S| \leq k$ and $|V(C)| \leq |S| \leq k$ for each $C \in cc(G - S)$. Hence, (G, k) is a yes-instance of FRACTURE NUMBER.

To show the if direction, assume that (G, k) is a yes-instance of FRACTURE NUMBER. Let $S \subseteq V$ be a set such that $|S| \leq k$ and $|V(C)| \leq k$ for each $C \in cc(G - S)$. We may assume without loss of generality that $|S| = k$ by adding arbitrary vertices if necessary. If $G[S]$ is connected, then S is indeed a connected safe set and we are done. Assume that S is not connected, and thus S does not contain the universal vertex u. Since u is a universal vertex, $G - S$ is connected, which implies that $|V \setminus S| \leq k$. Now we construct a set S' of size k by first picking u and then adding arbitrary $k - 1$ vertices. Since S' contains u, the graph $G[S']$ is connected. As $|V \setminus S'| = |V \setminus S| \leq k$, it holds for each $C \in cc(G - S')$ that $|C| \leq k = |S'|$. Thus, S' is a connected safe set of size k. $\square$

Lemma 4.4. *There is a polynomial-time reduction from* COMPONENT ORDER CONNECTIVITY *to* UNWEIGHTED VERTEX INTEGRITY *that increases pathwidth by at most 1.*

Proof. Let (H, ℓ, p) be an instance of COMPONENT ORDER CONNECTIVITY. We set $k = \ell p + \ell + p$. We construct a graph G by attaching p pendants (i.e., vertices of degree 1) to each vertex of H and then adding $k+1$ copies of $K_{1,k-p-1}$. Note that $\mathsf{pw}(G) \leq \mathsf{pw}(H) + 1$ since removing the degree-1 vertices in G decreases its pathwidth by at most 1 (see [4]) and $\mathsf{pw}(K_{1,k-p-1}) = 1$. We show that (G,k) is a yes-instance of UNWEIGHTED VERTEX INTEGRITY if and only if (H,ℓ,p) is a yes-instance of COMPONENT ORDER CONNECTIVITY.

To prove the if direction, assume that (H,ℓ,p) is a yes-instance of COMPONENT ORDER CONNECTIVITY and $S \subseteq V(H)$ satisfies that $|S| \leq p$ and $\max_{C \in \mathsf{cc}(H-S)} |V(C)| \leq \ell$. We show that S is a vi(k)-set of G. Since $|S| \leq p$, it suffices to show that $\max_{C \in \mathsf{cc}(G-S)} |V(C)| \leq k-p$. If $C \in \mathsf{cc}(G-S)$ contains no vertex of H, then C is one of the copies of $K_{1,k-p-1}$ or a single-vertex component corresponding to a pendant. Otherwise, $V(C) \cap V(H)$ induces a connected component of $H-S$, and thus $|V(C)| = (p+1)|V(C) \cap V(H)| \leq (p+1)\ell = k-p$.

To prove the only-if direction, assume that (G,k) is a yes-instance of UNWEIGHTED VERTEX INTEGRITY and S is an irredundant vi(k)-set of G. Since there are $k+1$ copies of $K_{1,k-p-1}$, there is one that does not intersect S. This implies that $\max_{C \in \mathsf{cc}(G-S)} |V(C)| \geq k-p$, and thus $|S| \leq p$. Let $S' = S \cap V(H)$ and let $C' \in \mathsf{cc}(H-S')$ be an arbitrary connected component of $H-S'$. As $|S'| \leq p$, it suffices to show that $|V(C')| \leq \ell$. Let C be the connected component of $G-S$ such that $V(C') \subseteq V(C)$. Observe that S contains no vertices of degree 1 since such vertices are redundant. This implies that, for each vertex of C', C contains all p pendants attached to it, and thus, $|V(C)| \geq (p+1)|V(C')|$. Since $|V(C)| \leq k = \ell p + \ell + p < (p+1)(\ell+1)$, we have $|V(C')| < \ell+1$. $\square$

Now Lemma 4.4 and Proposition 4.3 imply the following.

Corollary 4.5. UNWEIGHTED VERTEX INTEGRITY *is* W[2]-*hard parameterized by pathwidth.*

5 Concluding Remarks

We initiated the first systematic study of the problem of computing unweighted and weighted vertex integrity of graphs in terms of structural graph parameters. Our results show sharp complexity contrasts of the problem. (See Fig. 1.) There are still some cases where the complexity of the problem is unknown. For example, what is the complexity of UNWEIGHTED VERTEX INTEGRITY parameterized by treedepth or by feedback vertex set number, and what is the complexity of BINARY VERTEX INTEGRITY parameterized by modular-width?

References

1. Águeda, R., et al.: Safe sets in graphs: graph classes and structural parameters. J. Comb. Optim. **36**(4), 1221–1242 (2018). https://doi.org/10.1007/s10878-017-0205-2
2. Bagga, K.S., Beineke, L.W., Goddard, W., Lipman, M.J., Pippert, R.E.: A survey of integrity. Discret. Appl. Math. **37**(38), 13–28 (1992). https://doi.org/10.1016/0166-218X(92)90122-Q

3. Barefoot, C.A., Entringer, R.C., Swart, H.C.: Vulnerability in graphs – a comparative survey. J. Combin. Math. Combin. Comput. **1**, 13–22 (1987)
4. Belmonte, R., Hanaka, T., Katsikarelis, I., Lampis, M., Ono, H., Otachi, Y.: Parameterized complexity of safe set. J. Graph Algorithms Appl. **24**(3), 215–245 (2020). https://doi.org/10.7155/jgaa.00528
5. Bentert, M., Heeger, K., Koana, T.: Fully polynomial-time algorithms parameterized by vertex integrity using fast matrix multiplication. In: ESA 2023. LIPIcs, vol. 274, pp. 16:1–16:16 (2023). https://doi.org/10.4230/LIPIcs.ESA.2023.16
6. Clark, L.H., Entringer, R.C., Fellows, M.R.: Computational complexity of integrity. J. Combin. Math. Combin. Comput. **2**, 179–191 (1987)
7. Cygan, M., et al.: Parameterized Algorithms. Springer (2015). https://doi.org/10.1007/978-3-319-21275-3
8. Doucha, M., Kratochvíl, J.: Cluster vertex deletion: a parameterization between vertex cover and clique-width. In: MFCS 2012. Lecture Notes in Computer Science, vol. 7464, pp. 348–359. Springer (2012). https://doi.org/10.1007/978-3-642-32589-2_32
9. Drange, P.G., Dregi, M.S., van 't Hof, P.: On the computational complexity of vertex integrity and component order connectivity. Algorithmica **76**(4), 1181–1202 (2016). https://doi.org/10.1007/s00453-016-0127-x
10. Dvořák, P., Eiben, E., Ganian, R., Knop, D., Ordyniak, S.: The complexity landscape of decompositional parameters for ILP: programs with few global variables and constraints. Artif. Intell. **300**, 103561 (2021). https://doi.org/10.1016/j.artint.2021.103561
11. van Ee, M.: Some notes on bounded starwidth graphs. Inf. Process. Lett. **125**, 9–14 (2017). https://doi.org/10.1016/j.ipl.2017.04.011
12. Fellows, M.R., Stueckle, S.: The immersion order, forbidden subgraphs and the complexity of network integrity. J. Combin. Math. Combin. Comput. **6**, 23–32 (1989)
13. Fujita, S., Furuya, M.: Safe number and integrity of graphs. Discret. Appl. Math. **247**, 398–406 (2018). https://doi.org/10.1016/j.dam.2018.03.074
14. Gajarský, J., Lampis, M., Ordyniak, S.: Parameterized algorithms for modular-width. In: IPEC 2013. Lecture Notes in Computer Science, vol. 8246, pp. 163–176 (2013). https://doi.org/10.1007/978-3-319-03898-8_15
15. Garey, M.R., Johnson, D.S.: Computers and Intractability: A Guide to the Theory of NP-Completeness. Freeman, W. H (1979)
16. Gima, T., Hanaka, T., Kiyomi, M., Kobayashi, Y., Otachi, Y.: Exploring the gap between treedepth and vertex cover through vertex integrity. Theor. Comput. Sci. **918**, 60–76 (2022). https://doi.org/10.1016/j.tcs.2022.03.021
17. Gima, T., Otachi, Y.: Extended MSO model checking via small vertex integrity. In: ISAAC 2022. LIPIcs, vol. 248, pp. 20:1–20:15 (2022). https://doi.org/10.4230/LIPIcs.ISAAC.2022.20
18. Hliněný, P., Oum, S., Seese, D., Gottlob, G.: Width parameters beyond tree-width and their applications. Comput. J. **51**(3), 326–362 (2008). https://doi.org/10.1093/comjnl/bxm052
19. Hüffner, F., Komusiewicz, C., Moser, H., Niedermeier, R.: Fixed-parameter algorithms for cluster vertex deletion. Theory Comput. Syst. **47**(1), 196–217 (2010). https://doi.org/10.1007/s00224-008-9150-x
20. Kratsch, D., Kloks, T., Müller, H.: Measuring the vulnerability for classes of intersection graphs. Discret. Appl. Math. **77**(3), 259–270 (1997). https://doi.org/10.1016/S0166-218X(96)00133-3

420   T. Gima et al.

21. Lampis, M., Mitsou, V.: Fine-grained meta-theorems for vertex integrity. In: ISAAC 2021. LIPIcs, vol. 212, pp. 34:1–34:15 (2021). https://doi.org/10.4230/LIPIcs.ISAAC.2021.34
22. Lee, E.: Partitioning a graph into small pieces with applications to path transversal. Math. Program. **177**(1–2), 1–19 (2019). https://doi.org/10.1007/s10107-018-1255-7
23. Li, Y., Zhang, S., Zhang, Q.: Vulnerability parameters of split graphs. Int. J. Comput. Math. **85**(1), 19–23 (2008). https://doi.org/10.1080/00207160701365721
24. McConnell, R.M., Spinrad, J.P.: Modular decomposition and transitive orientation. Discret. Math. **201**(1–3), 189–241 (1999). https://doi.org/10.1016/S0012-365X(98)00319-7
25. Sorge, M., Weller, M.: The graph parameter hierarchy (2019). https://manyu.pro/assets/parameter-hierarchy.pdf

On the Complexity of List $\mathcal{H}$-Packing
for Sparse Graph Classes

Tatsuya Gima[1,5] 📷, Tesshu Hanaka[2] 📷, Yasuaki Kobayashi[3] 📷, Yota Otachi[1] 📷,
Tomohito Shirai[4], Akira Suzuki[4] 📷, Yuma Tamura[4(✉)], and Xiao Zhou[4]

[1] Graduate School of Informatics, Nagoya University, Nagoya, Japan
{gima,otachi}@nagoya-u.jp
[2] Department of Informatics, Kyushu University, Fukuoka, Japan
hanaka@inf.kyushu-u.ac.jp
[3] Graduate School of Information Science and Technology, Hokkaido University,
Sapporo, Japan
koba@ist.hokudai.ac.jp
[4] Graduate School of Information Sciences, Tohoku University, Sendai, Japan
{akira,tamura,zhou}@tohoku.ac.jp
[5] JSPS Research Fellow, Tokyo, Japan

Abstract. The problem of packing as many subgraphs isomorphic to
$H \in \mathcal{H}$ as possible in a graph for a class $\mathcal{H}$ of graphs is well studied in
the literature. Both vertex-disjoint and edge-disjoint versions are known
to be NP-complete for H that contains at least three vertices and at
least three edges, respectively. In this paper, we consider "list variants"
of these problems: Given a graph G, an integer k, and a collection $\mathcal{L}_{\mathcal{H}}$
of subgraphs of G isomorphic to some $H \in \mathcal{H}$, the goal is to compute
k subgraphs in $\mathcal{L}_{\mathcal{H}}$ that are pairwise vertex- or edge-disjoint. We show
several positive and negative results, focusing on classes of sparse graphs,
such as bounded-degree graphs, planar graphs, and bounded-treewidth
graphs.

Keywords: Packing problem · Parameterized complexity · Sparse
graph

1 Introduction

Packing as many graphs as possible into another graph is a fundamental problem in the field of graph algorithms. To be precise, for a fixed graph H, given
an undirected graph G and a non-negative integer k, the VERTEX DISJOINT
H-PACKING problem (resp. the EDGE DISJOINT H-PACKING problem) asks for
finding a collection $\mathcal{S}$ of k vertex-disjoint (resp. edge-disjoint) subgraphs of G

This work is partially supported by JSPS Kakenhi Grant Numbers JP20H00595,
JP20K11666, JP21K11752, JP21K21278, JP21K17707, JP21H05852, JP22H00513,
JP23H03344, JP23H04388, JP23KJ1066.

R. Uehara et al. (Eds.): WALCOM 2024, LNCS 14549, pp. 421–435, 2024.
https://doi.org/10.1007/978-981-97-0566-5_30

that are isomorphic to H. For a connected graph H, both problems are polynomially solvable if H has at most two vertices (resp. at most two edges) because they can be reduced to the MAXIMUM MATCHING problem, whereas the problems are shown to be NP-complete if H has at least three vertices (resp. at least three edges) [8,11]. Furthermore, both problems are naturally extended to VERTEX DISJOINT $\mathcal{H}$-PACKING and EDGE DISJOINT $\mathcal{H}$-PACKING [7,11], which respectively ask for finding a collection $\mathcal{S}$ of k vertex-disjoint and edge-disjoint subgraphs of G that are isomorphic to some graph in a (possibly infinite) fixed collection $\mathcal{H}$ of graphs. These problems are also well studied in specific cases of $\mathcal{H}$. In particular, when $\mathcal{H}$ is paths or cycles, it has received much attention in the literature because of the variety of possible applications [2,3,5,12,13]. In both cases, VERTEX DISJOINT $\mathcal{H}$-PACKING and EDGE DISJOINT $\mathcal{H}$-PACKING remain NP-complete for planar graphs [3,9].

Recently, Xu and Zhang proposed a new variant of EDGE DISJOINT $\mathcal{H}$-PACKING, which they call PATH SET PACKING, from the perspective of network design [18]. In the PATH SET PACKING problem, given an undirected graph G, a non-negative integer k, and a collection $\mathcal{L}$ of simple paths in G, we are required to find a subcollection $\mathcal{S} \subseteq \mathcal{L}$ of (at least) k paths that are mutually edge-disjoint. Notice that $\mathcal{L}$ may not be exhaustive: Some paths in G may not appear in $\mathcal{L}$. If $\mathcal{H}$ consists of a finite number of paths, EDGE DISJOINT $\mathcal{H}$-PACKING can be (polynomially) reduced to PATH SET PACKING because $\mathcal{H}$ is fixed and hence all paths in G isomorphic to some graph in $\mathcal{H}$ can be enumerated in polynomial time. Xu and Zhang showed that for a graph G with n vertices and m edges, the optimization variant of PATH SET PACKING is hard to approximate within a factor $O(m^{1/2-\epsilon})$ for any constant $\epsilon > 0$ unless NP = ZPP, while the problem is solvable in $O(|\mathcal{L}|n^2)$ time if G is a tree and in $O(|\mathcal{L}|^{\mathrm{tw}\Delta}n)$ time if G has treewidth tw and maximum degree Δ [18]. Very recently, Aravind and Saxena investigated the parameterized complexity of PATH SET PACKING for various parameters. For instance, PATH SET PACKING is W[1]-hard even when parameterized by pathwidth plus maximum degree plus solution size [1]. To the best of our knowledge, except for PATH SET PACKING, such a variant has not been studied for EDGE DISJOINT $\mathcal{H}$-PACKING, nor VERTEX DISJOINT $\mathcal{H}$-PACKING.

Our Contributions. In this paper, motivated by PATH SET PACKING, we introduce *list variants* of VERTEX DISJOINT $\mathcal{H}$-PACKING and EDGE DISJOINT $\mathcal{H}$-PACKING. In the VERTEX DISJOINT LIST $\mathcal{H}$-PACKING (resp. EDGE DISJOINT LIST $\mathcal{H}$-PACKING) problem, we are given a graph G, a non-negative integer k, and a collection (list) $\mathcal{L}_{\mathcal{H}}$ of subgraphs of G such that each subgraph in $\mathcal{L}_{\mathcal{H}}$ is isomorphic to some graph in $\mathcal{H}$. The problem asks whether there exists a subcollection $\mathcal{S} \subseteq \mathcal{L}_{\mathcal{H}}$ such that $|\mathcal{S}| \geq k$ and subgraphs of G in $\mathcal{S}$ are vertex-disjoint (resp. edge-disjoint). If $\mathcal{L}_{\mathcal{H}}$ contains all subgraphs of G isomorphic to some graph in $\mathcal{H}$, the problem is equivalent to VERTEX DISJOINT $\mathcal{H}$-PACKING (resp. EDGE DISJOINT $\mathcal{H}$-PACKING). Thus, the tractability of the list variants implies that of the original problems. If $\mathcal{H} = \{H\}$ for a fixed graph H, then we call these problems simply VERTEX (EDGE) DISJOINT LIST H-PACKING

Table 1. The complexity of VERTEX DISJOINT C_ℓ-PACKING (left) and EDGE DISJOINT C_ℓ-PACKING (right).

	$\Delta(G) = 3$	$\Delta(G) \geq 4$
$\ell = 3$	P [6]	NPC [6]
$\ell = 4$	P [Thm 4]	NPC [Thm 6]
$\ell = 5$		
$\ell \geq 6$	NPC [Thm 5]	

	$\Delta(G) = 3$	$\Delta(G) = 4$	$\Delta(G) \geq 5$
$\ell = 3$	P [6]		NPC [6]
$\ell = 4$	P [Thm 4]	NPC [Thm 7]	
$\ell = 5$			
$\ell \geq 6$	NPC [Thm 5]		

and VERTEX (EDGE) DISJOINT H-PACKING, respectively. For a positive integer ℓ, we denote by P_ℓ and C_ℓ the path and the cycle of ℓ vertices, respectively. (We assume $\ell \geq 3$ for C_ℓ.) When $\mathcal{P} = \{P_\ell : \ell \geq 1\}$, EDGE DISJOINT LIST $\mathcal{P}$-PACKING is equivalent to PATH SET PACKING. Therefore, EDGE DISJOINT LIST $\mathcal{H}$-PACKING generalizes both EDGE DISJOINT H-PACKING and PATH SET PACKING.

We first give sufficient conditions to solve VERTEX (EDGE) DISJOINT LIST H-PACKING on bounded degree graphs in polynomial time. These conditions directly indicate the polynomial-time solvability of VERTEX (EDGE) DISJOINT LIST C_ℓ-PACKING on graphs of maximum degree 3 for $\ell \in \{3, 4, 5\}$. It is worth noting that VERTEX DISJOINT P_3-PACKING remains NP-complete even for 2-connected bipartite planar cubic graphs [13]. In contrast, we show that VERTEX (EDGE) DISJOINT C_ℓ-PACKING on planar graphs of maximum degree 3 is NP-complete for any $\ell \geq 6$. As VERTEX (EDGE) DISJOINT C_ℓ-PACKING can be represented by its list variant, this result also indicates the hardness of VERTEX (EDGE) DISJOINT LIST C_ℓ-PACKING. We also give the NP-completeness of VERTEX (EDGE) DISJOINT C_ℓ-PACKING on planar graphs of maximum degree 4 for any $\ell \geq 4$. Therefore, we provide the complexity dichotomy of VERTEX (EDGE) DISJOINT C_ℓ-PACKING with respect to the maximum degree of a given graph and ℓ, as summarized in Table 1.

Second, we design a polynomial-time algorithm for VERTEX DISJOINT LIST $\mathcal{H}$-PACKING on bounded-treewidth graphs, provided that all graphs in $\mathcal{H}$ are connected. This implies that VERTEX DISJOINT LIST $\mathcal{H}$-PACKING belongs to XP parameterized by treewidth. Note that the connectivity condition on $\mathcal{H}$ is essential, because otherwise one can see that the problem is NP-complete even on forests (see Theorem 3). On the other hand, we show that VERTEX DISJOINT LIST $\mathcal{P}$-PACKING and VERTEX DISJOINT LIST $\mathcal{C}$-PACKING parameterized by pathwidth plus k are W[1]-hard, where $\mathcal{P} = \{P_\ell : \ell \geq 1\}$ and $\mathcal{C} = \{C_\ell : \ell \geq 3\}$. This result implies that there is probably no FPT algorithm for the problems parameterized by treewidth. One might think that XP algorithms parameterized by treewidth could also be designed for the edge-disjoint versions. We give a negative answer. We show that EDGE DISJOINT LIST $\mathcal{P}$-PACKING and EDGE DISJOINT LIST $\mathcal{C}$-PACKING parameterized by bandwidth plus k are W[1]-hard even for outerplanar and two-terminal series-parallel graphs, which have treewidth at most 2. In particular, the W[1]-hardness for EDGE DISJOINT LIST $\mathcal{P}$-PACKING, which is equivalent to PATH SET PACKING, strengthens the result of [1].

The above hardness results prompt us to further investigate the complexity of EDGE DISJOINT LIST P_ℓ-PACKING and EDGE DISJOINT LIST C_ℓ-PACKING on bounded-treewidth graphs. In this paper, we focus on series-parallel graphs, also known as graphs of treewidth at most 2. We show that EDGE DISJOINT LIST P_4-PACKING and EDGE DISJOINT LIST C_5-PACKING remain NP-complete even for series-parallel graphs. Since EDGE DISJOINT LIST P_3-PACKING is solvable in polynomial time for general graphs by reducing to MAXIMUM MATCHING, the former implies the complexity dichotomy of EDGE DISJOINT LIST P_ℓ-PACKING on series-parallel graphs with respect to ℓ. The remaining task is to settle the complexity of EDGE DISJOINT LIST C_ℓ-PACKING on series-parallel graphs for $\ell \leq 4$. We finally provide an algorithm that, given an n-vertex series-parallel graph and a collection $\mathcal{L}_\mathcal{H}$ of cycles with length $\ell \leq 4$, solves EDGE DISJOINT LIST C_ℓ-PACKING in $O(|\mathcal{L}_\mathcal{H}| + n^{2.5})$ time.

Due to the space limitation, several proofs, marked ♠, are omitted in this paper, which can be found in the full version.

2   Preliminaries

For a positive integer i, we denote $[i] = \{1, 2, \ldots, i\}$.

Let G be a graph. Throughout this paper, we assume that G is simple, that is, it has neither self-loops nor parallel edges. The sets of vertices and edges of G are denoted by $V(G)$ and $E(G)$, respectively. For $v \in V(G)$, we denote by $N_G(v)$ the set of neighbor of v and by $d_G(v)$ the degree of v in G. The maximum degree of a vertex in G is denoted by $\Delta(G)$ and the minimum degree of a vertex in G is denoted by $\delta(G)$. We may simply write uv to denote an edge $\{u, v\}$. For a positive integer t, we denote by tG the disjoint union of t copies of G. For a graph H, the H-vertex-conflict graph of G, denoted $I_H^V(G)$, is defined as follows. Each vertex of $I_H^V(G)$ corresponds to a subgraph isomorphic to H in G. Two vertices of $I_H^V(G)$ are adjacent if and only if the corresponding subgraphs in G share a vertex. The H-edge-conflict graph of G, denoted $I_H^E(G)$, is defined by replacing the adjacency condition in the definition of $I_H^V(G)$ as: two vertices of $I_H^E(G)$ are adjacent if and only if they share an edge in G.

A *claw* is a star graph with three leaves. A graph is said to be *claw-free* if it has no claw as an induced subgraph. Minty [14] and Sbihi [16] showed that the maximum independent set problem can be solved in polynomial time on claw-free graphs. This immediately implies the following proposition, which is a key to our polynomial-time algorithms.

Proposition 1. *If $I_H^V(G)$ is claw-free, then* VERTEX DISJOINT LIST H-PACKING *can be solved in* $n^{O(|V(H)|)}$ *time. Moreover, if $I_H^E(G)$ is claw-free, then* EDGE DISJOINT LIST H-PACKING *can be solved in* $n^{O(|V(H)|)}$ *time as well.*

We consider several width parameters of graphs, such as treewidth, pathwidth, and bandwidth. Due to the space limitation, we do not give their precise definitions in this paper. For the treewidth, pathwidth, and bandwidth of G, we denote them by $\mathrm{tw}(G)$, $\mathrm{pw}(G)$, and $\mathrm{bw}(G)$. We may simply write them as tw, pw, and bw without specific reference to G. It is well known that $\mathrm{tw}(G) \leq \mathrm{pw}(G) \leq \mathrm{bw}(G)$ for every graph G [4].

3   List C_ℓ-packing on bounded degree graphs

In this section, we focus on VERTEX DISJOINT LIST C_ℓ-PACKING and EDGE DISJOINT LIST C_ℓ-PACKING on bounded degree graphs.

Theorem 1. VERTEX DISJOINT LIST H-PACKING *can be solved in polynomial time if the following inequality holds:*

$$\Delta(G) \leq 2\delta(H) - \left\lfloor \frac{|V(H)|}{3} \right\rfloor.$$

Proof. By Proposition 1, it suffices to show that if $I_H^V(G)$ has a claw as an induced subgraph, then G has a vertex of degree more than $2\delta(H) - \lfloor|V(H)|/3\rfloor$. Let H^*, H_1, H_2, and H_3 be induced copies of H in G such that they correspond to an induced claw in $I_H^V(G)$ whose center is H^*. This implies that $V(H^*) \cap V(H_i) \neq \emptyset$ for $1 \leq i \leq 3$ and $V(H_i) \cap V(H_j) = \emptyset$ for $1 \leq i < j \leq 3$. This implies that some copy of H, say H_1, satisfies $|V(H^*) \cap V(H_1)| \leq \lfloor|V(H)|/3\rfloor$. Let $v \in V(H^*) \cap V(H_1)$. The vertex v has at least $\delta(H)$ neighbors in each of H^* and H_1 and at most $\lfloor|V(H)|/3\rfloor - 1$ of them belong to $V(H^*) \cap V(H_1)$. Thus,

$$d_G(v) \geq 2\delta(H) - (|V(H^*) \cap V(H_1)| - 1)$$
$$\geq 2\delta(H) - \left(\left\lfloor \frac{|V(H)|}{3} \right\rfloor - 1 \right)$$
$$> 2\delta(H) - \left\lfloor \frac{|V(H)|}{3} \right\rfloor,$$

which proves the claim.   □

Theorem 2 (♠). EDGE DISJOINT LIST H-PACKING *can be solved in polynomial time if the following inequality holds:*

$$\Delta(G) \leq 2\delta(H) - \left\lfloor \frac{|E(H)|}{3} \right\rfloor,$$

except that $H = tK_2$ for $3 \leq t \leq 5$.

Let us note that the exception in Theorem 2 is critical for the tractability of EDGE DISJOINT LIST H-PACKING. In fact, EDGE DISJOINT LIST H-PACKING is NP-complete even if $G = nK_2$ and $H = 3K_2$, which satisfy that $\Delta(G) = 1 \leq 2\delta(H) - \lfloor|E(H)|/3\rfloor$. This intractability is shown by reducing EXACT COVER BY 3-SETS, which is known to be NP-complete [10], to EDGE DISJOINT LIST H-PACKING. In EXACT COVER BY 3-SETS, we are given a universe U and a collection $\mathcal{C}$ of subsets of U, each of which has exactly three elements and asked

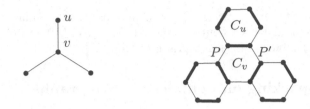

Fig. 1. An example of the construction of G' for $\ell = 6$. Bold lines indicate a matching of each cycle.

whether there is a pairwise disjoint subcollection C' of C that covers U (i.e., $U = \bigcup_{S \in C'} S$). This problem is a special case of EDGE DISJOINT LIST H-PACKING, where G has a copy of K_2 corresponding to each element in U and S consists of all copies of $3K_2$ corresponding to C. This reduction also proves that VERTEX DISJOINT LIST H-PACKING is NP-complete even if $G = nK_2$ and $H = 3K_2$.

Theorem 3. VERTEX DISJOINT LIST H-PACKING *and* EDGE DISJOINT LIST H-PACKING *are NP-complete even if* $G = nK_2$ *and* $H = 3K_2$.

As consequences of Theorems 1 and 2, we have the following positive results.

Theorem 4. *For* $\ell \in \{4, 5\}$, *there are polynomial-time algorithms for* VERTEX DISJOINT LIST C_ℓ-PACKING *and* EDGE DISJOINT LIST C_ℓ-PACKING *on graphs of maximum degree* 3.

Contrary to this tractability, for any $\ell \geq 6$, VERTEX DISJOINT C_ℓ-PACKING and EDGE DISJOINT C_ℓ-PACKING are NP-complete even on planar graphs of maximum degree 3.

Theorem 5. *For* $\ell \geq 6$, VERTEX DISJOINT C_ℓ-PACKING *and* EDGE DISJOINT C_ℓ-PACKING *are NP-complete even on planar graphs of maximum degree* 3.

Proof. Since VERTEX DISJOINT C_ℓ-PACKING and EDGE DISJOINT C_ℓ-PACKING are equivalent on graphs of maximum degree 3, we only consider VERTEX DISJOINT C_ℓ-PACKING.

To show NP-hardness, we perform a polynomial-time reduction from INDEPENDENT SET, which is known to be NP-complete even on planar graphs with maximum degree 3 and girth at least p for any integer p [15]. Here, the *girth* of G is the length of a shortest cycle in G.

Let G be a planar graph with $\Delta(G) \leq 3$ and girth at least $\ell + 1$. We construct a graph G' as follows. For each $v \in V(G)$, G' contains a cycle C_v of length ℓ. These cycles are called *primal cycles* in G'. Let M_v be an arbitrary matching in C_v with $|M_v| = 3$. For two adjacent vertices u, v in G, we identify one of the edges in M_u and in M_v as shown in Fig. 1.

The edges in M_v are called *shared edges*. By removing the shared edges from C_v, we obtain three paths, which we call *private paths* in C_v. Note that each private path belongs to exactly one primal cycle in G'. The construction of G' is done. It is easy to observe that G' is planar and has maximum degree 3. Also observe that the degree of each internal vertex of a private path is exactly 2.

We first claim that the length of cycles in G' except for primal cycles is greater than ℓ. To see this, let C be an arbitrary cycle that is not a primal cycle in G'. As shared edges form a matching in G', C must contain at least one private path P in C_v for some $v \in V(G)$. If C contains all the private paths in C_v, the length of C is greater than ℓ, except for the case $C = C_v$. Thus, we assume that C does not contain one of the private paths, say P', in C_v. Let $W = (v_0, v_1, \ldots, v_t)$ be a sequence of vertices of G defined as follows. We first contract all shared edges in C. Then, the contracted cycle can be partitioned into maximal subpaths $P_0, \ldots, P_t$ such that each P_i consists of edges of (possibly more than one) private paths in C_{v_i} for some $v_i \in V(G)$. Due to the maximality of P_i, we have $v_i \neq v_{i+1}$ for $0 \leq i \leq t$, where the addition in the subscript is taken modulo $t + 1$. Moreover, the sequence contains at least two vertices as C contains P and does not contain P', meaning that it must have a private path in $C_{v'}$ for some $v' \neq v$. For any pair of private paths in C_u and in C_w, they are adjacent with a shared edge if and only if u and w are adjacent in G. Thus, W is a closed walk in G.

Suppose that W contains a "turn", that is, $v_i = v_{i+2}$ for some i. This implies that C contains all private paths of $C_{v_{i+1}}$, which implies that the length of C is more than ℓ. Otherwise, W contains a cycle in G. Since the girth of G is at least $\ell + 1$, W has more than ℓ edges. Hence, C contains more than $\ell + 1$ private paths.

Now, we are ready to prove that G has an independent set of size at least k if and only if G' has a C_ℓ-packing of size at least k. From an independent set of G, we can construct a vertex-disjoint C_ℓ-packing by just taking primal cycles corresponding to vertices in the independent set. Since every cycle except for primal cycles has length more than ℓ, this correspondence is reversible: From a vertex-disjoint C_ℓ-packing of G' with size k, we can construct an independent set of G with size k.

Using a similar strategy of Theorem 5, we prove the following theorems. □

Theorem 6 (♠). *For $\ell \in \{4, 5\}$, VERTEX DISJOINT C_ℓ-PACKING is NP-complete even on planar graphs of maximum degree 4.*

Theorem 7 (♠). *For $\ell \in \{4, 5\}$, EDGE DISJOINT C_ℓ-PACKING is NP-complete even on planar graphs of maximum degree 4.*

4 VERTEX DISJOINT LIST $\mathcal{H}$-PACKING on bounded-treewidth graphs

In Sect. 5, we will see that EDGE DISJOINT LIST $\mathcal{H}$-PACKING is intractable even if $\mathcal{H}$ contains a single small connected graph and an input graph is series-parallel. In contrast to this intractability, VERTEX DISJOINT LIST $\mathcal{H}$-PACKING is polynomial-time solvable on series-parallel graphs and, more generally, bounded-treewidth graphs, if $\mathcal{H}$ consists of a finite number of connected graphs. More precisely, we show that VERTEX DISJOINT LIST $\mathcal{H}$-PACKING is XP parameterized by treewidth, provided that $\mathcal{H}$ consists of connected graphs. We also show that VERTEX DISJOINT LIST PATH PACKING is W[1]-hard parameterized by pathwidth.

Theorem 8 (♠). VERTEX DISJOINT LIST $\mathcal{H}$-PACKING *is solvable in* $n^{O(\text{tw})}$ *time, provided that all graphs in* $\mathcal{H}$ *are connected, where* n *is the number of vertices in the input graph.*

We would like to note that the connectivity of $\mathcal{H}$ is crucial as we have seen in Sect. 3 that VERTEX DISJOINT LIST H-PACKING is NP-complete even if $G = nK_2$ and $H = 3K_2$.

The following theorem complements the positive result of Theorem 8.

Theorem 9 (♠). VERTEX DISJOINT LIST P_ℓ-PACKING *is W[1]-hard parameterized by* pw $+ k$.

5   EDGE DISJOINT LIST $\mathcal{H}$-PACKING on series-parallel graphs

This section is devoted to showing several positive and negative results on *series-parallel graphs*. The class of series-parallel graphs is a well-studied class of graphs and is equivalent to the class of graphs of treewidth at most 2.

A *two-terminal labeled graph* is a graph G with distinguished two vertices called a *source* s and a *sink* t. Let $G_1 = (V_1, E_1)$ (resp. $G_2 = (V_2, E_2)$) be a two-terminal labeled graph with a source s_1 and a sink t_1 (resp. a source s_2 and a sink t_2). A *series composition* of G_1 and G_2 is an operation that produces the two-terminal labeled graph with a source s and a sink t obtained from G_1 and G_2 by identifying t_1 and s_2, where $s = s_1$ and $t = t_2$. A *parallel composition* of G_1 and G_2 is an operation that produces the two-terminal labeled graph with a source s and a sink t obtained from G_1 and G_2 by identifying s_1 and s_2, and identifying t_1 and t_2, where $s = s_1(= s_2)$ and $t = t_1(= t_2)$. We denote $G = G_1 \bullet G_2$ if G is created by a series composition of G_1 and G_2, and denote $G = G_1 \parallel G_2$ if G is created by a parallel composition of G_1 and G_2. We say that a two-terminal labeled graph G is a *two-terminal series-parallel graph* if one of the following conditions is satisfied: (i) $G = K_2$ with a source s and a sink t; (ii) $G = G_1 \bullet G_2$ for two-terminal series-parallel graphs G_1 and G_2; or (iii) $G = G_1 \parallel G_2$ for two-terminal series-parallel graphs G_1 and G_2.

We say that a graph G (without a source and a sink) is a *series-parallel graph* if each biconnected component is a two-terminal series-parallel graph by regarding some two vertices as a source and a sink[1].

---
[1] Some papers refer to a two-terminal series-parallel graph simply as a series-parallel graph. In this paper, we distinguish them explicitly to avoid confusion.

5.1 Hardness

A graph G is *outerplanar* if it has a planar embedding such that every vertex of G meets the unbounded face of the embedding. Every outerplanar graph is series-parallel but may not be two-terminal series-parallel. The following two theorems indicate that EDGE DISJOINT LIST $\mathcal{H}$-PACKING remains intractable even when a given graph is highly restricted.

Theorem 10 (♠).   EDGE DISJOINT LIST $\mathcal{P}$-PACKING *parameterized by* $bw(G) + k$ *is W[1]-hard even for outerplanar and two-terminal series-parallel graphs, where k is a solution size.*

Theorem 11 (♠).   EDGE DISJOINT LIST $\mathcal{C}$-PACKING *parameterized by* $bw(G) + k$ *is W[1]-hard even for outerplanar and two-terminal series-parallel graphs, where k is a solution size.*

We next focus on the case where $\mathcal{H}$ consists of a single graph and show that the problem remains hard. Let $K_{2,n}$ denotes the complete bipartite graph such that one side consists of two vertices and the other side consists of n vertices.

Theorem 12 (♠).   EDGE DISJOINT LIST P_4-PACKING *remains NP-complete even for the class of $K_{2,n}$.*

Obviously, $K_{2,n}$ is a two-terminal series-parallel graph. Since EDGE DISJOINT LIST P_3-PACKING is solvable for general graphs, Theorem 12 suggests that the complexity dichotomy with respect to path length still holds for very restricted graphs. Moreover, Theorem 12 immediately provides the following corollary, which strengthens the hardness result in [1] that PATH SET PACKING is W[1]-hard when parameterized by vertex cover number of G plus maximum length of paths in a given collection $\mathcal{L}$.

Corollary 1. PATH SET PACKING *is NP-complete even when a given graph has vertex cover number 2 and every path in $\mathcal{L}$ is of length 3.*

We also show the complexity of EDGE DISJOINT LIST C_5-PACKING, which highlights the positive result in Sect. 5.2.

Theorem 13 (♠).   EDGE DISJOINT LIST C_5-PACKING *remains NP-complete even for two-terminal series-parallel graphs.*

5.2 Polynomial-Time Algorithm of EDGE DISJOINT LIST C_ℓ-PACKING for $\ell \leq 4$

We design a polynomial-time algorithm for EDGE DISJOINT LIST C_ℓ-PACKING for $\ell \leq 4$ on two-terminal series-parallel graphs. Actually, we give a stronger theorem.

Theorem 14. *Let $C_{\leq 4} = \{C_3, C_4\}$. Given a series-parallel graph G with n vertices and a collection $\mathcal{L}_{\mathcal{H}}$ of cycles in G of length at most 4, EDGE DISJOINT LIST $C_{\leq 4}$-PACKING is solvable in $O(|\mathcal{L}_{\mathcal{H}}| + n^{2.5})$ time.*

We first note that we may assume that a given graph G is biconnected: the problem can be solved independently in each biconnected component. Moreover, from the definition of series-parallel graphs, every biconnected series-parallel graph can be regarded as a two-terminal series-parallel graph. We thus consider a polynomial-time algorithm that finds a largest solution of a given two-terminal series-parallel graph.

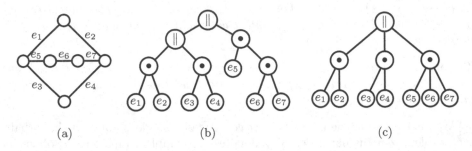

Fig. 2. (a) The graph G, (b) the decomposition tree of G, and (c) the layered decomposition tree of G.

The recursive definition of a two-terminal series-parallel graph G naturally gives us a rooted full binary tree T representing G, called the *decomposition tree* of G (see Fig. 2(a) and (b)). To avoid confusion, we refer to a vertex and an edge of T as a *node* and a *link*, respectively. For a node x of T, let T_x be a subtree of T rooted at x. Each leaf of T corresponds to an edge of G whose endpoints are labeled with a source s and a sink t. Each internal node x of T is labeled either • or ‖. Suppose that x has exactly two children x_1 and x_2. The label • indicates a series composition of two-terminal series-parallel graphs defined by T_{x_1} and T_{x_2}. The label ‖ indicates a parallel composition of two-terminal series-parallel graphs defined by T_{x_1} and T_{x_2}. We refer to nodes labeled • as •-*nodes* and to nodes labeled ‖ as ‖-*nodes*. We denote by G_x the graph composed by T_x. Let r be the root of T. Then, we have $G_r = G$. Note that, since $G_1 \bullet G_2$ and $G_2 \bullet G_1$ produce different two-terminal graphs, we assume that children of a •-node are ordered. In addition, since we have assumed G is 2-connected, the root r of T is labeled ‖ (assuming G has at least three vertices).

At the beginning of our algorithm, we construct a decomposition tree T' of a given graph G in linear time [17], and then transform it into a suitable form for our algorithm as follows (see also Fig. 2(c)). If a •-node x of T' has a child •-node x', then we contract a link xx' without changing the order of series compositions. For example, suppose that x has children x_1 and x'; x' has children x_2 and x_3; and $G_x = G_{x_1} \bullet G_{x'}$. Then, we contract the link xx' so that $G_x = G_1 \bullet G_2 \bullet G_3$. The contracted tree still tells how to construct G. Similarly, if a ‖-node x of T' has a child ‖-node x', then we contract the link xx'. We iteratively contract such links until each •-node has only leaves or ‖-nodes as its children, and ‖-node has only leaves or •-nodes as its children. Note that each ‖-node has at most one leaf of T' as its children because G has no multiple edges. The tree obtained in this way is called a *layered decomposition tree* of G and is denoted by T.

Let C be a cycle of a graph $G = (V, E)$. For a subgraph $G' = (V', E')$ of G, we say that C *enters* G' if C has both an edge in E' and an edge in $E \setminus E'$. Suppose that there is a •-node x of T such that x has $c \geq 4$ children $x_1, x_2, \ldots, x_c$. Then, no cycle of length at most 4 enters G_x; since G_x is created by series compositions of at least four two-terminal labeled graphs, every cycle entering G_x has length at least 5. Thus, the problem can be solved independently in each of G_x and the remaining part.

Assume that each •-node x of T has at most three children. Before explaining dynamic programming over T, we give the following key lemma.

Lemma 1 (♠). *Let x be any node of a layered decomposition tree T of a two-terminal series-parallel graph G and $\mathcal{L}_{\mathcal{H}}$ be a collection of cycles in G of length at most 4. For any solution $\mathcal{S} \subseteq \mathcal{L}_{\mathcal{H}}$ of G, there exists at most one cycle in $\mathcal{S}$ that enters G_x.*

Let $\mathcal{S}$ be a largest solution of G. Suppose that x is a •-node of T with c children $x_1, x_2, \ldots, x_c$. For an integer $i \in [c]$, let s_i and t_i denote a source and a sink of G_{x_i}, respectively. We distinguish the following two cases to consider:

(s_1) at least one cycle C in $\mathcal{S}$ enters G_x and $s_i t_i \in E(C)$ for every integer $i \in [c]$;
(s_2) at least one cycle C in $\mathcal{S}$ enters G_x and $s_i t_i \notin E(C)$ for some integer $i \in [c]$.

Similarly, for a ∥-node x with a source s and a sink t, we also distinguish the following two cases to consider:

(p_1) at least one cycle C in $\mathcal{S}$ enters G_x and $st \in E(C)$;
(p_2) at least one cycle C in $\mathcal{S}$ enters G_x and $st \notin E(C)$;

We note that the above cases are not exhaustive: there may be no cycle in $\mathcal{S}$ entering G_x. It is not necessary to consider such a case in the construction of our algorithm. We also note that by Lemma 1, there are no more than one (edge-disjoint) cycle satisfying these conditions.

Let $\mathcal{L}_{\mathcal{H}}^x$ be a restriction of $\mathcal{L}_{\mathcal{H}}$ to G_x, that is, $\mathcal{L}_{\mathcal{H}}^x = \{H \in \mathcal{L}_{\mathcal{H}} : E(H) \subseteq E(G_x)\}$. In our algorithm, for each node x of T, we compute the largest size of a subcollection $\mathcal{S}_x$ with $\mathcal{S}_x = \mathcal{S} \cap \mathcal{L}_{\mathcal{H}}^x$. Let $f^{\bullet}(x)$ be the largest size of $\mathcal{S}_x$ for a •-node x, and let $f^{\|}(x)$ be the largest size of $\mathcal{S}_x$ for a ∥-node x. Notice that, originally, leaves of T are labeled neither • nor ∥, and hence $f^{\bullet}(x)$ and $f^{\|}(x)$ cannot be defined for the leaves. For algorithmic simplicity, we consider a leaf x as a •-node if its parent is labeled ∥, and as a ∥-node if its parent is labeled •. This simplification allows us to define $f^{\bullet}(x)$ and $f^{\|}(x)$ for a leaf x accordingly.

We also define the truth values $b_j^{\bullet}(x)$ and $b_j^{\|}(x)$ for each $j \in \{1, 2\}$ and each node x of T. We set $b_j^{\bullet}(x) = 1$ (resp. $b_j^{\|}(x) = 1$) if and only if there exists a mutually edge-disjoint subcollection $\mathcal{S}_x'$ of $\mathcal{L}_{\mathcal{H}}^x$ that satisfies the following conditions:

$-$ $|\mathcal{S}_x'| = f^{\bullet}(x)$ (resp. $|\mathcal{S}_x'| = f^{\|}(x)$);

– there exists a cycle $C \in \mathcal{L}_\mathcal{H} \setminus \mathcal{L}_\mathcal{H}^x$ corresponding to the case (s_j) (resp. (p_j));
– all subgraphs in $\mathcal{S}'_x$ and C are edge-disjoint.

Intuitively speaking, $b_j^\bullet(x) = 1$ (and $b_j^\|(x) = 1$) if and only if we can further add a cycle C entering G_x into a partial solution at an ancestor of x.

We are ready to explain how to compute $f^\bullet(x)$, $f^\|(x)$, $b_j^\bullet(x)$, and $b_j^\|(x)$ for each node x of T and each $j \in \{1,2\}$.

Leaf Node. Suppose that x is a leaf of T. Let s and t be the source and the sink of G_x, respectively. One can verify that the following equalities hold: $f^\bullet(x) = f^\|(x) = 0$; $b_1^\bullet(x) = b_1^\|(x) = 1$ if and only if there exists a cycle C in $\mathcal{L}_\mathcal{H}$ such that $st \in E(C)$; and $b_2^\bullet(x) = b_2^\|(x) = 0$.

Internal •-node. Suppose that x is a •-node with c children $x_1, \ldots, x_c$. Since G_x consists of series compositions of $G_{x_1}, \ldots, G_{x_c}$, every cycle in G_x is contained in G_{x_i} for some i. We thus have

$$f^\bullet(x) = \sum_{i \in [c]} f^\|(x_i).$$

We next compute $b_j^\bullet(x)$ for each $j \in \{1,2\}$. Recall that x has at most three children.

Suppose that $c = 3$. If there exists a cycle $C \in \mathcal{L}_\mathcal{H} \setminus \mathcal{L}_\mathcal{H}^x$ that enters G_x, then it passes through s_1 and t_3, meaning that it enters G_{x_i} for all $i \in [3]$. Conversely, for every $i \in [3]$, if there is a cycle $C_i \in \mathcal{L}_\mathcal{H} \setminus \mathcal{L}_\mathcal{H}^{x_i}$ such that C_i enters G_{x_i}, then it must have $E(C_i) = \{s_1t_1, s_2t_2, s_3t_3, t_3s_1\}$, that is, the cycle is uniquely determined $C = C_i$ for $i \in [3]$. It is easy to observe that C is edge-disjoint from any cycles in $\mathcal{S}_x$ if and only if it is edge-disjoint from any cycles in $\mathcal{S}_{x_i}$ for all $i \in [3]$. Hence, we have $b_1^\bullet(x) = b_1^\|(x_1) \wedge b_1^\|(x_2) \wedge b_1^\|(x_3)$. This also implies that there is no cycle $C \in \mathcal{S}$ that enters G_x and $s_it_i \notin E(C)$, which yields that $b_2^\bullet(x) = 0$.

Suppose next that $c = 2$. By the similar argument to the case $c = 3$, we have $b_1^\bullet(x) = 1$ if and only if $b_1^\|(x_1) \wedge b_1^\|(x_2) = 1$ and there is a cycle $C \in \mathcal{L}_\mathcal{H} \setminus \mathcal{L}_\mathcal{H}^x$ such that $s_1t_1, s_2t_2 \in E(C)$. We explain how to decide $b_2^\bullet(x)$. If there is a cycle $C \in \mathcal{L}_\mathcal{H} \setminus \mathcal{L}_\mathcal{H}^x$ with $s_1t_1 \notin E(C)$ that enters G_x, then it enters both G_{x_1} and G_{x_2}, and it holds that $s_2t_2 \in E(C)$ because the length of C is at most 4. Conversely, for a cycle $C_1 \in \mathcal{L}_\mathcal{H} \setminus \mathcal{L}_\mathcal{H}^{x_1}$ such that C_1 enters G_{x_1} and $s_1t_1 \notin E(C_1)$, C_1 also enters G_{x_2} and G_x, and $s_2t_2 \in E(C_1)$ holds. The same argument is applied to a cycle $C \in \mathcal{L}_\mathcal{H} \setminus \mathcal{L}_\mathcal{H}^x$ with $s_2t_2 \notin E(C)$ that enters G_x. Thus, we have $b_2^\bullet(x) = (b_2^\|(x_1) \wedge b_1^\|(x_2)) \vee (b_1^\|(x_1) \wedge b_2^\|(x_2))$.

Internal ‖-node. Suppose that x is a ‖-node with c children $x_1, \ldots, x_c$. Let s and t be the source and the sink of G_x, respectively.

To compute $f^\|(x)$, we construct an *auxiliary graph* A_x whose vertex set is $\{a_1, a_2, \ldots, a_c\}$. We associate each child x_i of x with a vertex a_i. Let $i, j \in [c]$ be distinct integers. Suppose that x_i and x_j are internal nodes of T. Then, A_x

has an edge $a_i a_j$ if $b_1^\bullet(x_i) \wedge b_1^\bullet(x_j) = 1$ and there exists a cycle in $\mathcal{L}_{\mathcal{H}}^x$ that enters both G_{x_i} and G_{x_j}. Note that such a cycle C satisfies $|E(C) \cap E(G_{x_i})| = |E(C) \cap E(G_{x_j})| = 2$, which means that C must satisfy the case (s_1) for $\bullet$-nodes x_i and x_j. Suppose next that x_i is an internal node and x_j is a leaf of T. In this case, $st \in E(G_{x_j})$. Then, A_x has an edge $a_i a_j$ if at least one of the following conditions is satisfied:

1. x_i has exactly c children with $c \in \{2, 3\}$, $b_1^\bullet(x_i) = 1$, and there exists a cycle C in $\mathcal{L}_{\mathcal{H}}^x$ of length $c + 1$ that enters both G_{x_i} and G_{x_j}; or
2. $b_2^\bullet(x_i) = 1$.

Note that, in the second case $b_2^\bullet(x_i) = 1$, there is a cycle $C \in \mathcal{L}_{\mathcal{H}} \setminus \mathcal{L}_{\mathcal{H}}^{x_i}$ entering G_{x_i} such that x_i has a child y with $|E(C) \cap E(G_y)| \geq 2$. This implies that C has exactly three edges in G_{x_i} and hence we have $st \in E(C)$. Also note that there is no case that both x_i and x_j are leaves because G has no parallel edges. We complete the construction of A_x.

The intuition of the auxiliary graph A_x is as follows. If there is an edge $a_i a_j \in A_x$, then we can further add a cycle C in $G_{x_i} \parallel G_{x_j}$ that is edge-disjoint from any cycles in $\bigcup_{h \in [c]} \mathcal{S}_{x_h}$. We can simultaneously add such cycles for other edges in A_x. However, by Lemma 1, we cannot add more than one cycles entering G_{x_i}. Thus, in order to add as many such cycles as possible, the corresponding edges must form a matching in A_x. In fact, the following equality holds.

$$f^{\parallel}(x) = \sum_{i \in [c]} f^\bullet(x_i) + |M_x^*|, \tag{1}$$

where M_x^* be a maximum matching of A_x.

To compute $b_j^{\parallel}(x)$ for each $j \in \{1, 2\}$, we construct additional auxiliary graphs A_x^1 and A_x^2 from A_x. Let A_x^1 be the graph obtained from A_x as follows. We first add a vertex a'. Then, we add an edge $a' a_i$ if x has a leaf child x_i and there is a cycle $C \in \mathcal{L}_{\mathcal{H}} \setminus \mathcal{L}_{\mathcal{H}}^x$ such that C enters G_x and $st \in E(C)$.

Similarly, let A_x^2 be the graph obtained from A_x as follows. We first add a vertex a''. Then, for each $i \in [c]$, we add an edge $a'' a_i$ if x_i is an internal node of T, $b_1^\bullet(x_i) = 1$ and there is a cycle $C \in \mathcal{L}_{\mathcal{H}} \setminus \mathcal{L}_{\mathcal{H}}^x$ that enters G_{x_i}.

Let M_x^1 and M_x^2 be maximum matchings of A_x^1 and A_x^2, respectively. We let $b_1^{\parallel}(x) = 1$ if and only if $|M_x^1| > |M_x^*|$; and $b_2^{\parallel}(x) = 1$ if and only if $|M_x^2| > |M_x^*|$.

Finally, we conclude that $f^{\parallel}(r)$ is the size of a largest solution of G. Due to the space limitation, detailed proofs of the correctness and running time of our algorithm are postponed to the full version.

Acknowledgements. We thank the referees for their valuable comments and suggestions which greatly helped to improve the presentation of this paper.

References

1. Aravind, N.R., Saxena, R.: Parameterized complexity of path set packing. In: Lin, C.-C., Lin, B.M.T., Liotta, G. (eds.) the 17th International Conference and Workshops (WALCOM 2023), pp. 291–302. Springer Nature Switzerland, Cham (2023). https://doi.org/10.1007/978-3-031-27051-2_25
2. Bafna, V., Pevzner, P.: Genome rearrangements and sorting by reversals. In: Proceedings of 1993 IEEE 34th Annual Foundations of Computer Science, pp. 148–157 (1993). https://doi.org/10.1109/SFCS.1993.366872
3. Berman, F., Johnson, D., Leighton, T., Shor, P.W., Snyder, L.: Generalized planar matching. J. Algorithms 11(2), 153–184 (1990). https://doi.org/10.1016/0196-6774(90)90001-U
4. Bodlaender, H.L.: A partial k-arboretum of graphs with bounded treewidth. Theoret. Comput. Sci. 209(1), 1–45 (1998). https://doi.org/10.1016/S0304-3975(97)00228-4
5. Bontridder, K.M.J.D., et al.: Approximation algorithms for the test cover problem. Math. Program. 98(1–3), 477–491 (2003). https://doi.org/10.1007/s10107-003-0414-6
6. Caprara, A., Rizzi, R.: Packing triangles in bounded degree graphs. Inf. Process. Lett. 84(4), 175–180 (2002). https://doi.org/10.1016/S0020-0190(02)00274-0
7. Chung, F., Graham, R.: Recent results in graph decompositions. London Math. Soc. Lecture Note Ser. 52, 103–123 (1981)
8. Corneil, D.G., Masuyama, S., Hakimi, S.L.: Edge-disjoint packings of graphs. Discret. Appl. Math. 50(2), 135–148 (1994). https://doi.org/10.1016/0166-218X(92)00153-D
9. Dyer, M., Frieze, A.: On the complexity of partitioning graphs into connected subgraphs. Discret. Appl. Math. 10(2), 139–153 (1985). https://doi.org/10.1016/0166-218X(85)90008-3
10. Karp, R.M.: Reducibility among combinatorial problems. In: Miller, R.E., Thatcher, J.W. (eds.) Proceedings of a Symposium on the Complexity of Computer Computations, held March 20–22, 1972, at the IBM Thomas J. Watson Research Center, Yorktown Heights, New York, USA. The IBM Research Symposia Series, pp. 85–103. Plenum Press, New York (1972). https://doi.org/10.1007/978-1-4684-2001-2_9
11. Kirkpatrick, D.G., Hell, P.: On the complexity of general graph factor problems. SIAM J. Comput. 12(3), 601–609 (1983). https://doi.org/10.1137/0212040
12. Kosowski, A., Malafiejski, M., Zylinski, P.: Parallel processing subsystems with redundancy in a distributed environment. In: Wyrzykowski, R., Dongarra, J.J., Meyer, N., Wasniewski, J. (eds.) Parallel Processing and Applied Mathematics, 6th International Conference, PPAM 2005. LNCS, vol. 3911, pp. 1002–1009. Springer, Heidelberg (2005). https://doi.org/10.1007/11752578_121
13. Małafiejski, M., Żyliński, P.: Weakly cooperative guards in grids. In: Gervasi, O., et al. (eds.) Computational Science and Its Applications - ICCSA 2005, pp. 647–656. Springer, Berlin Heidelberg, Berlin, Heidelberg (2005). https://doi.org/10.1007/11424758_68
14. Minty, G.J.: On maximal independent sets of vertices in claw-free graphs. J. Comb. Theory, Ser. B 28(3), 284–304 (1980). https://doi.org/10.1016/0095-8956(80)90074-X
15. Murphy, O.J.: Computing independent sets in graphs with large girth. Discret. Appl. Math. 35(2), 167–170 (1992). https://doi.org/10.1016/0166-218X(92)90041-8

16. Sbihi, N.: Algorithme de recherche d'un stable de cardinalite maximum dans un graphe sans etoile. Discret. Math. **29**(1), 53–76 (1980). https://doi.org/10.1016/0012-365X(90)90287-R

17. Valdes, J., Tarjan, R.E., Lawler, E.L.: The recognition of series parallel digraphs. SIAM J. Comput. **11**(2), 298–313 (1982). https://doi.org/10.1137/0211023

18. Xu, C., Zhang, G.: The path set packing problem. In: Wang, L., Zhu, D. (eds.) COCOON 2018. LNCS, vol. 10976, pp. 305–315. Springer, Cham (2018). https://doi.org/10.1007/978-3-319-94776-1_26

Author Index

R. Uehara et al. (Eds.): WALCOM 2024, LNCS 14549, pp. 437–438, 2024.
https://doi.org/10.1007/978-981-97-0566-5

Printed in the United States
by Baker & Taylor Publisher Services